SUBCELLULAR MECHANISMS IN REPRODUCTIVE NEUROENDOCRINOLOGY

SUBCELLULAR MECHANISMS IN REPRODUCTIVE NEUROENDOCRINOLOGY

Edited by

F. NAFTOLIN
Department of Obstetrics and Gynecology,
Women's Pavilion,
Royal Victoria Hospital,
Montreal, Que. H3A 1A1 (Canada)

and

K.J. RYAN and I.J. DAVIES
Department of Obstetrics and Gynecology,
Harvard Medical School,
Boston, Mass. 02115 (U.S.A.)

ELSEVIER SCIENTIFIC PUBLISHING COMPANY
Amsterdam/Oxford/New York   1976

ELSEVIER SCIENTIFIC PUBLISHING COMPANY
335 Jan van Galenstraat
P.O. Box 211, Amsterdam, The Netherlands

AMERICAN ELSEVIER PUBLISHING COMPANY, INC.
52 Vanderbilt Avenue
New York, New York 10017

ISBN 0—444—41442—8

With 168 illustrations and 66 tables

Copyright © 1976 by Elsevier Scientific Publishing Company, Amsterdam

All rights reserved. No part of this publication may be reproduced, stored in a retrieval system, or transmitted in any form or by any means, electronic, mechanical, photocopying, recording, or otherwise, without the prior written permission of the publisher,
Elsevier Scientific Publishing Company, Jan van Galenstraat 335, Amsterdam

Printed in The Netherlands

List of contributors

Agnati, L.
  Department of Histology, Karolinska Institute, Stockholm, Sweden
Barden, N.
  Medical Research Council Group in Molecular Endocrinology, Centre Hospitalier de l'Université Laval, Quebec G1V 4G2, Canada
Beaulieu, M.
  Medical Research Council Group in Molecular Endocrinology, Centre Hospitalier de l'Université Laval, Quebec G1V 4G2, Canada
Beyer, C.
  Universidad Autonomia Metropolitana, Unidad Istapalapa, Mexico, D.F., Mexico
Borgeat, P.
  Medical Research Council Group in Molecular Endocrinology, Centre Hospitalier de l'Université Laval, Quebec G1V 4G2, Canada
Brawer, J.R.
  Departments of Obstetrics and Gynecology and Anatomy, McGill University Faculty of Medicine, and Royal Victoria Hospital, Montreal, Que., Canada
Brown-Grant, K.
  Medical Research Council External Scientific Staff, A.R.C. Institute of Animal Physiology, Babraham, Cambridge CB2 4AT, Great Britain
Brownstein, M.J.
  Laboratory of Clinical Science, National Institute of Mental Health, Bethesda, Md. 20014, U.S.A.
Challis, J.R.G.
  Department of Obstetrics and Gynecology, Women's Pavilion, Royal Victoria Hospital, Montreal, Que., Canada
Davies, I.J.
  Department of Obstetrics and Gynecology and Laboratory for Human Reproduction and Reproductive Biology, Harvard Medical School, Boston, Mass. 02115, U.S.A.
De Lean, A.
  Medical Research Council Group in Molecular Endocrinology, Centre Hospitalier de l'Université Laval, Quebec G1V 4G2, Canada
Drouin, J.
  Medical Research Council Group in Molecular Endocrinology, Centre Hospitalier de l'Université Laval, Quebec G1V 4G2, Canada
Dunlop, D.
  New York State Research Institute for Neurochemistry and Drug Addiction, Ward's Island, New York, N.Y. 10035, U.S.A.
Eneroth, P.
  Department of Obstetrics and Gynecology, Karolinska Hospital, Stockholm, Sweden
Enjalbert, A.
  Unité de Neurobiologie de l'INSERM, 2 ter rue d'Alésia, 75014 Paris, France

Epelbaum, J.
   Unité de Neurobiologie de l'INSERM, 2 ter rue d'Alésia, 75014 Paris, France
Everitt, B.
   Department of Histology, Karolinska Institute, Stockholm, Sweden
Fawcett, C.P.
   Department of Physiology, The University of Texas Health Science Center, Southwestern Medical School, Dallas, Texas 72535, U.S.A.
Ferland, L.
   Medical Research Council Group in Molecular Endocrinology, Centre Hospitalier de l'Université Laval, Quebec G1V 4G2, Canada
Fishman, J.
   Institute for Steroid Research and Department of Biochemistry, Montefiore Hospital and Medical Center, Albert Einstein College of Medicine, Bronx, N.Y. 10467, U.S.A.
Fuxe, K.
   Department of Histology, Karolinska Institute, Stockholm, Sweden
Goldstein, M.
   New York University Medical Center, New York, N.Y., U.S.A.
Gustafsson, J.-Å.
   Department of Chemistry, Karolinska Institute, Stockholm, Sweden
Harms, P.G.
   Department of Physiology, The University of Texas Health Science Center, Southwestern Medical School, Dallas, Texas 72535, U.S.A.
Hökfelt, T.
   Department of Histology, Karolinska Institute, Stockholm, Sweden
Jeffcoate, S.
   Department of Chemical Pathology, St. Thomas's Hospital Medical School, London, Great Britain
Johansson, O.
   Department of Histology, Karolinska Institute, Stockholm, Sweden
Karavolas, H.J.
   Department of Physiological Chemistry, The Endocrinology-Reproductive Physiology Program, and Waisman Center on Mental Development, University of Wisconsin, Madison, Wisc. 53706, U.S.A.
Kawakami, M.
   2nd Department of Physiology, Yokohama City University School of Medicine, Yokohama, Japan
Kimura, F.
   2nd Department of Physiology, Yokohama City University School of Medicine, Yokohama, Japan
Kordon, C.
   Unité de Neurobiologie de l'INSERM, 2 ter rue d'Alésia, 75014 Paris, France
Labrie, F.
   Medical Research Council Group in Molecular Endocrinology, Centre Hospitalier de l'Université Laval, Quebec G1V 4G2, Canada

Lajtha, A.
  New York State Research Institute for Neurochemistry and Drug Addiction, Ward's Island, New York, N.Y. 10035, U.S.A.

Lanman, T.
  Department of Obstetrics and Gynecology, Harvard Medical School, Boston, Mass. 02115, U.S.A.

Löfström, A.
  Department of Histology, Karolinska Institute, Stockholm, Sweden

Marks, N.
  New York State Research Institute for Neurochemistry and Drug Addiction, Ward's Island, New York, N.Y. 10035, U.S.A.

Martini, L.
  Department of Endocrinology, Institute of Endocrinology, University of Milan, 20129 Milan, Italy

McCann, S.M.
  Department of Physiology, The University of Texas Health Science Center, Southwestern Medical School, Dallas, Texas 72535, U.S.A.

McEwen, B.S.
  The Rockefeller University, 1230 York Avenue, New York, N.Y. 10021, U.S.A.

McKelvy, J.
  Department of Anatomy, University of Connecticut Health Center, Farmington, Conn. 06032, U.S.A.

Morin, O.
  Medical Research Council Group in Molecular Endocrinology, Centre Hospitalier de l'Université Laval, Quebec G1V 4G2, Canada

Naftolin, F.
  Department of Obstetrics and Gynecology, Women's Pavilion, Royal Victoria Hospital, Montreal, Que. H3A 1A1, Canada

Nuti, K.M.
  Department of Physiological Chemistry, The Endocrinology-Reproductive Physiology Program, and Waisman Center on Mental Development, University of Wisconsin, Madison, Wisc. 53706, U.S.A.

Ojeda, S.R.
  Department of Physiology, The University of Texas Health Science Center, Southwestern Medical School, Dallas, Texas 72535, U.S.A.

Poisner, A.M.
  Department of Pharmacology, University of Kansas Medical Center, Kansas City, Mo. 66103, U.S.A.

Reichlin, S.
  Endocrine Division, New England Medical Center Hospital, and Department of Medicine, Tufts University School of Medicine, Boston, Mass. 02111, U.S.A.

Robison, G.A.
  Department of Pharmacology, University of Texas, Health Science Center, Houston, Texas 77025, U.S.A.

Ruf, K.B.
: *Department of Physiology, University of Geneva, School of Medicine, 1211 Geneva 4, Switzerland*

Ryan, K.J.
: *Department of Obstetrics and Gynecology and Laboratory for Human Reproduction and Reproductive Biology, Harvard Medical School, Boston, Mass. 02115, U.S.A.*

Schiaffini, O.
: *Collegio Universitario, Division de Medicina, Departamento de Fisiologia, Las Palmas de Gran Canaria, Spain*

Short, R.
: *MRC Reproductive Biology Unit, Department of Obstetrics and Gynaecology, Edinburgh EH3 9ER, Great Britain*

Siu, J.
: *Department of Obstetrics and Gynecology, Harvard Medical School, Boston, Mass. 02115, U.S.A.*

Skett, P.
: *Department of Chemistry, Karolinska Institute, Stockholm, Sweden*

Sundberg, D.K.
: *Department of Physiology, The University of Texas Health Science Center, Southwestern Medical School, Dallas, Texas 72535, U.S.A.*

Van Houten, M.
: *Department of Anatomy, Tufts University School of Medicine, Boston, Mass., U.S.A.*

Wheaton, J.E.
: *Department of Physiology, The University of Texas Health Science Center, Southwestern Medical School, Dallas, Texas 72535, U.S.A.*

White, N.
: *Department of Chemical Pathology, St. Thomas's Hospital Medical School, London, Great Britain*

Wurtman, R.J.
: *Laboratory of Neuroendocrine Regulation, Department of Nutrition and Food Sciences, Massachusetts Institute of Technology, Cambridge, Mass. 02139, U.S.A.*

Yen, S.S.C.
: *Department of Reproductive Medicine (Obstetrics and Gynecology), School of Medicine, University of California, San Diego, Calif., U.S.A.*

Zimmerman, E.A.
: *Department of Neurology and the International Institute for the Study of Human Reproduction, Columbia University, New York, N.Y. 10032, U.S.A.*

Preface

In this book are found the contributions to a Symposium on Subcellular Mechanisms in Reproductive Neuroendocrinology which was held 13–15 October, 1975 at the House of the American Academy of Arts and Sciences on Jamaica Plain, near Boston, Mass. This International Symposium was sponsored by the Department of Obstetrics and Gynecology and the Laboratory for Human Reproduction and Reproductive Biology, Harvard Medical School. It was supported by a generous grant from the Upjohn Company (International Division).

The idea was to bring together investigators in diverse areas of subcellular neuroendocrine biology so that they might inform each other and the general scientific community of the current status of this area. Papers are arranged in the order in which they were presented on the program. In general, they are short technical reviews or somewhat longer topic oriented reviews. Each participant was chosen as an expert in an area and performed beautifully. The discussion was transcribed and is in the hands of the authors but was too bulky and general to burden the present book.

In the six months that have passed while our hardworking collaborators from Elsevier, Dr. P.S. Jackson and Mrs. E. Tjoa raced to prepare the book, little has changed to merit correction here. However, with the passage of time, it becomes clearer that studies in subcellular metabolism by central neuroendocrine tissues will continue to be very fruitful. It is hoped that this attempt to blend the elements of major topics and to cross pollinate between parallel disciplines and workers will be as useful to our readers as it was to the participants.

We should like to express the deepest appreciation to the Upjohn Company for the opportunity to construct such a well-balanced and representative symposium. Our thanks also to Ms. Lynn Resnick for many hours of administrative help and to Dr. Magda Marko and Dr. Nahid Parvizi for assistance with the manuscripts.

Finally, our thanks and admiration go to those professional colleagues in science who prepared these remarkable manuscripts and helped us in every way to make this a worthwhile venture.

Frederick Naftolin, Montreal
I. John Davies and Kenneth J. Ryan, Boston

Contents

List of contributors v

Preface ix

Chapter 1. Cellular organization of luteinizing hormone-releasing factor delivery systems
by J.R. Brawer and M. Van Houten (Montreal, Canada and Boston, Mass., U.S.A.).. 1

Chapter 2. Electrophysiological correlates of neuroendocrine tissues
by K.B. Ruf (Geneva, Switzerland) 33

Chapter 3. The role of calcium in neuroendocrine secretion
by A.M. Poisner (Kansas City, Kan., U.S.A.) 45

Chapter 4. Protein metabolism in neuroendocrine tissues
by A. Lajtha and D. Dunlop (New York, N.Y., U.S.A.) 63

Chapter 5. Localization of neurosecretory peptides in neuroendocrine tissues
by E.A. Zimmerman (New York, N.Y., U.S.A.) 81

Chapter 6. Biosynthesis and degradation of hypothalamic hypophysiotrophic factors
by S. Reichlin (Boston, Mass., U.S.A.) 109

Chapter 7. Biodegradation of hormonally active peptides in the central nervous system
by N. Marks (New York, N.Y., U.S.A.) 129

Chapter 8. Control of neurotransmitter synthesis by precursor availability and food consumption
by R.J. Wurtman (Cambridge, Mass., U.S.A.) 149

Chapter 9. Neurotransmitter interactions with neuroendocrine tissue
by C. Kordon, J. Epelbaum, A. Enjalbert and J. McKelvy (Paris, France and Farmington, Conn., U.S.A.) 167

Chapter 10. Hormonal regulation of the synthesis and metabolism of neurotransmitters
by M.J. Brownstein (Bethesda, Md., U.S.A.) 185

Chapter 11. On the role of neurotransmitters and hypothalamic hormones and their interactions in hypothalamic and extrahypothalamic control of pituitary function and sexual behavior
by K. Fuxe, T. Hökfelt, A. Löfström, O. Johansson, L. Agnati, B. Everitt, M. Goldstein, S. Jeffcoate, N. White, P. Eneroth, J.-Å. Gustafsson and P. Skett (Stockholm, Sweden, New York, N.Y., U.S.A. and London, Great Britain)...... 193

Chapter 12. Endogenous steroids in neuroendocrine tissues
by J.R.G. Challis, F. Naftolin, I.J. Davies, K.J. Ryan and T. Lanman (Boston, Mass., U.S.A. and Oxford, Great Britain) 247

Chapter 13. Specific binding of steroids by neuroendocrine tissues
by I.J. Davies, F. Naftolin, K.J. Ryan and J. Siu (Boston, Mass., U.S.A. and Montreal, Canada) 263

Chapter 14. Steroid receptors in neuroendocrine tissues: topography, subcellular distribution, and functional implications
  by B.S. McEwen (New York, N.Y., U.S.A.) 277
Chapter 15. Progesterone metabolism by neuroendocrine tissues
  by H.J. Karavolas and K.M. Nuti (Madison, Wisc., U.S.A.) 305
Chapter 16. Androgen reduction by neuroendocrine tissues: physiological significance
  by L. Martini (Milan, Italy) 327
Chapter 17. Androgen aromatization by neuroendocrine tissues
  by F. Naftolin, K.J. Ryan and I.J. Davies (Montreal, Canada and Boston, Mass., U.S.A.) 347
Chapter 18. Estrogen metabolism by neuroendocrine tissues
  by J. Fishman (Bronx, N.Y., U.S.A.) 357
Chapter 19. Oxidative metabolism in neuroendocrine tissue
  by O. Schiaffini (Las Palmas de Gran Canaria, Spain) 363
Chapter 20. Cyclic nucleotides in the limbic system
  by G.A. Robison (Houston, Texas, U.S.A.) 381
Chapter 21. Role of cyclic AMP in neuroendocrine control
  by F. Labrie, P. Borgeat, N. Barden, M. Beaulieu, L. Ferland, J. Drouin, A. De Lean and O. Morin (Quebec, Canada) 391
Chapter 22. Control of adenohypophyseal hormone secretion by prostaglandins
  by S.M. McCann, S.R. Ojeda, P.G. Harms, J.E. Wheaton, D.K. Sundberg and C.P. Fawcett (Dallas, Texas, U.S.A.) 407
Chapter 23. Limbic-preoptic responses to estrogens and catecholamines in relation to cyclic LH secretion
  by M. Kawakami and F. Kimura (Yokohama, Japan) 423
Chapter 24. The adenohypophysis; functional behavior of the gonadotrophs as target cells
  by S.S.C. Yen (San Diego, Calif., U.S.A.) 453
Chapter 25. Neuroendocrine mechanisms in sexual behavior
  by C. Beyer (Mexico, D.F., Mexico) 471
Chapter 26. Control of gonadotropin secretion
  by K. Brown-Grant (Cambridge, Great Britain) 485
Chapter 27. Special aspects — rhythms
  by R. Short (Edinburgh, Great Britain) 503

Subject index 515

Subcellular Mechanisms in Reproductive Neuroendocrinology, edited by
F. Naftolin, K.J. Ryan and J. Davies
© 1976 Elsevier Scientific Publishing Company—Amsterdam, The Netherlands

Chapter 1

Cellular organization of luteinizing hormone-releasing factor delivery systems

J.R. BRAWER and M. VAN HOUTEN

Departments of Obstetrics and Gynecology and Anatomy, McGill University Faculty of Medicine, and Royal Victoria Hospital, Montreal, Que. (Canada) and Department of Anatomy, Tufts University School of Medicine, Boston, Mass. (U.S.A.)

I. INTRODUCTION

Despite burgeoning interest in the hypothalamic control of gonadotrophin secretion and the accumulation of vast quantities of data on the physiology of this phenomenon, little is known about the structure of the final common pathway or pathways over which hypophysiotropic influence is exerted. Until recently, the final common gonadotrophotropic pathway was regarded in a largely conceptual fashion. Presumably, the luteinizing hormone-releasing factor (LRF) (and possibly the follicular hormone-releasing factor (FRF)) elaborating neurosecretory cells in the medial basal hypothalamus (MBH) projects axons to the perivascular space in the median eminence and, in response to the appropriate hormonal and neuronal cues, these axons release LRF into the hypothalamo-hypophysial portal system. Although this concept is undoubtedly correct, it is incomplete and lacking sufficient detail to elevate it from the category of theoretical necessity to that of anatomical reality.

It is now evident that the pathways and mechanisms by which the hypothalamus influences gonadotrophin secretion are complex and subtle. Recently, a variety of neuronal and non-neuronal cell types have been implicated in the final common pathway over which LRF is delivered to the portal system. These cells may not be restricted to the infundibular region of the hypothalamus as originally hypothesized, but may extend throughout the entire periventricular hypothalamus and even into non-hypothalamic structures such as the preoptic area, septum, and parolfactory area. Thus, the LRF delivery system may be comprised of several morphologically diverse and widely separated components, exhibiting different afferent and efferent connections and different cytological properties.

The following is a discussion of potential LRF delivery structures and an evaluation of present concepts concerning final common LRF pathways.

II. THE LOCATION OF NEUROSECRETORY LRF DELIVERY SYSTEMS

The MBH is generally considered to consist of premammillary arcuate, ventromedial anterior periventricular and suprachiasmatic nuclei (Fig. 1). The importance of the MBH

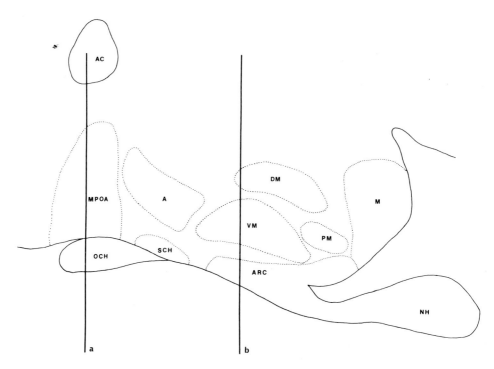

Fig. 1. Location of nuclei in medial basal hypothalamus. This schematic representation of a sagittal section of hypothalamus demonstrates the approximate relative positions and dimensions of some hypothalamic nuclei. The two solid lines traversing the diagram (a and b) indicate the levels represented by Figs. 2 and 3. Abbreviations: A, anterior hypothalamic nucleus; AC, anterior commissure; ARC, arcuate nucleus; DM, dorsomedial nucleus; M, mammillary body; MPOA, medial preoptic area; NH, neurohypophysis; OCH, optic chiasm; PM, premammillary nucleus; SCH, suprachiasmatic nucleus; VM, ventromedial nucleus.

in gonadotrophin control has long been suspected. A variety of classical lesion experiments (Bogdanove, 1954; Sawyer, 1956, 1959; Flerko and Bardos, 1959) indicate that an intact MBH is essential for ovulation and for maintenance of histologically normal gonads. Conversely, electrical stimulation of the MBH induces ovulation in atropine blocked rabbits (Saul and Sawyer, 1957) and in pentobarbital blocked rats (Critchlow, 1957, 1958). Furthermore, the hypophysiotropic area (Halász et al., 1962, 1965) coincides with the MBH. The hypophysiotropic area is defined as the region of the third ventricle that will maintain a transplanted pituitary gland in an active secretory state.

There are now considerable anatomical and electrophysiological data indicating that the MBH contains neurons that project axons to the median eminence, and therefore have access to the portal system. The greatest concentration of these neurons appears to be in the arcuate-ventromedial region (Figs. 1, 2). Axons have been traced from neuronal perikarya within the arcuate nucleus to the external layer of the median eminence in Golgi impregnated hypothalami (Spatz, 1951; Szentágothai et al., 1968). Dopamine containing axons have also been traced from arcuate nucleus to the external layer of the

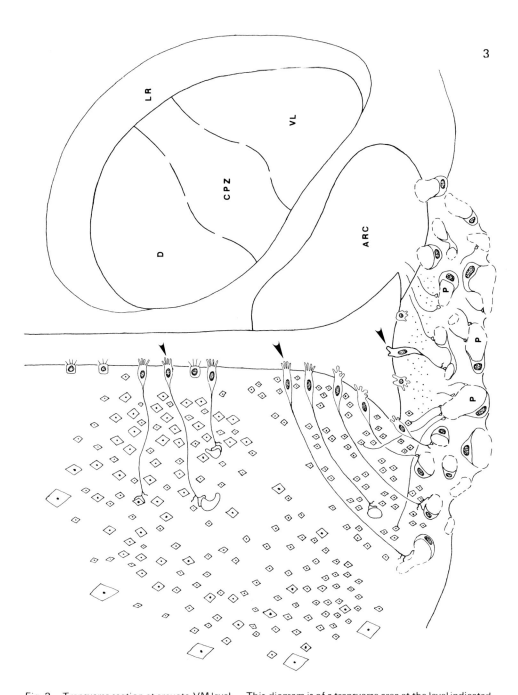

Fig. 2. Transverse section at arcuate-VM level. This diagram is of a transverse area at the level indicated by line b in Fig. 1. On the right side are outlined the areas occupied by the arcuate nucleus and the subdivisions of the ventromedial nucleus. On the left is a schematic representation of some of the histological aspects of this region. The lower region of the third ventricle and the median eminence are lined by tanycytes many of which project processes to the capillaries of the primary plexus (2 lower arrowheads). The tanycytes at the level of the ventromedial nucleus (upper arrowhead) project into the neuropil and often terminate on ordinary capillaries. Pericapillary nerve terminals in the median eminence derive in part from arcuate neurons as shown. Other sources of tuberoinfundibular axons are discussed at length in the text. The cytoarchitectonic subdivision of ventromedial nucleus discussed in the text is outlined on the right and on the left relative sizes of neurons are indicated. Abbreviations: ARC, arcuate nucleus; D, dorsomedial region; CPZ, cell poor zone; VL, ventrolateral region; LR, lateral rim.

median eminence by means of induced catecholamine fluorescence (Fuxe and Hökfelt, 1966).

Not only do neurons within the MBH project to the external layer of the median eminence, but they may be the only cells in the CNS that do so. Réthelyi and Halász (1970) have demonstrated that lesions within the MBH (hypophysiotropic area) result in degeneration of some of the perivascular axonal endings. If, however, the entire MBH-median eminence is surgically isolated from the rest of the brain, i.e., completely deafferented, no degeneration is detectable in the median eminence, indicating that the perivascular endings belong to neurons within the MBH island. Scott and Knigge (1970) have also reported absence of degeneration in the median eminence following complete deafferentation of arcuate nuclei-median eminence, suggesting that the vast majority of tuberoinfundibular axons originate in the arcuate nucleus. Such evidence must be considered cautiously since degeneration of tuberoinfundibular axons may be quite rapid (Raisman, 1972). Scott and Knigge (1970) let at least 9 postoperative days pass before examining the median eminence and any degenerative processes in the median eminence may have been completed long before.

In any event, electrophysiological studies support the view that the MBH projects extensively to the median eminence. Tuberoinfundibular neurons can be fired antidromically by application of a stimulating current to the median eminence (Harris et al., 1971; Makara et al., 1972; Sawaki and Yagi, 1973). In the rat, units from which antidromic spikes could be recorded subsequent to electrical stimulation of median eminence occur mostly in the arcuate nucleus and to a lesser extent in the ventromedial nucleus, particularly in its ventral region (Makara et al., 1972). Scattered tuberoinfundibular units were also localized in the suprachiasmatic nucleus, the anterior periventricular nucleus, and in the dorsal premammillary nuclei. Units projecting to the median eminence were also recently identified in the preoptic area (L. Renaud, personal communication). These various data document the existence of an extensive tuberoinfundibular system extending throughout the MBH albeit most concentrated in the arcuate nucleus.

To what extent this system consists of LRF delivery neurons is as yet somewhat unclear. The tuberoinfundibular system is functionally heterogeneous in that individual neurons comprising it may synthesize and transport one of a variety of adenohypophysiotropic hormones. It has also been demonstrated that many of the tuberoinfundibular neurons are dopaminergic (Hökfelt and Fuxe, 1972). It is not known whether these cells are simple dopaminergic neurons or are additionally involved in synthesis and transport releasing factors. Furthermore, although groups of cells that elaborate different regulating hormones may be deployed in roughly different regions of the MBH these populations are probably largely interblended, such that any region or nucleus of the MBH must contain cells serving a variety of functions. It has been suggested, for example, that the ventromedial nucleus (VM) may contain an element of the thyrotropin-releasing factor (TRF) system (Brown et al., 1974), the corticotropin-releasing factor (CRF) system (Halász et al., 1967; Feldman and Sarne, 1970; Palkovits and Stark, 1972), and the gonadotrophin-releasing factor (GRF) systems (Bernardis and Frohman, 1970; Pecile et al., 1972), as well as components involved in feeding behavior and in behavioral sexual receptivity (Carrer et al., 1973/74). Also, each of the releasing factor (RF) delivery systems may be fairly spread out. The TRF component of the tuberoinfundibular system, for example, may well consist of cells scattered throughout the MBH (Brown et al., 1974; Kruhlich et al., 1974).

Although the dispersed interblended organization of the various functional components of the tuberoinfundibular system renders the task of precise anatomical localization difficult, particularly at the cellular level, information is accumulating on the location, structure and behavior of the LRF delivery neurons.

Among the many experimental approaches applied to the localization of LRF cells, perhaps the most direct are immunological. Recently the chemical characterization and synthesis of LRF (Baba et al., 1971; Matsuo et al., 1971) has made possible the production of LRF antisera and has subsequently enabled the development of a specific radioimmunoassay. Palkovits et al. (1974), using radioimmunoassay, measured the relative amounts of LRF in individual nuclei removed from the hypothalamus by means of an ingenious micropunch technique (Palkovits, 1973). They found the highest concentrations of LRF in the median eminence. Although the arcuate nucleus contained considerably less LRF than the median eminence, it did have more than any other hypothalamic nucleus. The only other hypothalamic region to show significant concentration of LRF was the ventromedial nucleus (ventrolateral portion).

Although Palkovits et al. (1974) found only trace amounts of LRF in the suprachiasmatic nucleus and none in the preoptic area, other investigators have reported the presence of an anterior accumulation of LRF. The "anterior" LRF was first determined through bioassay of hypothalamic extracts and extracts of hypothalamic sections (Schneider et al., 1969; Crighton et al., 1970; Quijada et al., 1971). Recently radioimmunoassay was applied to extracts of serial hypothalamic sections cut in transverse, horizontal and sagittal planes (Wheaton et al., 1975). This study indicated an accumulation of LRF in the medial basal preoptic area particularly in the region of the organum vasculosum of the lamina terminalis. A second concentration peak occurred in the arcuate-median eminence region. The expanse of hypothalamus between these areas contained a somewhat uniform low level of LRF.

Attempts at LRF localization at the histological level by means of immunohistochemistry have also produced contradictory results. Barry et al. (1973, 1974) report LRF immunofluorescent staining in neuronal perikarya mostly in the preoptic, suprachiasmatic, and septal regions in castrated and colchicine or methanol treated guinea pigs. A small number of immunoreactive perikarya occur in normal untreated animals. Zimmerman et al. (1974) report LRF immunoperoxidase staining within neuronal perikarya only in the arcuate nucleus. Naik (1975a), using a double layer technique, reports perikaryal immunoreactivity in a large number of cells scattered throughout the preoptic area and MBH. The greatest concentration of immunoreactive cells, according to Naik (1975a), are in the preoptic-suprachiasmatic region and in the arcuate ventromedial region. Finally, there are a number of investigators who were unable to obtain perikaryal staining altogether (Baker et al., 1974; Kordon et al., 1974; Sétáló et al., 1975). Sétáló and his coworkers observed immunoreactive axons coursing through the anterior hypothalamus and funneling into the median eminence, but they observed no immunoreactivity in the preoptic area and no perikaryal staining even when castration and colchicine were used to increase the intrahypothalamic content of LRF. Barry et al. (1973, 1974) concur with Sétáló et al. (1975) to the extent that they also observed axons sweeping through the hypothalamus into the infundibulum but Barry and coworkers maintain that the cells of origin are mostly in the preoptic-suprachiasmatic area, the parolfactory area and in the septum. Preinfundibular lesions cause a backup of immunoreactive material anterior to the lesion confirming Barry's impression of anterior-posterior transport from preoptic-suprachiasmatic cells. Immunoreactive axons were also seen in such extrahypothalamic structures as the amygdala, zona incerta, and epithalamus (Barry et al., 1973), although LRF containing perikarya were not seen in these areas. Barry and Dubois (1973) were also able to obtain perikaryal staining in neonatal guinea pigs even without castration and colchicine treatment. The assertion of Barry and his coworkers that cells of the LRF delivery system are located largely in the preoptic-suprachiasmatic region and septum conforms with the results of Mess et al. (1967, 1970), who found that lesions in the area of the suprachiasmatic nucleus appeared to be most effective in reducing the hypothalamic content of LRF.

There is, therefore, considerable contradiction and discrepancy between the various studies on the anatomical localization of the LRF delivery system. However, several concrete concepts do emerge. It seems probable that the LRF delivery cells are scattered throughout the MBH and at least as far anterior as the medial preoptic area. One area of concentration of these cells is the arcuate-ventromedial region. The axons of these cells undoubtedly contribute to the LRF tuberoinfundibular tract. Another concentrated population of LRF cells is in the preoptic-suprachiasmatic region and it is not yet clear whether these contribute a significant projection to the median eminence or whether they project to other neurovascular structures and/or to the ventricular system. This will be discussed in detail later.

III. THE TUBEROINFUNDIBULAR LRF DELIVERY SYSTEM

III.1. Arcuate nucleus

Perhaps the most extensively studied source of LRF tuberoinfundibular fibers is the arcuate nucleus. This nucleus is comprised of a small cluster of small cells that nestle against the basal lateral wall of the third ventricle, just dorsal to the lateral recess (Fig. 2). These neurons commonly occur in clusters or in chains intercalated between the long tanycyte processes that course from the ventricular wall through the arcuate neuropil (Figs. 2, 3). Histologically, the nucleus appears quite ordinary in normal adult male and female rats. The neurons exhibit a narrow range of shapes and sizes and the standard varieties of glia are present.

Raisman (1972) describes a population of large neurons in the anterior region of the arcuate nucleus rostral to the median eminence. Most experimental cytological studies, however, are directed at the middle and posterior portions of the nucleus that inhabit the lower lateral ventricular wall adjacent to the median eminence. The present discussion is therefore confined to this region of arcuate nucleus.

The cytological appearance of arcuate neurons is unremarkable and is inconsistent with classical concepts of neurosecretory cytology, derived from observations of cells in paraventricular and supraoptic nuclei. For the most part, neurons of the arcuate nucleus in normal sexually mature rats exhibit the same variety of organelles and inclusions as other small neurons elsewhere in the CNS (Fig. 4). The endoplasmic reticulum is not unusually prominent although stacks of parallel cisternae comprising Nissl bodies do occur. More often, the rough endoplasmic reticulum is represented by short cisternae scattered randomly throughout the cell. Regular periodic variations in the conformation of the endoplasmic reticulum occur in normal female rats during the different stages of the estrous cycle. This will be discussed shortly.

Several small stacks of Golgi cisternae and associated vesicles usually appear in a perinuclear position in one pole of the cell. There are no large populations of specific granules although lysosomes are common.

Arcuate neurons commonly exhibit a variety of nucleolar satellites and nucleolus resembling cytoplasmic filamentous bodies (CFB) (Fig. 5). These features also occur in many other neuronal and non-neuronal cell types. Interestingly, the neurons in which they appear are often suspected neurosecretory cells (Brawer, 1971; LeBeux, 1971).

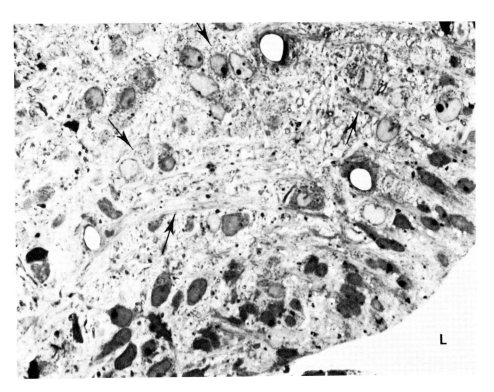

Fig. 3. Arcuate nucleus in cross-section. This is a micrograph of a thick Epon section stained with toluidine blue. In the lower right corner of the field is the lateral recess of the third ventricle (L). Lining the ventricle are tanycytes, the processes of which course through the arcuate neuropil (arrows). Neurons tend to occur in small clusters (arrowheads). x 1250.

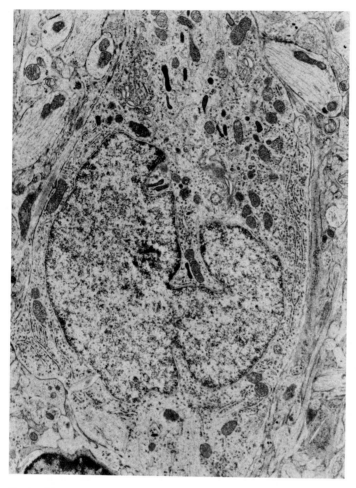

Fig. 4. Arcuate neuron. This is a cell from the arcuate nucleus of a normal sexually mature male rat. The nuclear profile is indented by groves and notches. The cytoplasm contains two small Nissl bodies on either side of the nucleus, although in the bottom of the field it is quite organelle poor. x 10,600.

Mark Van Houten of our laboratory has studied CFBs and intranuclear structures in the ventromedial nucleus. Although there is a greater variety of such structures in VM than in the arcuate nucleus, VM neurons do contain the type of nucleolus resembling CFB that typifies arcuate neurons (Fig. 5). In VM neurons, however, these particular CFBs are often encircled by cisternal rings. On rare occasions, the inclusion appears entrapped in a ring formed by one of a stack of annulate lamellae (Fig. 6). Since annulate lamellae derive from the nuclear envelope by delamination or "budding off", it is likely that this particular variety of CFB has its origin within the nucleus, possibly from the nucleolus. This is further suggested by nucleolar-nuclear envelope relationships observed in CFB containing cells (Van Houten and Brawer, unpublished observations).

It may be, therefore, that arcuate neurons engage in extrusion of nucleolar material into the cytoplasm. This process probably relates to some parameter of protein synthetic activity and the nucleolus-resembling CFB may ultimately prove to be a cytological marker for neurons primarily engaged in protein and/or polypeptide synthesis. In any event, it is an invariable cytoplasmic component of many arcuate neurons.

IV. NEGATIVE FEEDBACK

The arcuate nucleus is one source of the LRF tuberoinfundibular system that has been implicated as a direct target for negative gonadal steroid feedback. Implants of crystalline estradiol in the arcuate nucleus of male rats result in atrophy of the gonads and of accessory sexual structures. In female rats, estradiol implants in the arcuate nuclei or mammillary bodies result in ovarian and uterine atrophy. These atropic changes resemble those observed after hypophysectomy. Implants elsewhere are ineffective in this regard (Lisk, 1960). Estradiol implants in the arcuate region inhibit castration cell formation in the adenohypophysis following gonadectomy in the rat (Lisk, 1962) and in the rabbit (Kanematsu and Sawyer, 1963). Estradiol implants into the adenohypophysis did not prevent castration cell formation (Lisk, 1962). Furthermore, complete deafferentation of

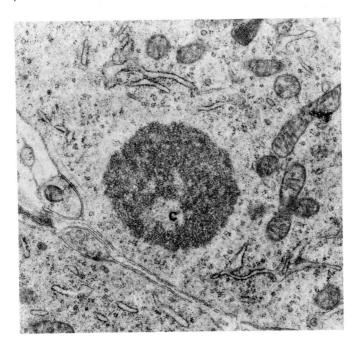

Fig. 5. Nucleolus resembling CFB in an arcuate neuron. This type of granulofilamentous inclusion is commonly observed in neurons of the arcuate and ventromedial nuclei. It is spheroidal, contains a clear area or cavitation (c) and is surrounded by a halo of cytoplasm that is relatively free of ribosomes. × 18,000.

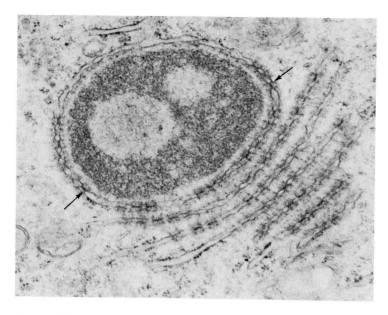

Fig. 6. CFB encircled by annulate lamella. This CFB is identical to that in Fig. 5, and represents one of the most common varieties of this inclusion. The CFB is in the vicinity of a stack of annulate lamellae, one of which forms a complete ring around the inclusion. Numerous nuclear pores (arrows) perforate the lamellae and attest to their nuclear origin. Although the cell in which this CFB appears is in the posterior dorsomedial region of the ventromedial nucleus, it is identical to those commonly seen in the arcuate nucleus as represented in Fig. 5. x 9000.

the MBH-median eminence does not prevent postgonadectomy castration cell formation (Halász and Gorski, 1967), suggesting that the castration cell effect is due to interruption of direct negative feedback to the MBH exclusive of any other CNS circuit. Ifft (1962) found a statistical decrease in nucleolar diameters of arcuate neurons in diestrous rats, and he interpreted this karyometric shift as an indication of reduced synthetic activity in these cells. Lisk and Newlon (1963) showed that estradiol implants in the arcuate nucleus result in statistical decreases in nucleolar diameters of arcuate neurons in addition to gonadal atrophy. Neither karyometric changes nor atrophic gonadal changes occurred after implantation of estradiol elsewhere in the hypothalamus.

It has further been shown that a proportion of the neurons of the arcuate nucleus exhibit label following injections of tritiated estradiol (Stumpf, 1970; Pfaff and Keiner, 1973), or testosterone (Sar and Stumpf, 1973). Neurons in several other hypothalamic and limbic areas also accumulated labeled steroid.

It is probable, therefore, that LRF tuberoinfundibular neurons within the arcuate nucleus are gonadal steroid targets and that the effect of the steroid is inhibitory to some facet of synthetic activity. This gonadal steroid inhibition of arcuate neurons ultimately translates into reduced gonadotrophin secretion.

Caution should be taken to avoid simplistic overinterpretation of steroid feedback effects on the arcuate nucleus. There may well also exist some sort of positive feedback at this level. Estradiol implants into the MBH in rabbits do not block ovulation immediately, but permit at least one reflex ovulation (Davidson and Sawyer, 1961). Ifft (1964) demonstrated a karyometric decrease in arcuate neurons of rats following gonadectomy, and Shin et al. (1974) have reported that castration lowered serum and hypothalamic LRF.

Although there are no cytological criteria for distinguishing a population of LRF tuberoinfundibular neurons in the arcuate nucleus, interference with negative gonadal steroid feedback produces consistent predictable cytological changes in a proportion of the cells. These changes implicate the neurons in which they occur in gonadotrophin regulation. Castration of male rats results in profound alteration in the organization of

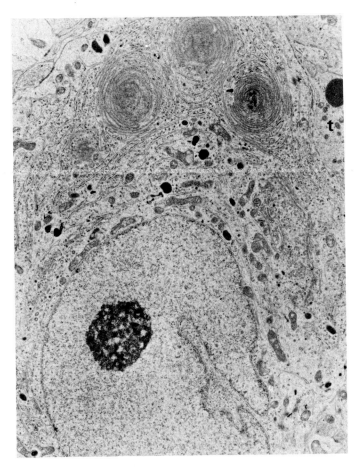

Fig. 7. Arcuate neuron of castrate. This is a cell in the arcuate nucleus of a sexually mature male rat that had been gonadectomized one month previously. At the top of the field are three whorled bodies situated in a complex network of RER. The whorled bodies consist of concentric cisternal shells. In the right upper corner, typical tanycyte processes (t) abutt the neuron. One of these processes contains a lipid droplet. x 9000.

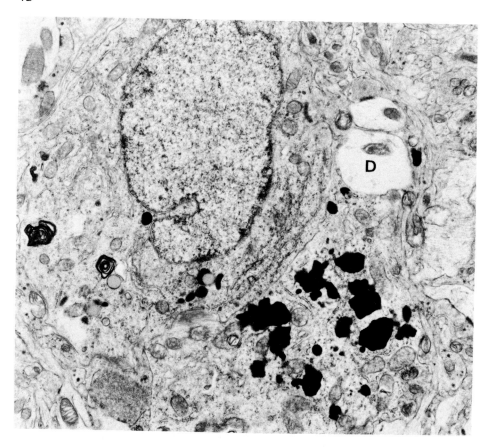

Fig. 8. Response of arcuate neuron and astrocyte to estradiol. This neuron (upper left) is from the arcuate nucleus of an adult female rat that received 2 mg estradiol valerate per month for 11 months. The cell contains lysosomes, lipofuscin and myelin figures indicative of cytopathology. The astrocytic process in the lower right is replete with large lipid pools also indicative of severe abnormality. Other neuronal processes in the neuropil also exhibit signs of deterioration such as the two empty dendritic profiles in the upper right. x 13,000.

the rough endoplasmic reticulum (RER) in approximately 20% of the cells in the arcuate nucleus. This consists of hypertrophy of the RER and the occurrence of rolls or whorles of cisternae (Fig. 7). The same effect also occurs in a percentage of arcuate neurons in gonadectomized female rats and to a lesser extent in normal females in diestrus (King et al., 1974). Cells exhibiting these post-castration whorles occur throughout the arcuate nucleus interspersed among ordinary unaffected neurons. There is, however, a tendency for whorled body containing neurons to occur in clusters.

Whorled RER formations are interpreted as a reflection of enhanced protein synthesis (Brawer, 1971; King et al., 1974), and the cells containing them are therefore probably the hypothalamic counterparts to the pituitary castration cells. This interpretation is supported by the finding that incorporation of radioactive lysine into arcuate neurons was significantly enhanced 5 weeks after castration (Litteria, 1973). The maximal whorled body effect occurred from 2 to 4 weeks post-castration (Brawer, 1971).

Hence, interruption of negative steroid feedback produces cytological hypertrophy in target cells. Conversely, large doses of estradiol produce pathological cytology in a small proportion of the arcuate neurons. The neuronal reaction to extensive estradiol treatment (2 mg estradiol valerate per month for up to 11 months) is characterized by large accumulations of lipofuscin, myelin figures, and a scant endoplasmic reticulum. These features are generally considered to indicate neuronal "shutdown" (Fig. 8). Although the effected cells look quite abnormal, they are few in number, and scattered among neurons that exhibit no response to the treatment. They do, however, tend to occur in clusters. Accompanying these degenerative perikaryal changes is a pronounced glial reaction involving astrocytes (Fig. 8) and phagocytic microglia.

Furthermore, scattered among the myriads of axons coursing through the arcuate nucleus in an anterior-posterior direction are clusters of degenerating fibers. These foci of axonal degeneration occur randomly in a vast field of normal axons. Engorged microglial cells often appear to be phagocytizing the degenerating fibers. Thus the estradiol may not only inhibit some aspect of cellular activity in target neurons, it may be toxic in large doses. The degenerating axons in the arcuate neuropil must originate more anteriorly. This may implicate extra-arcuate neurons in the anterior hypothalamus, possibly LRF tuberoinfundibular cells as targets for this steroid effect. However, glial reactions were never observed outside of the arcuate-VM area. Interestingly, small numbers of reactive glia are usually seen in this region in normally cycling female rats but rarely, if ever, in young mature males. Care must be taken in interpreting these data, since the estradiol could have a primary effect on some extrahypothalamic structure, such as the pituitary, and the lesion described above could result secondary to some alteration in pituitary activity.

Finally, the arcuate nucleus may be the critical structure that is modified by neonatal doses of testosterone resulting in androgenization of female rats. Nadler (1972, 1973) showed that implants of testosterone in the arcuate nucleus, but not in the preoptic area, were effective in androgenizing females. Furthermore, Litteria (1973) found that neonatal injections of testosterone propionate reduced the incorporation of [^{3}H]lysine in neurons of arcuate, paraventricular, periventricular and supraoptic nuclei in adults. No effect was observed in medial preoptic nucleus, lateral preoptic nucleus, ventromedial nucleus or suprachiasmatic nucleus. This suggests that some pattern of synthetic activity within the effected nuclei is masculinized by the treatment.

In our own laboratory, we have found that the arcuate nucleus of the rat is remarkably undeveloped at birth. The neurons are primitive in appearance and we have not been able to detect synapses in the arcuate nuclei of male or female rats in the first postnatal day. Therefore, development of the arcuate nucleus could easily be influenced by external factors such as testosterone, estrogen or castration during the neonatal period.

In summary then, there is a small population of neurons in the arcuate nucleus that form cisternal whorled bodies in response to gonadectomy in both male and female rats (Brawer 1971; King et al., 1974), there is a small population that exhibits cytopathology following extensive estradiol treatment (Brawer and Sonnenschein, 1975), and there exist a small number of neurons that selectively concentrate tritiated estradiol (Attramadal, 1970; Stumpf, 1970; Pfaff and Keiner, 1973), and testosterone (Sar and Stumpf, 1973). Also, according to some, a small proportion of arcuate neurons are immunoreactive for LRF (Zimmerman et al., 1974; Naik, 1975a). Although it has not yet been demonstrated that the populations of arcuate neurons reacting to these various treatments are one and the same, it is a strong possibility. Naik, for example (1975b), claims to have localized LRF by means of an immunoperoxidase technique at the fine structural level and he reports that the immunoreactivity often occurs in cells exhibiting whorled bodies. This suggests that the cells responding to steroid feedback are LRF delivery neurons.

Considering the small number of arcuate neurons implicated as LRF delivery cells and the large number of tuberoinfundibular terminals in the median eminence, it is likely that the arcuate nucleus is only one of several sources of the tuberoinfundibular pathway and may be only one of a number of contributors to the LRF delivery component.

V. VENTROMEDIAL NUCLEUS (VM)

The VM, particularly the ventrolateral portion (Fig. 2), has been implicated as a source of LRF tuberoinfundibular neurons. Radioimmunoassay of extracts of serial hypothalamic sections indicate an LRF concentration in the ventrolateral region of VM (Wheaton et al., 1975), and immunoreactive neurons have been identified in VM (Naik, 1975a). Also, tuberoinfundibular units have been identified electrophysiologically in the ventrolateral area as well as in other regions of VM.

Also, neurons in the ventrolateral region of VM have been shown to concentrate exogenous tritiated estradiol (Stumpf, 1970; Pfaff and Keiner, 1973). We have found whorled bodies in neurons of this region following gonadectomy in the male (unpublished results). Whorled bodies were not observed in normal or adrenalectomized rats. In addition, in female rats treated extensively with estradiol, neuronal and glial pathological responses identical to those seen in the arcuate nucleus occurred in the ventrolateral region of VM. It seems likely, therefore, that there is a collection of neurons in ventrolateral ventromedial nucleus that may belong to the LRF tuberoinfundibular system.

Interestingly, the cytology of many neurons in this region is somewhat similar to that of arcuate neurons. The principal cells in the ventrolateral part of VM are, for the most part, small and organelle poor (Fig. 9). This region is more heterogeneous than the arcuate nucleus in that large cells with extensive RER are interspersed among the small perikarya. The small cells of the ventrolateral VM also contain the same variety of nucleolus resembling CFBs that characterizes arcuate neurons. One distinguishing feature of the small neurons is the presence of numerous, unusually large multivesicular bodies. These were not observed in arcuate neurons. Since the small neurons of the ventrolateral part of VM share some common cytological and cytophysiological features with supposed LRF tuberoinfundibular neurons in the arcuate nucleus, they may also be suspected as LRF tuberoinfundibular cells.

In contrast, large neurons predominate in the other regions of VM. In the lateral rim, for example, the principal cells are very large and frequently contain vast quantities of

RER often arranged in parallel stacks of short cisternae comprising large Nissl bodies (Fig. 10). Golgi apparatus are also prominent in these cells which are often located in close apposition to capillaries. The large pericapillary neuronal profiles filled with RER and Golgi apparatus resemble the "synthetically active cells" seen in suprachiasmatic and medial preoptic nuclei (Clattenberg, 1974). The significance of this particular cell type will be discussed in relation to those nuclei.

Cells suspected of contributing a CRF delivery component to the tuberoinfundibular system may reside in the posterior dorsomedial part of VM (Fig. 2) (Palkovits and Stark, 1972), although there is as yet not much evidence to support this. Many of the cells of this region are fairly large and contain moderate RER which consists of long cisternae that meander randomly throughout the cytoplasm. Thus the RER of these cells forms large loosely organized Nissl bodies. This is in contrast to the dorsolateral region in which the RER is in the form of short cisternae arrayed in tight stacks. The posterior dorsomedial cells display a variety of CFBs and nucleus resembling bodies. They often also contain large complex myelin figures in relation to the Nissl bodies and diffuse CFBs are usually

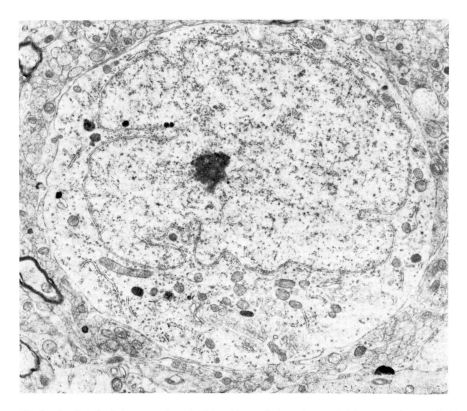

Fig. 9. Small cell of the ventrolateral VM. Most of the volume of this neuron is occupied by the nucleus which is typically indented. The RER consists of a few random cisternae. Outside of a small Golgi apparatus, mitochondria and lysosomes, the scant cytoplasm is rather sparsely populated. Nucleolus resembling CFBs often occur in these cells as do modest Nissl bodies. x 13,000.

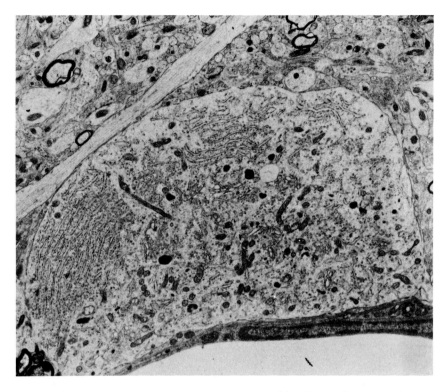

Fig. 10. Giant pericapillary neuron in the lateral rim of VM. This profile exhibits the extensive, well ordered stacks of RER that characterize cells of this region. The cytoplasm has a "crowded" appearance and contains in addition to the Nissl bodies, Golgi apparatus, randomly scattered free cisternae of the RER, lysosomes, and mitochondria. This cell conforms to the description of a "synthetically active" neuron (Clattenberg, 1974). x 5000.

present in the vicinity of such a Nissl body-myelin figure complex. Often CFBs are encircled by cisternal rings (Fig. 12).

Nucleolar satellites are common in these cells and can often be seen in contact with the inner leaflet of the nuclear envelope. The nucleolus frequently approaches the nuclear envelope as well, and the gap between the nucleolus and envelope is often bridged by granular filamentous material. The relationships between nucleus, nucleolus, CFBs and variant RER structures suggest that there is a constant disassembly and reassembly of the protein synthetic apparatus in these cells. This is the subject of an extensive study by Van Houten of our laboratory and will be reported at length elsewhere.

Suffice to say that many neurons of posterior dorsomedial VM are cytologically very different from the small cells that are concentrated in ventrolateral VM. They are also clearly different from the giant cells that characterize the lateral rim. Thus, within the same nucleus, there are neurons that are cytologically antipodal but that may perform similar functions, namely synthesis and delivery of a releasing factor (RF).

Neurosecretory cells may exhibit a wide range of cytological appearances and the proportion that conforms to the classical image typified by supraoptic and paraventricular neurons is probably small. It may be that structural variations in neurosecretory cells reflect differences in a variety of morphological and physiological parameters such as afferent connections, patterns of collateralization, extent and length of efferent axonal projections, differential responses to hormone feedback, etc. For example, as already mentioned, the principal cells of dorsomedial VM may be involved in CRF delivery, while those in ventrolateral VM are probably involved in LRF delivery. The cycle of ACTH secretion is diurnal as apposed to the 4- or 5-day rhythm luteinizing hormone (LH) secretion. CRF cells may therefore exhibit frequent manifestations of fluxes in protein synthetic activity such as disassembly and reassembly of protein synthetic apparatus. Furthermore, the cells involved in CRF delivery may well manifest acute cytological responses to the stress of preperfusion surgery since this increases ACTH secretion. LRF cells on a 4- or 5-day cycle exhibit gradual cytological changes consonant with the length of the estrous rhythm (King et al., 1974).

Thus it is as unrealistic to assume that all neurosecretory cells look alike as it is to pose that all neurons look alike and it is unwise to rely on normal fine structural criteria in identifying LRF delivery cells or for that matter, any neurosecretory cells.

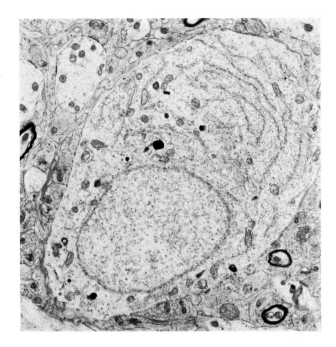

Fig. 11. Large neuron in the dorsomedial region of VM. In contrast to the giant cell in Fig. 10, this neuron has a rather "empty" appearance. Its most remarkable feature is a very loosely organized Nissl body consisting of more or less parallel long cisternae of the RER. This arrangement of RER typifies the principal cells of this region. x 8000.

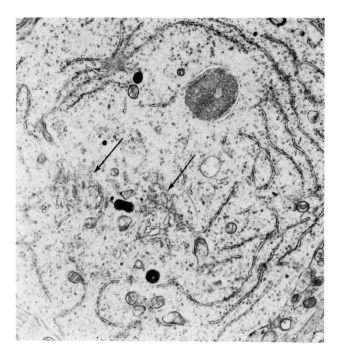

Fig. 12. Nissl body and cistern associated CFB in a principal cell of dorsomedial VM. This Nissl body consists of the characteristic long meandering cisternae loosely arranged in a large area of cytoplasm. In the upper right is a dense CFB partially encircled by cisternal elements. Often such CFBs are completely surrounded by a single closed cisternal ring. Golgi apparatus appears in the center of this profile (arrows). x 12,000.

VI. ANTERIOR HYPOTHALAMUS

There are regions of the anterior hypothalamus (in addition to the arcuate-ventromedial area) that undoubtedly contribute to the LRF tuberoinfundibular system. Evidence for this has already been discussed. Actually, very little is known about the small population of tuberoinfundibular LRF cells that are scattered throughout the anterior periventricular hypothalamus. It is clear, however, that they must be interblended with other RF neurons occupying the same regions. For example, somatostatin immunoreactive perikarya have been observed throughout the anterior periventricular region (Alpert et al., 1975). Somewhat more is known about the anteriormost hypothalamic extent of the tuberoinfundibular LRF system, i.e., the suprachiasmatic nucleus.

VII. SUPRACHIASMATIC NUCLEUS (SCH)

That the suprachiasmatic nucleus projects to the median eminence as part of the tuberoinfundibular pathway has been demonstrated electrophysiologically (Makara et al.,

1972; Sawaki and Yagi, 1973). This projection accounts, however, for only a small proportion of endings in the median eminence. It is probably comprised, at least partially, of LRF axons since there are a variety of immunological studies indicating the presence of LRF in suprachiasmatic cells (Barry et al., 1973; Naik, 1975a; Wheaton et al., 1975).

Suprachiasmatic neurons have been studied in the rabbit under different experimental conditions and endocrine states. In normal virgin rabbits, the cells of SCH are, for the most part, small and unremarkable in appearance. They resemble ordinary neurons that are, if anything, somewhat organelle poor (Clattenberg et al., 1972, 1975). A small number of suprachiasmatic neurons respond to castration by enlarging and developing extensive RER and Golgi apparatus (Clattenberg et al., 1975). These changes are interpreted as an indication of enhanced protein synthetic activity and presumably label the cells in which they occur as targets for negative gonadal steroid feedback. The suprachiasmatic nucleus is in fact included among the estrogen concentrating areas of the brain (Stumpf, 1970; Pfaff and Keiner, 1973). However, Litteria (1973b) examined the incorporation of [^{3}H]lysine into neurons in various hypothalamic nuclei and he was unable to detect any changes in incorporation in suprachiasmatic neurons under a variety of experimental conditions including castration. It is possible, however, that the rate of transport is equal to the rate of protein synthesis in these cells, and enhanced synthesis in castrates could be masked by accelerated transport. The cytological appearance of increased synthetic activity also occurs in a small percentage of suprachiasmatic neurons in female rabbits following coitus (Clattenberg et al., 1972).

VIII. SOME AFFERENT CONNECTIONS OF TUBEROINFUNDIBULAR NEURONS

A discussion of the structures that establish direct or indirect connections with the tuberoinfundibular system and are thus capable of modifying LH secretion is beyond the scope of this chapter. It should be noted, however, that the amygdala, hippocampus, and a number of mesencephalic nuclei all have anatomical access to the LRF delivery system (Raisman and Field, 1971). Furthermore, these areas receive a multiplicity of diverse afferent connections such that the role of extrahypothalamic inputs in the regulation of LRF secretion is fabulously complex.

Although limbic structures such as the amygdala establish direct connections with MBH nuclei such as VM, it seems that much of the extrahypothalamic input influencing LRF secretion is primarily by way of connections with the preoptic and suprachiasmatic areas which then contact LRF tuberoinfundibular neurons. Thus, much of the excitatory and inhibitory effects of extrahypothalamic structures on LH secretion are channeled through the preoptic area.

Furthermore, the preoptic area in and of itself exerts significant influence on the LRF system. For example, the preoptic area independent of any afferent connections is capable of driving the estrous cycle in rodents. Rats in which the preoptic area-MBH was completely deafferented exhibited normal ovulatory cycles (Halász, 1969; Taleisnik et al., 1970). If, however, the medial preoptic area (MPOA) (Fig. 13) is disconnected from the MBH, acyclicity characterized by persistent vaginal estrus and polyfollicular ovaries ensues.

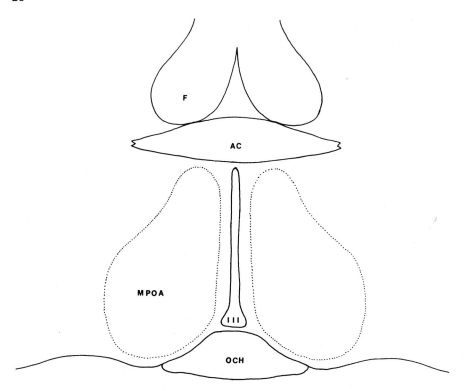

Fig. 13. Transverse section at the level of the preoptic nucleus. This diagram is of a transverse area at the level indicated by line a in Fig. 1. Abbreviations: AC, anterior commissure; III, third ventricle; F, fornix; MPOA, medial preoptic area; OCH, optic chiasm.

There is some controversy as to the physiological role of extrahypothalamic limbic structures in the control of gonadotrophin secretion. For example, electrical stimulation of the hippocampus has pronounced effects on gonadotrophin secretion and ovulation (Velasco and Taleisnik, 1969; Kawakami et al., 1973). Van Rees (1972) reports that anterior deafferentation of the preoptic area at 1:00 or 2:00 p.m. in proestrous rats blocks the first anticipated postsurgical ovulation. He suggests that subsequent ovulations depend on some sort of reorganization within the preoptic area. Brown-Grant and Raisman (1972) report that destruction of the fimbria and stria terminalis, tracts that connect the hippocampus and the amygdala with the POA, blocks the first anticipated ovulation but permits normal cycles subsequently. Animals sustaining such lesions could mate successfully and bear viable litters. It is difficult to imagine what sort of anatomical reorganization could occur in the POA so soon after deafferentation. One possible explanation of the acute effect of anterior preoptic deafferentation is neuronal "shock" of the sort that results in areflexia following acute transection of the corticospinal tracts (i.e., major deafferentation of the ventral horn of the spinal cord). Eventually the ventral horn recovers and deep tendon reflexes return and become hyperactive. In any event, it is apparent that limbic structures are not primarily essential to the initiation of the ovulatory cycle or to pituitary regulation of pregnancy and parturition in the rat, all of which the POA is capable of regulating. It is equally apparent, however, that limbic structures influence some reproductive functions such as development of puberty.

There is now good electrophysiological evidence that a direct POA-tuberoinfundibular pathway exists (Dyer and Cross, 1972; Cross, 1973; Dyer, 1973) and that this pathway is excitory to tuberoinfundibular neurons particularly within the arcuate-ventromedial area (Terasawa and Sawyer, 1969; Makara et al., 1972; Kawakami and Terasawa, 1974). It is also clear that the integrity of this pathway is essential for the maintenance of sexual cyclicity (Halász and Pupp, 1965; Halász and Gorski, 1967; Tejasen and Everett, 1967; Wuttke, 1974) and pregnancy or pseudopregnancy (Carrer and Taleisnik, 1970). Furthermore, the preoptic area is a target for gonadal steroid (Stumpf, 1970; Pfaff and Keiner, 1973; Sar and Stumpf, 1973), and this steroid feedback may be positive (Davidson, 1969). The preoptic area, therefore, integrates a variety of hormonal and neuronal inputs and under favorable conditions activates tuberoinfundibular neurons, resulting in increased LRF secretion.

Although the role of the preoptic area in gonadotrophin regulation is fairly clear in rodents, its function in primates is equivocal. There is evidence that a complete deafferentation of the pituitary stalk does not interfere with the menstrual cycle in the monkey (Knobil, 1974). On the other hand, there is evidence that disconnection of the POA from the MBH does abolish cyclicity in these animals (Spies et al., 1974).

In addition to its functions in regulating gonadotrophin secretion and in sexual behavior (Lisk, 1967), the POA exhibits sexual dimorphism and has been implicated in mechanisms of sexual differentiation. The area of dorsal POA medial to the stria terminalis exhibits many more non-strial synapses on dendritic spines in female rats than in males (Raisman and Field, 1973). Furthermore, neonatal androgen treatment causes the development in female rats of the male synaptic pattern and neonatal castration in males results in the female type synaptology. Exactly what role this dimorphic circuitry plays in sexual function is uncertain. It could be involved in influencing some parameter of reproductive function other than LRF release, since cyclical LRF delivery does not seem to depend upon afferent connections. Furthermore, although the corticomedial nucleus of the amygdala projects by way of the stria terminalis to strial POA and stimulation of the amygdala increases multiunit activity in the arcuate nucleus and enhances LH secretion (Kawakami and Terasawa, 1974), it is not the strial endings that are dimorphic. It is interesting in this regard that although the strial terminals do not exhibit dimorphism, stimulation of the amygdala results in enhanced LH secretion in female rats but not in males (Velasco and Taleisnik, 1969; Arai, 1971).

Possibly some complex component of sexual behavior is encoded in the non-strial POA circuit and it is this that changes with respect to alterations in POA circuitry following neonatal androgenization or castration. The effects of neonatal androgenization in cyclical LRF delivery may actually be mediated by the arcuate-VM (Nadler, 1972). On the other hand, the non-strial endings that exhibit sexual dimorphism may belong to an intrinsic MPOA circuit and thus may be essential to the regulation of the ovulatory cycle by the MPOA. One can only speculate on the function of these terminals until more is known about their origin.

Raisman and Field (1973) report that there is no sexual dimorphism in VM with respect to synaptology. This does not, however, preclude the possibility that dimorphism is present and reflected in some other, as of yet, unexamined aspect of neuronal morphology such as nuclear or cytoplasmic volume. In fact sexual dimorphism has been reported in the arcuate nucleus with respect to karyometric criteria (Staudt et al., 1973).

IX. MPOA-LRF DELIVERY SYSTEM

Although the MPOA has been implicated as an LRF delivery system, as already discussed, the evidence is somewhat contradictory and resistant to consistent interpretation. Most probably, there are LRF producing cells in the POA, but the histological localization and structural details of this LRF system are still obscure due to the size, heterogeneity, and complex connections and functions of this area.

The MPOA has been the subject of several experimental cytological investigations (Clattenberg, 1971, 1974). According to these studies, medial-POA and the suprachiasmatic nucleus are cytologically similar and both are composed of a more or less uniform population of rather unprepossessing neurons that are small in size and poor in RER. These cells are assumed to be synthetically inactive. Following mating, gonadectomy or even simple laparotomy, however, a small proportion of cells in MPOA as well as in suprachiasmatic nucleus appear to hypertrophy in that they become large and replete with RER, polysomes and lysosomes. This is regarded as an indication of enhanced protein synthetic activity and presumably the cells in which these changes are found, elaborate and deliver LRF. There are some problems with this interpretation. Although the change in RER and lysosomal content probably does reflect enhanced protein synthesis, there is sparse corroboration that the effected cells are indeed LRF delivery neurons. As previously explained, it is unwise to assume that all active neurosecretory cells share a common cytological appearance. Probable LRF delivery cells in the arcuate nucleus, for example, bear little resemblance to the classical neurosecretory cell (Scharrer and Scharrer, 1940), even when activated by gonadectomy. Furthermore, many neurons, e.g., alpha motor neurons, are very large and exhibit extensive RER, polysomes and Golgi apparatus are in no way neurosecretory.

Although hypertropic changes in POA neurons are consistent with coital activation of possible LRF cells by way of a transsynaptic route, the same changes seen after castration are paradoxical. Litteria (1973a), for example, reports a *decrease* in the incorporation of $[^3H]$lysine into MPOA cells in rats after castration. This is consistent with the positive feedback effect of gonadal steroids on the MPOA. Litteria's methods are, however, statistical and it is possible that the cells in MPOA, that are targets for negative feedback (Clattenberg, 1974), are so greatly outnumbered by positive feedback target cells that the negative feedback effect is masked. In any event, it is clear that more systematic morphological experimentation will be necessary to elucidate the cytophysiological organization of the MPOA and to identify and characterize its LRF component.

Until more is known about the MPOA LRF cells, one can only conjecture as to their pathways and mechanisms of delivery. At present, there seem to be three popular theories regarding MPOA-LRF delivery.

Barry et al. (1973) report the presence of a direct MPOA-infundibular tract in the guinea pig. The cells of origin populate the septum, MPOA, suprachiasmatic nucleus and parolfactory area and they give rise to axonal projections that sweep through the length of MBH to end in relation to the primary plexus in median eminence. As mentioned previously, this proposed pathway is the subject of considerable controversy. Although Barry and his coworkers demonstrate LRF immunofluorescent axons throughout the MPOA, anterior hypothalamus and median eminence, they have not demonstrated continuity between immunofluorescent infundibular axons and immunofluorescent MPOA neurons.

Although several electrophysiological studies have failed to disclose units in the POA that project to the median eminence (Makara et al., 1972; Sawaki and Yagi, 1973) others have (Renaud, personal communication). Furthermore, axonal degeneration was never observed in the median eminence following surgical disconnection of the POA (Réthelyi and Halász, 1970). It is possible that there are considerable species dependent differences in this system and the guinea pig may exhibit a major MPOA infundibular pathway that is of minor significance in the rat.

A second possible delivery route for preoptic LRF involving ependymal tanycytes has recently been the subject of much speculation. According to current conjecture, LRF neurosecretory cells in the MPOA may project into the third ventricle. Presumably, LRF is released from these intraventricular terminals into the cerebrospinal fluid (CSF) from whence it is transported to the primary capillary plexus by ependymal tanycytes.

There is diverse, albeit equivocal, evidence supporting this hypothesis. Supposed axon terminals have been observed within the ventricular system (Leonhardt, 1968; Wittowski, 1969). Of particular interest is the report that axons from the medial prechiasmatic region project into the preoptic recess of the third ventricle (LeBeux, 1972). The distribution of these intraventricular processes is not limited to the third ventricle, however, but includes lateral ventricles as well (Westergaard, 1972). Intraventricular axons are common in non-mammalian species (Oztan, 1967; Peute, 1969; Vigh-Teichman et al., 1970a). Matsui and Kobayashi (1968) indicate that axons in the third ventricle are rare in the rat, although this is certainly debatable.

Neurophysin (Robinson and Zimmerman, 1973), TRF (Shambaugh et al., 1975) and LRF (Morris and Knigge, 1975) have been found in CSF as would be expected if intraventricular neurosecretory projections truly exist. However, progesterone (Lurie and Weiss, 1967), insulin (Margolis and Altszular, 1967), metabolites of estradiol (Anand Kumar and Thomas, 1968) and human chorionic gonadotrophin (Bagshawe et al., 1968) also occur in CSF indicating that there is very little selectivity regarding compounds that have access to the ventricular system. Furthermore, Gunn et al. (1974) were unable to detect LRF in the CSF, and Cramer and Barraclough (1975) were unable to detect LRF in CSF even after electrical stimulation of the MPOA that resulted in an LH surge.

The capacity of tanycytes to transport substances, particularly RF from the CSF to the portal blood, has been examined using several experimental techniques. Infusion of LRF into the third ventricle stimulates a burst of LH secretion in female rats (Ondo et al., 1972). LRF injected into the ventricle can also be detected in portal blood within minutes after injection (Ben-Jonathon et al., 1974), but the amount recovered is rather small. It is possible that the intraventricular LRF could itself act upon LRF secreting neurosecretory cells and stimulate LRF secretion by the traditional tuberoinfundibular route. There is some evidence that such ultrashort feedback does occur (Kawakami and Sakuma, 1974). Uemura et al. (1975), however, demonstrated the presence of [^{125}I]LRF in tanycytes and primary plexus capillaries following intraventricular injection of the polypeptide. Thus, at least a proportion of the intraventricular LRF enters the portal blood by means of tanycyte transport. Weiner et al. (1972) found that LRF is capable of inducing a rise in serum LH when administered either intraventricularly or intravenously. The intravenous route, however, was more effective in this regard, even though the LRF was diluted in a huge volume of systemic blood. This, the authors feel, casts serious doubt as to the physiological significance of the CSF-transependymal pathway. However, Ben-Jonathon et al. (1974) demonstrated that LRF was effective in stimulating LH secretion when given either by intraventricular injection or intravenous injection. The intravenous route appeared to be more effective in producing a rapid rise, peak, and decline in plasma LH, whereas intraventricular injections produced a slow rise and slow decline. Uemura et al. (1975) found that intraventricular and intravenous injections of LRF produced roughly the same degree of elevation of plasma LH and FSH. So it appears that the intraventricular transependymal route is capable of delivering significant quantities of LRF from the CSF.

The ability of the median eminence, presumably the tanycyte component, to transport TRF has been documented (Knigge et al., 1974), however, tritiated corticosterone, [^{125}I]Na and [^{125}I]LH when injected into the CSF are also recovered in the portal blood (Ondo et al., 1972). Intraventricular injections of such non-physiologic compounds as horseradish peroxidase are also taken up and transported by tanycytes of the median eminence and the lower third ventricle (Kobayashi et al., 1972; Léranth and Schiebler, 1974). Also, although some investigations have demonstrated LRF in tanycytes (Zimmerman et al., 1974) others have not (Pelletier et al., 1974).

Thus, although some neurohormones may be present in the CSF and they may be subject to uptake and transport by tanycytes, this process appears to be extremely unselective. Considering the poor recovery of intraventricular LRF in portal blood (Ben-Jonathon et al., 1974), it is dangerous to assume that physiological amounts of LRF, when diluted in the vast reservoir of CSF, in the third ventricle, are responsible for pulsatile LH secretion. The possibility remains, however, that tanycytes may store LRF, and by doing so may be able to accumulate significant amounts of the substance over a period of several days. Thus far more studies have focused on the uptake and transport rather than the storage capacities of these cells, therefore this possibility deserves more attention.

As suggested by the experiments of Ben-Jonathon et al. (1974), the CSF transependymal route may provide a tonic pathway whereas release of LRF from neurosecretory ending could result in control of phasic aspects of LH secretion.

Both sexual dimorphism (Anand Kumar, 1968) and periodic structural modifications during the sexual cycle (Knowles and Anand Kumar, 1969; Brawer et al., 1974) occur in tanycytes. These phenomena involve the apical surface of the tanycyte and undoubtedly relate to the cells' differential absorptive capacities in the two sexes and at different stages of the sexual cycle. Interestingly, the region of ventricular wall in which the most pronounced cyclical changes occur during the estrous cycle in the rats overlies the ventromedial nucleus. Thus, many of the tanycytes that exhibit cyclical changes in morphology concurrent with the estrous cycle project to ordinary capillaries in VM or to the VM neuropil rather than to the portal system (Fig. 2).

The third possible delivery route for LRF synthesized in the POA may be by way of the organum vasculosum of the lamina terminalis (OVLT). Of all the circumventricular organs, the OVLT bears the strongest cytological resemblance to the median eminence (Weindl and Joynt, 1972). The OVLT is highly vascularized; its capillaries are fenestrated and the overlying ependyma is of the tanycyte variety. The capillaries are studded with myriads of nerve endings and tanycyte and feet (LeBeux, 1972; Weindl and Joynt, 1972). Thus the OVLT has all of the requirements for a neurohemal organ and it may be involved in neurovascular release of LRF. Lesley Alpert of our laboratory has observed LRF immunoreactive fibers in the region of the OVLT in rats as have others (Barry et al., 1973). Exactly how secretion of LRF into the capillaries of the OVLT could effect sexual function in mammals is unclear, since it seems that the capillaries of the OVLT simply empty into the systemic venous drainage. It is quite possible, however, that there are as of yet unknown vascular links between the OVLT and structures directly involved in control of sexual function. It is also possible that LRF acts on the anterior pituitary, in part, by the systemic circulation (Weiner et al., 1972; Ben-Jonathon et al., 1974) and that the OVLT provides the necessary link between LRF neurosecretory cells in the MPOA and the systemic circulation. Much more attention must be directed to the OVLT, particularly its neuronal and vascular architecture, in order to determine what role it may play as an LRF delivery system.

Finally, there is a strong possibility that LRF delivery cells function in capacities other than that of gonadotrophin control. It has been reported, for example, that LRF is capable of facilitating behavioral sexual receptivity in estrogen primed ovariectomized hypophysectomized rats (Pfaff, 1973). Furthermore, electrophysiologically identified tuberoinfundibular units in VM have been shown to project axon collaterals to the preoptic area and the dorsomedial nucleus of the thalamus (Renaud and Martin, 1975). Such collateralization may well account for the presence of LRF positive axons in areas generally thought to be devoid of LRF cells such as the zona incerta, epithalamus and amygdala (Barry et al., 1974). Barry and coworkers were actually able to observe collateralization of LRF immunoreactive axons.

It may well be that LRF delivery cells function as ordinary neurons involved in modulating and integrating sexual behavior, in addition to controlling gonadotrophin secretion by way of a neurosecretory mechanism. Such being the case, LRF may be considered one of a growing collection of chemical synaptic transmitters. A similar theory has been posed for TRF in order to account for its extrahypothalamic distribution and its behavioral effects.

ACKNOWLEDGEMENTS

The authors gratefully acknowledge the assistance of Ray Walsh and Lesley Alpert in the preparation of this manuscript.

REFERENCES

Alpert, L., Brawer, J., Patel, Y. and Reichlin, S. (1975) Somatostatinergic neurons in anterior hypothalamus: immunohistochemical localization. *Endocrinology*, in press.

Anand Kumar, T.C. (1968) Sexual differences in the ependyma lining the third ventricle in the area of the anterior hypothalamus of adult rhesus monkeys. *Z. Zellforsch.*, 90: 28–36.

Anand Kumar, T.C. and Thomas, G.H. (1968) Metabolites of ^{3}H-oestradiol-17-beta in the cerebrospinal fluid of the Rhesus monkey. *Nature (Lond.)*, 219: 628–629.

Arai, Y. (1971) Effect of electrochemical stimulation of the amygdala on induction of ovulation in different types of persistent estrous rats and castrated male rats with an ovarian transplant. *Endocr. jap.*, 18: 211–214.

Attramadal, A. (1970) Cellular localization if H^3-oestradiol in the hypothalamus. *Z. Zellforsch.*, 104: 572–581.

Baba, Y., Matsuo, H. and Schally, A. (1971) Structure of the porcine LH- and FSH-releasing hormone. 11. Confirmation of the proposed structure by conventional sequential analyses. *Biochem. biophys. Res. Commun.*, 44: 459–463.

Bagshawe, R.D., Hillary Orr, A., and Rushworth, A.G.J. (1968) Relationship between concentrations of human chorionic gonadotrophin in plasma and cerebrospinal fluid. *Nature (Lond.)*, 217: 950–951.

Baker, G., Dermody, W. and Reel, J. (1974) Localization of luteinizing hormone releasing hormone in the mammalian hypothalamus. *Amer. J. Anat.*, 139: 129–134.

Barry, J. et Dubois, M. (1973) La voie neurosécrétrice préoptico-infundibulaire à LH-RH chez le Cobaye au cours de la gestation. *Neuroendocrinologie*, 167: 1812.

Barry, J., Dubois, M. and Poulain, P. (1973) LRF producing cells of the mammalian hypothalamus (a fluorescent antibody study). *Z. Zellforsch.*, 146: 351–366.

Barry, J., Dubois, M. and Carette, B. (1974) Immunofluorescent study of the preoptico-infundibular LRF neurosecretory pathway in normal, castrated or testosterone-treated male guinea pig. *Endocrinology*, 95: 1416.

Ben-Jonathon, N., Mical, R. and Porter, J. (1974) Transport of LRF from CSF to hypophysial portal and systemic blood and release of LH. *Endocrinology*, 95: 18–25.

Bernardis, C. and Frohman, L. (1970) Effect of lesion size in the ventromedial hypothalamus on growth hormone and insulin levels in weanling rats. *Neuroendocrinology*, 6: 319–328.

Bogdanove, E.M. (1954) Location of hypothalamic lesions affecting the adenohypophysis. *Anat. Rec.*, 118: 282–283.

Brawer, J.R. (1971) The role of the arcuate nucleus in the brain-pituitary-gonad axis. *J. comp. Neurol.*, 143: 411–446.

Brawer, J. and Sonnenschein, C. (1975) Cytopathological effects of estradiol on the arcuate nucleus of the female rat. A possible mechanism for pituitary tumorigenesis. *Amer. J. Anat.*, 144: 57–87.

Brawer, J., Lin, P. and Sonnenschein, C. (1974) Morphological plasticity in the wall of the third ventricle during the estrous cycle in the rat: a scanning electron microscopic study. *Anat. Rec.*, 179: 481–490.

Brown, M., Palkovits, M., Saavedra, J., Bassiri, R. and Utiger, R. (1974) Thyrotropin-releasing hormone in specific nuclei of rat brain. *Science*, 185: 267–269.

Brown-Grant, K. and Raisman, G. (1972) Reproductive function in the rat following selective destruction of afferent fibers to the hypothalamus from the limbic system. *Brain Res.*, 46: 439–446.

Carrer, H. and Taleisnik, S. (1970) Induction and maintenance of pseudopregnancy after interruption of preoptic hypothalamic connections. *Endocrinology*, 86: 231—236.

Carrer, H., Asch, G. and Aron, C. (1973/74) New facts concerning the role played by the ventromedial nucleus in the control of estrous cycle duration and sexual receptivity in the rat. *Neuroendocrinology*, 13: 129—138.

Clattenberg, R. (1974) Ultrastructure of hypothalamic neurons and of the median eminence. *Canad. J. neurol. Sci.*, 1: 40—58.

Clattenberg, R., Singh, R. and Montemurro, D. (1971) Ultrastructural changes in the preoptic nucleus of the rabbit following coitus. *Neuroendocrinology*, 8: 289—306.

Clattenberg, R., Singh, R. and Montemurro, D. (1972) Postcoital ultrastructural changes in the neurons of the suprachiasmatic nucleus of the rabbit. *Z. Zellforsch.*, 125: 448—459.

Clattenberg, R., Montemurro, D. and Bruni, J. (1975) Neurosecretory activity within suprachiasmatic neurons of the female rabbit following castration. *Neuroendocrinology*, 17: 211—224.

Cramer, O. and Barraclough, C. (1975) Failure to detect luteinizing hormone-releasing hormone in the third ventricle cerebrospinal fluid under a variety of experimental conditions. *Endocrinology*, 96: 913—921.

Crighton, D., Schneider, H. and McCann, S. (1970) Localization of LH-releasing factor in the hypothalamus and neurohypophysis as determined by an in vitro method. *Endocrinology*, 87: 323—329.

Critchlow, B.V. (1957) Ovulation induced by hypothalamic stimulation in the rat. *Anat. Rec.*, 127: 283.

Critchlow, B.V. (1958) Ovulation induced by hypothalamic stimulation in the anaesthetized rat. *Amer. J. Physiol.*, 195: 171—174.

Cross, B. (1973) Towards a neurophysiological basis for ovulation. *J. Reprod. Fertil.*, Suppl. 20: 97—117.

Davidson, J.M. (1969) Feedback control of gonadotropin secretion. In *Frontiers in Neuroendocrinology*, W.F. Ganong and L. Martini (Eds.), Oxford University Press, New York, pp. 343—388.

Davidson, J. and Sawyer, C. (1961) Effects of localized intracerebral implantation of oestrogen on reproductive function in the female rabbit. *Acta endocr. (Kbh.)*, 37: 385.

Dyer, R. (1973) An electrophysiological dissection of the hypothalamic regions which regulate the pre-ovulatory secretion of luteinizing hormone in the rat. *J. Physiol. (Lond.)*, 234: 421—442.

Dyer, R. and Cross, B. (1972) Antidromic identification of units in the preoptic and anterior hypothalamic areas projecting directly to the ventromedial and arcuate nuclei. *Brain Res.*, 43: 254—258.

Feldman, S. and Sarne, Y. (1970) Effect of cortisol on single cell activity in hypothalamic islands. *Brain Res.*, 23: 65—67.

Flerko, B. und Bardos, V. (1959) Zwei verschiedene Effekte experimenteller Läsion des Hypothalamus auf die Gonadin. *Acta neuroveg. (Wien)*, 20: 248—262.

Fuxe, K. and Hökfelt, T. (1966) Further evidence for the existence of tuberoinfundibular dopamine neurons. *Acta physiol. scand.*, 66: 243—244.

Gunn, A., Fraser, H., Jeffcoate, S., Holland, D. and Jeffcoate, W. (1974) CSF and release of pituitary hormones. *Lancet*, 1: 1057.

Halász, B. (1969) The endocrine effects of isolation of the hypothalamus from the rest of the brain. In *Frontiers in Neuroendocrinology*, W.F. Ganong and L. Martini (Eds.), Oxford University Press, New York, pp. 307—342.

Halász, B. and Gorski, R. (1967) Gonadotrophic hormone secretion in female rats after partial or total interruption of neural afferents to the medial basal hypothalamus. *Endocrinology*, 80: 608—622.

Halász, B. and Pupp, L. (1965) Hormone secretion of the anterior pituitary gland after physical interruption of all nervous pathways to the hypophysiotropic area. *Endocrinology*, 77: 553—562.

Halász, B., Pupp, L. and Uhlarik, S. (1962) Hypophysiotrophic area in the hypothalamus. *J. Endocr.*, 25: 147—157.

Halász, B., Pupp, L., Uhlarik, S. and Tima, L. (1965) Further studies on the hormone secretion of the anterior pituitary transplanted into the hypophysiotrophic area of the rat hypothalamus. *Endocrinology*, 77: 343–345.

Halász, B., Slusher, M. and Gorski, R. (1967) Adrenocorticotrophic hormone secretion in rats after partial or total deafferentation of medial basal hypothalamus. *Neuroendocrinology*, 2: 43–55.

Harris, M.C., Makara, G.B. and Spyer, K.M. (1971) Electrophysiological identification of neurons of the tuberoinfundibular system. *J. Physiol. (Lond.)*, 218: 86–87.

Hökfelt, T. and Fuxe, K. (1972) On the morphology and neuroendocrine role of the hypothalamic catecholamine neurons. In *Brain–Endocrine Interaction. Median Eminence: Structure and Function, Int. Symp. Munich 1971*, K.M. Knigge, D.E. Scott and A. Weindl (Eds.), Karger, Basel, pp. 181–223.

Ifft, J. (1962) Evidence for gonadotropic activity of the hypothalamic arcuate nucleus in the female rat. *Anat. Rec.*, 142: 1.

Ifft, J.D. (1964) The effect of endocrine gland extirpations on the size of nucleoli in rat hypothalamic neurons. *Anat. Rec.*, 148: 599.

Kanematsu, S. and Sawyer, C.H. (1963) Effects of hypothalamic and hypophysial estrogen implants on pituitary gonadotrophic cells in ovariectomized rabbits. *Endocrinology*, 73: 687–695.

Kawakami, M. and Sakuma, Y. (1974) Responses of hypothalamic neurons to the microiontophoresis of LH-RH, LH and FSH under various levels of circulating ovarian hormones. *Neuroendocrinology*, 15: 290–307.

Kawakami, M. and Terasawa, E. (1974) Role of limbic structures on reproductive cycles. In *Biological Rhythms in Neuroendocrine Activity*, M. Kawakami (Ed.), Igaku Shoin Ltd., Tokyo, pp. 197–219.

Kawakami, M., Terasawa, E., Kimura, F. and Kubo, K. (1973) Correlated changes in gonadotropin release and electrical activity of the hypothalamus induced by electrical stimulation of the hippocampus in immature and mature rats. In *Hormones and Brain Function*, M. Lassati (Ed.), Plenum Press, New York, pp. 347–374.

King, J., Williams, T. and Gerall, A. (1974) Transformations of hypothalamic arcuate neurons. I. Changes associated with stages of the estrous cycle. *Cell Tiss. Res.*, 153: 497–515.

Kobayashi, H.M., Wada, M. and Uemura, H. (1972) Uptake of peroxidase from the third ventricle by ependymal cells of the median eminence. *Z. Zellforsch.*, 127: 545–551.

Kordon, C., Kerdelhve, B., Pattoo, E. and Jutisz, M. (1974) Immunocytochemical localization of LH-RH in axons and nerve terminals of the rat median eminence. *Proc. Soc. exp. Biol. (N.Y.)*, 147: 122.

Knobil, E. (1974) On the control of gonadotropin secretion in the Rhesus monkey. *Recent Progr. Hormone Res.*, 30: 1–46.

Knowles, F. and Anand Kumar, T. (1969) Structural changes related to reproduction in the hypothalamus and pris tuberalis of the Rhesus monkey. *Phil. Trans. B*, 256: 357–375.

Krulich, L., Quijada, M., Hefco, E. and Sundberg, D.K. (1974) Localization of thyrotropin-releasing factor (TRF) in the hypothalamus of the rat. *Endocrinology*, 95: 9–17.

LeBeux, Y. (1971) An ultrastructural study of the neurosecretory cells of the medial vascular prechiasmatic gland, the preoptic recess and the anterior part of the suprachiasmatic area. I. Cytoplasmic inclusions resembling nucleoli. *Z. Zellforsch.*, 114: 404–440.

LeBeux, Y. (1972) An ultrastructural study of the neurosecretory cells of the medial vascular prechiasmatic gland. II. Nerve endings. *Z. Zellforsch.*, 127: 439–461.

Leonhardt, H. (1968) Bukettformige Strukturen im Ependym der Regio hypothalamica des III. Ventrikels beim Kaninchen. Zur Neurosekretions und Rezeptorenfrage. *Z. Zellforsch.*, 88: 297–317.

Léranth, C. und Schiebler, T.H. (1974) Über die Aufnahme von Peroxidase aus dem 3. Ventrikel der Ratte. Elektronenmikroskopische Untersuchungen. *Brain Res.*, 67: 1–11.

Lisk, R.D. (1962) Inhibition of castration cell formation in the pituitary of the spayed rat by estradiol implants in the arcuate nucleus. *Amer. Zool.*, 2: 425.

Lisk, R.D. (1967) Sexual behavior: hormonal control. In *Neuroendocrinology*, Vol. 2, L. Martini and W.F. Ganong (Eds.), Academic Press, New York, pp. 197–240.

Lisk, R.D. and Newlon, M. (1963) Evidence for its direct effect on hypothalamic neurons. *Science*, 134: 223–224.

Litteria, M. (1973a) Inhibitory action of neonatal androgenization on the incorporation of (^{3}H)lysine in specific hypothalamic nuclei of the adult female rat. *Exp. Neurol.*, 41: 395–401.

Litteria, M. (1973b) Increased incorporation of ^{3}H-lysine in specific hypothalamic nuclei following castration in the male rat. *Exp. Neurol.*, 40: 309–315.

Lurie, P.O. and Weiss, J.B. (1967) Progesterone in cerebrospinal fluid during human pregnancy. *Nature (Lond.)*, 215: 1178–1179.

Makara, G., Harris, M. and Spyer, K. (1972) Identification and distribution of tuberoinfundibular neurones. *Brain Res.*, 40: 283–290.

Margolis, R.U. and Altszular, N. (1967) Insulin in the cerebrospinal fluid. *Nature (Lond.)*, 215: 1375–1376.

Matsui, T. and Kobayashi, H. (1968) Surface protrusion from the ependymal cells of the median eminence. *Arch. Anat. (Strasbourg)*, 51: 429–436.

Matsuo, H., Baba, Y., Nair, R., Arimura, A. and Schally, A. (1971) Structure of the porcine LH- and FSH-releasing hormone. 1. The proposed amino acid sequence. *Biochem. biophys. Res. Commun.*, 43: 1334–1339.

Mess, B., Fraschini, F., Motta, M. and Martini, L. (1967) The topography of the neurons synthesizing the hypothalamic releasing factors. In *Second int. Symp. Steroid Hormones*, L. Martini (Ed.), Excerpta Medica, Amsterdam, pp. 1004–1013.

Mess, B., Zanisi, M. and Tima, L. (1970). In *The Hypothalamus*, L. Martini, M. Motta and F. Fraschini (Eds.), Academic Press, New York, pp. 259–276.

Morris, M. and Knigge, K. (1975) Effect of ether anesthesia on LH-releasing hormone (LH-RH) secretion. *Fed. Proc.*, 23: 239.

Nadler, J. (1972) Intrahypothalamic locus for induction of androgen sterilization in neonatal female rats. *Neuroendocrinology*, 9: 349–357.

Nadler, J. (1973) Further evidence on the intrahypothalamic locus for androgenization of female rats. *Neuroendocrinology*, 12: 110–119.

Naik, D. (1975a) Immunoreactive LH-RH neurons in the hypothalamus identified by light and fluorescent microscopy. *Cell Tiss. Res.*, 157: 423–436.

Naik, D. (1975b) Immuno-electron microscopic localization of luteinizing hormone-releasing hormone in the arcuate nuclei and median eminence of the rat. *Cell Tiss. Res.*, 157: 437–455.

Ondo, J.G., Mical, R.S. and Porter, J.C. (1972) Passage of radioactive substances from CSF to hypophysial portal blood. *Endocrinology*, 91: 1239–1246.

Oztan, N. (1967) Neurosecretory processes projecting from the preoptic nucleus into the third ventricle of *Zoarces viviparus*. *Z. Zellforsch.*, 80: 458–460.

Palkovits, M. (1973) Isolated removal of hypothalamic or other brain nuclei of the rat. *Brain Res.*, 59: 444–450.

Palkovits, M. and Stark, E. (1972) Quantitative histological changes in rat hypothalamus following bilateral adrenalectomy. *Neuroendocrinology*, 10: 23–30.

Palkovits, M., Arimura, A., Brownstein, M., Schally, A. and Saavedra, J. (1974) Luteinizing hormone-releasing hormone (LH-RH) content of the hypothalamic nuclei in rat. *Endocrinology*, 95: 554–558.

Pecile, A., Müller, E.E. and Netti, C. (1972) Nervous system participation in growth hormone release from anterior pituitary gland. In *Growth and Growth Hormone*, A. Pecile and E.E. Müller (Eds.), Excerpta Medica, Amsterdam, pp. 261–270.

Pelletier, G., Labrie, F., Puviani, R., Arimura, A. and Schally, A. (1974) Immunohistochemical localization of luteinizing hormone-releasing hormone in the rat median eminence. *Endocrinology*, 95: 314–317.

Peute, J. (1969) Fine structure of the paraventricular organ of *Zenopus laeris* tadpoles. *Z. Zellforsch.*, 97: 564–575.

Pfaff, D. (1973) Luteinizing hormone-releasing factor potentiates lordosis behavior in hypophysectomized ovariectomized female rats. *Science*, 182: 1148–1149.

Pfaff, D. and Keiner, M. (1973) Atlas of estradiol-concentrating cells in the central nervous system of the female rat. *J. comp. Neurol.*, 151: 121–158.

Quijada, M., Krulich, L., Fawcett, C., Sundberg, D. and McCann, S. (1971) Localization of TSH-releasing factor (TRF), LH-RF and FSH-RF in the rat hypothalamus. *Fed. Proc.*, 30: 197–199.

Raisman, G. (1972) A second look at the parvicellular neurosecretory system. In *Brain–Endocrine Interaction. Median Eminence: Structure and Function, Int. Symp. Munich 1971*, K.M. Knigge, D.E. Scott and A. Weindl (Eds.), Karger, Basel, pp. 109–118.

Raisman, G. and Field, P. (1971) Anatomical considerations relevant to the interpretation of neuroendocrine experiments. In *Frontiers in Neuroendocrinology*, L. Martini and W.F. Ganong (Eds.), Oxford University Press, New York, pp. 3–44.

Raisman, G. and Field, P. (1973) Sexual dimorphism in the neuropil of the preoptic area of the rat and its dependence on neonatal androgen. *Brain Res.*, 54: 1–29.

Renaud, L. and Martin, J. (1975) Electrophysiological studies of connections of hypothalamic ventromedial nucleus neurons in the rat: evidence for a role in neuroendocrine regulation. *Brain Res.*, 93: 145–151.

Réthelyi, B. and Halász, B. (1970) Origin of nerve endings in the surface zone of the median eminence of the rat hypothalamus. *Exp. Brain Res.*, 11: 145–158.

Robinson, A.R. and Zimmerman, E.A. (1973) Cerebrospinal fluid and ependymal neurophysin. *J. clin. Invest.*, 52: 1260–1267.

Sar, M. and Stumpf, W. (1973) Autoradiographic localization of radioactivity in the rat brain after the injection of 1, 2-^{3}H-testosterone. *Endocrinology*, 92: 251–256.

Saul, G.D. and Sawyer, C.H. (1957) Atropine blockade of electrically induced hypothalamic activation of the rabbit adenohypophysis. *Fed. Proc.*, 16: 112..

Sawyer, C.H. (1956) Effects of central nervous system lesions on ovulation in the rabbit. *Anat. Rec.*, 124: 358.

Sawyer, C.H. (1959) Effects of brain lesions on estrous behavior and reflexogenous ovulation in the rabbit. *J. exp. Zool.*, 142: 227–246.

Sawaki, Y. and Yagi, K. (1973) Electrophysiological identification of cell bodies of the tubero-infundibular neurons in the cat. *J. Physiol. (Lond.)*, 230: 75–85.

Scharrer, E. and Scharrer, B. (1940) Secretory cells within the hypothalamus. *Res. Publ. Ass. nerv ment. Dis.*, 20: 170–194.

Schneider, A., Crighton, D. and McCann, S. (1969) Suprachiasmatic LH-releasing factor. *Neuroendocrinology*, 5: 271–280.

Scott, D.E. and Knigge, K.M. (1970) Ultrastructural changes in the median eminence of the rat following deafferentation of the basal hypothalamus. *Z. Zellforsch.*, 105: 1–32.

Sétáló, G., Vigh, A., Schally, A., Arimura, A. and Flerkó, B. (1975) LH-RH-containing neural elements in the rat hypothalamus. *Endocrinology*, 96: 135–142.

Shambaugh, G., Wilber, J., Montoya, E., Ruden, H. and Blonsky, E. (1975) Thyrotropin releasing hormone (TRH): measurements in human spinal fluid. *J. clin. Endocr.*, 41: 131–134.

Shin, S., Howitt, C. and Milligan, J. (1974) A paradoxical castration effect on LH-RH levels in male rat hypothalamus and serum. *Life Sci.*, 14: 2491–2496.

Spatz, H. (1951) Neues über die Verknupfung von Hypophyse und Hypothalamus. *Acta neuroveg. (Wien)*, 3: 1–23.

Spies, H., Resko, J. and Norman, R. (1974) Evidence of preoptic hypothalamic influence on ovulation in the Rhesus monkey. *Fed. Proc.*, 33: 222.

Staudt, J., Dorner, G., Doll, R. und Blose, J. (1973) Geschlechtsspezifische morphologische Unterschiede im Nucleus arcuatus und Paramamillaris ventralis der Ratte. *Endokrinologie*, 62: 234–236.

Stumpf, W.E. (1970) Estrogen neurons and estrogen-neuron systems in the periventricular brain. *Amer. J. Anat.*, 129: 207–218.

Szentágothai, J., Flerkó, B., Mess, B. and Halász, B. (Eds.) (1968) *Hypothalamic Control of the Anterior Pituitary*, Akadémiai Kiadó, Budapest, Ch. 2.

Taleisnik, S., Velasco, M. and Astrada, J. (1970) Effect of hypothalamic deafferentation on the control of luteinizing hormone secretion. *J. Endocr.*, 46: 1–7.

Tejasen, T. and Everett, J. (1967) Surgical analysis of the preoptico-tuberal pathway controlling ovulatory release of gonadotropins in the rat. *Endocrinology*, 81: 1387–1396.

Terasawa, E. and Sawyer, C. (1969) Changes in electrical activity in the rat hypothalamus related to electrochemical stimulation of adenohypophysial function. *Endocrinology*, 85: 143–149.

Uemura, H., Asai, T., Nozaki, M. and Kobayashi, H. (1975) Ependymal absorption of luteinizing hormone-releasing hormone injected into the third ventricle of the rat. *Cell Tiss. Res.*, 160: 443–452.

Van Houten, M. and Brawer, J. (1976) Evidence for nuclear extrusion in ventromedial neurons of the rat hypothalamus. *Brain Res.*, (Submitted).

Van Rees, G.P. (1972) Control of ovulation by the anterior pituitary gland. In *Topics in Neuroendocrinology, Progress in Brain Research, Vol. 38*, J. Ariëns Kappers and J.P. Schadé (Eds.), Elsevier, Amsterdam, pp. 193–210.

Velasco, M. and Taleisnik, S. (1969) Release of gonadotropins induced by amygdaloid stimulation in the rat. *Endocrinology*, 84: 132–139.

Vigh-Teichmann, I., Vigh, B. und Koritsanszky, S. (1970a) Liquorkontaktneurone im Nucleus paraventricularis. *Z. Zellforsch.*, 103: 483–501.

Vigh-Teichmann, I., Vigh, B. und Koritsanszky, S. (1970b) Liquorkontaktneurone im Nucleus lateralis tuberis von Fischen. *Z. Zellforsch.*, 105: 325–338.

Vigh-Teichmann, I., Vigh, B., Koritsanszky, S. und Aros, B. (1970c) Liquorkontaktneurone im Nucleus infundibularis. *Z. Zellforsch.*, 108: 17–34.

Weindl, A. and Joynt, R. (1972). In *Brain–Endocrine Interaction. Median Eminence: Structure and Function*, K. Knigge, D.E. Scott and A. Weindl (Eds.), Karger, Basel, pp. 280–297.

Weiner, R., Terkel, J., Blake, C., Schally, A. and Sawyer, C. (1972) Changes in serum luteinizing hormone following intraventricular and intravenous injections of luteinizing hormone-releasing hormone in the rat. *Neuroendocrinology*, 10: 261–272.

Westergaard, E. (1972) The fine structure of nerve fibers and endings in the lateral cerebral ventricles of the rat. *J. comp. Neurol.*, 144: 345–353.

Wheaton, J., Krulich, L. and McCann, S. (1975) Localization of luteinizing hormone-releasing hormone in the preoptic area and hypothalamus of the rat using radioimmunoassay. *Endocrinology*, 97: 30–38.

Wittowski, W. (1969) Ependymokrinie und Rezeptoren in der Wand des Recessus infundibularis der Maus und ihre Beziehung zum kleinzelligen Hypothalamus. *Z. Zellforsch.*, 93: 530.

Wuttke, W. (1974) Preoptic unit activity and gonadotropic release. *Exp. Brain. Res.*, 19: 205–216.

Zimmerman, E.A., Hsu, K.C., Ferin, M. and Kozlowski, G. (1974) Localization of gonadotropin-releasing hormone (Gn-RH) in the hypothalamus of the mouse by immunoperoxidase technique. *Endocrinology*, 95: 1–8.

Chapter 2

Electrophysiological correlates of neuroendocrine tissues

K.B. RUF

Department of Physiology, University of Geneva, School of Medicine, 1211 Geneva 4 (Switzerland)

I. INTRODUCTION

Mammalian neurosecretory neurons exhibit the essential properties of conventional nerve cells (Yagi et al., 1966). The study of the electrophysiological parameters of these "endocrine neurons" is therefore an obvious complement of morphological, biochemical and immunological methods employed in current neuroendocrine research. Although for this type of work a rather elaborate electronic set-up is required, the recording of electrical activity within the central nervous system is not particularly difficult, and this type of endeavor has therefore been popular ever since Cross and Green (1959) first reported on single-unit activity in the hypothalamus (for recent reviews, see Cross, 1973, 1974; Dyer, 1974). Electrophysiological studies in experimental animals have led to considerable advances in our understanding of the function of the neurohypophysis, but their contribution to the elucidation of the mechanisms of central control over the adenohypophysis and of the endocrine components of behavior have so far been disappointing. The main reason for this difference is the difficulty encountered in identifying components of and afferent pathways to the diffuse "parvicellular system" constituting the "hypophysiotrophic area" as opposed to the relative ease with which the more clearly delineated and simpler component of the "magnocellular system" can be studied.

This chapter should mainly concern itself with electrophysiological studies pertaining to the neural control of reproductive, i.e., adenohypophysial, function. However, much more solid evidence has accumulated on the neurohypophysis, and this system will be described first and in greater detail, since concepts and research strategies developed in this area are presently being extrapolated, with some success, to the study of anterior pituitary function. Other central autonomic mechanisms such as temperature control (Lomax and Green, 1975), regulation of food intake (Teitelbaum and Wolgin, 1975) and drinking behavior (Fitzsimons, 1975) have recently been reviewed and will not be discussed here.

II. ELECTROPHYSIOLOGICAL ANALYSIS OF THE NEUROHYPOPHYSIAL SYSTEM

It is of interest to note that the pioneering investigation by Kandel (1964) on the preoptic nucleus of the goldfish already contains two analytical features which are highly desirable in this field of study but which are generally lacking in other designs: antidromic identification of single units (for technical details see Cross, 1973; Dyer, 1974) and recording of intracellular potentials. With hindsight the technique of antidromic stimulation appears to be so perfectly suited for the purpose of identification of recording sites that it is difficult to see why investigators did not make use of this technique much earlier and why it was not applied to mammals before the study of Yagi et al. (1966). With respect to intracellular recordings, electrode penetration is of course facilitated by the slightly larger diameter of magnocellular preoptic cells and by the fact that the teleost brain is pulseless; the supraoptic nucleus (SON) and paraventricular nucleus (PVN) of higher vertebrates contain neurons which are smaller and thus more difficult to impale. Nevertheless, studies by Koizumi and Yamashita (1972) and by Koizumi et al. (1973) have shown that this approach is also feasible in mammals in vivo, and Sakai et al. (1974) have recently extended it to supraoptic neurosecretory tissue maintained in culture. Otherwise, studies have to rely on extracellular recordings of single or multiple units and on the frequency and pattern of discharge as their main parameters. It would appear, however, that for the assessment of modulatory hormonal influences on nervous tissue, the direct recording of changes in resting membrane potentials would provide an additional powerful and discriminatory tool for neuroendocrine research.

The principle of retrograde invasion of SON and PVN units by electrical shocks applied to nerve terminals in the neurohypophysis enables the investigator to identify true neurosecretory cells and to distinguish them from interneurons which are either unrelated to the system or connected with it by interposition of synapses. With this technique, it has also been possible to establish the existence of inhibitory recurrent collateral pathways, first in the goldfish (Kandel, 1964) and subsequently also in mammals (Kelly and Dreifuss, 1970). The demonstration of such pathways is of interest because it indicates that endocrine neurons may make true synaptic contacts with conventional nerve cells, and it raises the question as to the nature of the neurotransmitter substance released at these sites. More recently, evidence for the existence of facilitatory axon collaterals in the neurohypophysial system has been obtained (Koizumi et al., 1973).

The technique of antidromic identification of SON and PVN units does not, by itself, allow to identify their synaptic inputs or to state whether they secrete vasopressin or oxytocin. Fortunately, several experimental manipulations (e.g., teat stimulation, dilatation of the cervix uteri, hemorrhage, carotid occlusion, dehydration) are known to preferentially release one or the other of these neurohypophysial hormones, and investigators have concerned themselves with electrical recordings in these circumstances. Wakerley and Lincoln (1973) and Lincoln and Wakerley (1974, 1975) have obtained evidence for a particular type of neuronal discharge in the PVN and SON of the lactating rat (cf. Fig. 1), and these observations have led to precise calculations of oxytocin release induced by trains of action potentials (Lincoln, 1974). Single unit activity in the magnocellular system during the preferential release of vasopressin has been monitored by Dyball (1971), Vincent et al. (1972), Harris et al. (1975), among others, and phasically discharging neurons appear to be particularly implicated in the secretion of this hormone.

Since the neurons which elaborate these neurohypophysial hormones are densely clustered in the SON and PVN, it has been relatively easy to test their sensitivity to potential neurotransmitter substances applied locally by microelectrophoresis. Moss et al. (1972a) and Dreifuss and Kelly (1972) have obtained evidence for muscarinic and nicotinic excitation of these neurosecretory cells, whereas norepinephrine generally inhibits their rate of discharge. In contrast, the nature of the inhibitory transmitter liberated at recurrent collateral nerve endings is still unclear; the original suggestion by Nicoll and Barker (1971) that the neurohypophysial hormones might themselves be the chemical messengers released at these sites has not been generally accepted (Cross, 1974), although neuronal effects of such polypeptides on neurosecretory cells are clearly demonstrable in a non-mammalian system (Barker and Gainer, 1974). In the rat, the electrophoretic delivery of oxytocin leads to excitation rather than to inhibition of magnocellular cells (Moss et al., 1972b), and the phenomenon of recurrent inhibition is also observed in animals of the homozygous Brattleboro strain (Dreifuss et al., 1974; Dyball, 1974) which are unable

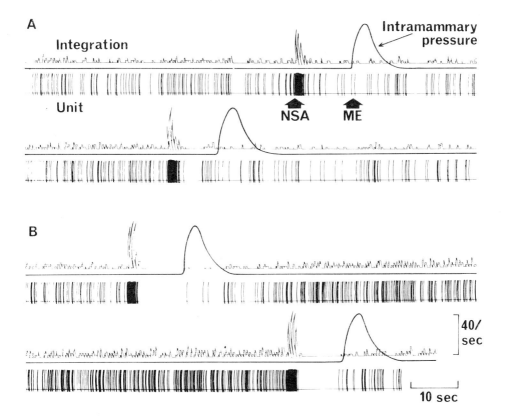

Fig. 1. Correlation between single-unit activity and changes in intramammary pressure in two identified supraoptic neurons (A and B) in the lactating rat during suckling. An increase in neurosecretory activity (NSA) is followed, with relatively uniform latency, by a milk ejection response (ME). (From Lincoln and Wakerley, 1975.)

to synthesize vasopressin. The question as to whether endocrine neurons are able to synthesize neurotransmitters which are distinct from their hormonal product assumes new significance in view of recent observations that neurons of the parvicellular system not only send axons to the neurovascular junction of the median eminence but also to neurons at more rostral sites (Harris and Sanghera, 1974; Makara and Hodács, 1975; cf. Barry and Dubois, 1974), that hypothalamic releasing factors exhibit observable behavioral effects (McCann and Moss, 1974) and that they influence single neurons when delivered by microelectrophoresis (Steiner, 1973; Dyer and Dyball, 1974; Renaud et al., 1975).

The application of the technique of antidromic identification has led investigators to realize that most endocrine cells discharge at low rates or are altogether silent. What then can be derived from the determination of the rate of neuronal discharge? Hayward and Jennings (1973), studying the activity of magnocellular neuroendocrine cells in the hypothalamus of the unanesthetized monkey, distinguished 3 types of firing patterns: (a) silent cells, (b) continuously active cells, (c) burster neurons, and they equated these functional states with, respectively (a) hormone synthesis, (b) tonic release of neurohypophysial peptides and carrier proteins, (c) pulsatile increased release of secretory products. It follows from this classification that the state of hormone synthesis has no electrophysiological equivalent at all and thus escapes detection unless antidromic invasion techniques are used. On the other hand, the observations by Wakerley and Lincoln (1973) and Lincoln and Wakerley (1974, 1975; cf. Fig. 1) on electrical activity during oxytocin release provide the clearest indication so far that depolarization by action potentials is quantitatively linked to hormone release from neurohypophysial terminals in vivo. As demonstrated by Harris et al. (1969) and by Nordmann and Dreifuss (1972), there is, however, no linear relationship between the two parameters, discharge of neurosecretory products being facilitated by higher frequency of stimulation of the terminals. Thus, the mere determination of the mean frequency of discharge of a given population of endocrine neurons is in itself an insufficient indication of the extent of hormone release, and the latter must be assessed directly, preferably on-line, by suitable assay systems. It is in this respect that investigators of the hypophysiotrophic area are most critically handicapped as compared with their colleagues studying input—output relationships in the magnocellular neuroendocrine transducer system.

III. ELECTROPHYSIOLOGICAL ANALYSIS OF GONADOTROPIN CONTROL MECHANISMS

Within the hypothalamus proper, the mediobasal hypothalamus and the medial preoptic area appear to be most prominently involved in the central control of gonadotropin release, at least in the female rat, whereas ascending neuronal pathways originating in adrenergic cells of the brain stem and the amygdala would seem to form the most important components of extrahypothalamic control. Antidromic identification of tuberoinfundibular neurons was achieved by Yagi and Sawaki (1970), by Makara et al. (1972) and by Sawaki and Yagi (1973). More recently, Yagi and Sawaki (1975) have obtained evidence, within this same system, for the existence of recurrent pathways mediating both inhibition and facilitation in the manner previously described for the neurohypophysial system of the rat (Koizumi and Yamashita, 1972; Koizumi et al., 1973). Differential sensitivity of identified tuberoinfundibular neurons to microelectrophoretically applied norepinephrine and dopamine has been reported (Moss et al., 1975b).

Projections of preoptic neurons to the mediobasal hypothalamus have been studied by Dyer and Cross (1972), Dyer (1973), Whitehead and Ruf (1974) and Wuttke (1974). There is agreement that such direct connections between these two areas implicated in reproductive biology in the female rat do indeed exist, and the chemical sensitivity of preoptic neurons projecting to the median eminence region has been tested with microelectrophoretic techniques. Dyball et al. (1974) and Whitehead and Ruf (1974) found mostly inhibitory effects of dopamine and of norepinephrine and either excitatory or no responses to acetylcholine. Only few electrophysiological studies of point-to-point connections of these areas with extrahypothalamic sites have been carried out as yet; particularly noteworthy are the report by Renaud and Martin (1975) on projections of the stria terminalis or the amygdala on identified tuberoinfundibular neurons within the ventromedial nucleus (Fig. 2), and a similar study by Fenske et al. (1975) on projections from the amygdala and the midbrain onto preoptic neurons whose axons terminate in the mediobasal hypothalamus.

If one accepts the premise that all nerve cells subjected to neuroendocrine investigations should be properly identified, judgment on all findings obtained prior to the era of

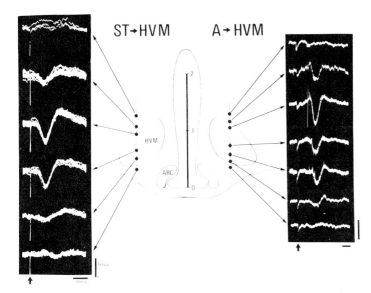

Fig. 2. Electrophysiological localization of recording sites within the ventromedial hypothalamic nucleus (HVM) based on field potentials evoked by stimulation of the stria terminalis (on the left) or amygdala (on the right). Sites of recording were either confined to the anatomical localization of the ventromedial nucleus or recorded potentials were of maximum amplitude within its boundaries. (From Renaud and Martin, 1975.)

antidromic invasion (reviewed by Beyer and Sawyer, 1969) must be reserved until it can be demonstrated that they are in keeping with results obtained in identified structures. This is of particular importance for observations which are based on the so-called "multi-unit background activity technique" (Beyer et al., 1967). Any investigator who has studied single unit activity and the responsiveness of single neurons to microelectrophoretically applied drugs cannot avoid being impressed by the large disparity of results obtained in neurons found in close vicinity to each other (cf. Fenske et al., 1975) and is led to wonder whether in a structure as heterogeneous as the hypothalamus integrated recordings from large pools of neurons might yield any interpretable results. In spite of certain improvements of the multi-unit technique (Johnson et al., 1972) and the undeniable advantage that it allows for recordings from unrestrained animals over extended periods of time, numerous technical objections have been raised (Cross, 1973; Dyer, 1974), and the discriminatory power of this approach would appear limited unless multi-unit can be separated reliably by advanced computer techniques (Roberts and Hartline, 1975). In cases where correlations between multi-unit activity and hormone concentrations have been attempted, results have been unsatisfactory (Brown-Grant, 1974); correlations with single unit activity appear more promising in this respect (Wuttke, 1974). The only area in which a comparison between findings obtained in identified structures and "blind" recordings can be made with even a modest degree of reliability would appear to be that of neuronal effects of estrogen, of gonadotropins and of luteinizing hormone releasing factor (LRF). The effects of androgens (cf. Pfaff and Gregory, 1971) and of progesterone (cf. Terasawa and Sawyer, 1970) on antidromically identified single neurons have yet to be described.

III.1. Influence of gonadal steroids on electrical activity of hypothalamic and preoptic neurons

Cross and Dyer (1970) and Dyer et al. (1972) established that the spontaneous activity of (unidentified) single units in these areas tends to be higher during proestrus than during any other stage of the estrous cycle of the rat. Unidentified estrogen-sensitive single neurons in these areas were also demonstrated by Yagi (1973). Yagi and Sawaki (1973) found an elevation of mean neuronal activity in the preoptic area during diestrus II and during proestrus (when plasma estrogen levels are rising) and an elevation of such activity in the basal hypothalamus during proestrus and after castration (i.e., in circumstances in which the secretion of luteinizing hormone is increased). These findings would appear to bear out earlier ones reported by Kawakami et al. (1970) using the multi-unit technique. However, when the sensitivity of identified preoptic neurons to injected estrogen was tested directly (Whitehead and Ruf, 1974), the steroid was found to exert a long-lasting depression of some units and to have no influence on their responsiveness to locally applied dopamine or norepinephrine (Fig. 3). These findings are hard to reconcile with the notion of a stimulatory estrogen effect on neurons situated in the presumed preoptic trigger zone for LH release; they are best explained by assuming a presynaptic site of action (ascending noradrenergic pathways?) of the steroid. Furthermore, while Yagi and Sawaki (1973) stressed similarities in the responsiveness of preoptic and hypothalamic neurons to estrogen, Bueno and Pfaff (1975) obtained evidence for reciprocal steroid effects in these areas. As long as single-unit studies disagree to such an extent on the direction of the predominant estrogen effect and as long as findings in antidromically identified neurons are not more abundant, comparisons with earlier results (summarized by Cross, 1973) are scarcely warranted.

III.2. Influence of gonadotropins and of releasing factors on electrical activity of hypothalamic and preoptic neurons

Earlier electrophysiological evidence supporting the concept that gonadotropins may exert direct effects on neuronal activity within the brain has been summarized by Beyer and Sawyer (1969). Kawakami and Sakuma (1974) have restudied the problem by electrophoretically applying luteinizing hormone, follicle-stimulating hormone and LRF to identified cells of the tuberoinfundibular tract or to preoptic neurons shown to be con-

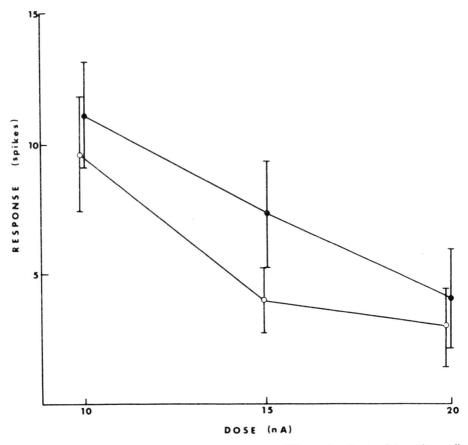

Fig. 3. Mean responses of 5 identified preoptic neurons at different dose levels of dopamine applied iontophoretically. The curve with the open circles is the dose—response curve recorded before an i.v. injection of estradiol benzoate, and the closed circles represent the dose—response curve recorded 30 min after the estrogen injection. (From Whitehead and Ruf, 1974.)

nected with the mediobasal hypothalamus. Single-unit responses to all 3 agents (mostly activation, but also instances of inhibition) were more frequent in the basal hypothalamus than in the preoptic area and were obtained with very small electrophoretic currents. However, the variable degrees of neuronal excitation, inhibition and non-responsiveness observed render the interpretation of these results difficult. Moss et al. (1975a) have reported preliminary results of the microelectrophoretic application of releasing factors and of neurotransmitters to identified preoptic neurons; the majority of cells showed little, if any, response to these agents.

IV. COMPARISON OF EXPERIMENTAL FEATURES OF THE NEUROHYPOPHYSIAL AND THE HYPOPHYSIOTROPHIC SYSTEM

As already outlined in Section II, the magnocellular system of the mammalian hypothalamus has a number of features which make electrophysiological investigations relatively easy. The anatomical distribution of neurosecretory cells in the SON and the PVN is well known, the cells are densely packed and their diameter is such that extracellular as well as intracellular recordings can be made. Their axons run in a well-defined bundle through the median eminence region and terminate in a well-defined structure, the posterior pituitary lobe. Their main secretory products, oxytocin and vasopressin, have been identified and have been available in synthetic form for a number of years, allowing investigators to develop immunohistochemical and radioimmunological methods for their precise determination. In addition, reliable, relatively simple and fairly specific bioassays, some of which can be run on-line during electrophysiological experiments, are available. Hormone release from the neurohypophysial system can readily be obtained, in vivo or in vitro, by electrical stimulation of the pituitary stalk or by the exposure of nerve endings to increased potassium concentrations. Thus the relationship between generation of action potentials, depolarization of terminals and hormone release can be studied in great detail. In contrast, the parvicellular system elaborating hypophysiotrophic substances (releasing factors) is poorly defined. These neurons are small, making recording of their electrical activity unrewarding. They appear to form diffuse systems; only a minority of their neural secretory products are known with a high degree of certainty, and even fewer have become available in synthetic form. This means that immunological methods for their detection in tissue or in the circulation are still in their infancy and that functional parameters have to be assayed by tests of anterior pituitary function which do not necessarily bear a linear relationship with the discharge of hypothalamic hormones. The secretory products of the parvicellular system are secreted, probably in their greater part, into the hypophysial portal circulation, which is notoriously inaccessible and difficult to monitor. Even if the neurosecretory elements could be identified, beyond doubt it would thus be difficult to establish input—output relationships to parallel electrical recordings.

V. SUMMARY AND CONCLUSIONS

The electrophysiological analysis of endocrine neurons which produce hypophysiotrophic substances must be preceded by the careful identification of their localization within the brain and of their afferent inputs from hypothalamic and extrahypothalamic sites. Antidromic identification of perikarya is a necessary requirement for valid electrophysiological work, but this technique must be supplemented by the demonstration, preferably by immunohistochemical means, of the presence of a particular releasing factor within a particular cell from which recordings have been obtained. When protein hormones are applied locally by microelectrophoresis, conclusive evidence must be presented that they are indeed ejected from the microelectrode tip. Biologically active water-soluble steroid conjugates should be similarly applied. Improved culture methods for the maintenance of parvicellular endocrine neurons in vitro and the development of reliable radioimmunoassay methods for the releasing factors they produce may eventually make it possible to study input—output relationships more directly and in the absence of anesthetic agents, and to correlate it more reliably with electrical events.

REFERENCES

Barker, J.L. and Gainer, H. (1974) Peptide regulation of bursting pacemaker activity in a molluscan neurosecretory cell. *Science*, 184: 1371—1373.

Barry, J. and Dubois, M.P. (1974) Study of the preoptico-infundibular LH-RH neurosecretory pathway in female guinea-pigs during gestation and the estrous cycle. In *Neurosecretions — the Final Neuroendocrine Pathway,* F. Knowles and L. Vollrath (Eds.), Springer, Berlin, pp. 148—153.

Beyer, C. and Sawyer, C.H. (1969) Hypothalamic unit activity related to control of the pituitary gland. In *Frontiers in Neuroendocrinology,* W.F. Ganong and L. Martini (Eds.), Oxford University Press, London, pp. 255—287.

Beyer, C., Ramirez, V.D., Whitmoyer, D.I. and Sawyer, C.H. (1967) Effects of hormones on the electrical activity of the brain in the rat and rabbit. *Exp. Neurol.*, 18: 313—326.

Brown-Grant, K. (1974) Steroid hormone administration and gonadotrophin secretion in the gonadectomized rat. *J. Endocr.*, 62: 319—332.

Bueno, J. and Pfaff, D.W. (1975) Single unit recording in hypothalamus and preoptic area of oestrogen-treated and untreated ovariectomized female rats. *Brain Res.*, 101: 67—78.

Cross, B.A. (1973) Unit responses in the hypothalamus. In *Frontiers in Neuroendocrinology,* W.F. Ganong and L. Martini (Eds.), Oxford University Press, New York, pp. 133—171.

Cross, B.A. (1974) The neurosecretory impulse. In *Neurosecretion — the Final Neuroendocrine Pathway,* F. Knowles and L. Vollrath (Eds.), Springer, Berlin, pp. 115—128.

Cross, B.A. and Dyer, R.G. (1970) Characterization of unit activity in hypothalamic islands with special reference to hormone effects. In *The Hypothalamus,* L. Martini, M. Motta and F. Fraschini (Eds.), Academic Press, New York, pp. 115—122.

Cross, B.A. and Green, J.D. (1959) Activity of single neurones in the hypothalamus: effect of osmotic and other stimuli. *J. Physiol. (Lond.),* 148: 554—569.

Dreifuss, J.J. and Kelly, J.S. (1972) The activity of identified supraoptic neurones and their response to acetylcholine applied by iontophoresis. *J. Physiol. (Lond.),* 220: 105—118.

Dreifuss, J.J., Nordmann, J.J. and Vincent, J.D. (1974) Recurrent inhibition of supraoptic neurosecretory cells in homozygous Brattleboro rats. *J. Physiol. (Lond.),* 237: 25P—27P.

Dyball, R.E.J. (1971) Oxytocin and ADH secretion in relation to electrical activity in antidromically identified supraoptic and paraventricular units. *J. Physiol. (Lond.),* 214: 245—256.

Dyball, R.E.J. (1974) Single unit activity in the hypothalamo-neurohypophysial system of Brattleboro rats. *J. Endocr.*, 60: 135—143.

Dyball, R.E.J., Dyer, R.G. and Drewett, R.F. (1974) Chemical sensitivity of preoptic neurones which project to the medial basal hypothalamus. *Brain Res.*, 71: 140–143.

Dyer, R.G. (1973) An electrophysiological dissection of the hypothalamic regions which regulate the pre-ovulatory secretion of luteinizing hormone in the rat. *J. Physiol. (Lond.)*, 234: 421–442.

Dyer, R.G. (1974) The electrophysiology of the hypothalamus and its endocrinological implications. In *Integrative Hypothalamic Activity, Progress in Brain Research, Vol. 41*, D.F. Swaab and J.P. Schadé (Eds.), Elsevier, Amsterdam, pp. 133–147.

Dyer, R.G. and Cross, B.A. (1972) Antidromic identification of units in the preoptic and anterior hypothalamic area projecting directly to the ventromedial and arcuate nuclei. *Brain Res.*, 43: 254–258.

Dyer, R.G. and Dyball, R.E.J. (1974) Evidence for a direct effect of LRF and TRF on single unit activity in the rostral hypothalamus. *Nature (Lond.)*, 252: 486–488.

Dyer, R.G., Pritchett, C.J. and Cross, B.A. (1972) Unit activity in the diencephalon of female rats during the oestrous cycle. *J. Endocr.*, 53: 151–160.

Fenske, M., Ellendorf, F. and Wuttke, W. (1975) Response of medial preoptic neurons to electrical stimulation of the mediobasal hypothalamus, amygdala and mesencephalon in normal, serotonin or catecholamine deprived female rats. *Exp. Brain Res.*, 22: 495–507.

Fitzsimons, J.T. (1975) The renin–angiotensin system and drinking behavior. In *Hormones, Homeostasis and the Brain, Progress in Brain Research, Vol. 42*, W.H. Gispen, Tj.B. van Wimersma Greidanus, B. Bohus and D. de Wied (Eds.), Elsevier, Amsterdam, pp. 215–233.

Harris, M.C. and Sanghera, M. (1974) Projections of medial basal hypothalamic neurones to the preoptic anterior hypothalamic areas and the paraventricular nucleus in the rat. *Brain Res.*, 81: 401–411.

Harris, G.W., Manabe, Y. and Ruf, K.B. (1969) A study of the parameters of electrical stimulation of unmyelinated fibres in the pituitary stalk. *J. Physiol. (Lond.)*, 203: 67–81.

Harris, M.C., Dreifuss, J.J. and Legros, J.J. (1975) Excitation of phasically firing supraoptic cells during vasopressin release. *Nature (Lond.)*, 258: 80–82.

Hayward, J.N. and Jennings, D.P. (1973) Activity of magnocellular neuroendocrine cells in the hypothalamus of unanesthetized monkeys. I. Functional cell types and their anatomical distribution in the supraoptic nucleus and in the internuclear zone. *J. Physiol. (Lond.)*, 232: 515–543.

Johnson, J.H., Clemens, J.A., Terkel, J., Whitmoyer, D.I. and Sawyer, C.H. (1972) Technique for recording multiple-unit activity from the brain of the freely-moving rat. *Neuroendocrinology*, 9: 90–99.

Kandel, E.R. (1964) Electrical properties of hypothalamic neuroendocrine cells. *J. gen. Physiol.*, 47: 691–717.

Kawakami, M. and Sakuma, Y. (1974) Responses of hypothalamic neurons to the microiontophoresis of LH-RH, LH and FSH under various levels of circulating ovarian hormones. *Neuroendocrinology*, 15: 290–307.

Kawakami, M., Terasawa, E. and Ibuki, T. (1970) Changes in multiple unit activity of the brain during the estrous cycle. *Neuroendocrinology*, 6: 30–48.

Kelly, J.S. and Dreifuss, J.J. (1970) Antidromic inhibition of identified rat supraoptic neurones. *Brain Res.*, 22: 406–409.

Koizumi, K. and Yamashita, H. (1972) Studies of antidromically identified neurosecretory cells of the hypothalamus by intracellular and extracellular recordings. *J. Physiol. (Lond.)*, 221: 683–705.

Koizumi, K., Ishikawa, T. and Brooks, C.McC. (1973) The existence of facilitatory axon collaterals in neurosecretory cells of the hypothalamus. *Brain Res.*, 63: 408–413.

Lincoln, D.W. (1974) Dynamics of oxytocin secretion. In *Neurosecretion – the Final Neuroendocrine Pathway*, F. Knowles and L. Vollrath (Eds.), Springer, Berlin, pp. 129–133.

Lincoln, D.W. and Wakerley, J.B. (1974) Electrophysiological evidence for the activation of supraoptic neurones during the release of oxytocin. *J. Physiol. (Lond.)*, 242: 533–554.

Lincoln, D.W. and Wakerley, J.B. (1975) Factors governing the periodic activation of supraoptic and paraventricular neurosecretory cells during suckling in the rat. *J. Physiol. (Lond.)*, 250: 443–461.

Lomax, P. and Green, M.D. (1975) Neurotransmitters and temperature regulation. In *Hormones, Homeostasis and the Brain, Progress in Brain Research, Vol 42*, W.H. Gispen, Tj.B. van Wimersma Greidanus, B. Bohus and D. de Wied (Eds.), Elsevier, Amsterdam, pp. 253–261.

Makara, G.B. and Hodács, L. (1975) Rostral projections from the hypothalamic arcuate nucleus. *Brain Res.*, 84: 23–29.

Makara, G.B., Harris, M.C. and Spyer, K.M. (1972) Identification and distribution of tuberoinfundibular neurones. *Brain Res.*, 40: 283–290.

McCann, S.M. and Moss, R.L. (1974) Putative neurotransmitters involved in discharging gonadotropin-releasing neurohormones and the action of LH-releasing hormone on the CNS. *Life Sci.*, 16: 833–852.

Moss, R.L., Urban, I. and Cross, B.A. (1972a) Microelectrophoresis of cholinergic and aminergic drugs on paraventricular neurons. *Amer. J. Pysiol.*, 223: 310–318.

Moss, R.L., Dyball, R.E.J. and Cross, B.A. (1972b) Excitation of antidromically identified neurosecretory cells of the paraventricular nucleus by oxytocin applied iontophoretically. *Exp. Neurol.*, 34: 95–102.

Moss, R.L., Kelly, M. and Dudley, C. (1975a) Responsiveness of medial-preoptic neurons to releasing hormones and neurohumoral agents. *Fed. Proc.*, 34: 219 (abstract).

Moss, R.L., Kelly, M. and Riskind, P. (1975b) Tuberoinfundibular neurons: dopaminergic and norepinephrinergic sensitivity. *Brain Res.*, 89: 265–277.

Nicoll, R.A. and Barker, J.L. (1971) The pharmacology of recurrent inhibition in the supraoptic neurosecretory system. *Brain Res.*, 35: 501–511.

Nordmann, J.J. and Dreifuss, J.J. (1972) Hormone release evoked by electrical stimulation of rat neurohypophyses in the absence of action potentials. *Brain Res.*, 45: 604–607.

Pfaff, D.W. and Gregory, E. (1971) Correlation between preoptic area unit activity and the cortical EEG: difference between normal and castrated male rats. *Electroenceph. clin. Neurophysiol.*, 31: 223–230.

Renaud, L.P. and Martin, J.B. (1975) Electrophysiological studies of connections of hypothalamic ventromedial nucleus neurons in the rat: evidence for a role in neuroendocrine regulation. *Brain Res.*, 93: 145–151.

Renaud, L.P., Martin, J.B. and Brazeau, P. (1975) Depressant action of TRH, LH-RH and somatostatin on activity of central neurones. *Nature (Lond.)*, 255: 233–235.

Roberts, W.M. and Hartline, D.K. (1975) Separation of multi-unit nerve impulse trains by a multi-channel linear filter algorithm. *Brain Res.*, 94: 141–149.

Sakai, K.K., Marks, B.H., George, J.J. and Koestner, A. (1974) The isolated organ-cultured supraoptic nucleus as a neuropharmacological test-system. *J. Pharmacol. exp. Ther.*, 190: 482–491.

Sawaki, Y. and Yagi, K. (1973) Electrophysiological identification of cell bodies of the tuberoinfundibular neurones in the rat. *J. Physiol. (Lond.)*, 230: 75–85.

Steiner, F.A. (1973) Effects of locally applied hormones and neurotransmitters on hypothalamic neurons. In *Proc. 4th Int. Congr. Endocrinol., Washington, D.C., 1972,* Int. Congr. Ser. No. 273, Excerpta Medica, Amsterdam, pp. 202–204.

Teitelbaum, P. and Wolgin, D.L. (1975) Neurotransmitters and the regulation of food intake. In *Hormones, Homeostasis and the Brain, Progress in Brain Research, Vol. 42*, W.H. Gispen, Tj.B. van Wimersma Greidanus, B. Bohus and D. de Wied (Eds.), Elsevier, Amsterdam, pp. 235–249.

Terasawa, E. and Sawyer, C.H. (1970) Diurnal variation in the effects of progesterone on multiple unit activity in the rat hypothalamus. *Exp. Neurol.*, 27: 359–374.

Vincent, J.D., Arnauld, E. and Nicolescu-Catargi, A. (1972) Osmoreceptors and neurosecretory cells in the supraoptic complex of the unanesthetized monkey. *Brain Res.*, 45: 278–281.

Wakerley, J.B. and Lincoln, D.W. (1973) The milk-ejection reflex of the rat: a 20- to 40-fold acceleration in the firing of paraventricular neurones during oxytocin release. *J. Physiol. (Lond.)*, 57: 477–493.

Whitehead, S.A. and Ruf, K.B. (1974) Responses of antidromically identified preoptic neurons in the rat to neurotransmitters and to estrogen. *Brain Res.*, 79: 185–198.

Wuttke, W. (1974) Preoptic unit activity and gonadotropin release. *Exp. Brain Res.*, 18: 205–216.

Yagi, K. (1973) Changes in firing rates of single preoptic and hypothalamic units following an intravenous administration of estrogen in the castrated female rat. *Brain Res.*, 53: 343–352.

Yagi, K. and Sawaki, Y. (1970) On the localization of neurosecretory cells controlling adenohypophysial function. *J. Physiol. Soc. Jap.*, 32: 621–622.

Yagi, K. and Sawaki, Y. (1973) Feedback of estrogen in the hypothalamic control of gonadotrophin secretion. In *Neuroendocrine Control*, K. Yagi and S. Yoshida (Eds.), Univ. Tokyo Press, Tokyo, pp. 297–325.

Yagi, K. and Sawaki, Y. (1975) Recurrent inhibition and facilitation: demonstration in the tuberoinfundibular system and effects of strychnine and picrotoxin. *Brain Res.*, 84: 155–159.

Yagi, K., Azuma, T. and Matsuda, K. (1966) Neurosecretory cell: capable of conducting impulse in rats. *Science*, 154: 778–779.

Chapter 3

The role of calcium in neuroendocrine secretion

ALAN M. POISNER

Department of Pharmacology, University of Kansas Medical Center, Kansas City, Kan. 66103 (U.S.A.)

I. INTRODUCTION

The purpose of this review is to examine the role of calcium on neuroendocrine secretion. Two types of cells will be discussed: the classical neuroendocrine cells of the hypothalamo-neurohypophysial system and the chromaffin cells of the adrenal medulla. The latter are endocrine cells (of neuroectodermal origin) under direct nervous control and a comparison of secretory mechanisms in the two types of cells provides a broader perspective. Similarities between the neurohypophysis (posterior pituitary gland) and the adrenal medulla have been discussed (Douglas, 1968; Rubin, 1970; Poisner, 1973a). In both secretory tissues hormone release occurs principally by exocytosis. Evidence for this view has been presented elsewhere (Rubin, 1970; Kirshner and Viveros, 1972; Poisner, 1973b; Dreifuss, 1975) and will not be the subject of the present discussion. Instead, attention will be directed to the role of calcium in hormone storage, transport and release.

II. CALCIUM IS REQUIRED FOR HORMONE RELEASE

II.1. *Posterior pituitary gland*

The release of hormones from the isolated posterior pituitary gland in response to electrical stimulation or to elevated potassium concentrations (high K) is dependent on extracellular calcium (Ca^{2+}) (Douglas and Poisner, 1964a; Haller et al., 1965; Ishida, 1968). No other cations besides Ca^{2+} are required in the case of electrical stimulation (Douglas and Sorimachi, 1971) or in high K-induced hormone release (Douglas and Poisner, 1964a). In fact, other physiological cations such as Na^+ and Mg^{2+} inhibit hormone release and this will be discussed later. The only cations found to be able to substitute for calcium are the related alkaline earth ions, Ba^{2+} and Sr^{2+} (Haller et al., 1965; Ishida, 1968; Buchs et al., 1972). Whether these cations replace calcium in critical steps or release intracellular calcium (or both) remains to be determined.

II.2. *Adrenal medulla*

The release of epinephrine and norepinephrine from the adrenal medulla in response to high K and a variety of secretogogues also depends on the presence of Ca^{2+} (Douglas and Rubin, 1961; Poisner and Douglas, 1966). Again, as in the case of the posterior pituitary, no other cations are required (Douglas and Rubin, 1961) and Na^+ and Mg^{2+} are inhibitory (Douglas and Rubin, 1963). Sr^{2+} and Ba^{2+} can substitute for Ca^{2+} (Douglas and Rubin, 1964).

III. HORMONE RELEASE IS GREATER AS EXTRACELLULAR CALCIUM IS RAISED

III.1. Posterior pituitary gland

The release of vasopressin in response to high K increases as Ca^{2+} is raised from 0.11 to 4.4 mM (Douglas and Poisner, 1964a) and a similar response is found for electrical stimulation (Mikiten, 1967). At higher Ca^{2+} concentrations, hormone release begins to fall.

III.2. Adrenal medulla

Catecholamine release from the perfused cat adrenal gland in response to acetylcholine and high K also increases as Ca^{2+} is raised (Douglas and Rubin, 1961, 1963). Under the experimental conditions employed, hormone release continues to rise as the Ca^{2+} is raised to 17.6 mM.

IV. HORMONE RELEASE IS ASSOCIATED WITH AN INCREASE IN CALCIUM INFLUX

IV.1. Posterior pituitary gland

When vasopressin release is evoked from the isolated posterior pituitary gland, there is an increased uptake of ^{45}Ca (used as a marker for Ca^{2+}) in the presence or absence of Na^+ (Douglas and Poisner, 1964b; Dreifuss et al., 1975). Calcium efflux is also increased by high K (Douglas and Poisner, 1964b). However, this effect is not seen in Na-free medium, even though hormone output is potentiated (Dreifuss et al., 1975). In studies on bovine posterior pituitary glands, no increase in ^{45}Ca uptake in response to electrical stimulation could be demonstrated although an increase was seen in response to high K (Russell and Thorn, 1974a). Nevertheless, further experiments by these authors suggested that electrical stimulation of the posterior pituitary gland is associated with calcium uptake (Russell and Thorn, 1974b). An increase in ^{45}Ca uptake associated with electrical stimulation of the posterior pituitary was demonstrated by other workers (Nordmann and Dreifuss, 1972). Strontium not only substitutes for Ca^{2+} in high K-induced oxytocin release, but also is taken up by the posterior pituitary gland during stimulation (Dreifuss et al., 1975).

IV.2. Adrenal medulla

^{45}Ca uptake is also increased in the adrenal medulla upon stimulation with high K or acetylcholine (Douglas and Poisner, 1961, 1962). There is also electrophysiological evidence for secretogogue-induced calcium influx. Thus, acetylcholine causes a depolarization of isolated chromaffin cells in Na-free medium which is linearly related to the logarithm of the calcium concentration (Douglas et al., 1967a, b).

V. AGENTS WHICH INCREASE CALCIUM INFLUX CAUSE HORMONE RELEASE

V.1. Posterior pituitary gland

The so-called calcium ionophores (A-23187 and X-537A) have been reported to cause an increase in calcium transport across biological membranes (Reed and Lardy, 1972; Pressman, 1973). Somewhat conflicting results have been reported concerning their effects on neurohypophysial hormone release. In two instances, X-537A has been reported to stimulate hormone release, while A 23187 was ineffective (Nakazato and Douglas, 1974; Nordmann and Currell, 1975). Under different experimental conditions, a calcium-dependent stimulation of vasopressin release by A-23187 was demonstrated (Russell et al., 1974a). In this latter case, hormone release was more pronounced when exposure to the ionophore occurred in the absence of calcium, followed by introduction of calcium without ionophore. Russell et al. (1974a) found an increase of ^{45}Ca efflux upon addition of the ionophore (A-23187). Nordmann and Currell (1975) reported that A-23187 had no effect on hormone release or ^{45}Ca uptake, while X-537A decreased ^{45}Ca uptake in concentrations which were shown by Nakazato and Douglas (1974) to increase hormone release. Since X-537A in high concentrations can induce hormone release in the absence of calcium, it has been postulated that it may also act by increasing the release of intracellular calcium (Nakazato and Douglas, 1974; Nordmann and Currell, 1975). It seems apparent that valid comparisons of ionophore actions must include a precise description of experimental techniques, including concentrations of ionophore, duration of exposure, simultaneous presence of cations and (when possible), fluxes of calcium.

V.2. Adrenal medulla

A-23187 causes a calcium-dependent release of catecholamines from the perfused cat adrenal gland is antagonized by Mg^{2+} (Garcia et al., 1975). In experiments on frog adrenals, X-537A stimulates catecholamine release in calcium-free medium, but A-23187 is ineffective with or without calcium (Ricci et al., 1975). Again, some of this discrepancy is no doubt related to the different experimental conditions (preparation, concentrations of ionophore, etc.).

VI. AGENTS WHICH INHIBIT CALCIUM INFLUX REDUCE HORMONE RELEASE

VI.1 Posterior pituitary gland

Magnesium inhibits ^{45}Ca uptake as well as vasopressin release, although the two effects are not completely parallel (Douglas and Poisner, 1964a, b). Magnesium also inhibits Sr^{2+} uptake in conditions where there is an Sr^{2+}-dependent release of oxytocin and neurophysin (Dreifuss et al., 1975). Other agents which block both hormone release and calcium uptake include cobalt, manganese, lanthanum and D-600, a calcium-channel blocker (Dreifuss et al., 1973; Russell and Thorn, 1974b).

The release of vasopressin or oxytocin in response to high K or electrical stimulation is retarded by the presence of sodium ions (Douglas and Poisner, 1964a; Mikiten, 1967; Dreifuss et al., 1971). It was shown by Dreifuss et al. (1971) that there is a linear correlation between hormone release and the ratio $[Ca/Na^2]$. This is similar to the relation between Na^+ and Ca^{2+} for myocardial contractility (Luttgau and Niedergerke, 1958), and for the facilitative effect of norepinephrine on acetylcholine release at the neuromuscular juntion (Kuba and Tomita, 1971). All these studies suggest a direct competition between sodium and calcium for a binding site which in fact could be at a calcium transport site (Birks et al., 1968; Poisner, 1973c). Some support for this comes from the finding that sodium inhibits high K-induced ^{45}Ca uptake by the isolated neurohypophysis (Dreifuss et al., 1975). A full discussion of sodium—calcium interaction in hormone release has been presented elsewhere (Poisner, 1973c).

VI.2. Adrenal medulla

The local anesthetic, tetracaine, blocks catecholamine release evoked by high K or acetylcholine (Rubin et al., 1967; Douglas and Kanno, 1967). This agent also blocks ^{45}Ca uptake (Rubin et al., 1967) and it blocks specifically the inward calcium current in isolated chromaffin cells (Douglas and Kanno, 1967). Magnesium blocks catecholamine release and ^{45}Ca exchange in perfused adrenal glands (Rubin et al., 1967). However, magnesium also blocks catecholamine release (exocytosis) induced by sodium deprivation in calcium-free medium, and thus some of the influence of magnesium may be exerted independent of calcium influx (i.e., at an intracellular site) (Lastowecka and Trifaro, 1974).

VII. AGENTS WHICH AFFECT INTRACELLULAR CALCIUM DISTRIBUTION STIMULATE OR POTENTIATE HORMONE RELEASE

VII.1. Posterior pituitary gland

Microsomal membranes prepared from the bovine neurohypophysis show an ATP-activated uptake of calcium (Poisner and Hong, 1974; Russell and Thorn, 1975). This calcium uptake is inhibited by low temperature (Poisner and Hong, 1974). Low temperature causes release of vasopressin from the isolated neurohypophysis (Douglas and Ishida, 1965; Hong and Poisner, 1974a) and also decreases ^{45}Ca efflux in 40 calcium-containing medium (Hong and Poisner, unpublished). It is therefore possible that one effect of low temperature is to increase intracellular calcium by decreasing its binding (or transport) by cellular membranes and that this increased level of cellular ionized calcium triggers hormone release. N-ethyl-maleimide (NEM) also causes vasopressin release from the isolated neurohypophysis (Douglas et al., 1965; Russell et al., 1974b) and inhibits calcium uptake by microsomal membranes (Russell and Thorn, 1975). The latter workers suggested that NEM acts by increasing intracellular calcium, since ^{45}Ca efflux was increased. However, possible direct effects of NEM on neurosecretory granules should be kept in mind in the absence of evidence for exocytosis with this agent.

Another agent known to increase calcium mobilization in various tissues is caffeine. Caffeine can initiate or potentiate cold-induced release of vasopressin from the isolated neurohypophysis (Hong and Poisner, 1972, 1974a). The release of vasopressin in response to cold temperature occurs in calcium-free medium, resembles exocytosis, and bears much resemblance to cold-induced muscle contraction (Hong and Poisner, 1974a; Poisner and Hong, 1974). In both cases, redistribution of intracellular calcium is the suggested mechanism. Cold temperature also causes the release of vasopressin from isolated neurosecretory granules, but this phenomenon shows many differences from cold-induced hormone release from the isolated neurohypophysis (Hong and Poisner, 1974a, b; Poisner and Hong, 1974).

VII.2. Adrenal medulla

An ATP-activated uptake of calcium by microsomal membranes from the adrenal medulla has also been demonstrated (Poisner and Hava, 1970). This uptake is inhibited by thiocyanate barium and sulfhydryl reagents, all of which can stimulate catecholamine release in the absence of extracellular calcium (Poisner and Hava, 1970). Moderate reduction in energy stores (by nitrogen or cyanide) potentiates calcium-evoked catecholamine release (Rubin, 1969) and this could be related to inhibition of calcium uptake by mitochondrial and microsomal membranes (Poisner and Hava, 1970).

Other agents known to stimulate catecholamine release in calcium-free medium include caffeine (Poisner, 1973d; Rahwan et al., 1973), aminophylline (Poisner, 1973e), chlorpromazine and amphetamine (Rahwan et al., 1973), thymol (Poisner, 1973a), vinblastine (Poisner, 1973g; Poisner and Cooke, 1975) and cyclic AMP (Peach, 1972). These agents have been shown to mobilize calcium from tissue stores in the adrenal medulla and/or skeletal muscle (for references, see Poisner, 1973a, b). In some cases, theophylline has been shown to potentiate acetylcholine-induced catecholamine release (Serck-Hanssen, 1974). The similarities in the utilization of membrane-bound calcium between muscle and secretory tissue have been commented on previously (Poisner, 1970, 1973a, b, c, d, e).

VIII. CALCIUM MAY BE IMPORTANT FOR STORAGE OF HORMONES IN SECRETORY GRANULES

VIII.1. Posterior pituitary gland

Calcium appears to be concentrated in neurosecretory granules. A value of 72 nEquiv./mg protein (36 nmoles/mg) was found by Thorn et al. (1975). In a study on ATP-activated uptake of ^{45}Ca by cell fractions of bovine neurohypophysis, only a very small uptake was found for the neurosecretory granules (Russell and Thorn, 1975). However, in these studies the glands were kept on ice and fractionated in the cold, maneuvers which are known to labilize secretory (Hong and Poisner, 1974a, b; Poisner and Hong, 1974). Whether the increased calcium taken up into the cell upon stimulation (Douglas and Poisner, 1964b) is partly stored in neurosecretory granules remains to be determined.

VIII.2. Adrenal medulla

Calcium is also concentrated in chromaffin granules. Values which have been reported are 65 nmoles/mg protein (Borowitz et al., 1965), 82 nmoles/mg (Borowitz, 1969), and 125 nmoles/mg (Serck-Hanssen and Christiansen, 1973). In perfused bovine adrenal glands, stimulated repeatedly with acetylcholine, there is an increased concentration of calcium in chromaffin granules which reaches 242 nmoles/mg (Serck-Hanssen and Christiansen, 1973). These authors suggest that the chromaffin granules are quantitatively important for calcium removal from the cytosol. A small component of ATP-activated ^{45}Ca uptake by chromaffin granules is resistant to the actions of mitochondrial inhibitors (Poisner and Hava 1970) and this may be relevant to the results just described. The calcium taken up by chromaffin granules is not released by EDTA (Borowitz et al., 1965; Serck-Hanssen and Christiansen, 1973) and is presumably inaccessible to the chelating agent, which does not readily penetrate the granule membrane (Kirshner et al., 1966).

Another aspect of intragranular calcium is the finding of Pletscher et al. (1974) that calcium participates with ATP and catecholamine in the formation of large molecular weight complexes. Thus, the retention of hormones within secretory granules appears to be, in part, due to the presence of calcium (and perhaps also magnesium) in large complexes which can not permeate the granule membrane.

IX. CALCIUM MAY BE IMPORTANT FOR TRANSPORT OF HORMONES

IX.1. Posterior pituitary gland

The movement of neurosecretory granules from the hypothalamus to the posterior pituitary gland appears to be due to the microtubule-dependent fast axoplasmic transport and is inhibited by colchicine (Norstrom and Sjöstrand, 1971; Norstrom et al., 1971; Norstrom, 1975). Colchicine also inhibits the release of vasopressin from the isolated neurohypophysis (Douglas and Sorimachi, 1972a) the potential role of calcium in microtubule-associated functions will be discussed below.

IX.2. Adrenal medulla

The distance traversed by a secretory granule from its site of formation to its site of discharge is much greater in the neurohypophysial system than in the adrenal medulla. Nevertheless, the chromaffin cells are very well endowed with microtubules and microtubule protein (Redburn et al., 1972; Poisner, 1973f; Poisner and Cooke, 1975). A combination of morphological, biochemical and pharmacological evidence suggests that microtubules are important for catecholamine secretion (Poisner and Bernstein, 1970; Poisner, 1973f; Poisner and Cooke, 1975). Since some secretogogues are relatively insensitive to the inhibitory effects of colchicine, it has been suggested that the influence of the antimitotic drugs is on cholinergic receptors (Douglas and Sorimachi, 1972b; Trifaro et al., 1972). However, there are many reasons for concluding that the effect of antimitotic agents is not simply exerted on cholinergic receptors (Poisner, 1973f; Poisner and Cooke, 1975). It is likely that the intracellular movement of granules toward the cell surface in the adrenal medulla, as in other cells, is related to an interaction with microtubules. Antagonism between the effects of calcium and antimitotic drugs has been shown for catecholamine release (Poisner and Cooke 1975). The potential sites of calcium—microtubule interaction include: (a) effects on tubulin polymerization, (b) effects on granule-tubule binding, and (c) effects on biochemical reactions associated with tubulin.

Calcium can be shown to inhibit tubulin polymerization in vitro (Borisy and Olmsted, 1972; Weisenberg, 1972) and possibly in situ (Schlapfer and Bunge, 1973). On the other hand, there is evidence that calcium may be required for fast axoplasmic transport (Dravid and Hammerschlag, 1975; Hammerschlag et al., 1975). More recent evidence suggests that even in vitro calcium may facilitate microtubule assembly (Borisy et al., 1975).

Evidence that calcium may bind to granules will be presented later. The possibility that calcium binds to tubulin has been discussed (Wilson et al., 1970; Poisner and Cooke, 1975). This type of evidence, as well as the close association between chromaffin in granules and microtubules (Poisner and Cooke, 1975), suggests that calcium may serve as a bridge between microtubules and secretory granules. Connection between microtubules and synaptic vesicles have been demonstrated (Smith, 1971; Smith et al., 1975).

The apparent energy requirement for axoplasmic transport has led to the suggestion that a Mg-Ca-ATPase may be important in energy transduction (Ochs, 1971). As will be discussed later, energy depletion inhibits the secretion of vasopressin (Douglas et al., 1965, Warberg and Thorn, 1969) and of catecholamine (Rubin, 1969).

Another cellular element that may be involved in intracellular transport is the microfilament, which is thought to be similar to muscle actin (Wessells et al., 1971). Microfilaments are present in chromaffin cells in association with a subplasmalemmal group of microtubules (Poisner, 1973f; Poisner and Cooke, 1975; Cooke and Poisner, unpublished). A similar distribution is found in beta cells of the pancreas, and a theory of secretion involving the microtubule-microfilament network has been presented (Malaisse et al., 1975). Although cytochalasin B, an agent known to disrupt microfilaments, can inhibit hormone release from the posterior pituitary gland (Douglas and Sorimachi, 1972a) and from the adrenal medulla (Douglas and Sorimachi, 1972a; Poisner, 1973f), this agent has other actions which may not be related to microfilaments.

Biochemical reactions associated with microtubules may be affected by calcium. Tubulin-associated proteins have been found to have protein kinase activity and/or to be phosphorylated in vitro or in situ (Lagnado et al., 1975; Rappaport et al., 1975; Sloboda et al., 1975). This activity is stimulated by cyclic AMP and inhibited by calcium (Rappaport et al., 1975). It is possible that the phosphorylated proteins may serve as calcium-binding sites which regulate microtubule—granule interaction. Neurosecretory granules also show endogenous, cyclic AMP-activated, phosphorylation which is inhibited by calcium (McKelvy, 1975). Furthermore, much of the phosphorylated component is present in granule membranes and the protein kinase activity is greater in mixtures of neurosecretory granules and tubulin (McKelvy, 1975).

Membranes of chromaffin granules are also phosphorylated in vitro from ^{32}ATP. Some of the phosphorylated membrane components are protein and some are lipid (Trifaro and Dworkind, 1971, 1975). Tubulin-like protein has also been found in isolated chromaffin granules (Redburn et al., 1972; Poisner, 1973f; Poisner and Cooke, 1975). The possible interactions of calcium, cyclic AMP, protein kinase and tubulin with chromaffin granules remain to be determined.

X. CALCIUM MAY ACT DIRECTLY ON THE SECRETORY GRANULE

Calcium could act directly on the granule (a) as a bridge between granule membrane and microtubules or plasma membrane, (b) as a co-factor in a granule membrane enzymatic reaction, or (c) as a direct physical effect.

Calcium is known to reduce the negative change on neurosecretory granules (Poisner and Douglas, 1968; Vilhardt and Jorgensen, 1972) and on chromaffin granules (Banks, 1966). This has led to the suggestion that calcium may serve as a bridge between granule and plasma membrane (Banks, 1966; Poisner and Trifaro, 1967; Poisner and Douglas, 1968). It has been argued that this approximation may permit the fusion reaction (Poisner and Trifaro, 1967; Poisner, 1973a) and/or an ATP-activated release process (Poisner and Trifaro, 1967; Poisner and Douglas, 1968). It is known that sites of attachment of synaptic vesicles with neuronal membrane increase in number only in the presence of calcium (Pfenninger and Rovainen, 1974). Calcium-stained particles can be seen on synaptic vesicles fixed in calcium-containing solutions (Politoff et al., 1974). The possible association between calcium, granules, and microtubules is discussed elsewhere (Section IX; Poisner, 1973a).

Poisner and Trifaro (1967) suggested that calcium serves as a co-factor in ATP-induced release of catecholamines in situ by bringing the granule membrane ATPase in proximity to mobilized plasma membrane ATP (Section XI). A similar suggestion was made concerning vasopressin release (Poisner and Douglas, 1968). Another possibility is that calcium acts on a tropomyosin-like complex to facilitate ATP-induced hormone release (Poisner, 1970, 1973a, b). This suggestion is strengthened by the finding that a protein extracted from the adrenal medulla conveys calcium sensitivity on the ATP-activated release reaction (Oka et al., 1972).

In reasonable concentrations, calcium alone has no direct effect on hormone release from isolated secretory granules (Poisner and Trifaro, 1967; Poisner and Douglas, 1968). The possible role of calcium in aggregating granule membranes has been considered (Poisner, 1970; Edwards et al., 1974).

XI. CALCIUM MAY BE CRITICAL FOR THE FUSION PROCESS

The mechanism by which membranes fuse with one another is still not well understood. This obviously applies to the fusion of secretory granule membranes with surface membranes which takes place during exocytosis. It has been suggested that calcium, along with ATP and ATPase, plays a central role in the fusion of chromaffin granules with the cell membrane (Poisner and Trifaro, 1967). Calcium ATP, and ATPase are central to many theories of fusion (Poste and Allison, 1971) and endocytosis, as well (Ben-Bassat et al., 1972; Penniston, 1972). More recent studies on calcium and membrane fusion have shown that calcium is critical for fusion of muscle membranes (Shainberg et al., 1970; Van der Bosch et al., 1972) and erythrocyte membranes (Ahkong et al., 1975). Furthermore it has been shown that Mg^{2+} inhibits the fusion-promoting effect of Ca^{2+} (Schudt et al., 1973). Thus the respective effects of Ca^{2+} and Mg^{2+} on membrane fusion parallel their effects on hormone release by exocytosis. It has been suggested that aggregation of intrinsic membrane proteins (or intramembranous particles) allows membrane fusion (Poste and Allison, 1973). Whether increased membrane fluidity is a consequence or a cause of particle aggregation is uncertain (Ahkong et al., 1975). It has been found that exocytosis in the neurohypophysis (a calcium-dependent process) is associated with a reduction in the number of intramembranous particles in the nerve membrane (Dempsey et al., 1973). The changes in the surface membrane may begin to occur prior to frank exocytosis, perhaps in the stage of attachment of the neurosecretory granule to the nerve membrane (Dempsey et al., 1973). When isolated chromaffin granules are exposed to Ca^{2+}, they adhere to one another in a reversible way and in some cases, appear to fuse (Edwards et al., 1974). However, Mg^{2+} acts like Ca^{2+}, high concentrations of Ca^{2+} have been employed, and the time sequence is relatively slow. Nevertheless, the reversible aggregation and fusion of granule membranes under the influence of calcium points to a potential role in exocytosis. It is interesting that in some secretory cells chains of granules are observed during secretory activity (Ichikawa, 1965; Amsterdam et al., 1969; Rohlich et al., 1971).

XII. CALCIUM MAY INTERACT WITH CYCLIC NUCLEOTIDES

XII.1. Posterior pituitary gland

Although cyclic AMP has been implicated in hormone release in a number of tissues, its role in neurohypophysial and adrenomedullary secretion is not clearly established. In isolated rat neurohypophysis, neither dibutyryl cyclic AMP nor dibutyryl cyclic AMP caused hormone release at a concentration of 10 mM (Thorn et al., 1975). The fact that caffeine can potentiate hormone release in some conditions (Hong and Poisner, 1972, 1974a) may simply reflect an action of caffeine on calcium mobilization independent of its inhibition of phosphodiesterase. As mentioned before, cyclic AMP stimulates the phosphorylation of neurosecretory granule membranes by endogenous protein kinase and this process is inhibited by calcium (McKelvy, 1975). The only positive information on a possible role of cyclic AMP in vasopressin release comes from studies on the isolated rat neurohypophysis. Vasopressin release is stimulated in this preparation by angiotensin II and this is associated with an accumulation of cyclic AMP (Gagnon et al., 1973; Gagnon and Heisler, 1974). The accumulation of cyclic AMP occurred only in the presence of theophylline. Since calcium has been shown to be critical for hormone release and since cyclic AMP can increase calcium efflux from mitochondria (Borle, 1974), it is possible that an accumulation of cyclic AMP could initiate hormone release through calcium mobilization and/or protein kinase activity. However, this has yet not been established and this speculation does not require cyclic AMP mediation of physiologically induced hormone release.

XII.2. Adrenal medulla

The possible role of cyclic nucleotides in catecholamine release has been considered previously (Poisner, 1970, 1973a, b). Most of the features of second messenger status for cyclic AMP have been found. Thus, cyclic AMP levels in the adrenal medulla are increased by maneuvers which cause hormone release (Guidotti et al., 1973), cyclic AMP itself can stimulate catecholamine secretion even in the absence of calcium (Peach, 1972); phosphodiesterase inhibitors initiate secretion and raise cyclic AMP levels (Guidotti et al., 1973; Poisner, 1973d, e) and cyclic AMP-activated protein kinase is also present in the adrenal medulla (Shima et al., 1974). However, activation of adenyl cyclase in cell fractions of the adrenal medulla by secretogogues has not been achieved (Serck-Hanssen et al., 1972; Hurko et al., 1974). In addition to this missing criterion of second messenger status, there are other reasons for questioning the importance of cyclic AMP in catecholamine release (Guidotti et al., 1973; Jaanus and Rubin, 1974).

The levels of cyclic GMP in the adrenal medulla are also increased by aminophylline (Guidotti et al., 1973) and the protein kinase activity in the medulla is also activated by cyclic GMP (Shima et al., 1974). Experiments in progress indicate that cyclic GMP can evoke catecholamine release from the isolated adrenal gland and that acetylcholine increases cyclic GMP levels (Poisner, 1973b; Shima et al., 1974). In many tissues, cyclic GMP accumulation is produced through muscarinic stimulation by a calcium-sensitive mechanism (Schultz et al., 1973).

Cyclic nucleotide levels in the adrenal medulla may also be regulated by a calcium-sensitive activator of phosphodiesterase (Uzunov et al., 1975).

XIII. CALCIUM MAY REGULATE CONTRACTILE ACTIVITY DURING HORMONE RELEASE

In early studies on secretory mechanisms in the adrenal medulla and the neurohypophysis, attention was drawn to the parallels to muscle physiology (Douglas and Rubin, 1961, 1963; Douglas and Poisner, 1961, 1964a). This parallelism was stressed repeatedly in further work, although the primary basis for this comparison was related to the similar critical requirement for calcium. A new reason for implicating a contractile process was presented in 1967 when it was shown that catecholamine release from isolated chromaffin granules could be evoked by ATP plus Mg^{2+} and was correlated with ATPase activity (Poisner and Trifaro, 1967). Subsequent studies showed that vasopressin release from neurosecretory granules is also induced by ATP and Mg^{2+} (Poisner and Douglas, 1968).

Since inhibitors of ATP synthesis suppress hormone release (Douglas et al., 1965; Kirshner and Smith, 1966; Rubin, 1969; Warburg and Thorn, 1969), it had been suspected that ATP was important for some step in hormone release. With the demonstration of the effect of ATP on isolated granules, it was suggested that the granule membrane ATPase functions as a mechanochemical transducer similar to the actomyosin system (Poisner and Trifaro, 1967, 1969; Poisner, 1970, 1973a). One missing link in this argument was the apparent lack of requirement for calcium. However, it was pointed out that this might be due to the lack of a tropomyosin—troponin complex in the in vitro test system (Poisner, 1970). Subsequent work has shown that a cytoplasmic factor can convey calcium sensitivity to the ATP- and Mg^{2+}-activated release of catecholamines (Oka et al., 1972). This further strengthens the argument for a truly contractile event (Poisner, 1970) since the parallel to isolated actomyosin contraction is apparent (Poisner, 1973a). A further step in establishing the contractile nature of hormone release would be to demonstrate the presence of actomyosin-like proteins in chromaffin granules. Initial studies along this line have been presented (Poisner, 1970, 1973a). It has recently been shown that chromaffin granules also bind exogenous actin and myosin (Burridge and Phillips, 1975). The presence of myosin-like protein in synaptic vesicles and a tropomyosin—troponin complex (calcium-sensitizing factor) in synaptosomes has also been described (Berl et al., 1973; Fine et al., 1973; Puszkin and Kochwa, 1974). These more recent findings strengthen the suggestion that release of hormones from secretory granules involves a contractile event utilizing actin, myosin and the tropomyosin—troponin complex. It is likely that actin and myosin are present in surface membranes and actin-like protein is also present in microfilaments in the adrenal medulla (Cooke and Poisner, unpublished).

XIV. SUMMARY AND CONCLUSIONS

The natural physiological stimulus to neurohypophysial hormone release is the action potential transmitted down the pituitary stalk from the hypothalamus. The subsequent depolarization triggers release by allowing calcium influx. Similarly, in the adrenal medulla, the natural stimulus is acetylcholine released from the splanchnic nerve endings which also causes depolarization and calcium influx. In both systems sodium ions are the principal charge carriers for depolarization but only Ca^{2+} is required for hormone release. A rise in intracellular calcium also seems to be required for release of neurotransmitters (Miledi, 1973; Llinás and Nicholson, 1975). Pharmacological or non-physiological means of inducing hormone release (for example, low temperature, caffeine, low sodium) may evoke hormone release with no depolarization or in fact with hyperpolarization. In some cases calcium may be mobilized from internal stores. Cytoplasmic Ca^{2+} is regulated in part by Na^+ (intra- and extracellular) (Baker, 1972; Blaustein, 1974). The common features in hormone release from these neuroendocrine glands are a requirement for calcium and a source of metabolic energy, probably ATP.

The secretory granules in the vicinity of the surface membrane are the first ones to discharge hormone. The transport of granules which are some distance away is probably facilitated by a mechanism involving microtubules. This transport may be regulated by calcium and cyclic nucleotides and may also involve microfilaments. It is proposed that both the movement of granules to the cell surface and the final ejection of hormone via exocytosis utilize a contractile process and, like muscle, this involves ATP and a calcium-regulated step. Finally, the fusion process of granule membrane with surface membrane may also utilize ATP and calcium. Following exocytosis, ATP must be regenerated and calcium levels must be returned to resting levels. This requires energy and includes surface and intracellular membranes, probably including the secretory granules.

This overview on neuroendocrine secretion stresses three points. (1) Hormone release from the neurohypophysis shares many similarities with secretion from the adrenal medulla. In both systems, hormone release occurs via exocytosis by calcium- and energy-dependent steps. These steps may utilize other intracellular organelles in addition to the secretory granules. (2) It is suggested that several steps in the overall secretory process (including the final discharge step) involve the same biochemical events associated with muscle contraction, utilizing actin, myosin and tropomyosin-like proteins. (3) Calcium, which regulates many functions in a variety of cells, modifies and controls hormone release by acting at multiple sites.

REFERENCES

Ahkong, Q.F., Fisher, D., Tampion, W. and Lucy, J.A. (1975) Mechanisms of cell fusion. *Nature (Lond.),* 253: 194—195.

Amsterdam, A., Ohad, I. and Schramm, M. (1969) Dynamic changes in the ultrastructure of the acinar cell of the rat parotid gland during the secretory cycle. *J. Cell Biol.*, 41: 753—773.

Baker, P.F. (1972) Transport and metabolism of calcium ions in nerve. *Progr. Biophys. molec. Biol.,* 24: 177—223.

Banks, P. (1966) An interaction between chromaffin granules and calcium ions. *Biochem. J.*, 101: 18C—20C.

Ben-Bassat, I., Bensch, K.G. and Schrier, S.L. (1972) Drug-induced erythrocyte membrane internalization. *J. clin. Invest.,* 51: 1833—1844.

Berl, S., Puszkin, S. and Nicklas, W.J. (1973) Actomyosin-like protein in brain. *Science,* 179: 441—446.

Birks, R.I., Burstyn, P.G.R. and Firth, D.R. (1968) The form of sodium-calcium competition at the frog myoneural junction. *J. gen. Physiol.,* 52: 887—907.

Blaustein, M.P. (1974) The inter-relationship between sodium and calcium fluxes across cell membranes. *Rev. Physiol. Biochem. Pharmacol.,* 70: 33—82.

Borisy, G.G. and Olmsted, J.B. (1972) Nucleated assembly of microtubules in porcine brain extracts. *Science,* 177: 1196—1197.

Borisy, G.G., Marcum, J.M., Olmsted, J.B., Murphy, D.B. and Johnson, K.A. (1975) Purification of tubulin and associated high molecular weight proteins from porcine brain and characterization of microtubule assembly *in vitro. Ann. N.Y. Acad. Sci.,* 253: 107—131.

Borle, A.B. (1974) Cyclic AMP stimulation of calcium efflux from kidney, liver and heart mitochondria. *J. Membrane Biol.,* 16: 221—236.

Borowitz, J.L. (1969) Effect of acetylcholine on subcellular distribution of ^{45}Ca in bovine adrenal medulla. *Biochem. Pharmacol.,* 18: 715—723.

Borowitz J.L., Fuwa, K. and Weiner, N. (1965) Distribution of metals and catecholamines in bovine adrenal medulla sub-cellular fractions. *Nature (Lond.),* 205: 42—43.

Buchs, M., Dreifuss, J.J., Grau, J.D. and Nordmann, J.J. (1972) Strontium as a substitute for calcium in the process leading to neurohypophysial hormone secretion. *J. Physiol. (Lond.),* 222: 168—169.

Burridge, K. and Phillips, J.H. (1975) Association of actin and myosin with secretory granule membranes. *Nature (Lond.),* 254: 526—529.

Dempsey, G.P., Bullivant, S. and Watkins, W.B. (1973) Ultrastructure of the rat posterior pituitary gland and evidence of hormone release by exocytosis as revealed by freeze-fracturing. *Z. Zellforsch.,* 143: 465—484.

Douglas, W.W. (1968) Stimulus-secretion coupling: the concept and clues from chromaffin and other cells. *Brit. J. Pharmacol.,* 34: 451—474.

Douglas, W.W. and Ishida, A. (1965) The stimulant effect of cold on vasopressin release from the neurohypophysis *in vitro. J. Physiol. (Lond.),* 179: 185—191.

Douglas, W.W. and Kanno, T. (1967) The effect of amethocaine on acetylcholine-induced depolariza-

tion and catecholamine secretion in the adrenal chromaffin cell. *Brit. J. Pharmacol.,* 30: 612–619.
Douglas, W.W. and Poisner, A.M. (1961) Stimulation of uptake of calcium-45 in the adrenal gland by acetylcholine. *Nature (Lond.),* 192: 1299.
Douglas, W.W. and Poisner, A.M. (1962) On the mode of action of acetylcholine in evoking adrenal medullary secretion increased uptake of calcium during the secretory response. *J. Physiol. (Lond.),* 162: 385–392.
Douglas, W.W. and Poisner, A.M. (1964a) Stimulus-secretion coupling in a neurosecretory organ: the role of calcium in the release of vasopressin from the neurohypophysis. *J. Physiol. (Lond.),* 172: 1–18.
Douglas, W.W. and Poisner, A.M. (1964b) Calcium movement in the neurohypophysis of the rat and its relation to the release of vasopressin. *J. Physiol. (Lond.),* 172: 19–30.
Douglas, W.W. and Rubin, R.P. (1961) The role of calcium in the secretory response of the adrenal medulla to acetylcholine. *J. Physiol. (Lond.),* 159: 40–57.
Douglas, W.W. and Rubin, R.P. (1963) The mechanism of catecholamine release from the adrenal medulla and the role of calcium in stimulus-secretion coupling. *J. Physiol. (Lond.),* 167: 288–310.
Douglas, W.W. and Rubin, R.P. (1964) The effects of alkaline earths and other divalent cations on adrenal medullary secretion. *J. Physiol. (Lond.),* 175: 231–241.
Douglas, W.W. and Sorimachi, M. (1971) Electrically evoked release of vasopressin from isolated neurohypophyses in sodium-free media. *Brit. J. Pharmacol.,* 42: 647P.
Douglas, W.W. and Sorimachi, M. (1972a) Effects of cytochalasin B and colchicine on secretion of posterior pituitary and adrenal medullary hormones. *Brit. J. Pharmacol.,* 45: 143P–144P.
Douglas, W.W. and Sorimachi, M. (1972b) Colchicine inhibits adrenal medullary secretion evoked by acetylcholine without affecting that to potassium. *Brit. J. Pharmacol.,* 45: 129–132.
Douglas, W.W., Ishida, A. and Poisner, A.M. (1965) The effect of metabolic inhibitors on the release of vasopressin from the isolated neurohypophysis. *J. Physiol. (Lond.),* 181: 753–759.
Douglas, W.W., Kanno, T. and Sampson, S.R. (1967a) Effects of acetylcholine and other medullary secretagogues and antagonists on the membrane potential of adrenal chromaffin cells: an analysis employing techniques of tissue culture. *J. Physiol. (Lond.),* 188: 107–120.
Douglas, W.W., Kanno, T. and Sampson, S.R. (1967b) Influence of the ionic environment on the membrane potential of adrenal chromaffin cells and on the depolarizing effect of acetylcholine. *J. Physiol. (Lond.),* 191: 107–121.
Dravid, A.R. and Hammerschlag, R. (1975) Axoplasmic transport of proteins *in vitro* in primary afferent neurons of frog spinal cord: effect of Ca^{++}-free incubation conditions. *J. Neurochem.,* 24: 711–718.
Dreifuss, J.J. (1975) A review on neurosecretory granules: their contents and mechanisms of release. *Ann. N.Y. Acad. Sci.,* 248: 184–199.
Dreifuss, J.J., Grau, J.D. and Bianchi, R.E. (1971) Antagonism between Ca and Na ions at neurohypophyseal nerve terminals. *Experientia (Basel),* 27: 1295–1296.
Dreifuss, J.J., Grau, J.D. and Nordmann, J.J. (1973) Effects on the isolated neurohypophysis of agents which affect the membrane permeability to calcium. *J. Physiol. (Lond.),* 231: 96P–98P.
Dreifuss, J.J., Grau, J.D. and Nordmann, J.J. (1975) Calcium movements related to neurohypophysial hormone secretion. In *Calcium Transport in Contraction and Secretion,* E. Carafoli, F. Clementi, W. Drabikowski, and A. Margreth (Eds.), North-Holland, Amsterdam, pp. 271–279.
Edwards, W., Phillips, J.H. and Morris, S.J. (1974) Structural changes in chromaffin granules induced by divalent cations. *Biochim. biophys. Acta (Amst.),* 356: 164–173.
Fine, R.E., Blitz, A.L., Hitchcock, S.E. and Kaminer, B. (1973) Tropomyosin in brain and growing neurones. *Nature New Biol.,* 245: 182–186.
Gagnon, D.J. and Heisler, S. (1974) Accumulation of cyclic AMP in neurohypophyses in response to angiotensin. *Biochim. biphys. Acta (Amst.),* 338: 394–397.
Gagnon, D.J., Cousineau, D. and Boucher, P.J. (1973) Release of vasopressin by angiotensin II and prostaglandin E_2 from the rat neuro-hypophysis *in vitro*. *Life Sci.,* 12: 487–497.

Garcia, A.G., Kirpekar, S.M. and Prat, J.C. (1975) A calcium ionophore stimulating the secretion of catecholamines from the cat adrenal. *J. Physiol. (Lond.)*, 244: 253–262.

Guidotti, A., Mao, C.C. and Costa, E. (1973) Transsynaptic regulation of tyrosine hydroxylase in adrenal medulla: possible role of cyclic nucleotides. In *Frontiers in Catecholamine Research*, E. Usdin and S. Snyder (Eds.), Pergamon Press, New York, pp. 231–236.

Haller, E.W., Sachs, H., Sperelakis, N. and Share, L. (1965) Release of vasopressin from isolated guinea pig posterior pituitaries. *Amer. J. Physiol.*, 209: 79–83.

Hammerschlag, R., Dravid, A.R. and Chiu, A.Y. (1975) Mechanism of axonal transport: a proposed role for calcium ions. *Science*, 188: 273–275.

Hong, J.S. and Poisner, A.M. (1972) Release of vasopressin from bovine posterior pituitary glands *in vitro* induced by low temperature, caffeine, high potassium and electrical stimulation. *Proc. 5th Int. Congr. Pharmacol.* (Abstr.): 105.

Hong, J.S. and Poisner, A.M. (1974a) Effect of low temperature on the release of vasopressin from the isolated bovine neurohypophysis. *Endocrinology*, 94: 234–240.

Hong, J.S. and Poisner, A.M. (1974b) Further studies on cold-induced release of vasopressin from isolated bovine neurosecretory granules. *Neuroendocrinology*, 16: 165–177.

Hurko, O., Elster, P. and Wurtman, R.J. (1974) Adenylate cyclase activity in bovine adrenal medulla. *Endocrinology*, 94: 592–593.

Ichikawa, A. (1965) Fine structural changes in response to hormonal stimulation of the perfused canine pancreas. *J. Cell Biol.*, 24: 369–385.

Ishida, A. (1968) Stimulus-secretion coupling on the oxytocin release from the isolated posterior pituitary lobe. *Jap. J. Physiol.*, 18: 471–480.

Jaanus, S.D. and Rubin, R.P. (1974) Analysis of the role of cyclic adenosine 3′,5′-monophosphate in catecholamine release. *J. Physiol. (Lond.)*, 237: 465–476.

Kirshner, N. and Smith, W.J. (1966) Metabolic requirements for secretion from the adrenal medulla. *Science*, 154: 422–423.

Kirshner, N. and Viveros, O.H. (1972) The secretory cycle in the adrenal medulla. *Pharmacol. Rev.*, 24: 385–398.

Kirshner, N., Hoolloway, C., Smith, W.J. and Kirshner, A.G. (1966) Uptake and storage of catecholamines. In *Mechanisms of Release of Biogenic Amines*, U.S. von Euler, S. Rosell and B. Uvnäs (Eds.), Pergamon Press, Oxford, pp. 109–123.

Kuba, K. and Tomita, T. (1971) Noradrenaline action on nerve terminal in the rat diaphragm. *J. Physiol. (Lond.)*, 217: 19–31.

Lagnado, J., Tan, L.P. and Reddington, M. (1975) The *in situ* phosphorylation of microtubular protein in brain cortex slices and related studies on the phosphorylation of isolated brain tubulin preparations. *Ann. N.Y. Acad. Sci.*, 253: 577–597.

Lastowecka, A. and Trifaro, J.M. (1974) The effect of sodium and calcium ions on the release of catecholamines from the adrenal medulla: sodium deprivation induces release by exocytosis in the absence of extracellular calcium. *J. Physiol. (Lond.)*, 236: 681–705.

Llinás, R. and Nicholson, C. (1975) Calcium role in depolarization-secretion coupling: an aequorin study in squid giant synapse. *Proc. nat. Acad. Sci. (Wash.)*, 72: 187–190.

Luttgau, H.C. and Niedergerke, R. (1958) The antagonism between Ca and Na ions on the frog's heart. *J. Physiol. (Lond.)*, 14: 486–505.

Malaisse, W.J., Malaisse-Lagae, F., Van Obberghen, E., Somers, G., Devis, G., Ravazzola, M. and Orci, L. (1975) Role of microtubules in the phasic pattern of insulin release. *Ann. N.Y. Acad. Sci.*, 253: 630–652.

McKelvy, J.F. (1975) Phosphorylation of neurosecretory granules by AMP-stimulated protein kinase and its implication for transport and release of neurophysin proteins. *Ann. N.Y. Acad. Sci.*, 248: 80–91.

Mikiten, T.M. (1967) *Electrically Stimulated Release of Vasopressin from Rat Neurohypophyses in vitro*, Ph.D. Thesis, Yeshiva University, New York.

Miledi, R. (1973) Transmitter release induced by injection of calcium into nerve terminals. *Proc. roy. Soc. B*, 193: 427–430.

Nakazato, Y. and Douglas, W.W. (1974) Vasopressin release from the isolated neurohypophysis induced by a calcium ionophore, X-537A. *Nature (Lond.)*, 249: 479–481.

Nordmann, J.J. and Currell, G.A. (1975) The mechanism of calcium ionophore induced secretion from the rat neurohypophysis. *Nature (Lond.)*, 253: 646–647.

Nordmann, J.J. and Dreifuss, J.J. (1972) Hormone release evoked by electrical stimulation of rat neurohypophyses in the absence of action potentials. *Brain Res.*, 45: 604–607.

Norstrom, A. (1975) Axonal transport and turnover of neurohypophyseal proteins in the rat. *Ann. N.Y. Acad. Sci.*, 248: 46–63.

Norstrom, A. and Sjöstrand, J. (1971) Axonal transport of proteins in the hypothalamo-neurohypophysial system of the rat. *J. Neurochem.*, 18: 29–39.

Norstrom, A., Hansson, H.A. and Sjöstrand, J. (1971) Effects of colchicine on axonal transport and ultrastructure of the hypothalamo-neurohypophyseal system of the rat. *Z. Zellforsch.*, 113: 271–293.

Ochs, S. (1971) Local supply of energy to the fast axoplasmic transport mechanism. *Proc. nat. Acad. Sci. (Wash.)*, 68: 1279–1282.

Oka, M., Izumi, F. and Kashimoto, T. (1972) Effects of cytoplasmic and microsomal fractions on ATP-Mg^{++} stimulated catecholamine release from isolated adrenomedullary granules. *Jap. J. Pharmacol.*, 22: 207–214.

Peach, M.J. (1972) Stimulation of release of adrenal catecholamine by adenosine 3′,5′-cyclic monophosphate and theophylline in the absence of extracellular Ca^{2+}. *Proc. nat. Acad. Sci. (Wash.)*, 69: 834–836.

Penniston, J.T. (1972) Endocytosis by erythrocyte ghosts; dependence upon ATP hydrolysis. *Arch. Biochem. Biophys.*, 153: 410–412.

Pfenninger, K.H. and Rovainen, C.M. (1974) Stimulation- and calcium-dependence of vesicle attachment sites in the presynaptic membrane: a freeze-cleave study on the lamprey spinal cord. *Brain Res.*, 72: 1–23.

Pletscher, A., Da Prada, M., Berneis, K.H., Steffen, H., Lutold, B. and Weder, H.G. (1974) Molecular organization of amine storage organelles of blood platelets and adrenal medulla. In *Advances in Cytopharmacology Vol. 2*, B. Ceccarelli, F. Clementi and J. Meldolesi (Eds.), Raven Press, New York, pp. 257–264.

Poisner, A.M. (1970) Release of transmitters from storage: a contractile model. In *Biochemistry of Simple Neuronal Models, Advances in Biochemical Psychopharmacology,* E. Costa and E. Giacobini (Eds.), Raven Press, New York, pp. 95–108.

Poisner, A.M. (1973a) Stimulus-secretion coupling in the adrenal medulla and posterior pituitary gland. In: *Frontiers in Neuroendocrinology,* W.F. Ganong and L. Martini (Eds.), Oxford Univ. Press, New York, pp. 33–59.

Poisner, A.M. (1973b) Mechanisms of exocytosis. In *Frontiers in Catecholamine Research*, E. Usdin and S. Snyder (Eds.), Pergamon Press, New York, pp. 477–482.

Poisner, A.M. (1973c) Sodium-calcium interaction as a trigger for the secretory process. In *Pharmacology and the Future of Man, Proc. 5th Int. Congr. Pharmacology, San Francisco, 1972, Vol. 4,* Karger, Basel, pp. 359–368.

Poisner, A.M. (1973d) Caffeine-induced catecholamine secretion: similarity to caffeine-induced muscle contraction. *Proc. Soc. exp. Biol. (N.Y.)*, 142: 103–105.

Poisner, A.M. (1973e) Direct stimulant effect of aminophylline of catecholamine release from the adrenal medulla. *Biochem. Pharmacol.*, 22: 469–476.

Poisner, A.M. (1973f) Microtubules and catecholamine secretion from the adrenal medulla. In *Proc. 4th int. Congr. Endocrinology, Washington, D.C., 1972, Int. Congr. Series No. 273,* Excerpta Medica, Amsterdam, pp. 299–304.

Poisner, A.M. (1973g) Cellular mechanisms of catecholamine release. In *Proc. 8th Midwest Conf. on Endocrinology and Metabolism, Columbia, Mo., 1972*, X.J. Musacchia and R.B. Breitenbach (Eds.), Univ. of Missouri Press, Columbia, pp. 121–129.

Poisner, A.M, and Bernstein, J. (1970) A possible role of microtubules in catecholamine release from the adrenal medulla: effect of colchicine, vinca alkaloids and deuterium oxide. *J. Pharmacol. exp. Ther.*, 177: 102–108.

Poisner, A.M. and Cooke, P. (1975) Microtubules and the adrenal medulla. *Ann. N.Y. Acad. Sci.*, 253: 653–669.

Poisner, A.M. and Douglas, W.W. (1966) The need for calcium in adrenal-medullary secretion evoked by biogenic amines, polypeptides and muscarinic agents. *Proc. Soc. exp. Biol. (N.Y.)*, 123: 62–64.

Poisner, A.M. and Douglas W.W. (1968) A possible mechanism of release of posterior pituitary hormones involving adenosine triphosphate and an adenosine triphosphatase in the neurosecretory granules. *Molec. Pharmacol.*, 4: 531–540.

Poisner, A.M. and Hava, M. (1970) The role of ATP and ATPase in the release of catecholamines from the adrenal medulla. IV. ATP-activated uptake of calcium by microsomes and mitochondria. *Molec. Pharmacol.*, 6: 407–415.

Poisner, A.M. and Hong, J.S. (1974) Storage and release of vasopressin from neurosecretory granules and the neurohypophysis. *Advanc. Cytopharmacol.*, 2: 303–310.

Poisner, A.M. and Trifaro, J.M. (1967) The role of ATP and ATPase in the release of catecholamines from the adrenal medulla. I. ATP-evoked release of catecholamines, ATP, and protein from isolated chromaffin granules. *Molec. Pharmacol.*, 3: 561–571.

Poisner, A.M. and Trifaro, J.M. (1969) The role of adenosine triphosphate and adenosine triphosphatase in the release of catecholamines from the adrenal medulla. III. Similarities between the effects of adenosine triphosphate on chromaffin granules and on mitochondria. *Molec. Pharmacol.*, 5: 294–299.

Politoff, A.L., Rose, S. and Pappas, G.D. (1974) The calcium binding sites of synaptic vesicles of the frog sartorius neuromuscular junction. *J. Cell Biol.*, 61: 818–823.

Poste, G. and Allison, A.C. (1971) Membrane fusion reaction: a theory. *J theor. Biol.*, 32: 165–184.

Poste, G. and Allison, A.C. (1973) Membrane fusion. *Biochim. biophys. Acta (Amst.)*, 300: 421–465.

Pressman, B.C. (1973) Properties of ionophores with broad range cation selectivity. *Fed. Proc.*, 32: 1698–1703.

Puszkin, S. and Kochwa, S. (1974) Regulation of neurotransmitter release by a complex of actin with relaxing protein isolated from rat brain synaptosomes. *J. biol. Chem.*, 249: 7711–7714.

Rahwan, R.G., Borowitz, J.L. and Miya, T.S. (1973) The role of intracellular calcium in catecholamine secretion from the bovine adrenal medulla. *J. Pharmacol. exp. Ther.*, 184: 106–118.

Rappaport, L., Leterrier, J.F. and Nunez, J. (1975) Protein-kinase activity, *in vitro* phosphorylation and polymerization of purified tubulin. *Ann. N.Y. Acad. Sci.*, 253: 611–629.

Redburn, D.A., Poisner, A.M. and Samson, Jr., F.E. (1972) Comparison of microtubule protein (tubulin) from adrenal medulla and brain. *Brain Res.*, 44: 615–624.

Reed., P.W. and Lardy, H.A. (1972) A-23187: a divalent cation ionophore. *J. biol. Chem.*, 247: 6970–6977.

Ricci, Jr., A., Sanders, K.M., Portmore, J. and Van der Kloot, W.G. (1975) Effects of the ionophores, X-537A and A-23187 on catecholamine release from the *in vitro* frog adrenal. *Life Sci.*, 16: 177–184.

Rohlich, P., Anderson, P. and Uvnäs, B. (1971) Electron microscopic observations on compound 48/80-induced degranulation in rat mast cells. Evidence for sequential exocytosis of storage granules. *J. Cell Biol.*, 51: 465–483.

Rubin, R.P. (1969) The metabolic requirements for catecholamine release from the adrenal medulla. *J. Physiol. (Lond.)*, 202: 197–209.

Rubin, R.P. (1970) The role of calcium in the release of neurotransmitter substances and hormones. *Pharmacol. Rev.*, 22: 389–428.

Rubin, R.P., Feinstein, M.B., Jaanus, S.D. and Paimre, M. (1967) Inhibition of catecholamine secretion and calcium exchange in perfused cat adrenal glands by tetracaine and magnesium. *J. Pharmacol. exp. Ther.*, 155: 463–471.

Russell, J.T. and Thorn, N.A. (1974a) Calcium and stimulus-secretion coupling in the neurohypophysis. I. 45-Calcium transport and vasopressin release in slices from ox neurohypophyses stimulated electrically or by a high potassium concentration. *Acta endocr. (Kbh.)*, 76: 449–470.

Russell, J.T. and Thorn, N.A. (1974b) Calcium and stimulus-secretion coupling in the neurohypophysis. II. Effects of lanthanum, a verapamil analogue (D600) and prenylamine on 45-calcium transport and vasopressin release of isolated rat neurohypophyses. *Acta endocr. (Kbh.)*, 76: 471–487.

Russell, J.T. and Thorn, N.A. (1975) Adenosine triphosphate dependent calcium uptake by subcellular fractions from bovine neurohypophyses. *Acta physiol. scand.*, 93: 364–377.

Russell, J.T., Hansen, E.L. and Thorn, N.A. (1974a) Calcium and stimulus-secretion coupling in the neurohypophysis. III. Ca^{2+} ionophore (A-23187)-induced release of vasopressin from isolated rat neurohypophyses. *Acta endocr. (Kbh.)*, 77: 443–450.

Russell, J.T., Warberg, J. and Thorn, N.A. (1974b) Calcium and stimulus-secretion coupling in the neurophypophysis. IV. Effect of N-ethylmaleimide on 45-calcium^{2+} transport and vasopressin release by isolated rat neurohypophyses. *Acta endocr. (Kbh.)*, 77: 691–698.

Schlapfer, W.W. and Bunge, R.P. (1973) Effects of calcium ion concentration on the degeneration of amputated axons in tissue culture. *J. Cell Biol.*, 59: 456–470.

Schudt, C., Van der Bosch, J. and Pette, D. (1973) Inhibition of muscle cell fusion *in vitro* by Mg^{2+} and K^+ ions. *FEBS Lett.*, 32: 296–298.

Schultz, G., Hardman, J.G., Schultz, K., Baird, C.E. and Sutherland, E.W. (1973) The importance of calcium ions for the regulation of guanosine 3′:5′-cyclic monophosphate levels. *Proc. nat. Acad. Sci. (Wash.)*, 70: 3889–3893.

Serck-Hanssen, G. (1974) Effects of theophylline and propranolol on acetylcholine-induced release of adrenal medullary catecholamines. *Biochem. Pharmacol.*, 23: 2225–2234.

Serck-Hanssen, G. and Christiansen, E.N. (1973) Uptake of calcium in chromaffin granules of bovine adrenal medulla stimulated *in vitro*. *Biochim. biophys. Acta (Amst.)*, 307: 404–414.

Serck-Hanssen, G., Christofferson, T., Morlano, J. and Osnes, J.B. (1972) Adenyl-cyclase activity in bovine adrenal medulla. *Europ. J. Pharmacol.*, 19: 297–300.

Shainberg, A., Yagil, G. and Yaffe, D. (1970) Control of myogenesis *in vitro* by Ca^{2+} concentration in nutritional medium. *Exp. Cell. Res.*, 58: 163–167.

Shima, S., Mitsunaga, M., Kawashima, Y., Taguchi, S. and Nakao, T. (1974) Studies on cyclic nucleotides in the adrenal gland. *Biochim. biophys. Acta (Amst.)*, 341: 56–64.

Sloboda, R.D., Rudolph, S.A., Rosenbaum, J.L. and Greengard, P. (1975) Cyclic AMP-dependent endogenous phosphorylation of a microtubules-associated protein. *Proc. nat. Acad. Sci. (Wash.)*, 72: 177–181.

Smith, D.S. (1971) On the significance of cross-bridges between microtubules and synaptic vesicles. *Phil. Trans. B*, 261: 395–405.

Smith, D.S., Jarlfors, U. and Cameron, B.F. (1975) Morphological evidence for the participation of microtubules in axonal transport. *Ann. N.Y. Acad. Sci.*, 253: 472–506.

Thorn, N.A., Russell, J.T. and Vilhardt, H. (1975) Hexosamine, calcium, and the role of calcium in hormone release. *Ann. N.Y. Acad. Sci.*, 248: 202–216.

Trifaro, J.M. and Dworkind, J. (1971) Phosphorylation of membrane components of adrenal chromaffin granules by adenosine triphosphate. *Molec. Pharmacol.*, 7: 52–65.

Trifaro, J.M. and Dworkind, J. (1975) Phosphorylation of the membrane components of chromaffin granules: synthesis of diphosphatidylinositol and presence of phosphatidylinositol kinase in granule membranes. *Canad. J. Physiol. Pharmacol.*, 53: 479–492.

Trifaro, J.M., Collier, B., Lastowecka, A. and Stern, D. (1972) Inhibition by colchicine and by vinblastine of acetylcholine-induced catecholamine release from the adrenal gland: an anticholinergic action, not an effect on microtubules. *Molec. Pharmacol.*, 8: 264–267.

Uzunov, P., Revuelta, A. and Costa, E. (1975) A role for the endogenous activator of 3′,5′-nucleotide phosphodiesterase in rat adrenal medulla. *Molec. Pharmacol.*, 11: 506–510.

Van der Bosch, J., Schudt, C. and Pette, D. (1972) Quantitative investigation of Ca^{++} and pH-dependence of muscle cell fusion *in vitro*. *Biochem. biophys. Res. Commun.*, 48: 326–332.

Vilhardt, H. and Jorgensen, T. (1972) Free flow electrophoresis of isolated secretory granules from bovine neurohypophyses. *Experientia (Basel)*, 28: 852–853.

Warburg, J. and Thorn, N.A. (1969) *In vitro* studies of the release mechanism for vasopressin in rats. III. Effect of metabolic inhibitors on the release. *Acta endocr. (Kbh.)*, 61: 415–424.

Weisenberg, R.C. (1972) Microtubule formation *in vitro* in solutions containing low calcium concentrations. *Science*, 177: 1104–1105.

Wessells, N.K., Spooner, B.S. and Ash, F. (1971) Microfilaments in cellular and developmental processes. *Science*, 171: 135–143.

Wilson, L., Bryan, J., Ruby, A. and Mazia, D. (1970) Precipitation of proteins by vinblastine and calcium ions. *Proc. nat. Acad. Sci. (Wash.)*, 66: 807–814.

Chapter 4

Protein metabolism in neuroendocrine tissues

ABEL LAJTHA and DAVID DUNLOP

New York State Research Institute for Neurochemistry and Drug Addiction, Ward's Island, New York, N.Y. 10035 (U.S.A.)

I. INTRODUCTION

One could ask, in an overview such as this, of protein metabolism of the nervous system, whether it is justifiable to deal with neural metabolism as being so very different from that in other organs. Indeed, most studies of the detailed mechanism of protein formation from amino acids, and of proteolysis to amino acids, indicate that the basic principles and mechanisms in brain are not different from those in other organs and cells (Roberts, 1971). There are, however, cogent reasons for considering the nervous system separately. First, some aspects of protein metabolism are specific for the nervous system. Important examples are the lack of cell regeneration in the adult brain and axonal flow in which proteins formed in the cell body are moved along the axon at various rates. These are indications that some of the structural specificity of the nervous system is reflected in its protein metabolism (Lajtha, 1964a; Droz, 1973). The second important reason for examining the characteristics of brain protein metabolism is the role played thereby in the function and pathology of the nervous system. Protein metabolism's role in such highly specific functions as excitation and information processing must necessarily receive special consideration.

In the early experiments investigating incorporation (in vivo) of radioactively labeled amino acids, the radioactivity found in brain proteins was much less than that in the proteins in the rest of the organism. Such a finding, in the organ that is the depository of long term memory and which has no cellular regeneration, suggested that most of the central nervous system proteins were metabolically stable. Subsequently it was shown that the main reason for the low rate of labeling of brain proteins was the lack of penetration of the precursor amino acid through cerebral barriers into the brain. Now the evidence is convincing that most brain proteins are rapidly metabolized (Lajtha, 1964a; Lajtha and Marks, 1971) which means that the role of protein metabolism in nervous function, including neuroendocrine regulation, can be studied.

II. METHODS FOR MEASURING PROTEIN METABOLISM

There is a rather extensive literature on protein turnover in various organs under various conditions. In brain, alterations of turnover have been studied with development, inhibition, drugs, and in pathology. However, the limitations of much of the methodology in these studies has to be emphasized and considered (Schimke, 1970; Lajtha and Marks, 1971). In general 3 methods are used for measuring rates of protein metabolism in vivo:

(a) Measuring the incorporation of labeled amino acid.

(b) Measuring the release of radioactivity from proteins that were previously labeled with an administered amino acid.

(c) Measuring the rate of change in a protein, such as restoration of enzyme activity following irreversible inhibition of the existing enzyme content.

While the third approach is applicable only to individual enzymes, the first two may also be used to study mixtures of proteins.

(a) To measure protein turnover from incorporation of an amino acid, the specific activity of the precursor must be known. With the administration of a pulse dose, the most frequently used approach, it is difficult to measure or estimate the true precursor specific activity in the tissue. The amino acid is rapidly exchanged with various tissue pools and often rapidly metabolized, resulting in gross changes in the precursor specific activity. Determinations at several different times are therefore required. This makes turnover measurements difficult to conduct and unfortunately subject to very great errors in many cases. In recent years several procedures have been developed in efforts to avoid these problems by establishing constant precursor specific activities. These methods include (1) the infusion of precursors at rates calculated to establish a constant blood specific activity (Garlick and Marshall, 1972; Seta et al., 1973), (2) multiple glucose injections to establish a reservoir of precursor for glutamic acid and glutamine, thereby imparting a fairly constant specific activity to those amino acids (Austin et al., 1972), (3) the flooding of the endogenous amino acid pools by the injection of gross amounts of labeled precursor, overwhelming any endogenous dilution (Dunlop et al., 1975a), and (4) the implantation of a pellet of precursor of low solubility, viz., tyrosine, which by slowly dissolving establishes a constant blood specific radioactivity (Lajtha et al., 1975). In our work we have used methods (1), (2) and (4) and found each to give reliable measures of protein turnover.

Methods (1), (2) and (3) maintain constant specific activities for several hours; (4) for several days. These procedures will undoubtedly allow much more accurate determinations of turnover rates. Only fragmentary results of applying some of these techniques to the study of neuroendocrine tissues are presently available.

(b) The decrease of a radioactive label in a protein is conceptually the easiest measure of turnover. Technically, however, this method suffers from the error of reincorporation, and, for mixture of protein, the uneven labeling as mentioned in the critique of Method a. Local recapture of label released by metabolism may result in underestimating turnover rates by a factor of 2–4, i.e., reincorporation may capture 50–75% of the released amino acid locally (Arias et al., 1969; Rannels et al., 1975).

(c) Methods based on measuring changes in enzyme activity may be suspect for two reasons. First, these changes may be due to activation or inhibition of the enzyme rather than turnover. While the use of specific antibodies may avoid this problem, the more serious question remains of whether the approaches usually employed (general inhibition of protein synthesis, induction, or irreversible inactivation of enzyme) may not themselves alter enzyme turnover rates as has been demonstrated in some cases.

Obviously when dealing with a single protein, the turnover rate can be calculated once the precursor and protein specific radioactivities are known over some time interval. But when dealing with groups of protein, such as whole brain protein, the turnover rate measured is an average value. That is, when the turnover rate is correctly measured we know that a certain number of μM of amino acid is incorporated per mg protein per hour, but with short periods of incorporation (a few hours when $t_{1/2}$ is in days) we have no way of distinguishing the different classes of proteins which produce this average. Only a very small fraction of the total protein is labeled and this of course is mostly in the rapidly metabolized fractions. The average turnover rate could for example reflect 10% of the protein with 10 times the average turnover rate. Thus the measured rate may be a perfectly accurate *average,* but not particularly representative of the bulk of the protein. If information is desired about the various classes of proteins yielding the average, then incorporation must be measured over a relatively long period, i.e., a time on the order of the average half-life. This has rarely been done (Lajtha et al., 1975).

III. THE FREE AMINO ACID POOL AND PROTEIN TURNOVER

To measure incorporation, the specific activity of the precursor free amino acid has to be known. It is therefore necessary to know if the average tissue specific activity reflects the specific activity of the true precursor. Also of importance are the questions of how the composition of the free amino acid pool influences protein metabolism, and whether changes in free amino acid levels alter brain metabolism. Interest in these latter questions is also in part methodological since two of the methods for turnover measurements which are based on establishing constant precursor specific activities, after the body concentration of percursor. The method of measuring the decay of enzyme activity after inhibition of protein synthesis also involves changes in the body levels of most amino acids.

III.1. The precursor pool for protein synthesis

Studies of pancreas, cartilage, and muscle (Hider et al., 1969; Adamson et al., 1972; Van Venrooij et al., 1974) indicated that incorporation preferentially occurs from an extraneous, presumably extracellular pool of amino acids instead of from the average intracellular pool. Two pools of free amino acids could be distinguished in liver, one of which had a fairly rapid equilibrium with amino acids in the circulation. This pool served as the precursor for incorporation. A second pool, perhaps derived from proteolysis in the lysosomes, was not in equilibrium (Mortimore et al., 1972). The precursor pool problem has not been clearly resolved for all systems because of numerous technical difficulties. Of the studies on brain, one dealing with the incorporation of amino acid metabolically converted within the brain (Fern and Garlick, 1974), and two others comparing the kinetics of incorporation to changes in the external and internal pool specific activities in brain slices (Jones and McIlwain, 1971, Dunlop et al., 1974), all indicated that, for brain at least, the average intracellular amino acid pool, rather than the external pool, is used for protein synthesis. Another in vivo study, comparing rates measured with traces of amino acid to rates measured with the flooding procedures, supports the conclusion that for brain the average specific radioactivity is probably a fairly good approximation of the true precursor-specific radioactivity (Dunlop et al., 1975a, b).

III.2. Compartmentation of amino acids

Although there are indications that the average free amino acid specific activity in the tissue can be used for calculations of protein turnover, this value does not indicate that free amino acids are homogenously distributed within the brain. The metabolic compartmentation of glutamic acid (Balázs and Cremer, 1973) is well established — separate pools of glutamate are in metabolic equilibrium with different precursors and are utilized in different ways. Measurements of exchange of cerebral amino acids with plasma amino acids in vivo or of exchange of amino acids in tissue with those in the medium in brain slice experiments indicated rapidly and slowly exchangeable compartments (Seta et al., 1973; Dunlop et al., 1974; Neidle et al., 1975). The rapidly exchanging pool shows a high flux between plasma and brain free amino acid. We calculate the half-life of most cerebral amino acids in minutes. The rate of amino acid movement between the free amino acid pool and brain proteins is also rapid. In most cases, protein metabolism incorporates an amount of amino acid equal to that present in the free brain pool within an hour.

The various compartments of amino acids reflect the structural heterogeneity of the organ. A number of amino acids, especially those with putative neurotransmitter activity, such as GABA, taurine, glycine, or glutamic acid, show a regional heterogeneity. Furthermore, there are indications that variations in the composition of the free pool occur between cellular structures to produce specialized nuclear, mitochondrial, and lysosomal free amino acid pools.

III.3. Changes in the free amino acid pool

The free amino acid pool of the brain is characteristic for this organ and is very different from that present in other organs. It contains a number of compounds, not components of proteins, that are present mainly or only in the brain. Among the protein precursor amino acids, glumatic and aspartic acids are present at particularly high concentrations in the brain free pool. The free amino acid pool changes during development: many of the non-essential amino acids increase, while the essential amino acids decrease with age (Davis and Himwich, 1973). A critically important question at present is whether such alterations in the free amino acid pool do affect protein metabolism. This is of great interest for conditions such as amino acidurias in which brain amino acid levels are strongly affected for long periods, recurrently or constantly, throughout development. This question is of particular importance since the pathological changes in phenylketonuria could be caused by the effect of high cerebral phenylalanine on the formation of specific proteins.

Unfortunately this problem is still unresolved. Elevation of the body level of an amino acid usually interferes with the transport of the labeled precursor into the brain and alters the brain concentrations of numerous other amino acids. The decreased access of the precursor to the brain can give a false impression of inhibition of protein formation. In our studies (Dunlop et al., 1975a, b) a number of amino acids over a considerable concentration range did not seem to influence protein metabolism in brain. More recent indications are that if changes occur due to increased cerebral phenylalanine they affect the metabolism of specific proteins, possibly myelin proteins, rather than the bulk of cerebral proteins (Agrawal and Davison, 1973).

IV. HALF-LIFE OF CEREBRAL PROTEINS

Early studies that measured the rate of amino acid incorporation into brain proteins showed decreased incorporation with increasing experimental time. This was interpreted as a demonstration of the heterogeneity of brain protein metabolism, some fractions being metabolized rapidly, others slowly (Lajtha, 1964a; Lajtha and Marks, 1971). More recently, a turnover rate of approximately 0.6% of brain protein/hr has been found when several different methods of administration were used with a variety of amino acid precursors (Table I). It has to be emphasized that these are short term experiments (a few hours) and that little more than 1% of the total proteins are labeled, which is more than sufficient for determining the average rate. An important question not answered by these experiments is whether the average rate is characteristic for the bulk of the total brain proteins. The average incorporation rate (0.6%/hr) is equivalent to an average half-life of approximately 4 days. In other words, if most brain proteins have half-lives close to the average, within 4 days half of the brain proteins would be broken down and replaced by freshly formed proteins.

We recently developed a method which extends the time for incorporation up to 5 days by utilizing the insolubility of tyrosine. An intraperitoneal injection of a labeled tyrosine suspension is combined with the subcutaneous implantation of a [^{14}C]tyrosine

TABLE I

Rates of brain protein synthesis determined by different procedures in the literature

The average rate of amino acid incorporation in experiments lasting up to a few hours is approximately 0.6%/hr. If turnover were to continue at this initial rate the half-life of total brain proteins would be estimated at less than 4 days.

Method	Species	Precursors	Rate (%/hr)	References
Adult				
Pulse	rat	Tyr	0.61	Oja (1967)
Pulse	rat	Tyr	0.40	Lindroos and Oja (1971)
Infusion	rat	Val, Lys, Arg, Gly, Tyr, Leu	0.55	Seta et al. (1973)
Infusion	mouse	Tyr	0.68	Garlick and Marshall (1972)
Glucose injections	rat	Glu-Gln	0.8	Austin et al. (1972)
Semi-gross injection	rat	Lys	0.4	Henshaw et al. (1971)
Gross injection	mouse	Val	0.69	Dunlop et al. (1975a, b)
Gross injection	rat	Val	0.62	Dunlop et al. (1975a, b)
Tyrosine subcut.	mouse	Tyr	0.60	Lajtha et al. (1975)

pellet. Slow, continuous absorbtion of the administered tyrosine keeps the specific activity of the free tyrosine in the plasma fairly constant throughout the experimental period. The constant specific activity in the plasma rapidly induces a similarly constant specific activity in the brain free tyrosine. Additional advantages of this method are that the specific activity of the tyrosine in the brain is similar to that of the plasma, and that during this extended incorporation period, any slowly equilibrating brain free amino acid pools should approach equilibrium. Thus the specific activity of precursor pools is likely to be homogenous. The constancy of the average specific activity in the free pool can be seen in Table II (Lajtha et al., 1976). One disadvantage of this method is that it increases plasma tyrosine concentrations above normal physiological levels, although it causes only a slight elevation of cerebral free tyrosine. These long term studies show how greatly the incorporation over the long term differs from the average incorporation rate (Table II). If the bulk of the brain proteins had turnover rates similar to the average value and only very small fractions varied from the average with moderately different rates, the incorporation over time would be substantially similar to that predicted by the average half-life. But when the actual per cent replacement is compared to that predicted on the basis of the average incorporation rate, discrepancies are evident by 14 hr and by 120 hr where the replacement is only 59% of the calculated value. It has always been known that the average turnover rate (half-life) derived from the incorporation rate (nm/mg protein/hr) might not adequately reflect the turnover of the bulk of the proteins. Obviously an average may differ greatly from a median. Here for the first time we have a quantitative measure of replacement over a relatively long period. If two or more fractions of protein with different half-lives are assumed to comprise the total, the closest fit is found with a small, fast pool and a much larger slow pool. The smaller fraction would contain 5.7% of the protein with an average half-life of 15 hr, while the remaining 94.3% would constitute the slow fraction with an average half-life of 10 days. Although these values are still approximations and are averages, they indicate that a large majority of brain proteins have metabolic rates within the range indicated: a much more rapid fraction (with a half-life of a few hours or less) or a much slower fraction (a half-life of 20 days, or more) could comprise only a small fraction of the total.

TABLE II

Incorporation of tyrosine into brain proteins

The incorporation rate of tyrosine into brain proteins decreases over the 5-day period of measurement. The "predicted" replacement is the per cent protein which would turnover if each of the brain proteins actually had half-lives equal to the average half-life (4.50 days).

Incorporation time (hr)	Specific activity of tyrosine (disint./min/μmole)		Per cent replacement of protein	
	Free pool	Protein bound	Predicted	Measured
1	6400	41	0.64	0.64
14	6200	410	8.60	6.60
50	6400	1100	27.00	17.00
120	6300	2000	54.00	32.00

High rates of turnover are not unique to brain. In our studies of long term labeling with tyrosine, the renewal rate of liver and kidney proteins exceeded that of brain, and within 3 days, 3 quarters of the total protein-bound tyrosine had been replaced in these organs, again demonstrating rapid and extensive protein turnover (Table III) (Lajtha et al., 1976).

TABLE III

Incorporation of tyrosine into tissue proteins
Rates of protein turnover are higher in liver and kidney than in brain: after 70 hr the per cent protein-bound tyrosine replaced is 21.78, and 71% in brain, liver, and kidney, respectively. The best fit to the incorporation curves was obtained by assuming the following average half-lives for the tissue proteins: brain, two compartments, 6% of brain proteins with a half-life of 15 hr, 94% with 10 days; liver, a single compartment, with all proteins having a half-life of 26 hr; kidney, two compartments, 71% of proteins with half-lives of 18 hr, 29% with half-lives of 63 hr.

Incorporation time (hr)	Protein bound as per cent of free tyrosine specific activity		
	Brain	Liver	Kidney
4	2.3	11	8
12	5.8	25	23
24	10	49	38
70	21	78	71
120	33		

V. HETEROGENEITY OF TURNOVER RATES

Although the calculations that fit the incorporation data best do not suggest a continuum of protein fractions, each with its own turnover rate, studies of isolated components clearly demonstrate a very wide range of turnover rates. In this, brain is similar to other tissues. One liver amino acid transferase (tyrosine-glutamic) was found to have a half-life of 20 hr, another (glutamic-alanine), 84 hr (Schimke, 1970). The half-life of cytochrome b_5 was 2.3 days; that of cytochrome c, 6.1 days (Druyan et al., 1969). Turnover rates of some relatively pure fractions from brain are: 4 days for neurotubulin, 16 days for S-100 (Cicero and Moore, 1970), 21 days for basic protein, 35 days for proteolipids, and 54 and 104 days for histones (Piha et al., 1966; Lajtha and Marks, 1971). Purification techniques have to be improved before final values for the half-lives of some brain protein fractions are established unequivocally, but differences may even now be perceived between the half-lives of neuronal and glial proteins (Hamberger et al., 1971; Satake, 1972) among the various proteins present in myelin (Sabri et al., 1974) and among the mitochondrial membrane components (Marks et al., 1970). There are differences then not only between structural elements (neurons and glia) but even within the elements of one structure, to wit the proteins in myelin specific subcellular membranes.

Even after 5 days of labeling only a third of the total brain protein was replaced. One may therefore ask whether all of the proteins are turning over. To answer this, mice were fed a diet containing radioactive lysine for approximately 2 weeks until the plasma specific activity approached that of the diet (Lajtha and Toth, 1966). The animals were then mated. The same diet was fed throughout pregnancy and to the issue throughout their growth. These offspring therefore had access only to radioactive lysine throughout their development. After the progeny matured, the diet was exchanged for one containing

TABLE IV

Decrease in protein-bound radioactivity in mouse brain with time

At 0 time, we exchanged the radioactively labeled diet (given to the experimental animal throughout development) for unlabeled diet. In the beginning of the experiment all brain proteins were radioactive; with time, 97% of the proteins were replaced with newly formed non-radioactive proteins.

Experimental time (days)	Counts/min/ μmoles lysine	% of original activity left
0	30	100
30	8.8	29
60	2.1	7.0
150	0.7	2.4

non-labeled lysine. In time, at least 97% of the radioactivity disappeared from the brain proteins (Table IV), showing that any permanently stable fraction, if present, must comprise only a minute portion of the total. The average half-life of this decay was about 18 days. As discussed, half-lives calculated from decay are underestimates and represent minimal values.

As we have seen the average values given in Tables I and III accurately measure the incorporation rate of free amino acid into protein but give no indication of what protein classes are responsible for that average rate. And though nervous tissue clearly contains many proteins with vastly different turnover rates, our recent data suggest that the bulk of the brain protein may be represented by the two fractions described above. It is probable there are fractions in brain with extremely high turnover rates (half-lives in minutes) and other with very slow rates, but the kinetics indicate that these must comprise only a very small fraction of the total. If more than 2% of the protein had a half-life of less than 2 hr, it would have been detected in the kinetic analysis. The feeding experiments limit the size of any permanent or semi-permanent fraction to less than 3%. Since proteins are characteristically functional, the factors controlling turnover are presumably related to the protein's functions. That the bulk of brain protein may be treated as two fractions in spite of the great diversity of individual proteins turnover rates will hopefully yield some insight into the general function of the nervous system. But for the present, what controls turnover and what the functional significance of high or low rates may be, is not clear.

VI. REGIONAL HETEROGENEITY

Regional aspects of cerebral protein metabolism are of particular relevance to the subject of this volume. The findings that monoaminoxidase has an 11-day half-life in brain, and a 4-day half-life in the superior cervical ganglion (Goridis and Neff, 1971), demonstrate the most simple, basic reason for regional heterogeneity in turnover rates: different rates for the same protein in different regions. There are, however, other discernible causes for this heterogeneity.

The numerous structural components of the brain exhibit distinct rates of protein turnover, and since these are heterogenously distributed, the result must be regional heterogeneity of turnover rates. For example, neuronal–glial differences should contribute to differences between neuron-rich and glial-rich regions. Differences have been found in various layers of the cerebellum. The half-lives of the proteins of the granule cell layer, the molecular layer and the white matter were 33, 59 and 136 hr, respectively (Austin et al., 1972). The higher rate of protein synthesis in cell bodies resulted in more bound radioactivity in areas rich in cell bodies, compared to areas rich in dendrites or fiber tracts.

Further differences in the protein turnover rates of various cells and structures have been indicated with high resolution autoradiography. These were differences between migratory cells, pyramidal cells, ganglion cells and sedentary cells (Droz and Koenig, 1970) and differences among neurons with larger vs. those with smaller axons (Koya and Friede 1969). Within the cells, as expected, the Nissl substances are labeled more rapidly than are the mitochondria (Droz and Koenig, 1970), while turnover in the nuclei is low, in addition to there being differences due to neuronal vs. glial nuclei.

The axonal flow of proteins, specific for the nervous system, is another cause of regional heterogeneity of protein metabolism. It has been well-established that proteins, synthesized within the cell body, flow down along the axon. Fast and slow components (over 100 mm/day, under 10 mm/day) can be clearly distinguished and some of the various protein components of each flow could be identified (Ochs, 1972, Hoffman and Lasek, 1975). The major flow occurs in proximodistal direction, although retrograde flow has also been observed. In addition to individual proteins, all structural elements, mitochondria and vesicles may participate in this flow. Such movements may account not only for different metabolic rates (dependent on flow rates), but also for separation of the site of synthesis from that of breakdown. It is generally accepted that the two facets of protein turnover, synthesis and breakdown, occur at different sites through different mechanisms, and are probably regulated through different controls in the cell. Axonal flow is an extreme example of the separation of these two processes. Although axonal flow has been studied in some detail, and similar processes are presumably important in neuroendocrine regions, the quantitative aspects of axonal flow of proteins have not been established. For instance, it is not known whether only a small fraction or a major portion of the total neuronal proteins participate in axonal flow, nor do we know how flow rates relate to protein turnover rates. The fate of transported protein in the terminals is also unknown. It must be dissipated through some mechanism such as secretion, degradation or retrograde transport.

Secretion of protein is another functional property which may complicate turnover measurements and, if occurring on a regional basis, introduce additional heterogeneity. Plasma proteins secreted by the liver have a very high "turnover rate" and of course are secreted rather than degraded. Incorporation into such fractions may merit separate consideration if the average incorporation rate is materially altered.

The pituitary undoubtedly constitutes a region of particular complexity. It would not be surprising if axonal transport contributed significantly to the total protein and if secretion of proteins or peptides accounted for significant losses of protein compared to degradation, situations which presumably do not exist in most brain regions. Long term incorporation studies should prove highly revealing. Our initial experiments point to a high rate of protein formation in the hypophysis that is unexplained by the greater permeability of the structure to free amino acid, and so may be related to the factors mentioned (Dunlop and Lajtha, in preparation). The rate of synthesis in this structure appears to be approximately 3 times the average rate in cerebral hemisphere.

VII. DEVELOPMENT AND PROTEIN TURNOVER

The rate of protein turnover in adult brain is sufficiently high that the net deposition of brain protein achieved during development could be attained simply by lowering the degradation rate. This, however, is not the case since numerous studies have shown a greater rate of protein formation in the young. Recent results from a number of laboratories indicate an average rate of incorporation of approximately 2% for cerebral hemisphere (Dunlop et al., 1975a), as compared to the 0.6%/hr in adults (Table V). From the gross incorporation rate and the net deposition rate, the rate of breakdown (or disappearance) can be calculated as is shown (Table VI) and, somewhat contrary to expectation, breakdown is also higher in the immature than in the mature brain. This increased breakdown during the phase of net growth is somewhat puzzling and it may be partially connected with the change in protein composition. In all brain areas and cells a higher

TABLE V

Incorporation rates in young rat brain proteins

Incorporation rates in immature brain are higher than in adult. For 2-day-old animals, the rate was 2.2%/hr as compared to 0.6% in adults. The rate is not influenced by the level of amino acid in brain, i.e., tracer levels or large loads of several amino acids yield similar results.

[^{14}C]amino acid	Per cent replacement/hr	
	Tracer levels	Large dose
Valine	2.4	2.1
Tyrosine	2.2	
Lysine	1.9	2.0
Histidine	2.4	2.1
Average	2.2	2.1

turnover in the immature brain was found (Johnson and Sellinger, 1971). Rates decreasing with ontogenetic age have also been found in experimental preparations, e.g., in brain cell suspensions. The increased incorporation in preparations from young animals was due to higher rates of synthesis, not to lower rates of degradation. Even greater decreases (25-fold) in incorporation rates occur in brain slices as development proceeds, but the evaluation of these changes has become rather difficult. It seems that incorporation in vivo and in slices from young brain are fairly close, while slices from adult brain show much lower rates of incorporation than does the tissue in vivo (Table VII). This divergence of in vitro from in vivo rates may not occur in systems other than the slice but there is little quantitative data suitable for comparing in vitro rates to the in vivo situation for other systems. Such systems have been used to show changes in the ribosomes during development (Roberts, 1971) and in the acceptor activity of tRNA (Johnson and Chou, 1973). To what extent these changes reflect in vivo developmental changes in turnover is not known.

TABLE VI

Changes in protein metabolism in rat brain with age

The rates of breakdown can be calculated from measurements of incorporation rates and the net increase in proteins over a time period. Protein breakdown in vivo is high in the immature brain and decreases during development.

Age (days)	mg protein metabolized/100 mg tissue protein/hr					
	Cerebral hemisphere			Cerebellum		
	Synthesis	Deposition	Breakdown	Synthesis	Deposition	Breakdown
2	2.1	0.5	1.6	2.7	1.3	1.4
8	2.0	0.5	1.5	2.3	1.3	1.0
18	1.0	0.2	0.8	1.2	0.3	0.9
37	0.6	0.01	0.6	0.7	0.01	0.7

TABLE VII

Incorporation of valine in vivo and in brain slices

Incorporation rates in brain slices are higher than in most other isolated systems (cells, homogenates, ribosomes) but are still considerably below rates measured in vivo. The difference is relatively greater in adult brain. Furthermore, in vitro, there is some net loss of protein with time by autolysis. Note that in vivo incorporation rates do not decrease between the 2nd and 9th days of age (a period of rapid growth). Incorporation was measured in the 30 min–2 hr period with high concentrations of [^{14}C]valine.

Age (days)	nMoles incorporated/mg protein/hr		
	In vivo	In slices	Slice% in vivo
2	9.9	7.2	73
7	9.8	5.6	57
9	9.9	5.3	54
14	7.0	3.4	49
Adult	2.8	0.4	14

In vitro systems have the advantage of allowing much closer control and ease manipulation of the environment of the tissue. In addition, with very small tissues such as pineal or pituitary the use of in vitro incubations could constitute an economy of considerable scale. For any given period of incorporation the minimum precursor specific radioactivity must be inversely related to the amount of protein counted. Unfortunately at this time the very wide discrepancy between in vitro and in vivo rates of synthesis renders this approach questionable for adult organs (Dunlop et al., 1976b).

We have preliminary data indicating that the large differences in protein synthesis rates found in all cases between young and adult brain regions may not pertain in the pituitary.

VIII. ALTERATIONS OF PROTEIN METABOLISM

The factors that alter protein turnover during growth and in the adult brain have been the subject of numerous studies. Because of the methodological complexity of measuring turnover in a meaningful manner, the voluminous results of these studies are all too frequently equivocal. Numerous mechanisms are involved. In many cases the factors studied do not alter protein metabolism at all but affect the access of amino acid to the brain. In such cases the apparent changes in protein turnover are artifactual and only reflect changes in pooling, transport, metabolism or exit of the labeled amino acid. In other cases factors influence cell division, and therefore, are effective in the brain only during the developmental period when active mitosis occurs. If cell division is arrested at an early stage, a permanent deficit can ensue without necessarily any change in adult protein turnover rates. In other situations a factor may influence protein metabolism but in an indirect way. Prenatal effects on energy supply, membrane structure, or enzyme distribution are examples. Even when factors directly affect protein metabolism, various mechanisms may be involved such as interference with amino acid activation, messenger or transfer RNA formation, mRNA stability, polypeptide chain elongation, the activation of proteinases or sensitization of protein substrates to proteinase action.

VIII.1. Effects of malnutrition

Lack of protein or amino acids in the diet has a major effect on cell division, since it is effective only during a sensitive period of development. The effects of amino acid or protein malnutrition on the adult brain are minimal (Zamenhof et al., 1971; Winick, 1974). Some neurons are formed postnatally but their number is small (Altman and Das, 1965). Thus possible nutritional effects occurring at later stages due to interference with their development may be undetectable at the present time. The resistance of the adult brain is somewhat specific. Nutritional influences on protein turnover were found in adult liver (Swick and Ip, 1974). This reflects the effects of excess amino acids, which affect protein formation in muscle while exhibiting minimal effects on brain (Morgan et al., 1971; Oja, 1972). Clearly, protein malnutrition affects many aspects of brain function besides curtailing cell division. The synthesis of specific proteins, enzymes, or membranes may be reduced, while lipid metabolism or the cerebral energy supply can be altered in various ways. If malnutrition persists for a brief period, recovery and rebound are possible. Regeneration and factors that would inhibit regeneration at a later time constitute an important area for research. The resistance of adult brain to malnutrition is of some interest. Preliminary data indicate that its composition is maintained not by decreased breakdown or binding of the amino acid content, but by efficient scavenging by cerebral amino acid transport systems (Banay-Schwartz and Lajtha, in preparation).

VIII.2. Endocrine influences

Hormones affect protein metabolism in various ways. In the developing nervous system the effect is observed mainly on cell division, and hormone effects therefore depend on the developmental stage of the brain or brain area. For example, a greater effect is found in the cerebellum in rodents since development of this area occurs at a later stage. A decreased number of cells was found following administration of thyroxine and of cortisol during maturation (Balázs and Richter, 1973). Thyroid hormones may affect protein synthesis in specific structures, e.g., mitochondria, and can be expected to have an indirect effect through interference with energy metabolism (Sokoloff, 1971). The sites at which hormone activity influences development are not known, but the period sensitive to hormone activity is similar to that during which malnutrition is effective.

Some studies have been conducted on the effects of hormones or lack of hormones on amino acid incorporation into various areas of the pituitary or pineal. In most cases the methodology was such that the changes seen may have resulted from altered transport, metabolism of precursor, blood flow or other factors. This should be a most fruitful area for the application of more accurate methods.

VIII.3. Stimulation and inhibition

Many studies have attempted to identify specific proteins, whose metabolism is altered during a specific functional load. For example, many have tried to identify a protein fraction that is altered during learning in order to study the mechanisms of memory storage. Some of the early literature has been summarized (Lajtha, 1964b; Jakoubek and Semiginovsky, 1970). A more recent example is the effect of light on the metabolism of specific protein fractions (Haywood et al., 1974). Effects of electric stimulation (Dunn et al., 1971) and convulsants were also studied. The identification of a specific protein fraction that may be responsible for some of the observed functional alterations remains to be demonstrated.

The effects of inhibition of brain activity on protein turnover are more clearcut than are those of excitation. A decrease in protein turnover during sleep and especially during hibernation has been indicated (Palladin et al., 1976). The greatest effects were observed after the brain temperature was lowered. This affected protein turnover drastically (Lajtha and Sershen, 1975). In goldfish, lowering the temperature from 34 to $10°C$ caused a 95% decrease in the rate of tyrosine incorporation (Fig. 1). This very strong inhibition, observed in most protein fractions, is of interest because, at low temperature, the learning ability of fish is not affected (Neale et al., 1973). It is crucial that the temperature effect be considered in drug studies since so many of those agents may alter body temperature.

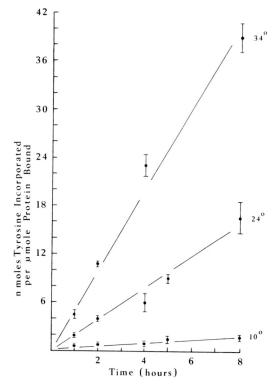

Fig. 1. Tyrosine incorporation into goldfish brain proteins is linear for an 8 hr period. This incorporation depends on temperature (in fish the brain temperature can be controlled by the tank water temperature): at 10°C incorporation is 5% of that at 34°C, and at 24°C, it is 36% of that at 34°C.

VIII.4. Pathological changes

Of great interest in investigations of alterations in protein metabolism are those changes connected with pathological alterations of brain function. As mentioned, in amino acidurias protein metabolism may or may not be altered, or perhaps after the levels of metabolites or of available energy undergo changes, then protein metabolism may be affected secondarily (Roberts, 1974; Gaull et al., 1975). Though there appears to be a deficit of brain protein, inhibition of synthesis has not been demonstrated.

Some of the most detailed information available is from studies of Wallerian degeneration, and in brain, changes accompanying multiple sclerosis and experimental allergic encephalitis. Increased proteinase activity was found around multiple sclerotic plaques (Benato and Gabrielescu, 1964; Einstein et al., 1972), and, in particular, increased acid protease activity. Changes either in the substrates (myelin proteins) making them more available to proteolytic attack, or in the surrounding proteinase activity could account for the initiation of myelin breakdown in some of the demyelinating conditions. That decreases in myelin basic protein are found in multiple sclerosis plaques and in allergic encephalitis, and that basic protein is a good substrate for brain cathepsin D, further supports this possibility (see also Dr. Marks' chapter in this volume).

There are very few factors which may be stated to unequivocally alter protein metabolism in the nervous system. Regrettably much of the work in this field has relied on techniques known to be deficient, and the conclusions of a vast segment of this literature are at best questionable.

IX. CONCLUSION

There is no reason to doubt now that protein metabolism is highly active in the nervous system. The great bulk of brain protein is in a dynamic state, undergoing continuous synthesis and breakdown. There could be a fraction that is stable, but it is probably very small. Relatively recent advances in methods appear to be resulting in much more reliable determination of turnover rates. Long term studies show that the average rate in brain may be treated as the result of two fractions with widely different turnover rates. For the large majority of proteins the average half-life is 10 days. Protein synthesis and breakdown are higher in the rapidly growing brain and decrease with maturation. Although the bulk of this turnover seems resistant to such outside influences as changes in levels of amino acids or other metabolites, stimulation, or drugs, specific changes in sensitive proteins or structures are possible under a variety of conditions. Decreased temperature greatly decreases protein turnover, about 6% decrease per °C. Some of the changes due to nutritional or endocrine deficiency cause permanent damage emphasizing the vulnerability of protein metabolism. On the other hand, the high synthetic rate may signify a possible regenerative capacity of the organ. At the least the high levels of continuous synthesis and breakdown do make possible rapid alterations in protein composition under a variety of conditions. Although the precise manner in which alterations of protein turnover participate in specific aspects of brain function has yet to be determined, improved methods should help solve these problems.

REFERENCES

Adamson, L.F., Herington, A.L. and Bornstein, J. (1972) Evidence for the selection by the membrane transport system of intracellular or extracellular amino acids for protein synthesis. *Biochim. biophys. Acta (Amst.)*, 282: 352–365.

Agrawal, H. and Davison, A.N. (1973) Myelination and amino acid imbalance in the developing brain. In *Biochemistry of Developing Brain*, W.A. Himwich (Ed.), Marcel Dekker, New York, pp. 143–186.

Altman, J. and Das, G.D. (1965) Post-natal origin of microneurones in the rat brain. *Nature (Lond.)*, 207: 953–956.

Arias, I.M., Doyle, D. and Schimke, R.T. (1969) Studies on the synthesis and degradation of proteins of the endoplasmic reticulum of rat liver. *J. biol. Chem.*, 244: 3303–3315.

Austin, L., Lowry, O.H., Brown, J.G. and Carter, J.G. (1972) The turnover of protein in discrete areas of the rat brain. *Biochem. J.*, 126: 351–359.

Baláz, R. and Richter, D. (1973) Effects of hormones on the biochemical maturation of the brain. In

Biochemistry of the Developing Brain, Vol. 1, W.A. Himwich (Ed.), Marcel Dekker, New York, pp. 253–299.

Benato, G. and Gabrielescu, E. (1964) Histochemistry of proteinases in allergic reactions. *Ann. Histochem.*, 9: 295–304.

Cicero, T.J. and Moore, B.W. (1970) Turnover of the brain specific protein S-100. *Science*, 169: 1333–1334.

Davis, J.M. and Himwich, W.A. (1973) Amino acids and proteins of developing mammalian brain. In *Biochemistry of the Developing Brain*, W.A. Himwich (Ed.), Marcel Dekker, New York, pp. 55–110.

Droz, B. (1973) Renewal of synaptic proteins. *Brain Res.*, 62: 383–394.

Droz, B. and Koenig, H.L. (1970) Localization of protein metabolism in neurons. In *Protein Metabolism of the Nervous System*, A. Lajtha (Ed.), Plenum, New York, pp. 93–108.

Druyan, R., De Bernard, B. and Rabinowitz, M. (1969) Turnover of cytochromes labeled with δ-aminolevulinic Acid-^{3}H in rat liver. *J. biol. Chem.*, 244: 5874–5878.

Dunlop, D.S., van Elden, W. and Lajtha, A. (1974) Measurements of rates of protein synthesis in rat brain slices. *J. Neurochem.*, 22: 821–830.

Dunlop, D.S., van Elden, W. and Lajtha, A. (1975a) A method for measuring brain protein synthesis rates in young and adult rats. *J. Neurochem.*, 24: 337–344.

Dunlop, D.S., van Elden, W. and Lajtha, A. (1975b) Optimal conditions for protein synthesis in incubated slices of rat brain. *Brain Res.*, 99: 303–318.

Dunn, A., Giuditta, A. and Pagliuca, N. (1971) The effect of electronconvulsive shock on protein synthesis in mouse brain. *J. Neurochem.*, 18: 2093–2099.

Einstein, E.R., Csejtey, J., Dalal, K.B., Adams, C.W.M., Bayliss, O.B. and Hallpike, J.F. (1972) Proteolytic activity and basic protein loss in and around multiple sclerosis plaques: combined biochemical and histochemical observations. *J. Neurochem.*, 19: 653–662.

Fern, E.B. and Garlick, P.J. (1974) The specific radioactivity of the tissue free amino acid as a basis for measuring the rate of protein synthesis in the rat *in vivo*. *Biochem. J.*, 142: 413–419.

Garlick, P.J. and Marshall, I. (1972) A technique for measuring brain protein synthesis. *J. Neurochem.*, 19: 577–583.

Gaull, G.E., Tallan, H.H., Lajtha, A. and Rassin, D.K. (1975) Pathogenesis of brain dysfunction in inborn errors of amino acid metabolism. In *Biology of Brain Dysfunction*, G.E. Gaull (Ed.), Plenum Press, New York, pp. 47–143.

Goridis, C. and Neff, N.H. (1971) Monoamine oxidase: an approximation of turnover rates. *J. Neurochem.*, 18: 1673–1682.

Hamberger, A., Blomstrand, C. and Yanagihara, T. (1971) Subcellular distribution of radioactivity in neuronal and glial-enriched fractions after incorporation of [^{3}H]leucine *in vivo* and *in vitro*. *J. Neurochem.*, 18: 1469–1478.

Haywood, J., Hambley, J. and Rose, S.P.R. (1974) Effects of early visual experience on [^{14}C]lysine incorporation into the chick brain. *Biochem. Soc. Trans.*, 2: 241–243.

Henshaw, E.C., Hirsch, C.A., Morton, B.E. and Hiatt, H.H. (1971) Control of protein synthesis in mammalian tissues through changes in ribosome activity. *J. biol. Chem.*, 246: 436–446.

Hider, R.C., Fern, E.B. and London, D.R. (1969) Relationship between intracellular amino acids and protein synthesis in the extensor digitorum longus muscle of rats. *Biochem. J.*, 114: 171–178.

Hoffman, P.N. and Lasek, R.J. (1975) The slow component of axonal transport. Identification of major structural polypeptides of the axon and their generality among mammalian neurons. *J. Cell Biol.*, 66: 351–366.

Jakoubek, B. and Semiginovsky, B. (1970) The effect of increased functional activity on the protein metabolism of the nervous system. *Int. Rev. Neurobiol.*, 13: 255–288.

Johnson, D.E. and Sellinger, O.Z. (1971) Protein synthesis in neurons and glial cells of the developing rat brain: an in vivo study. *J. Neurochem.*, 18: 1445–1460.

Johnson, T.C. and Chou, L. (1973) Level and acceptor activity of mouse brain tRNA during neural development. *J. Neurochem.*, 20: 405–414.

Jones, D.A. and McIlwain, H. (1971) Amino acid production and translocation in incubated and perfused tissues from the brain. *J. Neurobiol.*, 2: 311–326.

Koya, G. and Friede, R.L. (1969) Sequential incorporation of [^{3}H]leucine in the rat spinal cord. Is the protein metabolism of nerve cells related to the size of their axons? *J. Anat. (Lond.)*, 105: 47–57.

Lajtha, A. (1964a) Protein metabolism of the nervous system. In *International Review of Neurobiology, Vol. 6*, C.C. Pfeiffer and J.R. Smythies (Eds.), Academic Press, New York, pp. 1–98.

Lajtha, A. (1964b) Alteration and pathology of cerebral protein metabolism. In *International Review of Neurobiology, Vol. 7*, C.C. Pfeiffer and J.R. Smythies (Eds.), Academic Press, New York, pp. 1–40.

Lathja, A. and Marks, N. (1971) Protein turnover. In *Handbook of Neurochemistry*, A. Lajtha (Ed.), Plenum Press, New York, pp. 551–629.

Lajtha, A. and Sershen, H. (1975) Changes in the rates of protein synthesis in the brain of goldfish at various temperatures. *Life Sci.*, in press.

Lajtha, A. and Toth, J. (1966) Instability of cerebral proteins. *Biochem. biophys. Res. Commun.*, 23: 294–298.

Lajtha, A., Latzkovits, L. and Toth, J. (1976) Comparison of turnover rates of proteins of the brain, liver and kidney in mouse in vivo following long term labeling. *Biochim. biophys. Acta (Amst.)*, in press.

Lindroos, O.F.C. and Oja, S.S. (1971) Hyperphenylalaninemia and the exchange of tyrosine in adult rat brain. *Exp. Brain Res.*, 14: 48–60.

Marks, N., D'Monte, B., Bellman, C. and Lajtha, A. (1970) Protein metabolism in cerebral mitochondria. I. Hydrolytic enzymes and amino acid incorporation into mitochondrial membranes. *Brain Res.*, 18: 308–324.

Morgan, H.E., Earl, D.C.N., Broadus, A., Wolpert, E.B., Giger, K.E. and Jefferson, L.S. (1971) Regulation of protein synthesis in heart muscle I. Effect of amino acid levels on protein synthesis. *J. biol. Chem.*, 246: 2152–2162.

Mortimore, G.E., Woodside, K.H. and Henry, J.E. (1972) Compartmentation of free valine and its relation to protein turnover in perfused rat liver. *J. biol. Chem.*, 247: 2776–2784.

Neale, J.H., Klinger, P.D. and Agranoff, B.W. (1973) Temperature-dependent consolidation of puromycin-susceptible memory in the goldfish. *Behav. Biol.*, 9: 267–278.

Neidle, A., Kandera, J. and Lajtha, A. (1975) Compartmentation and exchangeability of brain amino acids: evidence from studies of transport into tissue slices. *Arch. Biochem. Biophys.*, 169: 397–405.

Ochs, S. (1972) Fast transport of materials in mammalian nerve fibers. *Science*, 176: 252–260.

Oja, S.S. (1967) Studies on protein metabolism in developing rat brain. *Ann. Acad. Sci. fenn. A5*, 131: 1–81.

Oja, S.S. (1972) Incorporation of phenylalanine, tyrosine and tryptophan into protein of homogenates from developing rat brain. Kinetics of incorporation and reciprocal inhibition. *J. Neurochem.*, 19: 2057–2069.

Palladin, A.V., Belik, Ya.V. and Polyakova, N.M. (1976) *Protein Metabolism of the Brain*, Plenum Press, New York, 315 pp.

Piha, R.S., Cuénod, M. and Waelsch, H. (1966) Metabolism of histones of brain and liver. *J. biol. Chem.*, 241: 2397–2404.

Rannels, D.E., Kao, R. and Morgan, H.E. (1975) Effect on insulin on protein turnover in heart muscle. *J. biol. Chem.*, 250: 1694–1701.

Roberts, S. (1971). Protein synthesis. In *Handbook of Neurochemistry, Vol. 5*, A. Lajtha (Ed.), Plenum Press, New York, pp. 1–48.

Roberts, S. (1974) Effects of animo acid imbalance on amino acid utilization, protein synthesis and polyribosome function in cerebral cortex. In *Aromatic Amino Acids in the Brain. CIBA Foundation Symposium 22*, American Elsevier, New York, pp. 299–324.

Sabri, M.I., Bone, A.H. and Davison, A.N. (1974) Turnover of myelin and other structural proteins in the developing rat brain. *Biochem. J.*, 142: 499–507.

Satake, M. (1972) Some aspects of protein metabolism of the neuron. *Int. Rev. Neurobiol.*, 15: 189–213.

Schimke, R.T. (1970) Regulation of protein degradation in mammalian tissues. In *Mammalian Protein Metabolism*, H.N. Munro (Ed.), Academic Press, New York, pp. 177–228.

Seta, K., Sansur, M. and Lajtha, A. (1973) The rate of incorporation of amino acids into brain proteins during infusion in the rat. *Biochim. biophys. Acta (Amst.)*, 294: 472–480.

Sokoloff, L. (1971) The action of thyroid hormones. In *Handbook of Neurochemistry*, A. Lajtha (Ed.), Plenum Press, New York, pp. 525–549.

Swick, R.W. and Ip, M.M. (1974) Measurement of protein turnover in rat liver with [^{14}C]carbonate. *J. biol. Chem.*, 249: 6836–6841.

Van Venrooij, W.J., Moonen, H. and Van Loon-Klaasen, L. (1974) Source of amino acids used for protein synthesis in HeLa cells. *Europ. J. Biochem.*, 50: 297–304.

Winick, M. (1974) Malnutrition and the developing brain. In *Brain Dysfunction in Metabolic Disorders*, F. Plum (Ed.), Raven Press, New York, pp. 253–261.

Zamenhof, S., Marthens, E. van and Grauel, L. (1971) DNA (cell number) and protein in neonatal rat brain: alteration by timing of maternal dietary protein restriction. *J. Nutr.*, 101: 1265–1270.

Chapter 5

Localization of neurosecretory peptides in neuroendocrine tissues

EARL A. ZIMMERMAN

Department of Neurology and the International Institute for the Study of Human Reproduction, Columbia University, New York, N.Y. 10032 (U.S.A.)

I. INTRODUCTION

In the last few years the availability of antisera to a number of specific neurosecretory peptides has provided new opportunities in neuroendocrinology. Application of these antisera in radioimmunoassays and immunocytochemical methods has allowed the quantitation of hypothalamic hormones in specific brain regions and identification of their cellular and subcellular distribution. A considerable amount of new data has already been generated concerning the pathways which secrete the octapeptides, oxytocin and vasopressin, and their carrier proteins, neurophysins, and the decapeptide, gonadotropin-releasing hormone (Gn-RH). Somatostatin, a tetradecapeptide, has also been localized by both approaches, and the tripeptide, thyrotropin-releasing hormone (TRH) by radioimmunoassay. In similar studies, antisera to synthesizing enzymes of biogenic amines have been used to localize pathways which may impinge on and regulate neurosecretory systems (Hökfelt et al., 1974b).

II. THE MAGNOCELLULAR NEUROSECRETORY SYSTEM

II.1. Historical aspects

Neurosecretion was first established in the early 1940's when Gomori stains, originally developed to study secretory cells in the pancreas (Gomori, 1941), were applied to brain tissue. Granular elements in the cytoplasm of large hypothalamic neurons located in the supraoptic (SON) and paraventricular (PVN) nuclei had a selective affinity for these stains (chrome-alum-hematoxylin, aldehyde-fuchsin) and it became apparent that these cells may have secretory function (Scharrer and Scharrer, 1954; Bargmann, 1966, 1968). Furthermore this "neurosecretory material" could be followed in long beaded axon processes from these cell bodies through the hypothalamus, in a tract in the zona interna of the median eminence, to endings on blood vessels in the posterior pituitary gland (Scharrer and Scharrer, 1954). Since interruption of the tract in the pituitary stalk caused a proximal buildup of neurosecretory material, it was correctly postulated that posterior pituitary hormone substances were not formed in the posterior pituitary itself, but formed in the perikarya in the hypothalamus and were transported to the gland (Scharrer and Scharrer, 1954). After the active hormones were purified from posterior pituitary by Du Vigneaud (1956) and bioassays were developed, both the hormones were found to be concentrated in the SON and the PVN (Lederis, 1961).

A nuclear theory of the regulation of oxytocin and vasopressin secretion arose at this time which stated that oxytocin was formed in the PVN and vasopressin in the SON based on differences in the concentrations of the hormones in the two nuclei in a number of mammals (Lederis, 1961) and experimental lesions of the respective nuclei (Olivecrona, 1957). This nuclear separation, however, was not complete as shown by later experimental lesions (Sokol, 1970) and studies of homozygous Brattleboro rats with diabetes insipidus (DI rat) with Gomori stains (Sokol and Valtin, 1967). Since part of the stainable material present in normal rat posterior pituitary was absent from DI rat, Sokol and Valtin (1967) concluded that the hormones were located in different terminals and proposed a one-cell-one-hormone theory of vasopressin and oxytocin secretion. Sloper (1966) had shown that the Gomori stains depend on the presence of the disulfide bonds in neurophysins as well as oxytocin and vasopressin. Since more such bonds are contributed by neurophysin, the staining pattern described in DI rat is now known (Zimmerman et al., 1975b; Zimmerman, 1976) to be due in great part to the presence of oxytocin-neurophysin and the absence of the neurophysin associated with the absent vasopressin (Burford et al., 1971; Sunde and Sokol, 1975).

II.2. The neurosecretory theory

The neurosecretory theory was further advanced when transmission electron microscopy demonstrated large ($>$ 150 nm) granules distributed throughout magnocellular neurons (Bargmann, 1968). Sachs et al. (1969), by using radiolabeled precursors, further established that the hormones and carrier proteins were formed on the perikaryal ribosomes, packaged in Golgi apparatus and transported by axonal flow in granules. They also postulated that neurophysin and vasopressin were formed from a larger precursor molecule as the granules travel away from the perikaryon.

Centrifugation studies of posterior pituitary homogenates (for review see Dreifuss, 1975) demonstrated that the hormones and neurophysins were concentrated in these large granules. There was also evidence that different granules had higher concentrations of the hormones (LaBella et al., 1962). The work of Dean et al. (1968) suggested that neurophysin I (NPI) and II (NPII) were located in oxytocin- and vasopressin-containing granules, respectively, in the ox. Fawcett et al. (1968) also found that radiolabeled neurophysin appeared in the systemic circulation providing support for the exocytosis mechanism of secretion in the posterior pituitary gland proposed by Douglas and Nagasawa (1970).

II.3. Neurophysins

Van Dyke et al. (1941) initially isolated a protein of about 30,000 mol. wt. from posterior pituitary which had oxytocic and vasopressor biological activity. The protein moiety was renamed neurophysin (Acher et al., 1956) after it was shown that the approx. 1000 mol. wt. hormones were loosely bound to the carrier protein by electrostatic bonds. Further purification revealed a number of similar proteins which bound both the hormones in vitro (Ginsburg and Jayasena, 1968; Hollenberg and Hope, 1968). It now appears that there are probably two major neurophysins in mammalian species (Robinson and Frantz, 1973) with a molecular weight of about 10,000 (Capra and Walter, 1975). One neurophysin has an intragranular association with oxytocin and another with vasopressin. The present nomenclature is confusing. In ox, NPI is associated with oxytocin and NPII with vasopressin (Vandesande et al., 1975a). In pig, NPII is found with oxytocin and NPI with vasopressin (Pickup et al., 1973). In rat, 2 bands, NPII and III, appear related to oxytocin and NPI with vasopressin (Sunde and Sokol, 1975). Robinson (1975), in an attempt to avoid confusion, named them in man and monkey (Robinson et al., 1976) by their rise in peripheral plasma after estrogen and nicotine administration as measured by radioimmunoassays. Estrogen-stimulated neurophysin (ESN) appears to be related to oxytocin (see below) and nicotine-stimulated neurophysin (NSN) to vasopressin. Specific assays for neurophysins have been of great value in predicting the behavior of the associated hormones in portal and peripheral plasma which are also more difficult to assay (Robinson and Frantz, 1973). The rise of ESN after peripheral administration of estradiol and its rise at mid-cycle in rhesus monkeys (Robinson et al., 1976) is probably accompanied by a rise in oxytocin as suggested by some preliminary data by radioimmunoassay and bioassay in our laboratories (Haldar, Ferin, Robinson, and Zimmerman, unpublished).

III. LOCALIZATION OF NEUROPHYSIN

III.1 The general distribution of neurophysins by immunocytochemistry using cross-species antiseras

Immunocytochemical studies of neurosecretion began with the immunofluorescence studies of Alvarez-Buylla et al. (1970) in which anti-porcine neurophysin II was used to localize NP in the dog hypothalamo-hypophysial system. Zimmerman et al. (1973b) applied antiserum to bovine NPI to localize the magnocellular system of ox and rhesus monkey by immunoperoxidase techniques. Subsequently, a number of additional mammalian species were studied by light microscopy using this antiserum which reacts with all the neurophysins in rat and probably all those present in other species except the ox (Zimmerman et al., 1975a). Watkins (1975) has carried out similar cross-species studies.

In summary, these light microscopic studies revealed that neurophysins are located in all the cell bodies of the SON (Fig. 1c) and in the majority of those (magnocellular) in the PVN (Fig. 1a), but never in all of them. Some PVN cells may secrete some other hormone or have another function.

III.2. The wider fields of the SON and the PVN

The immunocytochemical studies revealed that the magnocellular system is more extensive than outlined by previous methods (Zimmerman et al., 1975a). Some cell bodies of PVN appear to be distributed rostrally as far as the anterior commissure; dorsally, out of the hypothalamus, anterior and ventral to the ventral anterior thalamus along the internal capsule; and caudally, extend to the level of the mid-median eminence. They also

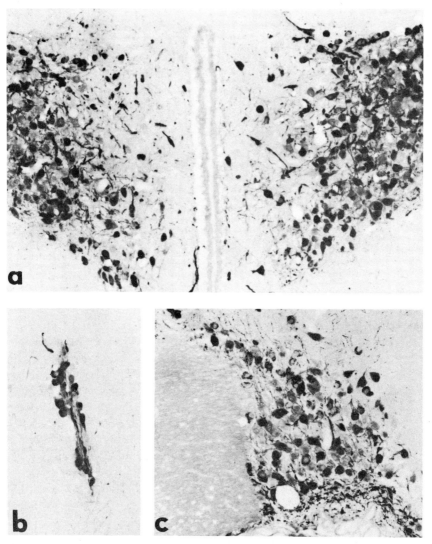

Fig. 1. a,b,c: magnocellular system of normal albino rat hypothalamus. Coronal section stained for neurophysin by immunoperoxidase technique with antiserum to bovine neurophysin. a: paraventricular nucleus; b: cells along the paraventricular tract between a and c; c: supraoptic nucleus. (From Zimmerman et al. (1975a) with permission.) x 200.

confirmed the earlier Gomori findings of Laquer (1954) that cells follow the tract of the PVN as they pass through and over the SON (Fig. 1b). It is no longer possible to clearly determine what is PVN and SON, and as shown below, the presence of oxytocin and vasopressin in both nuclei and in their extended fields makes the nuclear theory of hormone formation no longer tenable (Zimmerman, 1976). Although the majority of the cells of what is considered the SON are located over and around the optic tract lateral to the chiasm, neurophysin stains located some cells extending forward to the terminalis and caudally others reach the anterior median eminence where they line the ventral-lateral tuberal floor to the lateral hypothalamic sulcus.

III.3. Neurophysin in the suprachiasmatic nucleus (SCN)

The beaded axonal pathways of the SON and the PVN converge in a heavy tract in the zona interna of the median eminence on their way to posterior pituitary gland (Figs. 4a, 5). In the course of these studies the presence of neurophysin was discovered in a group of smaller (parvicellular) cell bodies in a portion of the SCN of the rat (Vandesande et al., 1974; Zimmerman et al., 1975a) and mouse (Zimmerman et al., 1975a) (Fig. 2). It was a surprise to find neurophysin outside of the magnocellular system and in a nucleus in which neurosecretory function was doubted on morphological grounds (Clementi and Ceccarelli, 1970). It was initially suggested that the neurophysin in these cells may be associated with corticotropin-releasing factor (CRF) and the source of neurophysin fibers which project to the portal system by Vandesande et al. (1974). Neurophysin increases in zona externa after adrenalectomy (Vandesande et al., 1974; Stillman et al., 1975; Zimmerman, 1976). Further studies demonstrated vasopressin, but not oxytocin in these cells which rim the dorsal portion of the SCN (Vandesande et al., 1975b; Zimmerman, 1976). Furthermore these cells are neurophysin and vasopressin negative in DI rat which have CRF activity (Zimmerman, 1976).

The function of vasopressin in these cells is, as yet, unknown, but it is tempting to think that they may mediate the effects of light on ACTH (Moore and Eichler, 1972). The retina innervates this nucleus directly (Moore, 1973). The function of vasopressin cells in SCN will be better understood when the terminal fields of these fine axons are

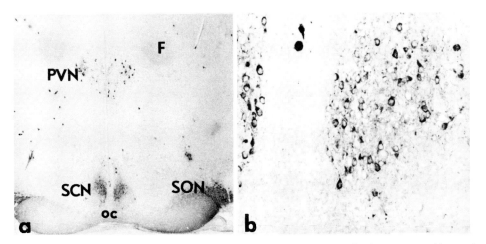

Fig. 2. a,b: normal albino rat coronal section stained for neurophysin by immunoperoxidase technique using antiserum to human neurophysin. Anterior to Fig. 1. a: note the position of the suprachiasmatic nucleus (SCN) over the optic chiasm (oc). SON, supraoptic nucleus; PVN, paraventricular nucleus; F, fornix. b: higher magnification of "a" showing neurophysin in parvicellular neurons in the dorsal portion of SCN. x 260. (From Zimmerman (1976) with permission.)

fully defined. It has been suggested that these fibers go to the infundibulum (Szentágothai et al., 1968) to be secreted into the portal system (Vandesande et al., 1974) where their secretory products may regulate ACTH. It is also possible that they end on cells of the medial-basal hypothalamus such as the dorsomedial or ventromedial nucleus (Swanson and Cowan, 1975) to stimulate presumed sites of the formation of CRF (Hedge et al., 1966).

Neurophysins or associated hormones have not been found in other parvicellular neurons. Zimmerman et al. (1975a) previously reported the presence of ESN in human arcuate cells, a tempting finding suggesting that ESN was associated with Gn-RH which is also found in this nucleus. This is now known to be an artefact (Zimmerman, 1976) as it was discovered that the staining was due to intrinsic peroxidase activity in arcuate cells in some aged brains.

III.4. Neurophysin and the ventricular system

Neurophysin was found in the perikarya of tanycytes lining the walls of the lower third ventricle and in their proximal processes in the median eminence of some rhesus monkeys by immunoperoxidase technique using antiserum to bovine neurophysin (Robinson and Zimmerman, 1973). These observations were also made in rats where the staining could be followed in long segments of tanycyte projections to the zona externa (Zimmerman et al., 1975c) (Fig. 4). Similar findings have been obtained in the duck (McNeill, Abel and Kozlowski, unpublished) and sheep (Watkins, 1975). Neurophysins have not been found in these cells at the electron microscopic (EM) level (Pelletier et al., 1974c; Silverman and Zimmerman, 1975). False negatives are a greater problem in immunoelectron microscopy (Silverman and Zimmerman, 1975). For example vasopressin has not been found in magnocellular perikarya at the EM level (LeClerc and Pelletier, 1974) but is easily demonstrated by light microscopy (Zimmerman, 1975a). Huge losses of the antigen from the tissue probably occur with both approaches to immunocytochemistry. Goldsmith and Ganong (1975) found that 98% of the Gn-RH was lost from the tissues during the preparation for immunoelectron microscopy.

The presence of neurophysin in tanycytes was an exciting finding. It was the first evidence that these specialized ependyma normally contain a neurosecretory substance and supported a wealth of data which suggested that tanycytes may participate in neuroendocrine function (Knigge, 1974) by transporting substances from or even to CSF (Porter et al., 1975). It has been considered that the CSF of the ventricular system may serve as a route for secretion of biologically active peptides to portal blood via tanycytes, an alternate route to the favored neurosecretory pathways via axon terminals on portal capillaries (Knigge, 1974). Zimmerman et al. (1974a) have also found Gn-RH in mouse tanycytes as discussed below.

Earlier reports of vasopressin in CSF (Vorherr et al., 1968) and the neurophysin staining in tanycytes led to measurements of neurophysin in CSF by Robinson and Zimmerman (1973). Slight but significantly more neurophysin was found in lumbar and cisternal CSF in monkey and man than in peripheral blood (Fig. 3).

How does NP get into CSF? Although it could be made in tanycytes or picked up from portal blood and secreted into CSF, the morphological characteristics of these cells suggest a transporting function (Knowles, 1972). It is more likely that vasopressin and neurophysin are secreted into CSF by nerve terminals. Transmission and scanning EM studies have shown numerous free axon endings projecting into the third ventricle some of which contain the larger 150 nm granules typical of magnocellular fibers and others with smaller granules thought to contain releasing factors and even ones with small granules associated with catecholamines (Scott et al., 1974b). As seen on light microscopic immunocytochemistry, neurophysin and vasopressin are often seen in fibers in many species which pass close to the floor of the third ventricle (Zimmerman, 1976). Goldsmith and Zimmerman (1975) have recently reported vasopressin and neurophysin in large granules in free axons in the floor of the rat median eminence by immunoelectron microscopy.

III.5. Neurophysin in zona externa of the median eminence

Large amounts of neurophysin were described in the zona externa of the median eminence in sheep by Parry and Livett (1973) and monkey by Zimmerman et al. (1973a) (Fig. 5). These light microscopic findings reopened long-standing arguments as to whether the magnocellular system projected to the portal system and whether vasopressin had any role in anterior pituitary function, particularly in ACTH release (Martini, 1966). The small amounts of aldehyde fuchsin material in zona externa which increased after adrenalectomy were attributed to CRF (Wittowski and Bock, 1972) or more recently to a CRF neurophysin (Watkins et al., 1974). Although CRF probably increases in zona externa after adrenalectomy, the NP that increases there is associated with immunostainable vasopressin in normal rats (Zimmerman, 1976). The small amount of neurophysin usually found in the zona externa of DI rats is not increased after adrenalectomy and is associated with oxytocin (Stillman et al., 1975).

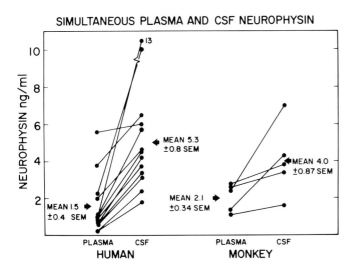

Fig. 3. Plasma and CSF neurophysin in man and monkey determined by radioimmunoassay. (From Robinson and Zimmerman (1973) with permission.)

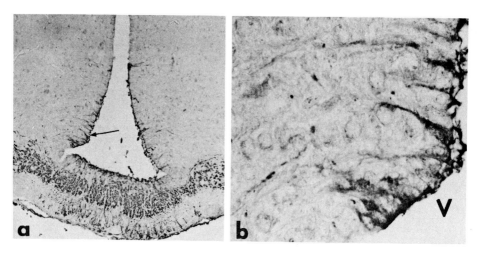

Fig. 4. a,b: normal rat mid-median eminence, coronal section, stained as in Fig. 2. a: neurophysin is seen in fibers in the hypothalamo-hypophysial tract in the zona interna and in zona externa as well as in tanycytes (arrow). x 110. b: at arrow in a, higher magnification (x 690) reveals neurophysin in the cytoplasm of tanycyte perikarya along the ventricle (V) and in their long processes. (From Zimmerman et al. (1975c) with permission.)

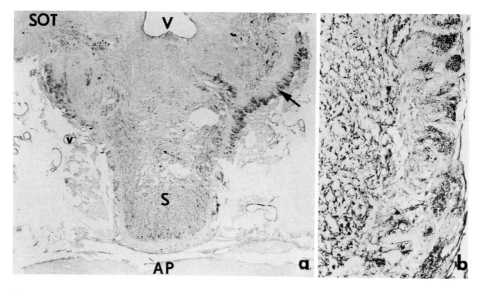

Fig. 5. Coronal section of monkey hypothalamus. Immunoperoxidase technique using anti-bovine neurophysin I. a: reaction products are found in the supraoptico-hypophysial tract (SOT) (= hypothalamo-hypophysial tract), pituitary stalk (S) and zona externa (arrow). V, third ventricle; AP, anterior pituitary gland; v, cross-section of long portal vein. b: higher magnification of "a" showing neurophysin around portal capillaries. x 504. (From Zimmerman et al. (1973a) with permission of the American Association for the Advancement of Science.)

An objection to magnocellular projections to the portal system in the zona externa was the paucity of terminals containing large granules in this region (Knigge and Scott, 1970), although there was some evidence of neurohypophysial hormones in granules in this region (Kobayashi et al., 1970). More recent transmission EM studies have demonstrated some typical magnocellular fibers near portal capillaries (Zimmerman et al., 1975c). Finally, neurophysin and vasopressin have been demonstrated in granules in axon terminals on or near portal capillaries by Silverman and Zimmerman (1975) in guinea pig (Fig. 6).

The large exogenous doses of vasopressin necessary to stimulate ACTH (Clayton et al., 1963) or GH release (Meyer and Knobil, 1966) were considered against a physiological

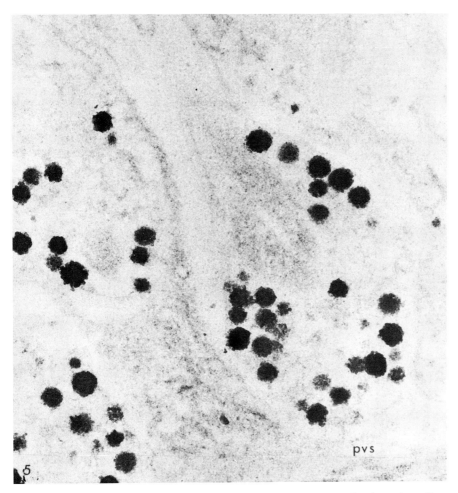

Fig. 6. Vasopressin in granules in an axon terminal in zona externa of guinea pig ending on the perivascular space (pvs) of a portal capillary. Immunoperoxidase PAP technique. x 56,000. (From Silverman and Zimmerman (1975) with permission.)

TABLE I

Estrogen-stimulated neurophysin (ESN), nicotine-stimulated neurophysin (NSN), and vasopressin (VP) measured by radioimmunoassays in simultaneous samples of monkey systemic venous blood (SB) and hypophysial portal blood (PB)

(From Zimmerman et al., 1973a; Zimmerman, 1976.)

Monkey	ES-NP (ng/ml)		NS-NP (ng/ml)		VP (pg/ml)	
	SB	PB	SB	PB	SB	PB
557	1.4	124	4.4	100	—	—
561	3.0	25	2.0	10	35	11,000
564	4.6	240	2.1	88	50	16,000
448	2.2	28	2.0	48	40	14,500

role for vasopressin in anterior pituitary function. However, more than adequate amounts of vasopressin to release ACTH and GH (approx. 13,500 pg/ml) were measured in monkey portal blood by Zimmerman et al. (1973a) (Table I). The radioimmunoassay used to detect vasopressin did not detect CRF (Zimmerman, 1976).

Convincing evidence is therefore accumulating that vasopressin normally has some role in tropic hormone secretion. What about oxytocin and its neurophysin which are also present in zona externa (Vandesande et al., 1975c)? Whether oxytocin might play a role in gonadotropin regulation will be discussed in Section IV.3.

IV. LOCALIZATION OF OXYTOCIN AND VASOPRESSIN

IV.1. Oxytocin, vasopressin and specific neurophysins

The association of a particular neurophysin with each of the two hormones was further established by radioimmunoassay and bioassay of brain regions (Zimmerman et al., 1974b) and by immunocytochemistry (Vandesande et al., 1975a). In the ox both neurophysins and both hormones were found in both the SON and the PVN, although 2–3 times more vasopressin and NPII were found in the SON than the PVN (Zimmerman et al., 1974b). As shown in Fig. 7 there was a correlation between the content of NPI and oxytocin and NPII and vasopressin in the two nuclei. This was confirmed by immunocytochemical studies which also show that NPI and oxytocin were found together in some cells and NPI and vasopressin in other cells of the SON and the PVN (Vandesande et al., 1975a). Zimmerman (1976) and coworkers found ESN and oxytocin in the same cells of man (Zimmerman et al., 1975b) and monkey (in preparation). In DI rat, the neurophysin present is oxytocin-neurophysin and is located in oxytocin-containing cells in the SON and the PVN (Zimmerman et al., 1975b) (Fig. 8). In all mammalian species studied, oxytocin cells primarily occupy the dorsal SON and parts of the PVN and are scattered throughout the rest of the neurophysin system described above.

The data presented thus far strongly support the one-cell-one-hormone theory including the studies of Vandesande et al. (1975a) on the ox showing a clean separation of oxytocin- and vasopressin-containing cells in the SON and the PVN. Studies in our labora-

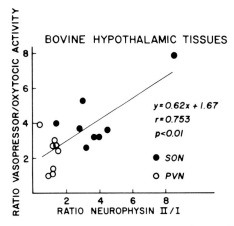

Fig. 7. Correlation of neurophysin II/I by radioimmunoassay with vasopressor/oxytocic activity by bioassay in acid extracts of bovine supraoptic (SON) and paraventricular (PVN) nuclei. (From Zimmerman et al. (1974b) with permission.)

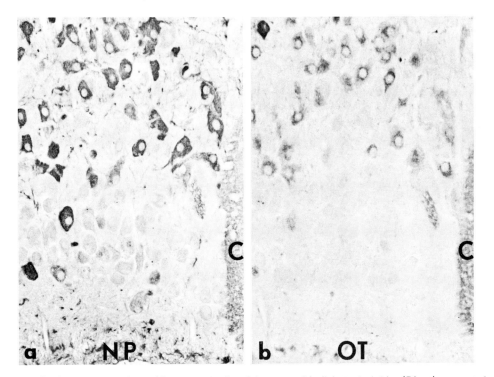

Fig. 8. Supraoptic nucleus of homozygous Brattleboro rat with diabetes insipidus (DI rat). a: reacted with antiserum to rat neurophysins by immunoperoxidase technique. b: adjacent 5 μm section reacted with anti-oxytocin. Both counterstained with dilute cresyl violet. The neurophysin (NP) present in DI rat is located in the same cells which contain oxytocin (OT) in the dorsal portion of the nucleus. C, edge of optic chiasm. × 260. (From Zimmerman (1976) with permission.)

tories, however, suggest that the separation may not be complete. In man (Zimmerman, 1976) and monkey (unpublished) the staining pattern for NSN and vasopressin coincided, but overlapped with cells containing ESN and oxytocin. A similar overlap was noted concerning the distribution of oxytocin and vasopressin in the rat (Zimmerman, 1976): vasopressin was found in most cells of the SON including the dorsal cells containing oxytocin. Despite elaborate absorption controls, we are forced at present to conclude that some cells contain both hormones (Sokol et al., 1976).

IV.2. Mid-cycle ESN

The remarkable rise of ESN in peripheral blood at mid-menstrual cycle in man and, as better demonstrated in monkey (Fig. 9), is of considerable interest. ESN, unlike NSN or vasopressin, appears very responsive to estrogen. ESN rises in pregnancy at mid-cycle (Robinson, 1975) and falls after ovariectomy (Robinson et al., 1976). After intramuscular administration of 300 µg of estradiol benzoate to a female rhesus monkey in the early follicular phase, ESN rises in peripheral plasma before LH, and stays elevated longer (Robinson et al., 1976) (Fig. 10). ESN in peripheral plasma appears to be a useful monitor of estradiol effects on the hypothalamus. The critical question at present, however, is

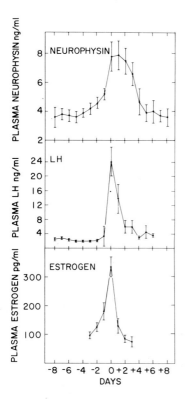

Fig. 9. Plasma estrogen-stimulated neurophysin, LH and estrogen at mid-cycle in 19 rhesus monkeys measured by radioimmunoassays. (From Robinson et al. (1976) with permission.)

whether ESN, or more likely the hormone it carries, has any real role in gonadotropin function. The presence of oxytocin in zona externa as described above and the large concentrations of ESN in portal blood (Table I) suggest that significant amounts of oxytocin could reach the anterior pituitary gland. Previous studies in which an effect of oxytocin on gonadotropin secretion was sought were disappointing (Reichlin and Mitnick, 1973). It may be possible, however, that a mid-cycle rise in oxytocin may not have a direct effect on gonadotropin secreting cells, but acts indirectly through Gn-RH by inhibiting its enzymatic destruction (Griffith et al., 1974). This appears to be a testable hypothesis. It seems less likely that ESN has hormone actions of its own or acts as a precursor molecule for an active yet unknown peptide fragment, but it must be considered.

Another dimension to the ESN-oxytocin story is where and how estradiol acts on this neurosecretory system. The immediate question that comes to mind is whether estradiol receptors are located on the cells of the SON and the PVN that form oxytocin and its neurophysin. Arcuate neurons avidly take up labeled estrogen as shown in autoradiography in rat (Pfaff and Keiner, 1973) and monkey (Keefer and Stumpf, 1975). Magnocellular neurons do incorporate estradiol in the squirrel monkey according to Keefer and

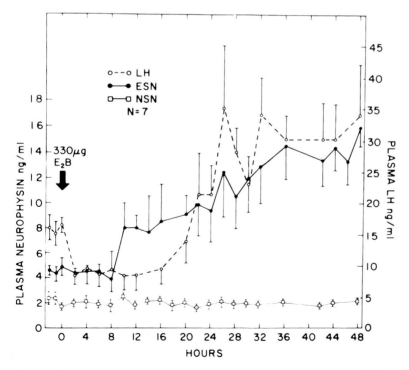

Fig. 10. Response of peripheral plasma estrogen-stimulated neurophysin (ESN) and luteinizing hormone (LH) to estradiol benzoate in 7 cycling rhesus monkeys. Nicotine-stimulated neurophysin (NSN) does not change while ESN rises before LH as measured by radioimmunoassay. (From Robinson et al., 1976.)

Stumpf (1975) but the uptake was scanty compared to arcuate nucleus as noted in our own studies of rhesus monkey (Gerlach et al., 1974). The 10-hr delay in the rise of ESN in peripheral blood of rhesus monkey after estradiol benzoate (Robinson et al., 1976) may involve new synthesis, transport and secretion of ESN stimulated by the steroid hormone. Alternatively, estradiol may act on other unknown neurons which project to the magnocellular system.

V. THE PARVICELLULAR SYSTEM

V.1. The parvicellular system and hypophysiotropins. Historical aspects

The vascular connection of the hypothalamus with the anterior pituitary gland was established in the 1940's (Harris, 1948). At the same time Scharrer and Scharrer (1954) were describing the magnocellular projections to the posterior pituitary with Gomori stains (Bargmann, 1968). The smaller cells (parvicellular) of the medial-basal hypothalamus or tuber which appeared to project to the portal system were Gomori negative and considered the source of hypothalamic hormones (Szentágothai et al., 1968). The arcuate nucleus appeared to be the major contributor to what became known as the tuberoinfundibular system. The magnocellular, Gomori positive, system forming vasopressin and oxytocin became known as the hypothalamo-hypophysial system. Most considered that the two systems were quite separate. As discussed earlier, electron microscopic studies revealed that smaller axon terminals with concentrations of smaller granules (about 100 nm) ending on or near portal capillaries in the zona externa were much more numerous than those containing the larger granules containing vasopressin and oxytocin (Knigge and Scott, 1970). The results of centrifugation studies supported the idea that the smaller granules contained releasing factors (Kobayashi et al., 1970).

Halász (1969) and coworkers found that the tonic secretion of anterior pituitary hormones could be maintained by hypothalamic islands containing the medial-basal hypothalamus in rats, lending further support to the concept that the tuberoinfundibular system contained in this region formed releasing factors. This region of the hypothalamus became known as the hypophysiotropic area and the releasing factors, hypophysiotropins (Szentágothai et al., 1968). The female rats described by Halász lacked cyclic changes in gonadotropins. The timing mechanism for the ovulatory surge of gonadotropins appeared to lie outside the island, perhaps in the preoptic area (Szentágothai et al.,1968). Stimulation of the preoptic area was shown to cause ovulation (Everett, 1965). McCann and coworkers (Crighton et al., 1970) also demonstrated that a significant portion of gonadotropin-releasing factor was located in the preoptic area by bioassay, although a larger amount was found in the medial-basal hypothalamus. Furthermore, lesions in the region of the suprachiasmatic nucleus abolished ovulation and reduced the releasing factor content in median eminence-arcuate region (Schneider et al., 1969). It was then considered that there was a neural connection between the two regions and that sex releasing factor was transferred from the preoptic area to the portal system for the surge of gonadotropin by a preoptico-infundibular tract.

Anatomical studies of this pathway and the other portions of the tuberoinfundibular system have been hampered by the fineness of the fibers, lack of Gomori stainability and smallness and sparsity of granulations compared to the magnocellular system. Granules have been found in some preparations in the SCN (Suburo and Pellegrino de Iraldi, 1969; Clattenberg et al., 1972) and, as mentioned earlier, the presence of neurophysin and vasopressin in some SCN cells also suggests neurosecretory function. Sparse granules have also been found in arcuate perikarya which change in number along with the ribosomes during the rat estrus cycle (Zambrano, 1969). At the present time, however, the tuberoinfundibular system is incompletely defined by traditional, basic anatomical tools. Further support for arcuate projections to zona externa have come from electrophysiological experiments in which arcuate cells were identified by antidromic stimulation of the median eminence (see review by Sawyer, 1975).

V.2. Localization of gonadotropin-releasing hormone

Barry et al. (1973) were the first to localize Gn-RH. Most of their studies were carried out in guinea pig using immunofluorescence. A great number of positive fibers were concentrated in zona externa of the median eminence. This finding has been confirmed in every subsequent immunocytochemical report (Baker et al., 1974; King et al., 1974; Kordon et al., 1974; Sétáló et al., 1975; Hökfelt et al., 1976; Kozlowski et al., 1976). In rat, mouse, and sheep (see Fig. 11a) these fibers are more concentrated in lateral zona externa while in guinea pig they are distributed more generally (Baker et al., 1974; Zimmerman, 1976). Somatostatin is also seen in zona externa (Hökfelt et al. 1974a; Pelletier et al., 1974a) but like vasopressin (Zimmerman, 1976) is found all along the sweep of the zona externa as viewed in mid-coronal section. Gn-RH is also concentrated in the posterior infundibulum (Kordon et al., 1974). Immunoelectron microscopic studies using immunoperoxidase technique have now established that Gn-RH is contained in granules of about 100 nm diameter in nerve terminals near portal capillaries in rat (Pelletier et al., 1974b; Goldsmith and Ganong, 1975). Centrifugation and radioimmunoassay of the granules confirmed their content of Gn-RH (Taber and Karavolas, 1975).

By light microscopic immunocytochemistry, Gn-RH is also found in scattered fibers in the internal portions of the median eminence. The staining appears axonal and is even found just under the ependymal lining of the floor of the third ventricle (Kordon and Ramirez, 1976). Whether these fibers actually enter the third ventricle like those containing vasopressin (Goldsmith and Zimmerman, 1975) is not known, but this might be found in future ultrastructural studies. The Gn-RH-containing fibers in the zona interna in rat do not appear to belong to the hypothalamo-hypophysial tract.

The critical question, however, which has not been fully answered by immunocytochemists concerns the location of cell bodies supposed to form the hypophysiotropins and transport them via their axons to portal capillaries according to the neurosecretory theory. Cells of the arcuate nucleus are the prime candidates for the sites of synthesis of Gn-RH since most of the hormone outside of the high concentration in median eminence is located here by radioimmunoassay (Palkovits et al., 1974; Wheaton et al., 1975). Most investigators have not found immunoreactive Gn-RH in perikarya (Kordon et al., 1974). Two laboratories, however, have published reports concerning Gn-RH positive perikarya. Barry et al. (1974), by using colchicine to stop axonal flow, found immunofluorescence

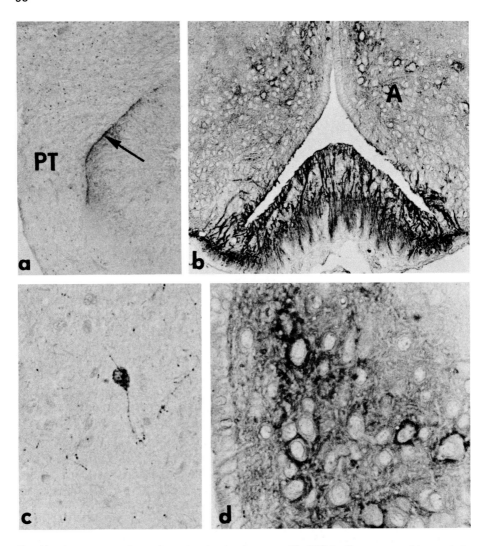

Fig. 11. Localization of gonadotropin-releasing hormone (Gn-RH) by immunoperoxidase technique. a: coronal section of sheep median eminence. Gn-RH is concentrated in lateral zona externa (arrow) along pars tuberalis (PT). x 65. (From Kozlowski et al. (1976) with permission.) b: Gn-RH in mouse median eminence and arcuate nucleus (A); coronal section. x 200. (From Zimmerman et al. (1974a) with permission.) c: medial preoptic area of guinea pig. Note Gn-RH in perikaryon and associated process of a neuron. (Silverman, unpublished.) d: higher magnification of b showing Gn-RH within the cytoplasm and on the surface of arcuate perikarya. x 600.

in cell bodies of the arcuate nucleus and others scattered around in the preoptic area including cells in the parolfactory cortex and around the anterior commissure in guinea pig. In both male and female New Zealand white mice, Zimmerman et al. (1974a) reported Gn-RH in arcuate perikarya by immunoperoxidase bridge technique and light microscopy. They found a number of positive perikarya scattered throughout the nucleus, but the majority were concentrated in the portion of the arcuate over the mid-median eminence, an area corresponding to the highest concentration found by assay of different parts of the nucleus (Palkovits et al., 1974). At this level, the positive cells were mostly concentrated in the dorsal-medial portion along the ventricle (Fig. 11b, d). Positive fibers could not be traced away from the cell bodies to the median eminence. Barry et al. (1974) also described a preoptico-infundibular pathway and suggested that the positive preoptic cell bodies send their Gn-RH to the portal bed in the median eminence.

These findings concerning perikaryal Gn-RH were questioned (Kordon et al., 1974) and considered as possible artefacts due to non-specific staining perhaps caused by antibodies to the protein used to form the conjugate in preparing the immunogen for Gn-RH. In the case of the studies of Zimmerman et al. (1974), bovine serum albumin (BSA) was conjugated to Gn-RH to produce the antiserum to Gn-RH. Prior to absorption with BSA, the antiserum did react with nerve cells in the immunoperoxidase technique, particularly the magnocellular system (Kozlowski et al., 1976). Absorption of the antiserum with BSA removed the staining in the magnocellular system but not the reactivity in arcuate perikarya. A more convincing control, absorption with Gn-RH, removed the findings in the arcuate nucleus, and tanycytes (Zimmerman et al., 1974a). Furthermore, the same staining system for Gn-RH used in the mouse produced what others reported in the rat (Zimmerman, unpublished): Gn-RH was only found in median eminence. Recently, A.J. Silverman (personal communication) has confirmed these findings in the mouse, including tanycyte Gn-RH described below in Swiss Albino mice using another antiserum to Gn-RH in an immunoperoxidase technique. With the same procedure, Silverman has also found positive perikarya in the guinea pig hypothalamus; Gn-RH was found in a few scattered perikarya in arcuate nucleus and in the medial preoptic area (Fig. 11c). The positive preoptic cells have not yet been seen in the mouse, but are under study. Silverman noted an interesting difference in the appearance of the immunostaining in the arcuate cell bodies in these two species. In the guinea pig the reactivity is always in the cytoplasm while in the mouse it appears both within and often outside on the surface of the cell bodies as shown in Fig. 11d. The latter finding is unexplained. If it is not a diffusion artefact, it may represent Gn-RH in axosomatic terminals or in ependymal endings which are known to occur in this region (Bleier, 1972). In this vein, it is interesting that radiolabeled Gn-RH put into the ventricle finds its way into arcuate cell bodies as well as into tanycytes of the median eminence as shown by autoradiography (Scott et al., 1974a).

V.3. Gn-RH in tanycytes

Gn-RH has only been reported in tanycytes of the mouse as shown by light microscopy (Zimmerman et al., 1974a). The staining is remarkable in its intensity and distribution throughout the cytoplasm of the entire course of these cells (Fig. 12a). The Gn-RH staining provides a good anatomical demonstration of the great numbers of these specialized ependymal cells whose cell bodies line the lower third ventricle and send numerous branching processes to the zona externa to end on portal vessels. Specificity of these findings was demonstrated by absorption with Gn-RH and they have been found in other strains

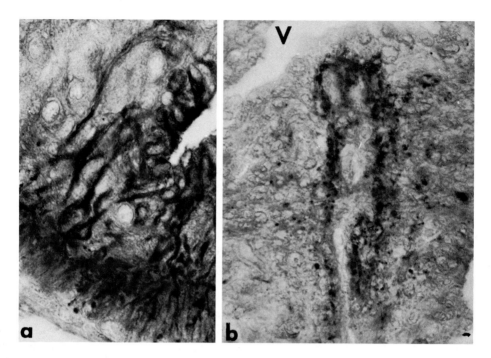

Fig. 12. Immunoreactive Gn-RH in mouse (a) tanycytes (higher magnification from Fig. 11b) and (b) organum vasculosum of the lamina terminalis in coronal sections. V, anterior tip of the third ventricle. Immunoperoxidase technique. × 860. (From Zimmerman et al. (1974a) with permission.)

of mice using other antisera on the light microscopic level (Kozlowski and Silverman, personal communications), but not at the electron microscopic level. Baker et al. (1975) recently reported immunoreactive Gn-RH associated with tanycytes of the rat median eminence. On the basis of the light microscopic pattern of staining, they interpreted their findings to mean that Gn-RH was located in axonal endings on tanycytes. Although such axonal endings on tanycytes may occur as shown by transmission electron microscopy in the past (Kobayashi and Matsui, 1969), immunoelectron microscopic studies may be needed to confirm this impression. The Gn-RH staining in mouse tanycytes is too extensive, is intracytoplasmic, and the surface too smooth to be adequately explained as Gn-RH in nerve terminals on them, although this may occur.

The findings in mouse tanycytes, like those for neurophysin discussed earlier, suggest that Gn-RH is normally found in these cells and that they have a role in reproduction. Anand Kumar and Knowles (1967) first suggested that the tanycytes of the median eminence of the rhesus monkey participate in reproduction. There are marked ultrastructural changes during the menstrual cycle (Knowles, 1972). At mid-cycle and after estrogen administration there is an increase in apical blebs on the ventricular surface and an increase in granules in the cytoplasm of these tanycytes. Most students of tanycytes consider their function primarily to be transport of materials from CSF to portal blood (Knigge and Scott, 1970; Knowles, 1972; Knigge, 1974). The microvilli and blebs on the ventricular surface suggest absorption at that end and the concentration of granules at their terminals, secretion into portal blood at the other (Knowles, 1972). That tanycytes can pick up Gn-RH from CSF has been demonstrated by Scott et al. (1974a) by autoradiography. When Gn-RH is infused into the ventricular system it appears in portal blood faster than it reaches the general circulation and it releases LH (Ben-Jonathan et al., 1974). The immunocytochemical data suggest that this actually occurs normally.

The presence of Gn-RH in CSF then is necessary to support the concept of tanycyte transport of Gn-RH. Cramer and Barraclough (1975) failed to detect it in rat ventricular CSF even after electrochemical stimulation of the preoptic area. On the other hand, Joseph et al. (1975) found significant concentrations in rat third ventricle which was about 10-fold higher 10 days after ovariectomy. They rapidly collected CSF after death by inserting a cannula through the base of the brain at the mammillary recess and posterior median eminence. Post mortem changes or contamination of the cannula by Gn-RH in the median eminence might have given falsely high values. They also found an increase in CSF levels in Gn-RH and a fall in the median eminence concentration after castration, however, which would not be fully explained by contamination. It seems more likely that the CSF concentrations were too high due to the radioimmunoassay used which detected peripheral plasma levels of about 50 pg/ml while other investigators generally could not detect it in peripheral blood (Araki et al., 1975). In our laboratories, Gn-RH was detected in third ventricular CSF of several monkeys during surgery by puncture of the median eminence of lamina terminalis (Ferin, Carmel and Zimmerman, unpublished). Values as high as 700 pg/ml were found, but again there is the question of contamination both from Gn-RH in median eminence tissue and the organum vasculosum of the lamina terminalis (OVLT) which also probably contains the hormone. It was undetectable in lumbar and cisternal CSF. Studies are now in progress to measure Gn-RH in CSF collected from indwelling cannulae in the third ventricle placed by a dorsal approach.

Vasopressin and neurophysin (Zimmerman et al., 1973a) and TRH (Knigge, 1974) are all found in CSF and it seems likely that Gn-RH will be found there as well. It may be very high in concentration at likely sites of secretion into the third ventricle such as the region of the OVLT and the median eminence and then it might be rapidly removed by ependyma so that it becomes undetectable in other ventricles or in the subarachnoid space.

V.5. Gn-RH in the organum vasculosum of the lamina terminalis (OVLT)

The OVLT is presently of considerable interest as both immunocytochemical (Barry et al., 1974; Zimmerman et al., 1974a; Kordon and Ramirez, 1976) and radioimmunoassay and bioassay studies (Wheaton et al., 1975) have now shown that most of all the Gn-RH in the preoptic area is concentrated in this structure. Leveque (1972) using histochemical stains had previously suggested that this structure may be important in reproduction.

If preoptic Gn-RH is important for the ovulatory surge of gonadotropins, then it might follow that the OVLT is important in this event. It is not known, however, how Gn-RH would get from the OVLT to the gonadotropic cells in the anterior pituitary gland if it does at all. Perhaps the OVLT has some other role and the surge is regulated by preoptico-infundibular neuronal pathways.

As seen in the mouse (Fig. 12b), light microscopic immunocytochemistry reveals Gn-RH concentrated around the vascular structure of the OVLT which appears to be located in terminals which could be those of tanycytes or axons. The studies of guinea pig (Barry et al., 1974) and rat (Baker et al., 1975; Kordon and Ramirez, 1976) suggest it is in axons, but ultrastructural studies may be needed to further establish the precise location. The source of possible Gn-RH bearing nerve fibers to the OVLT is also not certain, but they may come from perikarya in the preoptic region (Barry et al., 1974).

This circumventricular organ which lies in the lamina terminalis just in front of and over the optic chiasm at the anterior tip of the third ventricle (Fig. 12b) is very similar in structure to, though smaller than, the median eminence (Weindl and Joynt, 1972). It has fenestrated capillaries which receive tanycyte and axon terminals (Röhlich and Wenger, 1969). There are also nerve endings in this area which could also secrete into the third ventricle (Scott et al., 1974b; Weindl and Schinko, 1975). What is known about its venous drainage in man and monkey (Duvernoy et al., 1969) suggests that it goes into the systemic circulation and not the hypophysial portal system.

The OVLT, at present then, does not appear to be an accessory portal link to the anterior pituitary gland displaced forward in the hypothalamus. This idea is appealing, as somatostatin is also found here (Pelletier et al., 1975a, b) as well as the neurohypophysial peptides (Zimmerman, unpublished) as shown by light microscopic immunocytochemistry. Neurosecretion may be a general phenomenon at this site. It is also possible that these peptides may be absorbed from CSF at the OVLT or secreted into the ventricle by nerve endings converging into this region. In the latter case, Gn-RH for the ovulatory surge could be carried caudally by the CSF to the tanycytes of the median eminence and thence to the portal system, a possibility discussed but not favored by Wheaton et al. (1975). A common CSF pathway would not explain the specific point-to-point relationship between the portal system and the preoptic area described by Tejasen and Everett (1967), where unilateral frontal cuts abolished ovulatory responses to ipsilateral preoptic stimulation. However, they could be different pathways. A CSF pathway could explain the ovulatory behavior of the monkeys described by Krey et al. (1975) by transfer of Gn-RH from the preoptic area via CSF into the tanycytes of the island. Further experimentation on the preoptic area needs to be directed to the OVLT. For example, it would be helpful to know what its removal would do to reproduction. It was recently shown by destruction of the subfornical organ that this circumventricular organ is important for drinking (Dellmann and Simpson, 1975).

V.6. Pineal gland

Whether Gn-RH is located in pineal gland, yet another circumventricular organ is unsettled at this time. White et al. (1974) found larger concentrations in the pineal of sheep, oxen and pigs than the hypothalamus by bioassay and radioimmunoassay. Kozlowski and Zimmerman (1974) reported it in mouse and sheep pineals by immunocytochemistry. Further immunocytochemical studies have, however, failed to confirm it in our laboratories (Kozlowski and Zimmerman, unpublished) as have radioimmunoassay studies of sheep and monkey pineal extracts (Araki et al., 1975). Somatostatin has been reported in pineal and also in the subcommissural organ where it was found in ependyma by light microscopy (Pelletier et al., 1975a).

VI. TRH AND SOMATOSTATIN

TRH and somatostatin are present in highest concentration in median eminence as shown by radioimmunoassay (Brownstein et al., 1974, 1975). The highest nuclear concentration of somatostatin is found in the arcuate nucleus (Brownstein et al., 1975). Both peptides are, however, located in many other hypothalamic nuclei and, unlike the octapeptides and Gn-RH which are essentially limited in their distribution to the hypothalamus, they are found all over the brain.

There are no published reports concerning the cellular location of TRH. Perhaps due to its small size or lability, it is more easily lost from tissues. Somatostatin has been found in the zona externa of the median eminence both by light (Hökfelt et al., 1974a; Pelletier et al., 1975a, b) and electron microscopic immunocytochemistry (Pelletier et al., 1974a, 1975a, b) which demonstrated it in granules in parvicellular nerve terminals. It was also found in OVLT as outlined above (Pelletier et al., 1975a, b), and in a few fibers in the ventromedial nucleus (Hökfelt et al., 1974a). Gn-RH and somatostatin positive fibers were reported in the amygdala by light microscopy by Hökfelt et al. (1976). Somatostatin has not been found in perikarya except for the ependyma of the subcommissural organ (Pelletier et al., 1975a).

Although administration of somatostatin inhibits insulin, glucagon and gastrin secretion, it is apparently not located in the same cells of the pancreas and stomach which it inhibits (Dubois, 1975). It will be interesting to learn in the future how the inhibitor gets to the nearby hormone secreting cells in these organs.

VII. SUMMARY, CONCLUSIONS AND SPECULATIONS

The cytological and regional assay data obtained in the last 5 years have supported, modified, and expanded previous ideas about the participation of the brain in reproduction and other endocrine functions. Large amounts of rapidly changing data have accumulated in great part due to the availability of antisera to specific neurosecretory peptides. The data are incomplete, and at times contradictory, as the techniques are new. With improvements, the immunocytochemical techniques will probably result in more complete localization of hypophysiotropins at sites of synthesis, transport and secretion as well as in their targets (Sternberger and Petrali, 1975).

The new data have stimulated some new trends in our thinking. It has become apparent that vasopressin probably has a real role in anterior pituitary function. The responsiveness of oxytocin-neurophysin to estrogen stimuli suggests that the hypothalamo-hypophysial system, perhaps oxytocin itself, in some unknown way, may regulate gonadotropin secretion.

It has been shown that the synthetic hypophysiotropins, Gn-RH and somatostatin, are normally present in the small granules of parvicellular nerve terminals on portal capillaries, a major step forward in proving the neurosecretory theory for the tuberoinfundibular system as shown earlier for the hypothalamo-hypophysial system. There is also some evidence that Gn-RH is located in arcuate perikarya, and probably in other cell bodies in the preoptic as well, suggesting that these are sites of synthesis for this hormone. It is likely that improved immunocytochemical techniques will demonstrate the connections between Gn-RH positive arcuate perikarya and the known terminals in zona externa. Gn-RH-containing cell bodies in the preoptic area may be further established and their connections known. Although some of their fibers may be found to go to the infundibulum, they might connect to nuclei in the medial-basal hypothalamus, the SCN (Barry et al., 1974) or amygdala and play some transmitter-like role. Vasopressin might turn out to have a similar role in carrying messages from the SCN to CRF-forming neurons.

The demonstration of immunoreactive neurophysin and Gn-RH in tanycytes of the median eminence would support the notion that these peculiar cells *normally* play a role in neuroendocrine function. In this line of thinking, that large concentration of Gn-RH and other neurocircumventricular organs would suggest that the ventricular fluid may yet be an important communication medium. Vasopressin may be secreted into the third ventricle to be carried to other brain areas to affect learning (Van Wimersma Greidanus et al., 1975) and perhaps Gn-RH is secreted at the OVLT to be carried by CSF to the portal system in the median eminence or to other CNS regions to affect sexual behavior (Pfaff, 1973). It may also be established that the OVLT is an accessory hypophysial portal system, though it seems unlikely at present, or it may have functions which are unknown.

ACKNOWLEDGEMENTS

Supported by NIH Career Development (NINDS Teacher-Investigator Award), NS 11008 and a Ford Foundation Grant to the International Institute for the Study of Human Reproduction, Columbia University, New York. The author thanks his coworkers for many valuable contributions: A.J. Silverman, G.P. Kozlowski, P.C. Goldsmith, H.W. Sokol, R. Defendini, A.G. Robinson, P.W. Carmel, M. Ferin, L. Recht, P. Stillman, K.C. Hsu, M. Tannenbaum and S. Rosario.

REFERENCES

Acher, R., Chauvet, J. et Olivry, G. (1956) Sur l'existence d'une hormone unique hypophysaire. Relation entre l'oxytocin, le vasopressine et la protein de van Dyke extradites de la neurohypophyse du boeuf. *Biochim. biophys. Acta (Amst.)*, 22: 428–433.
Alvarez-Buylla, R., Livett, B.G., Uttenthal, L.O., Milton, S.H. and Hope, D.B. (1970) Immunohistochemical evidence for the transport of neurophysin in the neurosecretory neurons of the dog. *Acta physiol. scand.*, Suppl. 357: 5.
Anand Kumar, T.C. and Knowles, F.G.W. (1967) A system linking the third ventricle with the pars tuberalis of the rhesus monkey. *Nature (Lond.)*, 215: 54–55.

Araki, S., Ferin, M., Zimmerman, E.A. and VandeWiele, R.L. (1975) Ovarian modulation of immunoreactive gonadotropins-releasing hormone (Gn-RH) in the rat brain: evidence for a differential effect on the anterior and mid-hypothalamus. *Endocrinology*, 96: 644–650.

Baker, B.L., Dermody, W.C. and Reel, J.R. (1974) Localization of luteinizing hormone-releasing hormone in the mammalian hypothalamus. *Amer. J. Anat.*, 139: 129–134.

Baker, B.L., Dermody, W.C. and Reel, J.R. (1975) Distribution of gonadotropin-releasing hormone in the rat brain as observed with immunocytochemistry. *Endocrinology*, 97: 125–135.

Bargmann, W. (1966) Neurosecretion. *Int. Rev. Cytol.*, 19: 183–201.

Bargmann, W. (1968) Neurohypophysis: structure and function. In *Handbook of Experimental Pharmacology, Vol. XXIII, Neurohypophyseal Hormones and Similar Polypeptides*, B. Berde (Ed.), Springer, New York, pp. 1–39.

Barry, J., Dubois, M.P. and Poulain, P. (1973) LRF producing cells of the mammalian hypothalamus. *Z. Zellforsch.*, 146: 351–366.

Barry, J., Dubois, M.P. and Carette, B. (1974) Immunofluorescence study of the preoptico-infundibular LRF neurosecretory pathway in the normal, castrated or testosterone-treated male guinea pig. *Endocrinology*, 95: 1416–1423.

Ben-Jonathan, N., Mical, R.S. and Porter, J.C. (1974) Transport of LRF from CSF to hypophysial portal and systemic blood and the release of LH. *Endocrinology*, 95: 18–25.

Bleier, R. (1972) Structural relationships of ependymal cells and their processes within the hypothalamus. In *Brain–Endocrine Interaction. Median Eminence: Structure and Function*, K.M. Knigge, D.E. Scott and A. Weindl (Eds.), Karger, Basel, pp. 306–318.

Brownstein, M.J., Palkovits, M., Saavedra, J.M., Bassiri, R. and Utiger, R.D. (1974) Thyrotropin-releasing hormone in specific nuclei of rat brain. *Science*, 185: 267–269.

Brownstein, M., Arimura, A., Sato, H., Schally, A.V. and Kizer, J.S. (1975) The regional distribution of somatostatin in the rat brain. *Endocrinology*, 96: 1456–1461.

Burford, G.D., Jones, C.W. and Pickering, B.T. (1971) Tentative identification of a vasopressin-neurophysin and an oxytocin-neurophysin in the rat. *Biochem. J.*, 124: 809–813.

Capra, J.D. and Walter, R. (1975) Primary structure and evolution of neurophysins. *Ann. N.Y. Acad. Sci.*, 248: 92–111.

Clattenberg, R.E., Singh, R.P. and Montemurro, D.G. (1972) Post-coital ultrastructural changes in neurons of the suprachiasmatic nucleus in the rabbit. *Z. Zellforsch.*, 125: 448–459.

Clayton, G.W., Librik, L., Gardner, R.L. and Guillemin, R. (1963) Studies on the circadian rhythm of pituitary adrenocorticotropic release in man. *J. clin. Endocr.*, 23: 975–980.

Clemente, F. and Ceccarelli, B. (1970) Fine structure of rat hypothalamic nuclei. In *The Hypothalamus*, L. Martini, M. Motta and F. Fraschini (Eds.), Academic Press, New York, pp. 17–44.

Cramer, O.M. and Barraclough, C.A. (1975) Failure to detect luteinizing hormone-releasing hormone in third ventricle cerebral spinal fluid under a variety of experimental conditions. *Endocrinology*, 96: 913–921.

Crighton, D.G., Schneider, H.P.G. and McCann, S.M. (1970) Localization of LH-releasing factor in the hypothalamus and neurohypophysis as determined by in vitro assay. *Endocrinology*, 87: 323–329.

Dean, C.R., Hope, D.B. and Kazic, T. (1968) Evidence for the storage of oxytocin with neurophysin-I and vasopressin with neurophysin-II in separate neurosecretory granules. *Brit. J. Pharmacol.*, 24: 192–193.

Dellman, H.-D. and Simpson, J.B. (1975) Comparative ultrastructure and function of the subfornical organ. In *Brain–Endocrine Interaction. The Ventricular System in Neuroendocrine Mechanisms*, K.M. Knigge, D.E. Scott, H. Kobayashi and S. Ishii (Eds.), Karger, Basel, pp. 166–189.

Douglas, W.W. and Nagasawa, J. (1970) Membrane vesiculation at sites of exocytosis in the neurohypophysis and adrenal medulla: device for membrane conservation. *J. Physiol. (Lond.)*, 218: 94P–95P.

Dreifuss, J.J. (1975) A review of neurosecretory granules: their contents and mechanism of release. *Ann. N.Y. Acad. Sci.*, 248: 184–201.

Dubois, M.P. (1975) Immunoreactive somatostatin is present in discrete cells of the endocrine pancreas. *Proc. nat. Acad. Sci. (Wash.)*, 72: 1340–1343.
Duvernoy, H., Koritke, J.G. et Monnier, G. (1969) Sur la vascularisation de la lame terminale humaine. *Z. Zellforsch.*, 102: 49–77.
Du Vigneaud, V. (1956) Hormones of the posterior pituitary gland. In *The Harvey Lectures 1954–55*, Academic Press, New York, pp. 1–26.
Everett, J.W. (1965) Ovulation in rats from pre-optic stimulation through platinum electrodes. Importance of duration and spread of stimulus. *Endocrinology*, 76: 1195–1201.
Fawcett, C.P., Powell, A.E. and Sachs, H. (1968) Biosynthesis and release of neurophysin. *Endocrinology*, 83: 1299–1310.
Gerlach, J.L., McEwen, B.S., Pfaff, D.W., Ferin, M. and Carmel, P.W. (1974) Rhesus monkey brain binds radioactivity from ^{3}H-estradiol and ^{3}H-corticosterone, demonstrated by nuclear isolation and radioautography. *Endocrinology*, 94: A-370.
Ginsburg, M. and Jayasena, K. (1968) The distribution of proteins that bind neurohypophysial hormones. *J. Physiol. (Lond.)*, 197: 65–76.
Goldsmith, P.C. and Ganong, W.F. (1975) Ultrastructural localization of luteinizing hormone-releasing hormone in the median eminence of the rat. *Brain Res.*, 97: 181–193.
Goldsmith, P.C. and Zimmerman, E.A. (1975) Ultrastructural localization of neurophysin and vasopressin in rat median eminence. *Endocrinology*, 96: A-377.
Gomori, G. (1941) Observations with different stains on human islet of Langerhans. *Amer. J. clin. Pathol.*, 20: 665–666.
Griffith, E.C., Hooper, K.C., Jeffcoate, S.L. and Holland, D.T. (1974) The presence of peptidases in the rat hypothalamus inactivating LH-RH. *Acta endocr. (Kbh.)*, 77: 435–441.
Halász, B. (1969) The endocrine effects of isolation of the hypothalamus from the rest of the brain. In *Frontiers in Neuroendocrinology*, W.F. Ganong and L. Martini (Eds.), Oxford University Press, London, pp. 307–342.
Harris, G.W. (1948) Neural control of the pituitary gland. *Physiol. Rev.*, 28: 139–179.
Hedge, G.A., Yates, M.B., Marcus, R. and Yates, F.E. (1966) Site of action of vasopressin causing corticotropin release. *Endocrinology*, 79: 328–340.
Hökfelt, T., Efendic, S., Johansson, O., Luft, R. and Arimura, A. (1974a) Immunohistochemical localization of somatostatin (growth hormone-release inhibiting factor) in the guinea pig brain. *Brain Res.*, 80: 165–169.
Hökfelt, T., Fuxe, K., Goldstein, M. and Johansson, O. (1974b) Immunohistochemical evidence for the existence of adrenaline neurons in the rat brain. *Brain Res.*, 66: 235–251.
Hökfelt, T., Fuxe, K., Goldstein, M., Johansson, O., Fraser, H. and Jeffcoate, S.L. (1976) Immunofluorescence mapping of central monoamine and releasing hormone (LRH) systems. In *Anatomical Neuroendocrinology*, W.E. Stumpf and L.D. Grant (Eds.), Karger, Basel, in press.
Hollenberg, M.D. and Hope D.B. (1968) The isolation of the native hormone-binding proteins from bovine posterior pituitary lobes: crystallization of neurophysin I and II as complexes with (8-arginine)-vasopressin. *Biochem. J.*, 106: 557–564.
Joseph, S.A., Sorrentino, S., Jr. and Sundberg, D.K. (1975) Releasing hormones, LRF and TRF, in the cerebrospinal fluid of the third ventricle. In *Brain–Endocrine Interaction. II. The Ventricular System*, K.M. Knigge, D.E. Scott, H. Kobayashi and S. Ishii (Eds.), Karger, Basel, pp. 306–312.
Keefer, D.A. and Stumpf, W.E. (1975) Atlas of estrogen-concentrating cells in the central nervous system of the squirrel monkey. *J. comp. Neurol.*, 160: 419–441.
King, J.C., Parsons, J.A., Erlandsen, S.L. and Williams, T.H. (1974) Luteinizing hormone-releasing hormone (LH-RH) pathway of the rat hypothalamus revealed by the unlabeled antibody peroxidase-antiperoxidase method. *Cell Tiss. Res.*, 153: 211–217.
Knigge, K.M. (1974) Role of the ventricular system in neuroendocrine processes: initial studies on the role of catecholamines in transport of thyrotropin releasing factor. In *Frontiers in Neurology and Neuroscience Research*, P. Seeman and G.M. Brown (Eds.), University of Toronto Press, Toronto, pp. 40–47.

Knigge, K.M. and Scott, D.E. (1970) Structure and function of the median eminence. *Amer. J. Anat.*, 129: 223—228.

Knowles, F. (1972) Ependyma of the third ventricle in relation to pituitary function. In *Topics in Neuroendocrinology, Progr. Brain Res., Vol. 38*, J. Ariëns Kappers and J.P. Schadé (Eds.), Elsevier, Amsterdam, pp. 255—270.

Kobayashi, H. and Matsui, T. (1969) Fine structure of the median eminence and its functional significance. In *Frontiers in Neuroendocrinology*, W.F. Ganong and L. Martini (Eds.), Oxford University Press, London, pp. 3—46.

Kobayashi, H., Matsui, T. and Ishii, S. (1970) Functional electron microscopy of the hypothalamic median eminence. *Int. Rev. Cytol.*, 29: 281—381.

Kordon, C. and Ramirez, V.D. (1976) Recent developments in neurotransmitter—hormone interactions. In *Anatomical Neuroendocrinology*, W.E. Stumpf and L.D. Grant (Eds.), Karger, Basel, in press.

Kordon, C., Kerdelhué, B., Pattou, E. and Jutisz, M. (1974) Immunocytochemical localization of LHRH in axons and nerve terminals of the rat median eminence. *Proc. Soc. exp. Biol. (N.Y.)*, 147: 122—127.

Kozlowski, G.P. and Zimmerman, E.A. (1974) Localization of gonadotropin-releasing hormone (Gn-RH) in sheep and mouse brain. *Anat. Rec.*, 178: 396.

Kozlowski, G.P., Nett, T.M. and Zimmerman, E.A. (1976) Immunocytochemical localization of Gn-RH and neurophysin. In *Anatomical Neuroendocrinology*, W.E. Stumpf and L.D. Grant (Eds.), Karger, Basel, in press.

Krey, L.C., Butler, W.R. and Knobil, E. (1975) Surgical disconnection of the medial basal hypothalamus and pituitary function in the rhesus monkey. I. Gonadotropin secretion. *Endocrinology*, 96: 1073—1087.

LaBella, F.S., Beaulieu, G. and Reiffenstein, R.J. (1962) Evidence for the existence of separate vasopressin and oxytocin-containing granules in the neurohypophysis. *Nature (Lond.)*, 193: 173—174.

Laquer, G.L. (1954) Neurosecretory pathways between the hypothalamic paraventricular nucleus and the neurohypophysis. *J. comp. Neurol.*, 101: 543—554.

LeClerc, R. and Pelletier, G. (1974) Electron microscope localization of vasopressin in the hypothalamus and neurohypophysis of the normal and Brattleboro rat. *Amer. J. Anat.*, 140: 583—588.

Lederis, K. (1961) Vasopressin and oxytocin in the mammalian hypothalamus. *Gen. comp. Endocrinol.*, 1: 80—89.

Leveque, T.F. (1972) The medial prechiasmatic area in the rat and LH secretion. In *Brain—Endocrine Interaction. Median Eminence: Structure and Function*, K.M. Knigge, D.E. Scott and A. Weindl (Eds.), Karger, Basel, pp. 298—305.

Martini, L. (1966) Neurohypophysis and anterior pituitary activity. In *The Pituitary Gland, Vol. 3*, G.W. Harris and B.T. Donovan (Eds.), University of California Press, Berkeley, pp. 535—577.

Meyer, V. and Knobil, E. (1966) Stimulation of growth hormone secretion by vasopressin in the rhesus monkey. *Endocrinology*, 79: 1016—1020.

Moore R.Y. (1973) Retinohypothalamic projections in mammals: a comparative study. *Brain Res.*, 49: 403—409.

Moore, R.Y. and Eichler, V.B. (1972) Loss of a circadian adrenal corticosterone rhythm following suprachiasmatic lesions in the rat. *Brain Res.*, 42: 201—206.

Olivecrona, H. (1957) Paraventricular nucleus and pituitary gland. *Acta physiol. scand.*, 40, Suppl. 136: 1—178.

Palkovits, M., Brownstein, M., Saavedra, J.M. and Axelrod, J. (1974) Luteinizing hormone-releasing hormone (LH-RH) content of the hypothalamic nuclei in rat. *Endocrinology*, 96: 554—558.

Parry, H.B. and Livett, B.G. (1973) A new hypothalamic pathway to the median eminence containing neurophysin and its hypertrophy in sheep with natural scrapie. *Nature (Lond.)*, 242: 63—65.

Pelletier, G., Labrie, F., Arimura, A. and Schally, A.V. (1974a) Electron microscopic immunohistochemical localization of growth hormone-release inhibiting hormone (somatostatin) in the rat median eminence. *Amer. J. Anat.*, 140: 445—450.

Pelletier, G., Labrie, G., Puviani, R., Arimura, A. and Schally, A.V. (1974b) Electron microscope localization of luteinizing hormone-releasing hormone in the rat median eminence. *Endocrinology*, 95: 314–315.

Pelletier, G., Le Clerc, R., LaBrie, F. and Puviani, R. (1974c) Electron microscopic immunohistochemical localization of neurophysin in the rat hypothalamus and pituitary. *Molec. Cell. Endocrinol.*, 1: 157–166.

Pelletier, G., Le Clerc, R. and Dubé, D. (1975a) Immunohistochemical localization of somatostatin in the rat brain. *Endocrinology*, 96: A-153.

Pelletier, G., Le Clerc, R., Dubé, D., Labrie, F., Puviani, R., Arimura, A. and Schally, A.V. (1975b) Localization of growth hormone-release inhibiting hormone (somatostatin) in the rat brain. *Amer. J. Anat.*, 142: 497–500.

Pfaff, D.W. (1973) LHRH potentiates lordosis behavior in hypophysectomized ovariectomized rats. *Science*, 182: 1148–1150.

Pfaff, D.W. and Keiner, M. (1973) Atlas of estradiol-concentrating cells in the central nervous system of the female rat. *J. comp. Neurol.*, 151: 121–158.

Pickup, J.C., Johnston, C.I., Nakamura, S., Uttenthal, L.O. and Hope, D.B. (1973) Subcellular organization of neurophysins, oxytocin, (8-lysine)-vasopressin and adenosine triphosphatase in porcine posterior pituitary lobes. *Biochem. J.*, 132: 316–317.

Porter, J.C., Ben-Jonathan, N., Oliver, C. and Eskay, R.L. (1975) Secretion of releasing hormones and their transport from CSF to hypophysial portal blood. In *Brain–Endocrine Interaction. II. The Ventricular System,* K.M. Knigge, D.E. Scott, H. Kobayashi and S. Ishii (Eds.), Karger, Basel, pp. 295–312.

Reichlin, S. and Mitnick, M. (1973) Biosynthesis of hypothalamic hypophysiotropic factors. In *Frontiers in Neuroendocrinology*, W.F. Ganong and L. Martini (Eds.), Oxford University Press, London, pp. 61–88.

Robinson, A.G. (1975) Isolation, assay, and secretion of individual human neurophysins. *J. clin. Invest.*, 55: 360–367.

Robinson, A.G. and Frantz, A.G. (1973) Radioimmunoassay of posterior pituitary peptides: a review. *Metabolism*, 22: 1047–1057.

Robinson, A.G. and Zimmerman, E.A. (1973) Cerebrospinal fluid and ependymal neurophysin. *J. clin. Invest.*, 52: 1260–1267.

Robinson, A.G., Ferin, M. and Zimmerman, E.A. (1976) Neurophysin in monkeys: emphasis on the hypothalamic response to estrogen and ovarian events. *Endocrinology,* in press.

Röhlich, P. und Wenger, T. (1969) Elektronenmikroskopische Untersuchungen am Organon vasculosum laminae terminalis der Ratte. *Z. Zellforsch.*, 102: 483–506.

Sachs, H., Fawcett, C.P., Portanova, R. and Takabatake, Y. (1969) Biosynthesis and release of vasopressin and neurophysin. *Recent Progr. Hormone Res.*, 25: 447–491.

Sawyer, C.H. (1975) First Geoffrey Harris memorial lecture. Some recent developments in brain-pituitary-ovarian physiology. *Neuroendocrinology*, 17: 97–124.

Scharrer, E. and Scharrer, B. (1954) Hormones produced by neurosecretory cells. *Recent Progr. Hormone Res.*, 10: 183–240.

Schneider, H.P.G., Crighton, D.B. and McCann, S.M. (1969) Suprachiasmatic LH-releasing factor. *Neuroendocrinology*, 5: 271–380.

Scott, D.E., Dudley, G.K., Knigge, K.M. and Kozlowski, G.P. (1974a) In vitro analysis of the cellular localization of luteinizing hormone-releasing factor (LRF) in the basal hypothalamus of the rat. *Cell Tiss. Res.*, 149: 371–378.

Scott, D.E., Kozlowski, G.P. and Sheridan, M.N. (1974b) Scanning electron microscopy in the ultrastructural analysis of the mammalian cerebral ventricular system. *Int. Rev. Cytol.*, 37: 349–388.

Sétáló, G., Vigh, S., Schally, A.V., Arimura, A. and Flerkó, B. (1975) LH-RH-containing neural elements in the rat hypothalamus. *Endocrinology*, 96: 135–142.

Silverman, A.J. and Zimmerman, E.A. (1975) Ultrastructural immunocytochemical localization of neurophysin and vasopressin in the median eminence and posterior pituitary of the guinea pig. *Cell Tiss. Res.*, 159: 291–301.

Sloper, J.C. (1966) The experimental and cytopathological investigation of neurosecretion in the hypothalamus and pituitary. In *The Pituitary Gland, Vol. 3*, G.W. Harris and B.T. Donovan (Eds.), University of California Press, Berkeley, pp. 131–239.

Sokol, H.W. (1970) Evidence for oxytocin synthesis after electrolytic destruction of the paraventricular nucleus in rats with hereditary diabetes insipidus. *Neuroendocrinology*, 6: 90–97.

Sokol, H.W. and Valtin, H. (1967) Evidence for the synthesis of oxytocin and vasopressin in separate neurones. *Nature (Lond.)*, 214: 314–316.

Sokol, H.W., Zimmerman, E.A., Sawyer, W.H. and Robinson, A.G. (1976) The hypothalamo-hypophysial system of the rat: localization and quantitation of neurophysin by light microscopic immunocytochemistry in normal rats and in Brattleboro rats deficient in vasopressin and neurophysin. *Endocrinology*, in press.

Sternberger, L.A. and Petrali, J.P. (1975) Quantitative immunohistochemistry of pituitary receptors for luteinizing hormone-releasing hormone. *Endocrinology*, 96: A-105.

Stillman, M., Recht, L.D., Sokol, H.W., Seif, M. and Zimmerman, E.A. (1975) Neurophysin, vasopressin and the hypophysial portal system. Evidence that adrenalectomy selectively increases the neurophysin associated with vasopressin. *Soc. Neurosci.*, abstract.

Suburo, A.M. and Pellegrino de Iraldi, A. (1969) An ultrastructural study of the rat's suprachiasmatic nucleus. *J. Anat. (Lond.)*, 105: 439–446.

Sunde, D. and Sokol, H.W. (1975) Quantification of rat neurophysins by polyacrylamide gel electrophoresis: applications to the rat with hereditary hypothalamic diabetes insipidus. *Ann. N.Y. Acad. Sci.*, 248: 345–364.

Swanson, L.W. and Cowan, W.M. (1975) The efferent connections of the suprachiasmatic nucleus of the hypothalamus. *J. comp. Neurol.*, 16: 1–12.

Szentágothai, J., Flerkó, B., Mess, B. and Halász, B. (Eds.) (1968) *Hypothalamic Control of the Anterior Pituitary*, Akadémiai Kiadó, Budapest.

Taber, C.A. and Karavolas, H.J. (1975) Subcellular localization of LH releasing activity in the rat hypothalamus. *Endocrinology*, 96: 446–452.

Tejasen, T. and Everett, J.W. (1967) Surgical analysis of the preoptico-tuberal pathway controlling ovulatory release of gonadotropins in the rat. *Endocrinology*, 81: 1387–1396.

Vandesande, F., DeMey, J. and Dierickx, K. (1974) Identification of neurophysin-producing cells. I. The origin of the neurophysin-like substance-containing nerve fibers of the external region of the median eminence of the rat. *Cell Tiss. Res.*, 151: 187–200.

Vandesande, F., Dierickx, K. and DeMey, J. (1975a) Identification of the vasopressin-neurophysin II and the oxytocin-neurophysin I producing neurons in the bovine hypothalamus. *Cell Tiss. Res.*, 156: 189–200.

Vandesande, F., Dierickx, K. and DeMey, J. (1975b) Identification of vasopressin-neurophysin producing neurons of the rat suprachiasmatic nuclei. *Cell Tiss. Res.*, 156: 337–342.

Vandesande, F., Dierickx, K. and DeMey, J. (1975c) Identification of separate vasopressin-neurophysin II and oxytocin-neurophysin I containing nerve fibers in the external region of the bovine median eminence. *Cell Tiss. Res.*, 158: 509–516.

Van Dyke, H.B., Chow, B.F., Greep, R.O. and Rothen, A. (1941) The isolation of a protein from pars neuralis of the ox pituitary with constant oxytocic, pressor and diuresis-inhibiting effects. *J. Pharmacol. (Kyoto)*, 74: 190–209.

Van Wimersma Greidanus, Tj. B., Dogterom, J. and De Wied, D. (1975) Intraventricular administration of anti-vasopressin serum inhibits memory consolidation in rats. *Life Sci.*, 16: 637–644.

Vorherr, H., Bradbury, M.W.B., Hoghoughi, M. and Kleeman, C.R. (1968) Anti-diuretic hormone in cerebrospinal fluid during endogenous and exogenous changes in its blood level. *Endocrinology*, 83: 246–250.

Watkins, W.B. (1975) Immunohistochemical demonstration of neurophysin in the hypothalamo-neurohypophysial system. *Int. Rev. Cytol.*, 41: 241–284.

Watkins, W.B., Schwabedal, P. and Bock, R. (1974) Immunohistochemical demonstration of a CRF-associated neurophysin in the external zone of the rat median eminence. *Cell Tiss. Res.*, 152: 411–421.
Weindl, A. and Joynt, R.J. (1972) The median eminence as a circumventricular organ. In *Brain–Endocrine interaction. Median Eminence: Structure and Function,* K.M. Knigge, D.E. Scott and A. Weindl (Eds.), Karger, Basel, pp. 280–297.
Weindl, A. and Schinko, I. (1975) Vascular and ventricular neurosecretion in the organum vasculosum of the lamina terminalis of the Golden hamster. In *Brain–Endocrine Interaction. The Ventricular System in Neuroendocrine Mechanisms*, K.M. Knigge, D.E. Scott, H. Kobayashi and S. Ishii (Eds.), Karger, Basel, pp. 190–203.
Wheaton, J.E., Krulich, L. and McCann, S.M. (1975) Localization of luteinizing hormone-releasing hormone (LRH) in the preoptic area and hypothalamus of the rat using radioimmunoassay. *Endocrinology*, 97: 30–38.
White, W.F., Hedlund, M.T., Weber, G.F., Rippel, R.H., Johnson, E.S. and Wilber, J.F. (1974) The pineal gland: a supplemental source of hypothalamic releasing hormones. *Endocrinology*, 94: 1422–1426.
Wittowski, W. and Bock, R. (1972) Electron microscopical studies of the median eminence following inference with the feedback system, anterior pituitary–adrenal cortex. In *Brain–Endocrine Interaction. Median Eminence: Structure and Function*, K.M. Knigge, D.E. Scott and A. Weindl (Eds.), Karger, Basel, pp. 171–180.
Zambrano, D. (1969) The arcuate complex of the female rat during the sexual cycle. *Z. Zellforsch.*, 93: 560–570.
Zimmerman, E.A. (1976) Localization of hypothalamic hormones by immunocytochemical techniques. In *Frontiers in Neuroendocrinology, Vol. 4*, L. Martini and W.F. Ganong (Eds.), Raven Press, New York, in press.
Zimmerman, E.A., Carmel, P.W., Husain, M.K., Tannenbaum, M., Frantz, A.G. and Robinson, A.G. (1973a) Vasopressin and neurophysin: high concentrations in monkey hypophyseal portal blood. *Science*, 182: 925–927.
Zimmerman, E.A., Hsu, K.C., Robinson, A.G., Carmel, P.W., Frantz, A.G. and Tannenbaum, M. (1973b) Studies of neurophysin secreting neurons with immunoperoxidase techniques employing antibody to bovine neurophysin. I. Light microscopic findings in monkey and bovine tissues. *Endocrinology*, 92: 931–940.
Zimmerman, E.A., Hsu, K.C., Ferin, M. and Kozlowski, G. (1974a) Localization of gonadotropin-releasing hormone (Gn-RH) in the hypothalamus of the mouse by immunoperoxidase technique. *Endocrinology*, 95: 1–8.
Zimmerman, E.A., Robinson, A.G., Husain, M.K., Acosta, M., Frantz, A.G. and Sawyer, W.H. (1974b) Neurohypophysial peptides in the bovine hypothalamus: the relationship of neurophysin I to oxytocin and neurophysin II to vasopressin in supraoptic and paraventricular regions. *Endocrinology*, 95: 931–936.
Zimmerman, E.A., Defendini, R., Sokol, H.W. and Robinson, A.G. (1975a) The distribution of neurophysin-secreting pathways in the mammalian brain: light microscopic studies using the immunoperoxidase technique. *Ann. N.Y. Acad. Sci.*, 248: 92–111.
Zimmerman, E.A., Defendini, R., Sokol, H.W. and Robinson, A.G. (1975b) The intracellular distributions of oxytocin, vasopressin and respective neurophysins in mammalian hypothalamus. *Endocrinology*, 96: A-95.
Zimmerman, E.A., Kozlowski, G.P. and Scott, D.E. (1975c) Axonal and ependymal pathways for the secretion of biologically active peptides into hypophysial portal blood. In *Brain–Endocrine Interaction. II. The Ventricular System*, K.M. Knigge, D.E. Scott, H. Kobayashi and S. Ishii (Eds.), Karger, Basel, pp. 123–134.

Chapter 6

Biosynthesis and degradation of hypothalamic hypophysiotrophic factors*

SEYMOUR REICHLIN

Endocrine Division, New England Medical Center Hospital, and Department of Medicine, Tufts University School of Medicine, Boston, Mass. 02111 (U.S.A.)

I. INTRODUCTION

A wide variety of neurotransmitters, hormones and other biologically active molecules is found in the hypothalamus. These include the monoamine neurotransmitters norepinephrine, dopamine and serotonin, other active amines including histamine, epinephrine and polyamines, the neurohypophysial hormones, vasopressin and oxytocin with their associated neurophysin carrier, substance P and the hypophysiotrophic hormones of which three, thyrotrophic releasing hormone (TRH), luteinizing hormone releasing hormone (LRH) and somatostatin (SRIF) have been identified chemically. A number of other active compounds are present in the hypothalamus, identified mainly by their biological activities in regulating pituitary function, and a number of peptide substances have also been identified chemically for whom no function has as yet been ascribed.

The monoamines and other active amines are synthesized by a non-ribosomal mechanism within cell bodies of aminergic neurons by well characterized enzyme systems. On the other hand, knowledge of biosynthesis of the peptidergic neurons of the hypothalamus is remarkably scanty. The best understood of the peptidergic systems is that for the biosynthesis of vasopressin, and this system may serve as the model for the synthesis of the other peptide hormones.

II. BIOSYNTHESIS OF VASOPRESSIN

Of the peptide constituents of the hypothalamus, the biosynthesis of only vasopressin has been elucidated to any satisfactory extent, largely by the work of Sachs and collaborators.

As outlined in several publications (Sachs and Takabatake, 1964; Takabatake and Sachs, 1964; Sachs et al., 1969, 1971, 1973, 1974), the following features of vasopressin synthesis can be summarized.

(1) Prolonged infusion of high specific activity [^{35}S]cysteine into the hypothalamus of the dog leads to labeling of vasopressin after a lag period of at least 1.5 hr.

---

* Studies from this laboratory alluded to in this paper were supported by U.S.P.H.S. Grant No. 16684.

(2) In vivo synthesis of vasopressin is blocked by prior administration of puromycin, a drug that acts by blocking ribosomal protein synthesis.

(3) Guinea pig hypothalamic-neurohypophysial fragments, grown in tissue culture for several weeks, retain the ability to incorporate [^{35}S]cysteine into vasopressin. The incorporation is enhanced by a factor derived from fetal hypothalamus, and is blocked by inhibitors of protein biosynthesis including puromycin ($5 \times 10^{-4} M$), or cycloheximide ($2 \times 10^{-4} M$). Actinomycin D, an inhibitor of RNA synthesis, abolished vasopressin synthesis but only after a long lag period, suggesting a slow turnover of messenger.

(4) Neural lobes, separated from the hypothalamus do not form vasopressin. Treatment wich colchicine, an agent that inhibits axonal transport, enhances incorporation of precursor into vasopressin. It can be shown that after a pulse period of 18–24 hr, 7–15% of counts appeared in the neural lobe, suggesting transport of synthesized product.

(5) There is considerable evidence to support the view that vasopressin is synthesized through a precursor molecule. (a) Puromycin must be present from the initial exposure of the isotope to prevent incorporation into hormone. (b) A lag of at least 1.5 hr precedes appearance of labeled molecule. (c) During early synthesis, formed vasopressin is not associated with the ribosomal fraction. Instead it is found in the granular fraction. (d) Formation of vasopressin is blocked by 3 amino acid analogues, alpha-methyl leucine, α-methyl methionine, and histidinol. Since neither methionine, leucine nor histidine are constituents of vasopressin, it is likely that "the biosynthesis of the octapeptide hormone requires the simultaneous synthesis of protein."

As a general statement, Sachs points out the problem in demonstrating biosynthesis in this system. These include the extremely low rate of formation of vasopressin (estimated to be $10^{-2}-10^{-4}$ µg/hr), the high degree of contamination of product with both precursor and a large number of other labeled peptides. These same problems vex investigators of biosynthesis of all hypothalamic peptides. Sachs initially depended upon a large number of chromatographic steps for purification of product, but at the present time, the procedure has been simplified by use of affinity systems using coupled neurophysin and specific antibodies.

III. BIOSYNTHESIS OF HYPOPHYSIOTROPHIC HORMONES

Until the discovery of the chemical nature of the releasing hormones, scientific studies of their biosynthesis and of the factors controlling their synthesis were inaccessible to analysis. As proposed by Geschwind (1971) at least two possible routes of synthesis are reasonable to consider. One, analogous to the synthesis of vasopressin, would imply synthesis of peptide as a prohormone in the classical way, followed by enzymatic cleavage of an active fragment. On a theoretical basis at any rate this mechanism could well explain the biosynthesis of LRH, a decapeptide, and somatostatin, a cyclic peptide consisting of 14 amino acids. On the other hand, in the case of TRH, a tripeptide amide, alternative mechanisms of biosynthesis are theoretically possible. The most attractive alternative hypothesis as outlined by Reichlin and Mitnick (1973a, b) is that TRH is made by a non-ribosomal enzymatic mechanism analogous to the formation of several other biologically significant small peptides of mammalian tissues including glutathione (γ-Glu-Cys-Gly), and ophthalmic acid γ-glutamyl-α-amino-n-butyrylglycine. In bacteria, at least three small peptides are known to be synthesized by enzymatic function (gramicidin, tyrothrycin (see Reichlin and Mitnick (1973a) for review) and the pentapeptide nucleotide of *Staphylococcus aureus* (Ito and Strominger, 1960). An additional theoretical possibility for peptide bond formation applicable to the problem of TRH biosynthesis has been proposed by McKelvy and Grimm-Jørgensen (1975). This is the formation of interpeptide bonds by the participation of aminoacyl-tRNA transferases which has been demonstrated in bacterial cell wall biosynthesis (Kamiryo and Matsuhashi, 1972), and in mammalian brain (Soffer, 1973).

A critical analysis of presently available data fails to disclose which of these mechanisms for TRH synthesis is correct.

IV. ORGAN INCUBATES

IV.1. TRH biosynthesis

As initially reported by Mitnick and Reichlin (1971) and Reichlin et al. (1972), fragments of rat hypothalamic tissue incubated in amino acid precursors were found to incorporate label into peptides that had the same chromatographic behavior as authentic TRH. Many other peptides were also found. Systems used for separation of TRH were electrophoresis, thin-layer chromatography (TLC), Sephadex chromatography and CMC column chromatography. Grimm-Jørgensen and Reichlin (1973) also found that mouse hypothalamic fragments, incubated in labeled histidine, incorporated this amino acid into a substance with the mobility of TRH on thin-layer chromatography, and that the product, labeled with [^{14}C]histidine co-chromatographed to constant specific activity with [^{3}H]TRH.

Subsequently, McKelvy (1974) and McKelvy and Grimm-Jørgensen (1975) criticized the earlier work from my laboratory on rat TRH biosynthesis on the basis that the proof of identity of the labeled product had not been sufficiently rigorous and that only a small fraction of the radioactivity moving with TRH on electrophoresis or thin-layer chromatography was authentic TRH. Their work was carried out with much more sophisticated chemical separation methods than our original studies, and I believe that their observations may well be correct. Using guinea pig hypothalamic culture, McKelvy (1974) showed that [^{3}H]proline was incorporated into a compound that moved with TRH in at least 9 separate chromatographic steps, and formed an n-trifluoroacetyl n-butyl ester derivative that had chromatographic properties identical with TRH. Using the whole newt brain, Grimm-Jørgensen and McKelvy (1974) and McKelvy and Grimm-Jørgensen (1975) conducted a similar study by extensive chromatographic procedures including derivatization and proved identity of the compound. From these experiments it appears reasonable to conclude that the organ incubate and culture is capable of forming new TRH, but that the most rigorous separation methods are needed to establish the identity of the compound in view of the large number of new peptides that are formed. Parenthetically, in the electrophoretic system used extensively in my laboratory to separate TRH prior to the chemical elucidation of other releasing hormones, we find identical R_F values for LRH and somatostatin.

In rat brain fragment incubations, we found (Reichlin et al., 1972) that puromycin did not block the incorporation of amino acids into a peptide separable as TRH by thin-layer chromatography and electrophoresis. Since the identity of the isolated compound as TRH is now in doubt because of more recent findings alluded to above, the significance of this result can be questioned. On the other hand, Grimm-Jørgensen and McKelvy (1975) found that newt brain TRH formation neither was blocked by puromycin, nor by diphtheria antitoxin, a potent ribosomal poison also indicating a non-ribosomal form of biosynthesis.

IV.2. LRH biosynthesis

Several groups have reported that hypothalamic fragments or cultures incubated in the presence of labeled amino acid precursors of the LRH decapeptide incorporate precursor amino acids into a substance with the chromatographic behavior of carrier LRH decapeptide. In the studies of Johansson et al. (1972), 125 rat hypothalamic fragments were incubated for 4 hr in the presence of labeled glutamic acid, mixed amino acids, Mg ion, KCl, EDTA, mannitol, ADP and succinate. After sequential separation steps including polyacrylamide gel electrophoresis, cellulose thin-layer chromatography, and electrophoresis, radioactivity in small amounts was detected in the LRH spot. Moguilevsky et al. (1974) incubated hypothalamic tissue from 10 normal rats in Krebs solution containing glucose, [^{3}H]tyrosine, and separated the product on Sephadex G-25 columns and CM-sephadex. The area corresponding to LRH was counted and found to contain radioactivity. The significant fraction contained 82 counts, and when the source of the hypothalamus was the castrate rat, a larger number of counts were detected.

In a preliminary note Sachs et al. (1974) reported the isolation of a labeled peptide from 7-day guinea pig hypothalamic cultures that moved with LRH in several separation systems (CMC column, TLC). [^{3}H]Leucine and [^{3}H]tyrosine were added simultaneously for a 20-hr pulse.

In my laboratory, the effort to identify LRH biosynthesis began in 1970 prior to the isolation and identification of the LRH decapeptide and experiments were carried out in part in an attempt to determine the constituent amino acids of LRH, then unknown. The distribution of LRF activity was determined by bioassay in 3 chromatographic systems: TLC, CMC, and Sephadex G-10. Rat hypothalamic fragments incubated in KRB for 1 hr in the presence of each of 20 labeled amino acids were extracted into methanol and chromatographed in the 3 systems. Only 3 amino acids consistently moved with LRH. These were glutamic acid, threonine and alanine. Just as these findings became available, the structural formula of the LRH decapeptide was reported by Matsuo et al. (1971) and quickly confirmed by Burgus et al. (1972). Comparison of the LRH decapeptide constituents with our incorporation studies indicated that there were a number of discrepancies in one system or another. For example, labeled leucine, tyrosine and tryptophan were not associated with LRH biological activity in TLC and serine, glycine, proline and histidine did not follow LRH on CMC separation.

Two alternative explanations offered themselves to these observations. The one entertained first was that there was more than one LRF, an issue now almost completely settled in favor of a unitary hypothesis of GnRH, although reports persist of distinct types of LRF and of FSH (Fawcett et al., 1975) and of FSH-RF with LRH properties (Currie et al., 1973). The other explanation was that the incorporation studies simply had failed to indicate the synthesis of LRH, and that the chromatographic separations had been of other non-LRH labeled peptides.

This experience, together with the earlier problem of identifying the TRH product has been the basis of my reservations about the reports of amino acid incorporation into LRH activity alluded to above. In an effort to circumvent this problem an affinity column technique was developed in my laboratory to purify the product of incubations. Using standard techniques of coupling LRH to TBG, immunization in rabbits, purification of

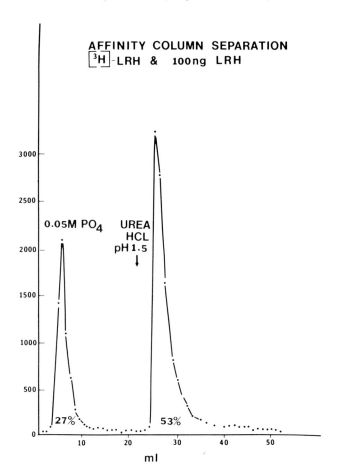

Fig. 1. Application of an affinity column for LRH separation prepared by coupling anti-LRH antibody to Sepharose. The initial peak represents unbound radioactive [^{3}H]LRH. The second peak is eluted by a mixture of urea and 1.5 M HCl. The separation was carried out in the presence of 100 ng LRH, indicating that the capacity of the column was at least this amount.

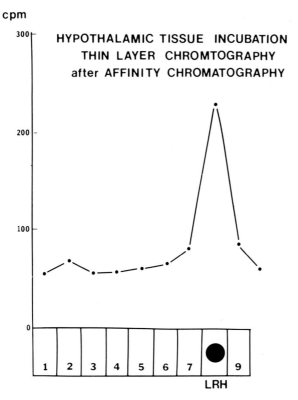

Fig. 2. Isolation of product of hypothalamic incubation in [^{3}H]proline on thin-layer chromatography after initial purification by electrophoresis followed by affinity chromatography. The peak containing LRH corresponded with Pauly-stained marker. Thin-layer chromatography carried out on silica gel plates in chloroform—methanol—distilled water (5:4:1).

immune gamma-globulin, and preparation of a Sepharose-stabilized antibody, we have, with the collaboration of Drs. Ivor M.D. Jackson, George Baum and Miss Stella Mothon, been able to isolate labeled LRH from rat hypothalamic incubates. The column used traps approximately 60% of [^{3}H]TRH in the presence of 100 ng of carrier LRH, an amount far in excess of the LRH load applied (Fig. 1). The methanol extract from 3 rat hypothalamic fragments incubated for 1 hr in Krebs—Ringer bicarbonate media together with 25 μCi of [^{3}H]proline was purified by electrophoresis in pyridine buffer, the region corresponding to LRH marker applied to the affinity column and the eluate of the column separated by both TLC (Fig. 2) and electrophoresis (Fig. 3) with cold LRH as carrier. The principle areas of radioactivity corresponded to authentic LRH. Hypothalamic tissue from castrated rats shows a higher degree of incorporation than in tissues from normal animals (Fig. 4) in confirmation of Moguilevsky's finding. I believe that these observations provide definitive proof of LRH decapeptide biosynthesis, but they are still in a preliminary stage, and we have not as yet had time to determine incorporation of other precursor amino acids, the effects of metabolic inhibitors, sex steroids and neurotransmitters on

LRH biosynthesis. Nevertheless it appears to me that this system offers considerable promise to unraveling the mechanism of LRH biosynthesis and its control.

In separate studies, Seyler et al. (1974) found that rat hypothalamic fragments incubated in vitro together with their media contained more bioassayable LRH activity at the end of incubation than at the beginning, suggesting that biosynthesis had occurred. Tissue from castrated rats apparently synthesized more LRH than tissues from normal animals. Similar experiments have recently been reported by Moguilevsky et al. (1975).

IV.3. The problem of TRH biosynthesis in hypothalamic extracts

Following the observation that puromycin did not block the incorporation of amino acids into a material that had the mobility of TRH in several chromatographic systems, the possibility was considered that TRH might be formed by a non-ribosomal enzymatic synthetic process analogous to that responsible for the formation of glutathione (Mitnick and Reichlin, 1972; Reichlin et al., 1972; Reichlin and Mitnick, 1973a, b). Procedures adapted from those used to study glutathione were used to study TRH biosynthesis. These included incubation of the soluble fraction of hypothalamic tissue in the presence

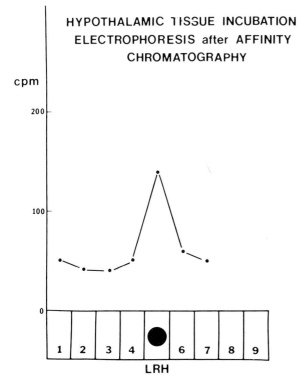

Fig. 3. Isolation of product of hypothalamic incubation in [^{3}H]proline on electrophoresis following an initial purification by electrophoresis followed by affinity chromatography. The principle peak of radioactivity corresponded to the behavior of marker LRH. Electrophoretic system carried out on Seprophore III strips run in 2% pyridine in 1% acetic acid for 70 min.

of Mg^{2+}, ATP, and precursor amino acids, glutamic acid, histidine and proline. A product was isolated from the incubate by a series of chromatographic steps, including charcoal adsorption and elution, TLC, CMC and Sephadex chromatography and electrophoresis all showing correspondence to synthetic TRH and it was proposed therefore that TRH was formed by a "synthetase" analogous to the enzyme that synthesizes glutathione. Further studies were made of the influence of a number of factors, including prior thyroid status on "TRH" formation, using electrophoresis to separate the product and the conclusion was drawn that thyroid deficiency decreased TRH synthesis, and that thyroxine feeding enhanced TRH synthesis. The observations were interpreted to indicate that thyroxine exerted a "positive" feedback effect on TRH synthesis, and by inference, on TRH secretion.

Unfortunately, these results that appeared so clear-cut have not been confirmed by a number of other workers who have either published their studies (Bauer, 1974; Bauer and

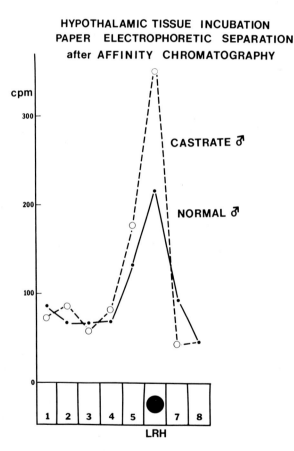

Fig. 4. Comparison of labeled LRH formed by hypothalamic fragments from castrated male rats as compared with normals. Product purified by initial separation by electrophoresis, affinity chromatography and as illustrated here, electrophoresis.

Kleinkauf, 1974; Dixon and Acres, 1975) or who have communicated these results to me in personal communications (McKelvy, LaBrie, Rosenberg, Barnea). Bauer and coworkers (Bauer et al., 1973; Bauer, 1974; Bauer and Kleinkauf, 1974) report that the incubation procedure led to the enzymatic degradation of TRH. In an attempt to circumvent this problem an inhibitor of TRH degradation was added to the reaction mixture. In the presence of the inhibitor substantial amounts of labeled TRH were isolated from the reaction mixture but Bauer believes that the product arose by a "substitution reaction" consequent to enzymatic degradation. In the studies of Dixon and Acres (1975) initial separation of product on charcoal and electrophoresis revealed counts corresponding to TRH, but the material did not behave like TRH in other separation systems. They also were unable to identify synthesis of TRH by isolated rat hypothalamic fragments incubated in vitro. McKelvy and Grimm-Jørgensen also failed to identify labeled TRH in freeze-dried porcine or rat hypothalamic tissue, or in fresh rat hypothalamic extract supernatants separated by low speed centrifugation (personal communication).

The latter workers have used whole newt brain for studies of TRH synthesis by extracts. Newt brain has a high concentration of extrahypothalamic TRH, and for this reason, it was presumed that it would provide a richer source of TRH synthesizing structures. In their studies (Grimm-Jørgensen and McKelvy, 1975; McKelvy and Grimm-Jørgensen, 1975) 5000 g supernatant was incubated in amphibian Ringer Hepes, at $22°C$ in the presence of 0.1 mM ATP and [^{3}H]proline, 25 μCi/ml showed progressive incorporation of radioactivity into "TRH" over a 30-min period. The identity of the product was confirmed by electrophoresis in acid and in alkaline media, and by dinitrophenylation. The DNP derivative was separated by electrophoresis and shown to have identical chromatographic behavior as authentic DNP-TRH. In the newt extract preparation, formation of TRH was blocked by *RNAse* pretreatment of the extract. On the basis of this observation they concluded that biosynthesis was by an RNA-dependent mechanism, but since they have previously shown in whole newt brain that biosynthesis was neither blocked by cycloheximide nor by diphtheria toxin, they proposed that synthesis might take place by an active transfer acyl-tRNA mechanism mediating peptide bond formation. Similar mechanisms have been identified in bacterial cell wall synthesis (Kamiryo and Matsuhashi, 1972) and in brain (Soffer, 1973).

Reports that TRH formation by soluble hypothalamic extracts could not be confirmed in a number of laboratories began to emerge 3 years ago, at the time that my laboratory moved to Boston. In collaboration with Dr. Richard Saperstein and Miss Stella Mothon, efforts were made to set up the originally described experiments as close as possible to the original technique. In a number of trials using fresh rat hypothalamic extracts, freeze-dried rat and porcine hypothalamic extracts, we were unable to demonstrate consistently and in a convincing way that new counts appeared in "TRH" as isolated by charcoal adsorption, electrophoresis and TLC. This failure to confirm earlier findings led to the reappraisal of the properties of the incubation system.

The three major areas of concern were the problem of degradation of product, the problem of degradation of ATP by ATPases of hypothalamic tissue and the unambiguous demonstration of product as authentic TRH. These problems apply to all studies of hypothalamic hormone biosynthesis by soluble systems. The first issue addressed ap-

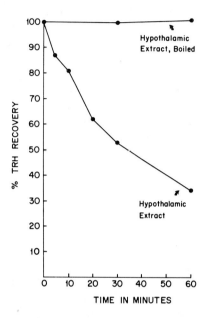

Fig. 5. Degradation of TRH by hypothalamic incubates. The product was identified by electrophoresis.

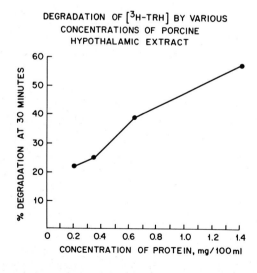

Fig. 6. Degradation of TRH by hypothalamic extracts is a function of extract concentration.

peared to be the problem of enzymatic degradation of product, as initially pointed out by Bauer et al. (1973) who used labeled TRH substrate and by Bassiri and Utiger (1974) who used radioimmunoassay methods. The more general problem of hypothalamic degradation of the hypophysiotrophic hormones has recently been reviewed in detail by Reichlin et al. (1976). The hypothalamic incubation system used by us was found to degrade TRH actively (Fig. 5), and the rate of inactivation proved to be a function of the concentration of extract added (Fig. 6). A series of compounds were studied in an attempt to find one that would protect TRH from degradation and would not interfere with biosynthesis. As reported by Saperstein et al. (1975) (Fig. 7, Table I) TRH degradation was blocked by a number of amidated and non-amidated peptides including TRH, deamido TRH (kindly supplied by Dr. Will White, Abbott Laboratories), LRH, deamido LRH, angiotensin amide, angiotensin I, angiotensin II and somatostatin (kindly provided by Dr. M. Goetz, Ayerst Laboratories). The reaction was not blocked by vasopressin, oxytocin nor by so-called MIF (Pro-Leu-Gly-NH_2). Our conclusion with respect to the degradation system is that hypothalamic tissues degrade TRH by one or more peptidases capable of attacking

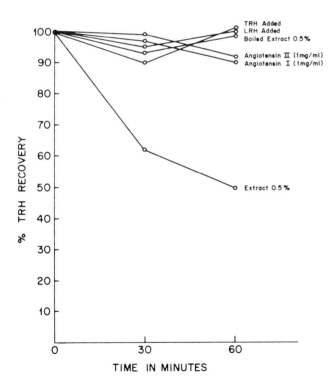

Fig. 7. Degradation of TRH is abolished by addition of certain peptides. This figure shows the effects of addition of LRH and angiotensin.

TABLE I

Summary of inhibitors of TRH and LRH degradation by hypothalamic extracts

	TRH	LRH
Enzyme inhibitors		
Trasylol	−	−
Benzamidine	+	+
Analogues		
TRH	+	Partial
Deamido TRH	+	Partial
LRH	+	+
Deamido LRH	+	+
Amidated peptides		
Pro-Leu-Gly-NH$_2$ (MIF)	−	−
Angiotensin amide	+	+
Oxytocin	−	Partial
Vasopressin	−	Partial
Non-amidated peptides		
Angiotensin I	+	+
Angiotensin II	+	+
Somatostatin	+	?

the amide bond as well as the peptide bond as has been shown for chymotrypsin. Similar studies carried out with LRH are illustrated in Fig. 8 and summarized in Table I.

The second problem is that of ATPase concentration. We found that ATP was rapidly destroyed by the incubation system — within 10 min virtually all was gone. Even with the addition of an ATP generating system concentration of ATP, initially 3.2 mM, fell to 1.6 mM by 15 min and 0.1 mM at the end of 0.5 hr. If in fact ATP is a necessary cofactor for the formation of TRH by a soluble system, its rapid degradation would require either the addition of even larger amounts of ATP or separation of ATPase from other extract components.

The third problem we have attacked is that of specificity of product. Immunological techniques have been applied to this problem using immunoprecipitation by a double antibody method and we are currently developing an affinity system analogous to that used for LRH separation.

In the last few months we have carried out 8 series of incubation studies in the presence of 2 μCi [^{3}H]proline per incubation. These have utilized freeze-dried porcine hypothalami (Oscar Meyer), freeze-dried rat hypothalamic fragments (courtesy of the Pituitary Distribution Program, NIH), fresh rat homogenates, all prepared as the 5000 g supernatant (and hence including particulate subfractions) and two "squish preparations" prepared by extruding fresh rat hypothalamic fragments through a 21-gauge needle. Based on the report by Montoya et al. (1975) that stress increased the concentration of whole brain TRH in the rat, we have also studied on one occasion a "squish" preparation of brain removed from rats subjected 24 hr earlier to laparotomy under ether anesthesia. In several experiments, an ATP generating system was added, with additional ATP added through the incubation, and in all studies, LRH, 100 μg/ml, and in some cases angiotensin II, was added to block degradation product. Initial separations were in most cases by adsorption to activated charcoal in minicolumns, elution in acid methanol and then electrophoresis in pyridine buffer at an acid pH. The spot corresponding to TRH (run in parallel strips) was eluted and counted. In 5 of 8 experiments counts appeared in the TRH region whereas incubation with boiled hypothalamic extracts showed no counts or a decreased number of counts as compared with the incubation material at zero time. Thus far, we have been able to carry out immunoprecipitation of TRH on the product of only one of these incubations, and no labeled material was precipitated.

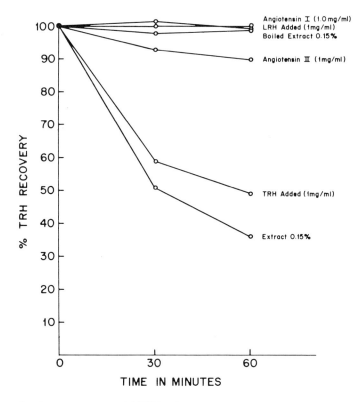

Fig. 8. Degradation of LRH by hypothalamic extracts is blocked by high concentrations of angiotensin.

TABLE II

Summary TRH biosynthesis experiments — January to June, 1975

Tissue		Conditions of incubation 37°C		After charcoal
Freeze dried				
1/16/75	Porcine 2.5%	0.05 M PO$_4$ Glu, His, Pro ATP 10^3, Mg^{2+}	Control no add. +Angio II +LRF	+ + + +
1/31	Porcine 2.5%	" " 1 hr preincubation	Control no add. +Angio II +LRF	+ I + +
Freeze dried				
4/2	Rat 5%	Tris · HCl Glu, His, Pro ATP 2.5 mM, Mg^{2+} ATP regenerating syst.	Control +LRF	+ +
4/16	Rat 5%	ATP 5 mM + addition at 15 sec	Control no add. +LRF	+ + +
Fresh rat hypothalamus				
5/16	5% wet weight	Tris · HCl Glu, His, Pro ATP 2.5 mM Mg^{2+} ATP regenerating	Control no add. +LRF	+ + +
6/17	"squished" 8% wet weight	" " Nitrogen atmosphere	Control no add. +LRF	— — —
Fresh rat hypothalamus				
6/24	"squished" whole brain 12% wet weight	Nitrogen atmosphere sensitized rat	Control no add. +LRF	— — —
6/24	"squished" whole brain 12%	Room atmosphere sensitized rat	Control no add. +LRF	— — —

Paper electrophoresis separation of TRH			
Counts/min (sec)			Counts/min/mg protein
0	30	30.0	
107	65	− 42	− 382
90	79	− 11	− 100
97	85	− 12	− 109
99	98	− 1	− 9
73	43	− 30	− 143
38	44	+ 6	+ 29
42	48	+ 6	+ 29
36	51	+ 15	+ 71
62	104	+ 42	+ 168
49	80	+ 31	+ 124
69	62	− 7	− 28
52	73	+ 21	+ 84
39	66	+ 27	+ 108
57	54	− 3	− 25
57	63	+ 6	+ 50
49	53	+ 4	+ 33
			Counts/min/mg tissue
81	75	− 6	− 14
59	72	+ 13	+ 13
97	153	+ 56	+ 15
19	19	0	0
20	19	− 1	− 2
37	36	− 1	− 2
23	22	1	− 2
81	22	− 59	− 103
58	71	+ 13	+ 23

IV.4. Biosynthesis of other releasing hormones by soluble hypothalamic extracts

In previous studies from this laboratory it was reported that a soluble hypothalamic extract dialyzed to remove preformed releasing hormone, upon incubation with a mixture of 20 common amino acids, ATP and Mg^{2+} showed an increase in bioassayable LRF activity (Reichlin and Mitnick, 1973b), prolactin releasing activity (PRF) (Mitnick et al., 1973) and growth hormone releasing activity (Reichlin and Mitnick, 1973c).

It has been reported by Johansson et al. (1973) that [^{14}C]glutamic acid was incorporated into chromatographically separated LRH-like material by a subcellular fraction of fresh porcine hypothalamic tissue. Batches of 250 fragments were homogenized, and the so-called "mitochondrial" fraction isolated, incubated for 3 hr at $37°C$ under an atmosphere of 95% air and 5% CO_2. Incubation media consisted of phosphate buffer, $MgCl_2$, KCl, EDTA, mannitol, ADP, succinate, amino acid mixture and 20 μCi [^{14}C]glutamic acid. Further separations included chromatography on Sephadex G-25, electrophoresis. The total number of counts recovered from the incubate was 85 counts/min.

Except for these experiments of Johanssen, I am not aware of other attempts to show cell-free synthesis of other releasing hormones. Experiments recently carried out on the stability of PRF, PIF, SRF and somatostatin in plasma indicate that the interpretation of de novo biosynthesis of new activity based on increased biological activity cannot be accepted without further study. We find (Patel and Reichlin, unpublished; Boyd, Spencer, Jackson and Reichlin, unpublished) that preparations of porcine hypothalamic extract, prepared by methanol or boiling acetic acid and hence free of tissue enzymes, show increased PRF activity and increased GH-RF activity upon incubation with rat plasma. Under the conditions of this incubation, immunoassayable somatostatin concentration declines extremely rapidly. We interpret these results to mean that the incubation procedure leads to a greater loss in the amount of somatostatin and of PIF than of the respective releasing hormone, and this leads to a spurious increase in releasing factor activity. That this may have occurred in the soluble hypothalamic extracts reported previously cannot be excluded at this time.

V. CONCLUDING REMARKS

In this paper an attempt has been made to summarize the current state of knowledge with respect to the mechanism of biosynthesis of the hypophysiotrophic hormones. It is reasonable to conclude that both LRH and TRH biosyntheses have been demonstrated to occur in whole hypothalamic incubation preparations from several species of animals including slices and cultures. The use of inhibitors of protein synthesis applied to the problem of TRH biosynthesis appears to indicate that classical ribosomal synthesis is not the underlying mechanism, but in limited studies of crude extracts of newt brain the finding that RNAse blocks the reaction suggests the possibility of an unusual form of peptide bond formation that involves an RNA-mediated process, rather than a non-ribosomal enzymatic process. The problem of LRH biosynthesis may prove to be easier to solve than that of TRH because purification of the product is on sounder footing, and the larger molecule is most likely formed by classical protein synthetic pathways analogous to those of vasopressin and oxytocin. It has already been shown in slice preparations that the rate of LRH biosynthesis is enhanced in tissues removed from castrated animals.

Problems in analysis of the mechanism of TRH biosynthesis by soluble hypothalamic extracts have been summarized. Earlier studies from my laboratory reporting the incorporation of precursor amino acids into a TRH product by particulate free extracts have not been confirmed by other workers, in some cases because rigorous purification procedures indicate that the newly labeled peptides resulting from this incubation are not TRH, or that the TRH formed from amino acid precursors may be the result of exchange reactions following enzymatic breakdown of TRH. We have ourselves been unable to demonstrate consistently and convincingly that tritiated proline is incorporated into TRH in a number of incubation preparations including those in which peptide inhibitors of TRH breakdown and ATP-generating systems have been added. For this reason, the status of the non-ribosomal biosynthesis of TRH is still uncertain.

REFERENCES

Bassiri, R. and Utiger, R.D. (1974) Thyrotropin-releasing hormone in the hypothalamus of the rat. *Endocrinology*, 94: 188–197.

Bauer, K. (1974) Degradation of thyrotropin releasing hormone (TRH). Its inhibition by Pyroglu-His-OCH$_3$ and the effect of the inhibitor in attempts to study the biosynthesis of TRH. In *Lipmann Symposium: Energy, Biosynthesis and Regulation in Molecular Biology*, Walter De Gruyter, Berlin, pp. 53–62.

Bauer, K. and Kleinkauf, H. (1974) Degradation of thyrotropin-releasing hormone (TRH) by serum and hypothalamic tissue and its inhibition by analogues of TRH. *Z. Physiol. Chem.*, 355: 1173–1176.

Bauer, K., Sy, J. and Lipmann, F. (1973) Degradation of thyrotropin releasing hormone (TRH) by extracts of hypothalamus. *Fed. Proc.*, 32: 489 (abstract).

Burgus, R., Butcher, M., Amoss, M., Ling, N., Monahan, M., Rivier, J., Fellows, R., Blackwell, R., Vale, W. and Guillemin, R. (1972) Primary structure of the ovine hypothalamic luteinizing hormone-releasing factor. *Proc. nat. Acad. Sci. (Wash.)*, 69: 278–282.

Currie, B.L., Johansson, K.N.G., Folkers, K. and Bowers, C.Y. (1973) On the chemical existence and partial purification of the hypothalamic follicle stimulating hormone releasing hormone. *Biochem. biophys. Res. Commun.*, 50: 14–19.

Dixon, J.E. and Acres, S.C. (1975) The inability to demonstrate the non-ribosomal biosynthesis of thyrotropin releasing hormone in hypothalamic tissue. *Fed. Proc.*, 34: 658 (abstract).

Fawcett, C.P., Beezley, A.E. and Wheaton, J.E. (1975) Chromatographic evidence for the existence of another species of luteinizing hormone releasing factor (LRF). *Endocrinology*, 96: 1311–1314.

Geschwind, I.I. (1971) Biochemical mechanisms in hormone storage and secretion. *Mem. Soc. Endocr.*, 19: 945–950.

Grimm-Jørgensen, Y. and McKelvy, J.F. (1974) Biosynthesis of thyrotropin releasing factor by newt (*Triturus viridescens*) brain *in vitro*. Isolation and characterization of thyrotropin releasing factor. *J. Neurochem.*, 23: 471–478.

Grimm-Jørgensen, Y. and McKelvy, J.F. (1975) Biosynthesis of TRH by newt brain homogenate *in vitro. Brain Res. Bull.*, in press.

Grimm-Jørgensen, Y. and Reichlin, S. (1973) Thyrotropin releasing hormone (TRH): neurotransmitter regulation of secretion by mouse hypothalamic tissue *in vitro*. *Endocrinology*, 93: 626–631.

Ito, E. and Strominger, J.L. (1960) *J. biol. Chem.*, 235: PC5. Cited by Sachs, H., Fawcett, P., Takabatake, Y. and Portanova, R. (1969) *Recent Progr. Hormone Res.*, 25: 447–484.

Johansson, N.G., Hooper, F., Sievertsson, H., Currie, B.L., Folkers, K. and Bowers, C. (1972) Biosynthesis in vitro of the luteinizing releasing hormone by hypothalamic tissue. *Biochem. biophys. Res. Commun.*, 49: 656–660.

Johansson, N.G., Currie, B.L., Folkers, K. and Bowers, C. (1973) Biosynthesis of the luteinizing hormone releasing hormone in mitochondrial preparations and by a possible pantetheine-template mechanism. *Biochem. biophys. Res. Commun.*, 53: 502–507.

Kamiryo, T. and Matsuhashi, M. (1972) The biosynthesis of the cross linking peptides in the cell wall peptidoglycan of *Staphylococcus aureus. J. biol. Chem.*, 247: 6306–6311.

Matsuo, H., Baba, Y., Nair, R.M.G., Arimura, A. and Schally, A.V. (1971) Structure of the porcine LH and FSH releasing hormone. 1. The proposed amino acid sequence. *Biochem. biophys. Res. Commun.*, 43: 1334–1339.

McKelvy, J.F. (1974) Biochemical neuroendocrinology. I. Biosynthesis of thyrotropin releasing hormone (TRH) by organ cultures of mammalian hypothalamus. *Brain Res.*, 65: 489–502.

McKelvy, J.F. and Grimm-Jørgensen, Y. (1975) Studies on the biosynthesis of thyrotropin releasing hormone in vitro. In *Hypothalamic Hormones: Chemistry, Physiology, Pharmacology and Clinical Uses*, M. Motta, P.T. Crosignani and L. Martini (Eds.), Academic Press, New York, pp. 13–26.

Mitnick, M. and Reichlin, S. (1971) Biosynthesis of TRH by rat hypothalamic tissue in vitro. *Science*, 172: 1241–1243.

Mitnick, M. and Reichlin, S. (1972) Enzymatic synthesis of thyrotropin releasing hormone (TRH) by hypothalamic "TRH synthetase". *Endocrinology*, 91: 1145–1153.

Mitnick, M., Valverde, R.C. and Reichlin, S. (1973) Enzymatic synthesis of prolactin releasing factor (PRF) by rat hypothalamic incubates and by extracts of rat hypothalamic tissue: evidence for PRF synthetase. *Proc. Soc. exp. Biol. (N.Y.)*, 143: 418–421.

Moguilevsky, J.A., Enero, M.A. and Szwarcfarb, B. (1974) Luteinizing hormone releasing hormone-biosynthesis by rat hypothalamus in vitro, influence of castration. *Proc. Soc. exp. Biol. (N.Y.)*, 147: 434–437.

Moguilevsky, J.A., Scacchi, P., Debeljuk, L. and Faigon, M.R. (1975) Effect of castration upon hypothalamic luteinizing hormone releasing factor (LH-RF). *Neuroendocrinology*, 17: 189–192.

Montoya, E., Wilber, J., White, W. and Gendrich, R. (1975) The influences of sex, gonadal ablations, and light upon GnRH and TRH in the rat brain (CNS). *Program 57th Meeting, Endocrine Soc., June, 1975*, Abstract 88.

Reichlin, S. and Mitnick, M.A. (1973a) Biosynthesis of hypothalamic hypophysiotropic hormones. In *Frontiers in Neuroendocrinology*, W.F. Ganong and L. Martini (Eds.), Oxford University Press, New York, pp. 61–88.

Reichlin, S. and Mitnick, M.A. (1973b) Biosynthesis of thyrotropin releasing hormone and its control by hormones, central monoamines and external environment. In *Hypothalamic Hypophysiotropic Hormones. Proc. int. Congr. Ser. 263, Acapulco*, Excerpta Medica, Amsterdam, pp. 124–135.

Reichlin, S. and Mitnick, M.A. (1973c) Enzymatic synthesis of growth hormone releasing factor (GH-RF) by rat incubates and by extracts of rat and porcine hypothalamic tissue. *Proc. Soc. exp. Biol. (N.Y.)*, 142: 497–501.

Reichlin, S., Martin, J.B., Mitnick, M., Boshans, R., Grimm-Jørgensen, Y., Bollinger, J., Gordon, J. and Malacara, J. (1972) The hypothalamus in pituitary-thyroid regulation. *Recent Progr. Hormone Res.*, 28: 229–277.

Reichlin, S., Sapirstein, D., Jackson, I.M.D., Boyd, A.E., III and Patel, Y. (1976) Hypothalamic hormone. *Ann. Rev. Physiol.*, in press.

Sachs, H. and Takabatake, Y. (1964) Evidence for a precursor in vasopressin biosynthesis. *Endocrinology*, 75: 943–948.

Sachs, H., Fawcett, P., Takabatake, Y. and Portanova, R. (1969) Biosynthesis and release of vasopressin and neurophysin. *Recent Progr. Hormone Res.*, 25: 447–484.

Sachs, H., Goodman, R., Osinchak, J. and McKelvy, J. (1971) Supraoptic neurosecretory neurons of the guinea pig in organ culture. *Proc. nat. Acad. Sci. (Wash.)*, 68: 2782–2786.

Sachs, H., Goodman, R., Shin, S., Shainberg, A. and Pearson, D. (1973) Vasopressin and neurophysin biosynthesis in hypothalamic organ cultures. In *Endocrinology, Proc. 4th int. Congr. Endocrinology*, R.O. Scow (Ed.), Excerpta Medica, Amsterdam, pp. 573–578.

Sachs, H., Pearson, D., Shainberg, A., Shin, S., Bryce, G., Malamed, S. and Mowles, T. (1974) Studies on the hypothalamo-neurohypophysial complex in organ culture. In *Recent Studies of Hypothalamic Function*, K. Lederis and K.E. Cooper (Eds.), Karger, Basel, pp. 50–66.

Saperstein, R., Mothon, S. and Reichlin, S. (1975) Enzymatic degradation of TRH and LRH by hypothalamic extracts. *Fed. Proc.*, 34: 239.

Seyler, L.E., Jr., Canalis, E. and Reichlin, S. (1974) Effect of stilbestrol on LH secretory response to LRH in men and women. *56th Meeting, Endocrine Society, Atlanta, June, 1974, abstract.*

Soffer, R.L. (1973) Post-translational modification of proteins catalyzed by amino acyl-tRNA-protein transferases. *Molec. Cell. Biochem.*, 2: 3–14.

Takabatake, Y. and Sachs, H. (1964) Vasopressin biosynthesis. III. *In vitro* studies. *Endocrinology*, 75: 934–942.

Chapter 7

Biodegradation of hormonally active peptides in the central nervous system

N. MARKS

New York State Research Institute for Neurochemistry and Drug Addiction, Ward's Island, New York, N.Y. 10035 (U.S.A.)

I. INTRODUCTION

Turnover (synthesis and breakdown) of neurosecretory peptides represents an important but frequently overlooked facet of the mechanisms regulating reproductive neuroendocrinology. Progress has been limited by the difficulties experienced in early studies on the chemical characterization of such peptides, but based on the available evidence, two processes with respect to breakdown can be differentiated: (1) formation of active hormones or factors by cleavage of inactive (protein or polypeptide) precursors, and (2) inactivation of active peptides by a separate set of degradative enzymes. This area has captured considerable interest following the many reports that hypothalamic releasing factors, and other neurosecretory materials, occur widely in the CNS, suggesting a pharmacological role in addition to their known endocrinological functions. These peptides have specific actions suggesting that they represent a new class of chemical messengers with properties similar to those of conventional neurotransmitters. If such is the case, they are presumably inactivated at their target sites by specific hydrolases. It is the purpose of this account to focus attention on the appropriate brain enzymes that may be involved in these processes, with the major emphasis placed on contribution from the author's laboratory. The peptides selected for detailed description are those with known structure and which appear to play a direct or indirect role in reproduction. Substance P, kinins, and angiotensins are included since they occur in neurosecretory areas and illustrate a number of unusual points concerning mechanisms of formation and/or degradation that are pertinent to possible subcellular mechanisms in regulation of gonadal function.

Brain, like other tissues, contains a large number of exo- and endopeptidases involved in the sequential breakdown of proteins to yield oligopeptides and free amino acids (Table I) (Marks, 1968; Marks and Lajtha, 1971). In several cases these brain enzymes show unexpected specificities, as illustrated by the limited cleavage of biologically active materials (Marks and Stern, 1974a; Benuck and Marks, 1975; Benuck et al., 1975). This is particularly relevant to hypothalamic releasing factors, which are characterized in most cases by the presence of unusual N-(pyroglutamyl) and C-(acylated amides) protected groups, implying the existence in brain of enzymes with novel specificities (Marks and Stern, 1974b). It might be noted that these end-groups along with the presence of

TABLE I

Comparison of exo- and endopeptidases in brain and neurosecretory areas

Enzymes were assayed by the procedures of Serra at al. (1972) and Marks and Stern (1974a, b). For hormone assays, 50 nmoles of substrate were incubated in a volume of 0.5 ml of 10 mM Tris—HCl buffer containing 0.5 mM Cleland's reagent and 0.25 mg homogenate protein for 4 hr at 37°C and then fixed with 3% w/v sulfosalicylic acid. Breakdown was measured by the appearance of free amino acids as detected in an analyzer modified to detect glycinamide. Results are the means of 6 determinations agreeing within 10%. See Fig. 1 for the structures of MIF, LH-RF, SRIF (somatostatin), and Substance P.

Substrate	Enzyme activity (μmoles/g fresh wt./hr)			
	Cortex	Hypothalamus	Pituitary	
			Anterior	Posterior
Hb (pH. 7.6)	9.3	9.1	12	11
Hb (pH 3.2)	1.5	17	44	46
Leu-Gly-Gly	360	420	700	700
Arg-β-NA	140	109	460	450
Arg-Arg-β-NA	27	18	29	29
MIF	3.2	3	4.3	3.5
TRF	0.6	2.3	0.9	tr
LH-RF	1.0	4.6	4.3	3.5
Substance P	13.2	13.2	15	21.6
Somatostatin	4.3	—	—	—

disulfide bridges exist in a large number of tissue components and may represent a fine control mechanism for regulating breakdown. At the present time very little is known about the mechanisms regulating protein turnover and breakdown in cells (Lajtha and Marks, 1971; and see Lajtha and Dunlop, Ch. 4 in this volume). The possibility also exists that some hormones are formed by a mechanism of de novo synthesis from precursor amino acids (see Reichlin, Ch. 6 in this volume). In most other cases, active peptides are formed by breakdown of precursor proteins, as will be described below for kinins and angiotensins.

The fact that the clinical usefulness of hypothalamic releasing factors has been severely limited by their very short half-lives has prompted an intensive and costly search for longer acting analogs (agonists and in some cases antagonists). There is also a search for analogs with useful pharmacological properties but with attenuated endocrinological side-effects. As will be discussed, a knowledge of biodegradation can facilitate the preparation of appropriate analogs with a more lasting action in tissues and blood, and which thereby have greater clinical potential. The stability of an active peptide in tissues, or in blood (for example in the hypophyseal circulation), is a very important consideration, frequently ignored in studies attempting to relate structure to receptor occupancy (Cuatrecasas, 1974; Marks and Stern, 1974a, 1975). Provided there is no interference with the minimum chemical structure required for binding, the introduction of residues (such as D-isomers) that hinder the action of proteolytic enzymes can be expected to lead to enhanced biological activity. This is particularly important in the case of peptide hormones that are readily destroyed in the gut and can be administered only by intravenous infusion or by other systemic routes.

Finally, an added justification for these studies is that a knowledge of the mechanisms involved in formation and breakdown of peptides can provide new insights into the pathways involved in the general mechanisms of protein turnover. The available data in this regard will be discussed in relation to each individual peptide as described below.

SITES OF CLEAVAGE OF HYPOTHALAMIC PEPTIDES BY BRAIN ENZYMES

1. *LH-RF (luliberin)*

    pGlu-His-Trp-Ser-Tyr-Gly-Leu-Arg-Pro-Gly-NH$_2$

    cathepsin M   carboxyamide peptidase

2. *SRIF (somatostatin)*

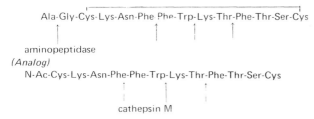

    aminopeptidase
    (Analog)
    N-Ac-Cys-Lys-Asn-Phe-Phe-Trp-Lys-Thr-Phe-Thr-Ser-Cys

    cathepsin M

3. *TRF (thyroliberin)*

    pyroglutamyl peptidase   aminopeptidase   deamidase (?)

4. *MIF (melanostatin)*

    aminopeptidase

5. *Substance P*

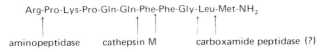

    aminopeptidase   cathepsin M   carboxamide peptidase (?)

Fig. 1. Established points of cleavage (↑), postulated points of cleavage (↓). Terminology taken from IUPAC-IUB recommendations (1974), *Biochemistry*, 14 (1975) 2559, and for enzymes from the recommendations of the *2nd Int. Conf. Intracellular Catabolism* (Marks, 1976). See text under the appropriate sub-headings for references and other details.

II. LULIBERIN (LH-RF)

This particular decapeptide is of interest in terms of inactivation since it is characterized by an N-terminal pyroglutamyl group and a C-terminal glycinamide (Fig. 1). On incubation with brain or extracts prepared from the neurosecretory regions or with serum it is inactivated with release of free amino acids (Tables II, III). It was estimated by Koch et al. (1974) using a bioassay procedure that the supernatant from one rat hypothalamus inactivated 1 nmole within 5 min, or based on the average weight for this region, 1–2 μmoles/g wet wt./hr. This rate is within the range of those obtained by a biochemical method for monitoring breakdown as summarized in Table III (Marks and Stern, 1974a, b). The rationale of this chemical procedure is based on the presence of a mixture of proteolytic enzymes in crude extracts such that the timed appearance of different residues supplies information as to the primary sites of cleavage. For example, incubation of 50 nmoles of LH-RF at pH 7.6 with 0.1 ml of a 10% supernatant of rat hypothalamus prepared with 0.32 M sucrose (containing 0.25 mg protein) led to the appearance within 3 hr of 60% of the internal amino acids Leu, Gly, and Tyr and only trace levels of glycinamide (Gly-NH_2). This is a rate equivalent to the inactivation of 4 μmoles/g wet wt./hr sufficient to degrade the known content of LH-RF in the rat hyperphysiotropic areas (30–50 nmoles) within 1 hr assuming a wet weight of about 10 mg for this region). The data on the preferential release of Leu and adjacent residues can be interpreted as showing a rate-limiting cleavage at an internal site, probably the Gly^5-Leu^6 bond, followed by secondary cleavages of the split fragments by exopeptidases to release other amino acids. This mechanism was confirmed by the use of analogs that blocked the internal site(s) of cleavage (Marks and Stern, 1974a) and by the isolation of the peptidyl product pyroGlu-His-Trp-Ser-Tyr-Gly (Koch et al., 1974).

II.1. Cleavage of analogs

Studies with analogs are of particular interest since several have higher activities with clinical potential (Fujino et al., 1974). In an analog in which Gly was replaced by D-Ala in position 6 the released amino acids adjacent to this site were blocked, but not the release of glycinamide (Table III). This was interpreted as showing the presence in brain of a second but slower inactivating enzyme acting on the C-terminus. Such C-terminal cleaving enzymes have been reported as present in brain and neurosecretory tissue and shown capable of inactivating oxytocin (see below). The presence of this enzyme (tentatively referred to as a carboxyamide peptidase) was confirmed by the use of a second analog, LH-RF-ethylamide (replacement of glycinamide by C_2H_5-NH_2), which blocked the release of glycinamide and the adjacent amino acids but not internal residues. Finally, both inactivating enzymes can be blocked using the double substituted analog desGly-D-Ala^6 (LH-RF)-ethylamide. It is an impressive fact that analogs (that blocked the action of these inactivating enzymes) were more potent in the release of LH and FSH hormones in vivo (Fujino et al., 1974; Marks and Stern, 1974a; Coy et al., 1975) (Table III). The low yield of His in all cases would tend to exclude a pyroglutamyl peptidase as a major factor in the inactivation process (by way of contrast see TRH below). It can be inferred, also, that the absence of free Gly as a product of degradation excludes the action of deamidases (see also Section VII, Substance P). The above results show that blocking degradation can lead to an improved biological activity. Predications can be made now with confidence as to the type of analog most likely to resist digestion (see also Section V, somatostatin).

II.2. Enzyme distribution

Inactivating enzymes occur in all regions of the brain studied to date, including the cortex, thalamus, hypothalamus, median eminence and pituitary (Table I) (Kochman et al., 1975; Griffiths et al., 1975a, b). In a recent study, the highest activity was shown for the pineal gland (Benuck and Marks, 1976); this is of interest since this region appears to be a repository for a number of active peptides (the values of LH-RF, however, may be lower than those previously reported) (Wilfred F. White, personal communication). Enzymes inactivating LH-RF are present largely in a supernatant fraction (Koch et al.,

TABLE II

Inactivation of hypothalamic releasing factors in serum of rat and man

The incubation mixture and experimental procedure were similar to Table I except for the use of 0.2 ml serum. See Fig. 1 for the structure of the substrates.

	Per cent inactivation			
	Rat		Man	
	4 hr	24 hr	4 hr	24 hr
MIF	100	100	0	5
LH-RF	52	100	14	60
SRIF	—	—	52	100
N-Ac-Cys^1-SRIF	—	—	28	—
TRF	25	80	5	30

TABLE III

Cleavage of LH-RF analogs by rat brain homogenate

Breakdown of LH-RF was monitored by the same procedures as described in Tables I and II. The numbers adjacent to the amino acids represent the sequence order as shown in Fig. 1. tr = trace quantities, nd = not determined.

Analog	nmoles per cent released								Activity*
	His^2	Ser^4	Tyr^5	Gly^6	Leu^7	Arg^8	Pro^9	$Gly\text{-}NH_2^{10}$	
LH-RF	20	30	45	46	50	50	30	15*	100
Ethylamide^{10}	18	41	26	38	40	30	tr	(0)*	nd
D-Trp^3	0	tr	5	38	43	38	23	29	nd
3-Me-His^2	(0)	0	22	18	28	17	15	18	nd
D-Ala^6	5	tr	tr	(0)	0	18	15	15	670
D-Ala^6, ethylamide	5	tr	tr	tr	0	tr	0	(0)*	8,000

* Ovulatory activity in vivo (LH-RF = 100).

TABLE IV

Partial purification of neutral endopeptidase (cathepsin M) inactivating Substance P, LH-RF and SRIF

Enzyme was purified batchwise using 10 ml of a supernatant prepared from 10% rat brain homogenate made with 0.32 M sucrose and containing 0.1 mM Cleland's reagent. The supernatant was applied to a column of DEAE-cellulose equilibrated with 40 mM Tris–HCl buffer pH 7.6 containing Cleland's reagent and eluted with an NaCl gradient as illustrated in Fig. 3. Breakdown was based on the liberation of free amino acids of the respective peptides as described in Table I. See Fig. 1 for the structure of the peptides.

	nmoles/mg fresh wt. or protein					
	Substance P		LH-RF		SRIF	
	Act.	Spec. act.	Act.	Spec. act.	Act.	Spec. act
Homogenate	2.7	25	1.0	7	4.3	39
100,000 × g supt.	1.3	52	0.70	26	3.7	161
DEAE cellulose	0.8	160	0.25	100	1.4	280

1974; Griffiths et al., 1974, 1975a, b) and this has been confirmed in subcellular fraction studies (Marks and Stern, 1974a; Benuck and Marks, 1976). More than 95% of the activity resided in the brain supernatant with less than 5% associated with synaptosomes and the mitochondrial enriched fractions. The finding of trace activities in synaptosomes and other particulates may be of interest since LH-RF is reported to occur at such sites (Shin et al., 1974a; Taber and Karavolas, 1975) and can be released following stimulation (Edwardson et al., 1972).

There has been only one attempt to purify the "neutral endopeptidase" inactivating LH-RF (Benuck and Marks, 1976). The enzyme is present in the cytosol fractions and requires rapid processing because of lability. It can be purified 15-fold by extraction with 0.32 M sucrose followed by DEAE-cellulose chromatography (Table IV). Enzyme from the active peak (eluted by 0.2 M NaCl) degraded histones and hemoglobin and showed properties similar to that of other brain neutral proteinases (cathepsin M) (Marks, 1976). Inhibition of LH-RF inactivating enzyme as a method of enhancing activity may be of clinical interest, but studies to date on crude fractions using tryptic inhibitors are inconclusive (Koch et al. 1974; Kochman et al., 1975).

Inactivation of LH-RF can occur also in serum and may partially account for its relatively short half-life in man and animals (Redding et al., 1973; Dupont et al., 1974). In our studies, the rates of inactivation were higher in rat than in human serum; this may be of some therapeutic interest (Table II). The rate of inactivation in man, however, was higher than MIF and TRF. In a study by Redding et al. (1973) using tritiated LH-RF the major end-product in man was reported to be pyroGlu-His. Based on appearance of free amino acids, however, we have concluded that inactivation in serum was by a mechanism similar to that in brain, with evidence for a rate-limiting internal cleavage (unpublished findings). In brain pyroGlu-His was rapidly cleaved to release His (Marks and Stern, 1974a).

Reports have appeared over the last decade that steroids and other hormones can directly affect peptidase activity and other enzymes in neurosecretory or other brain areas (see Marks, 1968). With respect to LH-RF, Griffiths and Hooper (1974) and Griffiths et al. (1975b) recently observed a higher rate of inactivation for LH-RF in hypothalamic tissue obtained from males as compared to females with some gradation of activity within the different regions of the hypothalamus neurosecretory region. It was reported that gonadectomy decreased inactivating enzyme present in the supernatant in some but not all areas of the hypothalamus, and these changes were reversed by estradiol in females and testosterone in males. These results were taken as implying possible steroid feedback affecting rates of breakdown in the hypothalamus. In this respect Shin et al. (1974a) observed an increase in LH-RF content in hypothalamus of male rats following castration that could be reversed by steroids. Kuhl and Taubert (1975) reported that a cystinyl aminopeptidase associated with inactivation in the hypothalamus was elevated on treatment with LH-RF. There are other examples in the literature of gonadectomy altering enzymatic activity (glucose, γ-glutamate and catecholamine metabolism) of neurosecretory areas of the brain and/or other organs, reversible in some cases by the administration of the appropriate steroid (Delap et al., 1975; Luine et al., 1975). It is probable that many factors can account for an alteration of enzyme activity, including the role of accessory factors (inhibitors or activators), the relative rates of degradation and synthesis of the enzymes or peptides involved, and the localization of enzyme versus substrate and transport factors. Many of the hormonal changes also may be mediated by secondary alterations in biogenic amines and the cAMP-adenylcyclase system; such questions have to be clarified before an association between breakdown and hormonal regulation in reproduction can be defined.

III. THYROLIBERIN (TRF)

Although TRF was the first hypothalamic releasing structure to be characterized, its mechanisms of inactivation in brain and serum are far from clear. Its structure, pyroGlu-His-Pro-NH$_2$, would exclude most of the known proteolytic enzymes. In our studies, incubation with brain extracts or serum led to the release of free His, showing that the pyroHis and the His-Pro-NH$_2$ bonds were cleaved (Fig. 1) (Marks, unpublished findings). Serum differed from brain, since free Pro also accompanied the appearance of His, indicating that deamidation at some point had occurred. Activity in brain was lower than the other hypothalamic peptides tested but comparable to that of LH-RF (Table I). In serum, the activity in rat was significantly higher as compared to that of man (Table II). The mechanisms for inactivation of TRF in brain appear to differ from those in serum.

To determine the mechanisms of inactivation we studied breakdown of a number of intermediates: pyroGlu-His itself was a good substrate with more rapid breakdown as compared to His-Pro-NH$_2$ with release of His in both cases but with no detectable levels of Pro, indicating that the primary cleavage site probably occurred at the pyroGlu-His bond. Neurosecretory tissue is known to contain a pyroglutamyl peptidase (EC 3.4.11.9), and this enzyme has been partially purified by Mudge and Fellows (1973) from bovine pituitaries. The enzyme is highly unstable but has similarities to a bacterial enzyme since it can be stabilized partially by addition of the inhibitor 2-pyrrolidone. With pyroGlu-Ala as the substrate the pH optimum was 7.3 with a K_m of $2.3 \times 10^{-4} M$ and the partially purified enzyme could cleave the pyroglutamyl moiety from TRF, its deamido form, and the O-methyl ester. In studies with TRF, His-Pro-NH$_2$ was formed, confirming our belief that the primary site of cleavage is the pyroGlu bond followed by subsequent hydrolysis at a lower rate of the remaining dipeptide. We have attempted to purify this enzyme from whole brain using a fluorimetric substrate pyroGlu-β-NA; enzyme activity is very low with this material but some activity is detectable and can be partially purified. It may be that the preferred substrate is the tripeptide itself; studies using the native hormone are in progress. If pyroglutamyl peptidase is involved in the breakdown of TRF, this is in contrast to LH-RF, which also has the N-protected group (Fig. 1). In the case of LH-RF, however, the lactam ring is adjacent to bulky histidyl and tryptophanyl groups and this may structurally hinder cleavage by a pyroglutamyl peptidase. There are some conflicting data pointing to the participation by a brain deamidase. In a brief report, Bauer et al. (1973) observed the formation of deamido TRF (and Pro-NH$_2$) on incubation of TRF with an extract of lyophilized porcine hypothalami. In studies by Nair et al. (1971) and Redding and Schally (1972) the end-products following an injection of tritiated TRF included the deamido forms, His-Pro-NH$_2$ and Glu-His-Pro. Visser et al. (1975), however, were unable to detect the deamido form in serum using a RIA for pyroGlu-His-Pro-OH. In our own studies, TRF breakdown in serum was accompanied by the production of His and Pro, indicating that deamidation played a role. By analogy to Substance P (below), deamidation may have occurred after the release of the C-terminal amide; in the case of TRF this would mean release of Pro-NH$_2$ prior to the appearance of Pro.

The role of the pyroglutamyl moiety in terms of breakdown and its regulations is unexplored. Pyroglutamate itself along with pyroglutamyl peptidase is reported to occur in many mammalian tissues (Szewezuk and Kwiatkowski, 1970). The level of pyroGlu in brain is 0.13 μmoles/g wet wt. and is higher than those in all other organs except kidney (Wilk and Orlowski, 1973). It has been shown that pyroGlu is an intermediate in the γ-glutamyl cycle, which is thought to play a role in amino acid transport (Orlowski et al., 1974). The presence of free pyroGlu in brain may be of interest in the synthesis of hormones with N-protected groupings (TRF, LH-RF, neurostenin).

IV. MELANOSTATIN (MIF)

Although there is doubt whether this tripeptide is an authentic releasing factor, there is a considerable literature on its formation, inactivation, and pharmacological activity. Its structure, Pro-Leu-Gly-NH$_2$, is identical to the C-terminus of oxytocin (Fig. 1) and there is some evidence that hypothalamic but not whole brain extracts can degrade oxytocin, leading to its release (Marks et al., 1973; Walter et al., 1973). In our studies on whole brain extracts, using labeled oxytocin, we failed to detect MIF as a degradation product, but incubation led to the release of Leu-Gly-NH$_2$ and Gly-NH$_2$ as the end-products (see below). Failure to detect MIF may arise from its rapid degradation by brain aminopeptidase leading to the formation of Pro, Leu, and Gly-NH$_2$ (Marks and Walter, 1972). Enzymes inactivating MIF have been purified from whole brain: in one case the enzyme was purified with Leu-Gly-Gly as the substrate (Simmons and Brecher, 1973) and in the other with Arg-β-NA (Marks and Walter, 1972). Since the intermediate dipeptides are hydrolyzed at very high rates we were unable to decide on the primary point of cleavage, namely, Pro-Leu or Leu-Gly-NH$_2$.

In view of the pronounced pharmacological effects in man there has been some interest in its biological half-life (Nair et al., 1973; Kastin et al., 1974). In a comparison of breakdown in various species we found that MIF is rapidly inactivated by sera of rat but only very slowly by human sera (Table II) (Walter et al., 1975). The explanation for this difference is unclear but may be related to the vestigial presence of the *pars intermedia* in man as compared to adult rats.

V. SOMATOSTATIN (SRIF)

The cyclic tetradecapeptide present in the hypothalamus inhibits the release of GH from the pituitary, and also glucagon and insulin from the pancreas of fasted animals. Its very short biological half-life, however, limits its clinical usefulness. In studies on inactivation in serum and brain we have used the same strategy as adopted for breakdown of LH-RF, namely, the timed appearance of amino acids. Results showed a preferential release of internal amino acids with a rate-limiting cleavage at the Trp^8-Lys^9 linkage and the bonds adjacent to the Phe residues (Fig. 1) (Marks and Stern, 1975). The slower release of Ala and Cys largely excludes the action of an aminopeptidase and a carboxypeptidase as the primary routes of inactivation. In many cases it is now known that the first two residues Ala-Gly are not required for biological activity (Brown et al., 1975). Conclusive evidence that internal cleavage is responsible for inactivation was provided by studies with analogs with a blocked N-terminal at position 3, namely, by the use of the linear and cyclic forms of desAla-Gly (Ac-Cys^3) somatostatin (Fig. 1). Incubation of this N-blocked analog as in the case of the native hormone led to the release of internal amino acids and a low yield of Cys (Marks and Stern, 1975). Little difference was observed between the linear and cyclic forms of this hormone in terms of breakdown (in contrast, see oxytocin below). There appears also to be little difference between the biological activity of these two forms.

Predications based on sites of cleavage as a method to prepare longer acting and/or more potent derivatives have been confirmed. Thus N-protected derivatives such as desAla-Gly-(N-Ac-Cy^3)-SRIF have in general a more durable action (Brown et al., 1975). In a recent report Rivier et al. (1975) showed that D-Trp^8-SRIF was 8 times more potent than the native analog but was not long-acting in vivo. The latter result is consistent with the fact that -Trp^8-Lys^9 bond is only one of three potential bonds cleaved by cathepsin M (Fig. 1) and additional studies are planned to determine if other analogs can be prepared with an increase in both potency and half-life. In a systematic study on substitution of each individual residue with Ala, no major increase in activity was found except in the case of those with L-Ala at positions 2 and 8 (Rivier et al., 1975). These results can be attributed to receptor occupancy rather than biodegradation since L-Ala amino acids probably do not affect breakdown to the same extent as the D-isomer. In studies on the proposed tertiary structure of somatostatin it was recently shown that there is a β-structure stabilized by stacking of the three aromatic rings Phe^7, Trp^8, and Phe^{11} (see Rivier et al., 1975). It is of interest that these three residues appear to be the ones most involved in the action of the neutral endopeptidase (Fig. 1) and may partially account for the increased potency of the D-Trp^8-somatostatin analog.

The enzyme inactivating somatostatin is present largely in brain supernatant fractions. It can be purified about 10-fold using the identical procedures for purification of other hormonal inactivating enzymes (Table IV). Further studies are required to differentiate this enzyme from those inactivating LH-RF, Substance P, and bradykinin.

VI. OXYTOCIN AND VASOPRESSIN

These hormones are characterized by a hexapeptidyl ring attached to a tripeptidyl alicyclic tail (Fig. 2). Both hormones are resistant to breakdown on prolonged exposure to crude brain extracts, and purified brain aminopeptidases (Marks et al., 1973). Incubation is accompanied, however, by the release of C-terminal fragments, notably Leu-Gly-NH$_2$ for oxytocin or Arg-Gly-NH$_2$ for vasopressin and Gly-NH$_2$ in both cases. This mechanism of inactivation is in contrast to the action of "oxytocinase" of sera that can

SITES OF CLEAVAGE OF ANGIOTENSINS, KININS, AND OXYTOCIN BY BRAIN ENZYMES

1. Pro- and angiotensin

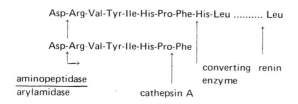

2. Kinin 9

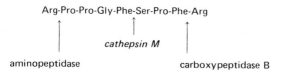

3. Oxytocin

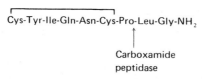

Fig. 2. See Fig. 1 for details and text for references.

remove N-terminal residues from oxytocin (see Marks and Lajtha, 1971). Product-precursor studies showed that dipeptide release precedes the formation of glycinamide, implying that cleavage occurred at the Cys^6-Pro^6 bonds. The enzyme involved has novel specificity, and since it acts on polypeptidyl amides it was named carboxamide peptidase. The enzyme inactivating oxytocin can be differentiated on the basis of its specificity from chymotryptic-like enzymes and from brain cathepsin M; carboxamide peptidase is incapable of splitting analogs with a free C-terminal carboxyl group (arginine vasopressinoic acid). Studies with other analogs show that the ring structure prevents the action of peptidases, since on its removal to yield linear peptides there is a rapid degradation by a brain aminopeptidase (purified on the basis of Arg-β-naphthylamide), for example, with S-bz-Cys-Tyr-Ile-Gln-Asn-S-bz-Cys-Pro-Leu-Gly-NH_2. Replacement of the terminal amino group by hydrogen and of the disulfide by an ethylene bridge — (1,6)-aminosuberic acid-lysine-vasopressin — prevented arylamidase action (Marks et al., 1973).

The role of a disulfide bridge in breakdown is an important topic since it is a characteristic of a very large number of proteins and polypeptides. In the case of oxytocin it appears that the hexapeptidyl ring confers a stable conformation unfavorable to the action of aminopeptidases and proteinases, in contrast to somatostatin, which has a disulfide bridge spanning 12 residues, and is susceptible to endopeptidase attack (see above) (Figs. 1,2). Tissues also contain transdehydrogenases capable of cleaving disulfides to yield linear polypeptides, which are known to be better substrates for degradation. Thus in the case of some hormones (insulin), the mechanism of breakdown appears to require two distinct catalytic mechanisms: scission of the ring structure followed by the action of intracellular exo- and endopeptidase. The question of a specific proteolytic enzyme recognizing a complex hormone such as insulin (insulinase) is still, however, an unresolved matter (Ansorge et al., 1973; Duckworth et al., 1975).

The enzyme(s) releasing the C-terminal fragments of oxytocin and vasopressin have been purified partially from rat brain cytosol fractions (Marks and Stern, 1974b). On elution from DEAE-cellulose columns two peaks of activity were obtained; one gave a higher release of the dipeptide Leu-Gly-NH_2 as compared to glycinamide.

VII. SUBSTANCE P

Polypeptides capable of causing contraction of the gut have been known for several decades, but their structures have been clarified only recently. Substance P is a unidecapeptide characterized by a C-terminal methionamide (Chang et al., 1971) capable of inducing a transient hypotension and depolarizing spinal motoneurons (Henry et al., 1975). These facts combined with its differential distribution in the nervous system have led to suggestions that it may play a role in neurotransmission. A function of this nature would necessitate rapid inactivation at its site of action. Early studies indicated that brain tissue could inactivate Substance P, but the questionable nature of the substrate used at that time did not permit characterization of the proteolytic enzyme(s) involved (see review Marks and Lajtha, 1971). Recently we showed that incubation of synthetic Substance P with crude or partially purified extracts of rat brain led to the preferential release of internal amino acids, implying the existence of a rate-limiting cleavage by an endopeptidase (Benuck and Marks, 1975; Fig. 2). In addition, there was a slower release of N- and C-terminal groups, indicating the presence of peptide hydrolases or other enzymes with the potential for inactivation. At longer periods of incubation we observed also the release of free Met. Studies with model peptides revealed that deamidation occurred only after release of methioninamide from peptide linkage. Thus, with the dipeptide Leu-Met-NH_2, the cleavage products at short incubation periods were Leu-Met-NH_2, and a trace of free Met. The formation of Met from methioninamide was confirmed by incubation of this residue with brain extracts, and this was shown to require extended incubation periods (Benuck and Marks, 1975). Deamidation is a known property of some aminopeptidases (notably leucine amino peptidase), but there have been few studies on polypeptidyl amides (Marks and Walter, 1972). This may represent an important pathway for regulation since this group is common, e.g., gastrin, secretin, α-MSH, cerulein, calcitonin, LH-RF, TRH, and MIF, to mention just a few. Current studies, however, on inactivation by brain of Substance P, LH-RF, and MIF suggest that deamidation does not play a role in the inactivation. There is some evidence as noted above, however, that deamidation might be involved in the inactivation of TRF in tissues and more especially in serum.

The enzyme inactivating Substance P by internal cleavage has been purified approximately 10-fold using simple extraction procedures and column chromatography (Fig. 3; Benuck and Marks, 1975). The major peak of activity was coincident with enzyme degrading hemoglobin and histones, and exhibited the properties associated with a brain neutral proteinase (cathepsin M). Further evidence that internal cleavage preceded release of free amino acids was provided by a bioassay utilizing contraction of guinea-pig ileum.

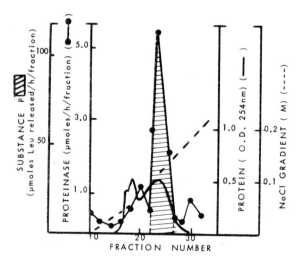

Fig. 3. Elution pattern of brain enzymes inactivating Substance P (hatched area) following application of a rat brain cytosol fraction to a DEAE-column (25 cm × 2 cm) and elution with NaCl as indicated in the diagram. Proteinase activity was measured with histones and hemoglobin as the substrate at pH 7.6. For details of assay see Table I, and for purification details see Table IV.

VIII. ANGIOTENSINS

Angiotensin formation represents an excellent example of a sequential breakdown process forming a peptidyl hormone from a protein precursor (plasma 2-globulin) by action of renin, to liberate a decapeptide (proangiotensin), followed by a converting enzyme (dipeptidyl hydrolase, EC 3.4.15.1) to liberate the active octapeptide and His-Leu (Fig. 1). Renin was first regarded as a renal hormone associated with vasoconstriction and ADH release, although the renin-angiotensin system is now reported to occur in many tissues, suggesting a broader role for this enzyme. CNS tissue, for example, is known to contain a renin-like enzyme acting on a plasma component, or on a brain protein, to release proangiotensin-like peptides. Renin-like enzymes occur in the hypothalamus and brain stem areas associated with synaptosomes, and a dipeptidyl hydrolase (converting enzyme) is present in microsomes (Fisher-Ferraro et al., 1971; Ganten et al., 1971a, b; Daul et al., 1975; Poth et al., 1975).

Ablation experiments show that the actions of angiotensins are mediated partly through a central mechanism. From this it has been inferred that angiotensin, despite its size and negative charge, can penetrate selected areas of the brain–blood or CSF–blood barriers (Severs and Daniels-Severs, 1973). Angiotensin has many diverse biological actions both in vitro and in the intact animal. Among these are changes in the transport and synthesis of catecholamines in nerve ending preparations from different tissues, a stimulation of vasopressin release from the posterior pituitary, and the transport of Na in arterial tissues. Among the inactivating enzymes reported as present in tissues are angiotensinases A_1 and A_2 (depending on whether the synthetic Asn'-amide or native hormone are used as substrates). These are Ca^{2+} dependent and inhibited by chelating reagents and DFP with similarities to aminopeptidases hydrolyzing α-Glu-β-NA as the preferred substrate (Marks, 1968). A second enzyme, angiotensinase B, has properties of an endopeptidase but has not been fully characterized (see Erdos, 1971). Angiotensin C (or polycarboxypeptidase) is present in lysosomes of swine kidney cortex, and is reported present in urine, leukocytes, and platelets of man; it is inhibited by DFP, its pH optima is 5.8, and it can be differentiated from cathepsin A by its preference for Z-Pro-Phe (Erdos and Yang, 1970; Yang et al., 1970a). Among potential enzymes for inactivation are cathepsin C (dipeptidyl transferase), which can remove N-terminal dipeptidyl fragments (see Marks and Lajtha, 1971), a lysosomal carboxypeptidase A (cathepsin A) removing neutral amino acids (Grynbaum and Marks, 1975), and cathepsin B_2 recently shown to remove all C-terminal amino acids from a number of peptides (Otto, 1976). Inactivation will depend on the localization of these enzymes in relation to the presence or absence of accessory factors.

In brain there is evidence for the existence of an aminopeptidase that hydrolyzes α-Glu-β-NA (angiotensin A), for a cathepsin C, and for lysosomal carboxypeptidases (Marks, 1968, 1970). Only two enzymes inactivating angiotensin in brain have been characterized, an aminopeptidase (based on Arg-β-NA as the substrate) shown capable of removing the first five N-terminal amino acids, and a lysosomal carboxypeptidase A that removes only the C-terminal Phe (Abrash et al., 1971; Grynbaum and Marks, 1976). The aminopeptidase (arylamidase) present in the soluble supernatant fraction is inhibited by

DFP; incubation of enzyme with angiotensin led to the release of Asp, Arg, Val, Tyr, Ile and the tripeptide His-Pro-Phe. The second enzyme releasing only C-terminal Phe is present in lysosomes, and it can be extracted by an acid buffer in the presence of detergents and purified 200-fold by use of DEAE-Sephadex and Sephadex G-200 (Grynbaum and Marks, 1976).

IX. KININS

Kinins are formed by the action of kallikrein on protein precursors (kinogens), present in plasma and tissue. Plasma kallikrein (or kininogenin, EC 3.4.31.8) yields kinin-9 (bradykinin), and that present in tissues yields kinin-11 (Met-bradykinin); kinin-11 (Met-Lys) can be formed by the action of kallikrein on tryptic fragments of globulins. Kallikrein itself is formed from a precursor (prekallikrein) by several mechanisms: enzymatic as noted above, a change in pH, contact with glass surfaces (activates the Hageman factor), heat, blood coagulation. Alterations in the release of kinins occur as a result of shock, or cerebral apoplexy, pregnancy, and arteriosclerosis (see review, Marks and Lajtha, 1971). Release of plasma kinins is retarded by inhibitors identical in most cases to those known to inhibit trypsin. These are reported to alleviate some of the symptoms accompanying stroke and postoperative hemorrhages (see Erdos and Yang, 1970; Erdos, 1971).

Kinin-9 has a very short half-life in blood in the range of minutes, pointing to the presence of potent kininases in body fluids and tissues. A number of enzymes have been described that can inactivate kinins. The first to be described was a carboxypeptidase N (kininase I) that removes the C-terminal Arg at pH 7.5; this enzyme can be differentiated from pancreatic carboxypeptidase B (which also inactivates) by its properties with respect to chelating agents and the effect of inorganic ions (Erdos and Yang, 1970). A second carboxypeptidase (kininase II) extracted from plasma and kidney inactivates kinin-9 by cleaving the Pro^6-Phe^7 bond with release of Phe^7-Arg^8 from the C-terminus. There are reported to be similarities between kininase-II and proangiotensin converting enzyme since it can release His-Leu to form angiotensin (Yang et al., 1970a, b). The identity of the enzymes has not been completely resolved and must await better methods of purification (Overturf et al., 1975).

Nerve tissue is reported to contain kininogen-like proteins, the enzyme kallikrein, and kinins (see Marks and Lajtha, 1971). There are several reports that brain and neurosecretory material contain carboxypeptidase-like enzymes that inactivate bradykinin and are present in high concentration in dorsal roots and ganglia. An enzyme present in the supernatant fraction was purified about 25-fold by Shikimi et al. (1970); it appeared to be a mixture of amino- and carboxypeptidases since it released N- and C-terminal Arg at pH 7.6. Peptide hydrolases, however, purified on the basis of Leu-Gly-Gly (aminopeptidase) and Arg-β-NA as substrate failed to affect kinins, implying the existence of other, more specific peptidases capable of releasing end-groups. Brain contains another potent inactivating enzyme that splits kinin-9 at the Phe^5-Ser^6 bond to release two polypeptide intermediates, Arg-Pro-Pro-Gly-Phe and Ser-Pro-Phe-Arg (Marks and Pirotta, 1971; Camargo et al., 1973). This enzyme can be partially purified and was shown to have properties akin to that of brain neutral proteinase (cathepsin M).

X. CONCLUSIONS

The short half-life of hypothalamic releasing factors and other peptides in vivo, together with their destruction in vitro, provides a good prima facie case that breakdown can play a significant role in endocrine control. A knowledge of the mechanisms of biodegradation can provide a good basis for the preparation of longer acting derivatives, as illustrated already in the case of LH-RF. Further progress in this area will depend on the nature of the enzymes involved and their characterization. A number of conclusions can be derived in this respect from present studies.

Brain and neurosecretory areas contain a neutral endopeptidase (cathepsin M) capable of inactivating LH-RF, SRIF, Substance P, and kinin-9 by cleavage at an internal site with release of polypeptidyl intermediates that can be split further by exopeptidases. In some cases, the internal cleavage is rate-limiting, and blocking this site with D-amino acids leads to an enhanced biological activity. Other enzymes involved also in breakdown appear to act at lower rates and play a less significant role in inactivation. Peptides with free N- and C-terminal groups, for example, are potential substrates for exopeptidases, and C-terminal amides can be removed by a C-terminal cleaving enzyme tentatively called "carboxamide peptidase". There is little evidence that LH-RF is inactivated by a pyroglutamyl peptidase; this peptide and Substance P also are not inactivated by deamidation.

The tripeptide MIF and TRF are degraded by CNS extracts and by serum with the release of free amino acids or acylated amides. Insufficient information is available in both cases to decide on the primary points of cleavage. MIF is split by CNS extracts and serum to yield all three amino acids, and this can be reproduced by purified brain aminopeptidases. Incubation of TRF with brain extracts leads to the release of His, showing that cleavage occurred at both of the adjacent bonds. One of the bonds can be cleaved by a pyroglutamyl peptidase purified from the pituitary gland. There is evidence that TRF and its breakdown products but not MIF can be deamidated. The rates of inactivation of these two tripeptides are strikingly lower in humans as compared to rat serum.

Oxytocin and vasopressin are inactivated by a carboxyamide peptidase releasing C-terminal dipeptide amides. The hexapeptidyl ring structure of these hormones hinders the action of known amino- and endopeptidases present in crude and purified brain extracts. There is no evidence that deamidation plays a role in the inactivation of these two nonapeptides.

Kinins and angiotensins are formed in brain from endogenous substrates and are inactivated by a variety of brain exo- and endopeptidases. Kinins are inactivated by a neutral endopeptidase (cathepsin M) at an internal bond, and also by specific exopeptidases; angiotensins can be inactivated by an arylamidase with release of the first 5 N-terminal amino acids, and also by a lysosomal carboxypeptidase A (cathepsin A) with release of C-terminal Phe.

ACKNOWLEDGEMENTS

The author is deeply indebted to Wilfred F. White for the gift of LH-RF and its analogs, and for many helpful discussions. I am also indebted to Jean Rivier (The Salk Institute, San Diego, Calif.) for the gift of SRIF and its analogs. The skilled help of Frederic Stern and Alice Grynbaum is also gratefully acknowledged.

A portion of this work was supported by a grant from the U.S.P.H.S. NB-03226.

REFERENCES

Abrash, L., Walter, R. and Marks, N. (1971) Inactivation studies of angiotensin II by purified enzymes. *Experientia (Basel)*, 27: 1352–1353.

Ansorge, S., Bohley, P., Kirschke, H., Langner, T., Wiederanders, B. and Hanson, H. (1973) Metabolism of insulin and glucagon: glutathiono-insulin transdehydrogenase from microsomes of rat liver. *Europ. J. Biochem.*, 32: 27–35.

Bauer, K., Jose, S. and Lipmann, F. (1973) Degradation of TRH by extracts of hypothalamus. *Fed. Proc.*, 32: 489.

Benuck, M. and Marks, N. (1975) Enzymatic inactivation of Sustance P by a partially purified enzyme from rat brain. *Biochem. biophys. Res. Commun.*, 65: 153–160.

Benuck, M. and Marks, N. (1976) Distribution and partial purification of an endopeptidase inactivating luliberin (LH-RF). *Neurochem. Res.*, in press.

Benuck, M., Marks, N. and Hashim, G. (1975) Metabolic instability of myelin proteins: breakdown of basic protein induced by brain cathepsin D. *Europ. J. Biochem.*, 52: 615–621.

Brown, M., Rivier, T., Vale, W. and Guillemin, R. (1975) Variability of the duration of inhibition of growth hormone release by N-acylated-de (Ala'Gly^2)-somatostatin analogs. *Biochem. biophys. Res. Commun.*, 65: 752–756.

Camargo, A.C.M., Shapanka, R. and Greene, L.T. (1973) Preparation, assay and partial characterization of a neutral endopeptidase from rabbit brain. *Biochemistry*, 12: 1838–1844.

Chang, M.M., Leeman, S.E. and Niall, H.D. (1971) Amino acid sequence of Substance P. *Nature New Biol.*, 232: 86–87.

Coy, D.H., Coy, E.T. and Schally, A.V. (1975) Structure–activity relationship of LH and FSH releasing hormone. In *Methods in Neurochemistry*, N. Marks and R. Rodnight (Eds.), Plenum, New York, pp. 393–496.

Cuatrecasas, P. (1974) Commentary: insulin receptors, cell membranes and hormone action. *Biochem. Pharmacol.*, 23: 2353–2361.

Daul, C.B., Heath, R.G. and Garey, R.E. (1975) Angiotensin-forming enzyme in human brain. *Neuropharmacology*, 14: 75–80.

Delap, L.W., Tate, S.S. and Meister, A. (1975) γ-Glutamyl transpeptidase of rat seminal vesicles. *Life Sci.*, 16: 691–709.

Duckworth, W.C., Heinemann, M. and Kitabachi, A.F. (1975) Proteolytic degradation of insulin and glucagon. *Biochim. biophys. Acta (Amst.)*, 377: 421–430.

Dupont, A., Labrie, F., Pelletier, G., Puriani, R., Coy, D.H., Coy, E.J. and Schally, A.V. (1974) Organ distribution of radioactivity and disappearance of radioactivity from plasma after administration of [^{3}H]LH-RH in mice and rats. *Neuroendocrinology*, 16: 65–73.

Edwardson, J.A., Bennett, G.W. and Bradford, H.F. (1972) Release of amino acids and neurosecretory substances after stimulation of nerve endings (synaptosomes) isolated from the hypothalamus. *Nature (Lond.)*, 240: 554–556.

Erdos, E.G. (1971) Enzymes that inactivate vasoactive peptides. In *Handbook of Experimental Pharmacology*, Springer, Berlin, pp. 620–653.

Erdos, E.G. and Yang, H.Y.T. (1970) kininases. In *Handbook for Experimental Pharmacology*, Springer, Berlin, pp. 289–323.

Fisher-Ferraro, C., Nahmod, V.E., Goldstein, J.J. and Finkelman, S. (1971) Angiotensin and renin in rat and dog brain. *J. exp. Med.*, 133: 353–361.

Fujino, M., Ramazak, I., Kobayashi, S., Fukuda, T., Shinagawa, S., Nakayama, R., White, W.F. and Rippel, R.H. (1974) Some analogs of LH-RF having intense ovulation-inducing activity. *Biochem. biophys. Res. Commun.*, 57: 1248–1256.

Ganten, D., Minnich, J.L., Granger, P., Haydu, K., Brecht, H.M., Barbeau, A., Boucher, R. and Genest, J. (1971a) Angiotensin-forming enzyme in brain tissue. *Science*, 173: 64–65.

Ganten, D., Boucher, R. and Genest, J. (1971b) Renin activity in brain tissue of puppies and adult dogs. *Brain Res.*, 33: 557–559.

Griffiths, E.C. and Hopper, K.C. (1974) Peptidase activity in different areas of the rat hypothalamus. *Acta endocr. (Kbh.)*, 77: 10–18.

Griffiths, E.C., Hooper, K.C., Jeffcoate, S.L. and Holland, D.T. (1974) Presence of peptidases in the rat hypothalamus inactivating luteinizing hormone releasing hormone (LH-RH). *Acta endocr. (Kbh.)*, 77: 435–442.

Griffiths, E.C., Hooper, K.C., Jeffcoate, S.L. and Holland, D.T. (1975a) Peptidases in different areas of the rat brain inactivity LH-RH. *Brain Res.*, 85: 161–164.

Griffiths, E.C., Hooper, K.C., Jeffcoate, S.L. and Holland, D.T. (1975b) The effects of gonadectomy and gonad steroids on the activity of hypothalamic peptidases inactivating LH-RH. *Brain Res.*, 88: 384–388.

Grynbaum, A., and Marks, N. (1976) Characteristics of a rat brain catheptic carboxypeptidase (cathepsin A) inactivating angiotensin II. *J. Neurochem.*, 26: in press.

Henry, J.L., Krnjević, K. and Morris, M.E. (1975) Substance P and spinal neurons. *Canad. J. Physiol. Pharmacol.*, 53: 423–432.

Kastin, A.J., Nissen, C., Redding, T.W., Nair, R.M.G. and Schally, A.V. (1974) Delayed appearance of ^{14}C labeled Pro-Leu-Gly·NH_2 from blood of hypophysectomized rats. *Neuroendocrinology*, 16: 36–42.

Koch, Y., Baram, T., Chobsieng, P. and Fridkin, M. (1974) Enzymic degradation of LH-RH by hypothalamic tissue. *Biochem. biophys. Res. Commun.*, 61: 95–103.

Kochman, K., Kerdelhué, B., Zor, U. and Jutisz, M. (1975) Studies of enzymatic degradation of luteinizing hormone-releasing hormone by different tissues. *FEBS Lett.*, 50: 190–194.

Kuhl, H. and Taubert, H.D. (1975) Shirt-loop feedback mechanism of luteinizing hormone. *Acta endocr. (Kbh.)*, 78: 649–663.

Lajtha, A. and Marks, N. (1971) Protein turnover. In *Handbook of Neurochemistry, Vol. 5*, A. Lajtha (Ed.), Plenum Press, New York, pp. 551–629.

Luine, V.N., Khitchevskaya, R.I. and McEwen, B.S. (1975) Effect of gonadal hormones on enzyme activities in brain and pituitary of male and female rats. *Brain Res.*, 86: 283–292.

Marks, N. (1968) Exopeptidases in the nervous system. *Int. Rev. Neurobiol.*, 11: 57–90.

Marks, N. (1970) Peptide hydrolases. In *Handbook of Neurochemistry, Vol. 3*, A. Lajtha (Ed.), Plenum Press, New York, pp. 133–171.

Marks, N. (1976) Specificity of breakdown based on the inactivations of active proteins and peptides by brain proteolytic enzymes. In *Proc. 2nd Int. Conf. on Intracellular Protein Catabolism*, V. Turk (Ed.), in press.

Marks, N. and Lajtha, A. (1971) Protein and polypeptide breakdown. In *Handbook of Neurochemistry, Vol. 5A*, A. Lajtha (Ed.), Plenum Press, New York, pp. 49–139.

Marks, N. and Pirotta, M. (1971) Breakdown of bradykinin and its analogs by rat brain neutral proteinase. *Brain Res.*, 33: 565–567.

Marks, N. and Stern, F. (1974a) Enzymatic mechanisms for the inactivation of luteinizing hormone-releasing hormone (LH-RH). *Biochem. biophys. Res. Commun.*, 61: 1458–1463.

Marks, N. and Stern, F. (1974b). In *Psychoneuroendocrinology*, N. Hatotani (Ed.), Karger, Basel, pp. 153–162.

Marks, N. and Stern, F. (1975) Inactivation of somatostatin (GH-RIH) and its analogs by crude and partially purified rat brain extracts. *FEBS Lett.*, 55: 220–224.

Marks, N. and Walter, R. (1972) MSH-release inhibiting factor: inactivation by proteolytic enzyme. *Proc. Soc. exp. Biol. (N.Y.)*, 140: 673–676.

Marks, N., Abrash, L. and Walter, R. (1973) Degradation of neurohypophyseal hormones by brain extracts and purified brain enzymes. *Proc. Soc. exp. Biol. (N.Y.)*, 142: 455–460.

Mudge, A.W. and Fellows, R.E. (1973) Bovine pituitary pyrrolidone-carboxyl peptidase. *Endocrinology*, 93: 1428–1434.

Nair, R.M.G., Redding, T.W. and Schally, A.V. (1971) Site of inactivation of TRH by human plasma. *Biochemistry*, 10: 3621–3624.

Nair, R.M.G., Redding, T.W., Kastin, A.J. and Schally, A.V. (1973) Site of inactivation of MIF by human plasma. *Biochem. Pharmacol.*, 22: 1915–1919.

Orlowski, M., Sessa, G. and Green, J.P. (1974) γ-Glutamyl transpeptidase in brain capillaries: possible site of a blood brain barrier for amino acids. *Science*, 184: 66–68.

Otto, K. (1976) Specificity of cathepsin B 2. In *2nd Symposium on Intracellular Protein Catabolism, Ljubljana*, in press.

Overturf, M., Wyatt, S. and Fitz, A. (1975) Angiotensin I (Phe^8-His^8) hydrolase and bradykininase from human lung. *Life Sci.*, 16: 1669–1682.

Poth, M.M., Heath, R.G. and Ward, M. (1975) Angiotensin-converting enzyme in human brain. *J. Neurochem.*, 25: 83–85.

Redding, T.W. and Schally, A.V. (1972) On the half-life of TRF in rats. *Neuroendocrinology*, 9: 250–256.

Redding, T.W., Kastin, A.J., Gonzalez-Barcena, D., Coy, D.H., Coy, E.J., Schalch, D.S. and Shally, A.V. (1973) The half-life, metabolism and excretion of tritiated LH-RH in man. *J. clin. Endocr.*, 37: 626–663.

Rivier, J., Brown, M. and Vale, W. (1975) D-Trp^8-somatostatin: an analog of somatostatin more potent than the native molecule. *Biochem. biophys. Res. Commun.*, 65: 746–751.

Serra, S., Grynbaum, A., Lajtha, A. and Marks, N. (1972) Peptide hydrolases in spinal cord and brain of the rabbit. *Brain Res.*, 44: 529–592.

Severs, W.B. and Daniels-Severs, A.E. (1973) Effect of angiotensin on the CNS. *Pharmacol. Rev.*, 25: 415–449.

Shikimi, T., Shigemitsu, H. and Heitanoh, I. (1970) Substrate specificity and amino acid composition of partially purified enzyme inactivating bradykinin in brain. *Jap. J. Pharmacol.*, 20: 169–170.

Shin, S.H., Howitt, C. and Milligan, J.V. (1974a) A paradoxical castration effect on LH-RH levels in male rat hypothalamus and serum. *Life Sci.*, 14: 2491–2496.

Shin, S.H., Morris, A., Snyder, J., Hymer, W.C. and Milligan, J.V. (1974b) Subcellular localization of LH releasing hormone in the rat hypothalamus. *Neuroendocrinology*, 16: 191–201.

Simmons, W.H. and Brecher, A.S. (1973) Inactivation of MIF by a Mn^{2+}-stimulated bovine brain aminopeptidase. *J. biol. Chem.*, 248: 5780–5784.

Szewezuk, A. and Kwiatkowska, J. (1970) Pyrrolidonyl peptidase in animal, plant and human tissues. Occurrence and some properties of the enzyme. *Europ. J. Biochem.*, 15: 92–96.

Taber, C.A. and Karavolas, H.J. (1975) Subcellular localization of LH releasing activity in the rat hypothalamus. *Endocrinology*, 96: 446–452.

Visser, J.T., Kluotwijk, W., Docter, R. and Hennemann, G. (1975) RIA for the measurement of Pyro-Glu-His-Pro, a proposed TRF metabolite. *J. clin. Endocr.*, 40: 742–745.

Walter, R., Griffiths, E.C. and Hooper, K.C. (1973) Production of MIF by a particulate preparation of hypothalami: mechanism of oxytocin inactivators. *Brain Res.*, 60: 449–508.

Walter, R., Neidle, A. and Marks, N. (1975) Significant differences in the degradation of Pro-Leu-Gly·NH$_2$ by human serum and that of other species. *Proc. Soc. exp. Biol. (N.Y.)*, 148: 98–103.

Wilk, S. and Orlowski, M. (1973) The occurrence of free L-pyrrolidine carboxylic acid in body fluids and tissues. *FEBS Lett.*, 33: 157–160.

Yang, H.Y.T., Erdos, E.G., Chiang, T.G., Jenssen, T.A. and Rodgers, J.G. (1970a) Characteristics of an enzyme that inactivates angiotensin II (angiotensinase C). *Biochem. Pharmacol.*, 19: 1201—1211.

Yang, H.Y.T., Erdos, E.G. and Levin, Y. (1970b) A dipeptidyl carboxypeptidase that converts angiotensin I and inactivates bradykinin. *Biochim. biophys. Acta (Amst.)*, 214: 374—376.

Subcellular Mechanisms in Reproductive Neuroendocrinology, edited by
F. Naftolin, K.J. Ryan and J. Davies
© 1976 Elsevier Scientific Publishing Company—Amsterdam, The Netherlands

Chapter 8

Control of neurotransmitter synthesis by precursor availability and food consumption*

RICHARD J. WURTMAN

Laboratory of Neuroendocrine Regulation, Department of Nutrition and Food Science, Massachusetts Institute of Technology, Cambridge, Mass. 02139 (U.S.A.)

I. INTRODUCTION

Mammalian brains employ as neurotransmitters three primary monoamines synthesized from the amino acids tryptophan and tyrosine; they are the indoleamine serotonin and the catecholamines dopamine and norepinephrine. Studies summarized in this report show that the rate at which brain neurons synthesize serotonin is, to a surprising extent, under "open-loop" control, that is, normally occurring variations in the availability of tryptophan to the brain (e.g., after eating) cause corresponding changes in serotonin synthesis and (probably) release. While catecholamine synthesis can also be controlled by brain tyrosine levels under experimental conditions, the effects of changed tyrosine tend to be buffered by closed feedback loops, which cause inverse changes in the activity of tyrosine hydroxylase as soon as catecholamine release into synapses is affected (tyrosine hydroxylase being the enzyme that rate-limits catecholamine synthesis). Serotoninergic neurons are clearly able to function as "sensors" of plasma amino acid patterns; catecholaminergic neurons may or may not fulfill a similar sensing function.

Tryptophan is an essential amino acid; hence, the tryptophan within brain neurons can derive only from the circulating amino acid or from tryptophan liberated by the lysis of intracellular proteins (Fig. 1). Circulating tryptophan, in turn, must come directly from food or from tryptophan present in tissues (Wurtman and Fernstrom, 1972).

The tyrosine in brain neurons may derive from the circulation, from the breakdown of brain proteins, or, potentially, from an additional source: the intraneuronal hydroxylation of brain phenylalanine (which may be catalyzed by tyrosine hydroxylase (Fig. 2; Shiman et al., 1971). Circulating tyrosine is obtained from tissue tyrosine or from dietary protein (as tyrosine or as phenylalanine, which is converted to tyrosine by hepatic phenylalanine hydroxylase). Hormones control the fluxes of tyrosine and most other amino acids between the plasma and such tissues as skeletal muscle; insulin, in particular, facilitates their uptake and thus decreases their plasma concentrations. The mechanism that

---

* These studies were supported in part by grants from the John A. Hartford Foundation and the United States Public Health Service (AM-14228).

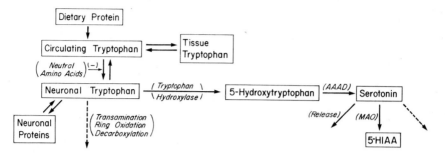

Fig. 1. Control of serotonin synthesis in brain neurons. AAAD, aromatic L-amino acid decarboxylase; MAO, monoamine oxidase; ——— indicates unproved pathway. (From: Wurtman and Fernstrom, 1975.)

controls the fluxes of tyrosine and tryptophan between the brain and the extracellular fluid apparently differs from that operating in most other tissues: it depends not primarily on direct effects of insulin or other hormones, but on plasma concentrations of tyrosine, tryptophan, and other neutral amino acids that compete for brain uptake (Blasberg and Lajtha, 1965; Fernstrom and Wurtman, 1972b). This competition is described in greater detail below. Both tyrosine and tryptophan are utilized within brain neurons to form peptides and proteins, as well as to synthesize monoamines; to a lesser extent, both compounds may also be catabolized within brain (by transamination, decarboxylation, and — for tryptophan — ring cleavage) (Wolstenholme, 1974).

The initial steps in the biosyntheses of serotonin and the catecholamines involve the hydroxylations of tryptophan and tyrosine to form the corresponding amino acids, 5-hydroxytryptophan (5-HTP) and dihydroxyphenylalanine (DOPA) (Figs. 1 and 2). These reactions are catalyzed by two specific enzymes, tyrosine hydroxylase (Nagatsu et al., 1964) and tryptophan hydroxylase (Lovenberg et al., 1968), which are highly localized within catecholamine-producing and serotonin-producing cells, respectively. Both enzymes utilize a reduced pteridine as a cofactor (Lovenberg et al., 1968; Shiman et al., 1971). Catecholamines can compete with this cofactor for attachment to a binding site on the tyrosine hydroxylase molecule, and may thereby inhibit the hydroxylation of tyrosine in vivo (Ikeda et al., 1966; Weiner, 1970); serotonin apparently does not compete for this site and does not significantly suppress its own biosynthesis by end-product inhibition (Jacoby et al., 1975). DOPA and 5-HTP are probably converted to their corresponding monoamines by a single decarboxylase enzyme (Lovenberg et al., 1962) (Figs. 1 and 2). Normally, very little DOPA and 5-HTP collect in significant concentrations within mammalian brain (Lindqvist et al., 1975); hence, their decarboxylations must be very rapid, even though their tissue concentrations are well below the K_ms of the decarboxylating enzyme. (It should be noted that, although tyrosine hydroxylase and tryptophan hydroxylase are confined to cells that are capable of synthesizing the catecholamines and serotonin, respectively, the decarboxylase enzyme is ubiquitously distributed, both inside (Lytle et al., 1972) and outside (Romero et al., 1973) the brain. Thus, when animals or humans receive exogenous DOPA or 5-HTP, most cells in the body at least transiently contain dopamine or serotonin; for this reason, the physiological responses elicited by the administration of these amino acids do not prove that such responses normally involve dopaminergic or serotoninergic mechanisms.)

Some brain neurons utilize dopamine as their neurotransmitter; others, which contain the enzyme dopamine β-oxidase, use dopamine simply as an intermediate in the biosynthesis of another transmitter — norepinephrine. Phenylethanolamine-N-methyl transferase is the enzyme that catalyzes the conversion of norepinephrine to the hormone epinephrine within the adrenal medulla; it, too, has been identified in mammalian brain (Pohorecky et al., 1969) and, more recently, has been found to be localized within neurons (Hökfelt et al., 1974). Hence, epinephrine may also function as a brain neurotransmitter.

II. CONTROL OF BRAIN SEROTONIN SYNTHESIS BY TRYPTOPHAN LEVELS, PLASMA AMINO ACID CONCENTRATIONS, AND NUTRITIONAL STATE

Abundant evidence suggests that the activities of the hydroxylase enzymes can control the rates at which brain neurons synthesize monoamine neurotransmitters. For example, short-term increases in the physiological activity of catecholaminergic neurons (caused either by direct electrical stimulation or by drug- or stress-induced changes in presynaptic activity) can accelerate the conversion of labeled tyrosine to a catechol; this effect indicates either that norepinephrine's end-product inhibition of tyrosine hydroxylase has decreased (Weiner, 1970) or that allosteric changes have occurred in the enzyme (Roth et al., 1974). Longer periods of enhanced presynaptic input increase the activity of tyrosine hydroxylase in vitro (Mueller et al., 1969) and may also increase the formation rate of the tyrosine hydroxylase enzyme protein, as assayed immunochemically (Joh et al., 1973). Tryptophan hydroxylase does not seem to be affected by end-product inhibition (Jacoby et al., 1975); however, the activity of the enzyme is apparently associated with neuronal

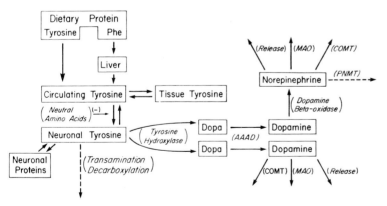

Fig. 2. Control of catecholamine synthesis in brain neurons. Phe, phenylalanine; MAO, monoamine oxidase; COMT, catechol-O-methyl transferase; PNMT, phenylethanolamine-N-methyl transferase; AAAD, aromatic L-amino acid decarboxylase; ——— indicates unproved pathway. (From: Wurtman and Fernstrom, 1975.)

firing rate, as has been shown in vivo by Andén and Modigh (1972) and in vitro by Zivkovic et al. (1973).

My associates and I have been interested in the possibility that monoamine synthesis in mammalian brains might depend not only on the activities of rate-limiting enzymes, but also on the tissue concentrations of the amino acid precursors. A series of experiments described below (performed largely in collaboration with Dr. John Fernstrom) has led us to conclude that tryptophan availability is the *major* physiological factor controlling the rate at which serotonin-containing brain neurons synthesize their neurotransmitter; indeed, the functional significance of these neurons may be related to their ability to serve as "sensors", thereby coupling serotonin synthesis with brain tryptophan levels and, moreover, with the pattern of amino acids in the plasma (Fernstrom and Wurtman, 1974).

II.1. Daily rhythms in amino acid metabolism

Our suspicion that amino acid availability might affect monoamine synthesis arose initially from observations on temporal changes in plasma amino acid concentrations. We found that if plasma samples were collected from untreated humans or rats at various times of day or night, their concentrations of tryptophan and of most other amino acids fluctuated characteristically during each 24-hr period (Wurtman et al., 1968; Wurtman, 1970). Among human subjects who ate 3 meals per day, tryptophan levels were lowest from 2 to 5 a.m.; they then rose 50–80% to attain a plateau in the late morning or early afternoon. In rats the daily nadir and peak occurred 8–10 hr later (Fernstrom et al., 1971) — a difference consistent with the rat's nocturnal feeding behavior. The plasma amino acid rhythms were shown not to be generated simply by the cyclic ingestion of dietary protein, inasmuch as they persisted (albeit with characteristic changes in amplitude and in times of peaks and nadirs) in human volunteers who ate essentially no protein for 2 weeks. Subsequent studies showed that the rhythms did disappear when subjects were placed on a total fast (Marliss et al., 1970); hence, such rhythms are not truly endogenous and circadian, but largely of nutritional origin; they result from the cyclic postprandial overflow of amino acids from the portal to the systemic circulations, and also from the postprandial release of insulin and other hormones that modify amino acid uptake into tissues.

The existence of amino acid rhythms in plasma suggested that the availability of these compounds to the brain and other tissues might also vary diurnally (perhaps in response to food consumption) and that such variations might, in turn, determine the rates at which they could be used for the synthesis of proteins and low-molecular weight derivations. To explore the significance of plasma amino acid rhythms, we decided to produce in rats fluctuations of the same amplitude as those occurring diurnally (i.e., by injecting an amino acid at a time of day when plasma levels are low) and to see whether such fluctuations were associated with increased in vivo utilization of the amino acid. The amino acid whose plasma concentration seemed most likely to influence its metabolic fate was tryptophan, since it is the least abundant amino acid in most tissues and foods (Wurtman and Fernstrom, 1972). Daily rhythms in the ingestion of tryptophan-containing proteins (and, presumably, in the concentration of tryptophan delivered to the liver via the portal venous circulation) had previously been shown to generate parallel rhythms both in the aggregation of hepatic polysomes (Fishman et al., 1969) and in the synthesis of the enzyme tyrosine transaminase in the liver (Wurtman, 1970).

After administering tryptophan to rats, we measured brain concentrations of serotonin and its chief metabolite, 5-hydroxyindole acetic acid (5-HIAA), in order to determine changes in brain serotonin synthesis — the dependent variable in our studies. Three lines of evidence suggested that the amount of tryptophan available to the brain might control serotonin synthesis: (1) the existence of roughly parallel daily rhythms in the brain concentrations of both tryptophan and serotonin (Albrecht et al., 1956; Wurtman and Fernstrom, 1972); (2) the likelihood, based on K_m measurements in vitro, that the concentrations of tryptophan in brain would not be sufficient to saturate tryptophan hydroxylase (Lovenberg et al., 1968); and (3) the great increases in brain concentrations of serotonin and 5-HIAA that occurred in animals receiving very large doses of tryptophan (50–1600 mg/kg, i.p.) (Moir and Eccleston, 1968; Wurtman and Fernstrom, 1972; Wolstenholme, 1974).

Initial experiments were designed to determine whether brain serotonin concentrations could be increased by raising the level of brain tryptophan from its daily nadir to values just below peak nocturnal levels. The noontime administration of L-tryptophan (12.5 mg/kg, i.p. — less than 5% of a 200-g rat's normal daily consumption) produced peak elevations in plasma and brain tryptophan that were within the nocturnal range of untreated rats (Fig. 3); in addition, brain serotonin levels rose 20–30% ($P < 0.01$) within 1 hr of treatment (Fernstrom and Wurtman, 1971a). Doses of 25 mg/kg caused proportionately greater increases in both brain tryptophan and brain serotonin; however, both concentrations remained within their normal dynamic ranges. Larger doses of tryptophan,

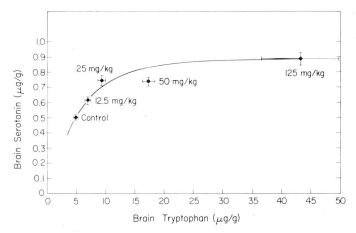

Fig. 3. Dose-response curve relating brain tryptophan and brain serotonin. Groups of 10 rats received L-tryptophan (12.5, 25, 50, or 125 mg/kg, i.p.) at noon and were killed 1 hr later. Horizontal bars represent standard errors of the mean for brain tryptophan; vertical bars represent standard errors of the mean for brain serotonin. All brain tryptophan levels in treated animals were significantly higher than the control levels ($P < 0.001$); all brain serotonin levels were also significantly higher ($P < 0.01$). (From: Fernstrom and Wurtman, 1971a.)

which caused brain tryptophan concentration to rise well beyond its physiological range, produced only minor additional increments in brain serotonin (Fig. 3).

II.2. Insulin, carbohydrate consumption, and brain serotonin synthesis

The increases in brain serotonin and 5-HIAA following the very small doses of tryptophan showed that changes in plasma and brain tryptophan (identical to those occurring diurnally) could influence brain serotonin synthesis. We next attempted to determine whether physiological *decreases* in the plasma amino acid could similarly depress serotonin synthesis. We anticipated that such a decrease could be produced by giving fasted rats enough insulin (2 IU/kg, i.p.) to lower plasma concentrations of glucose and most other amino acids. To our surprise, the hormone did not lower plasma tryptophan; it instead increased the tryptophan concentration by 30–40% (Fernstrom and Wurtman, 1972a).

TABLE I

Effects of consuming a carbohydrate-fat meal on serotonin and 5-hydroxyindole acetic acid concentrations in various brain regions and spinal cord

Animals were treated and tissues prepared as described, except that septum was included with "rest of brain", and cerebellum was not assayed for 5-hydroxyindoles. Cerebellar tryptophan concentrations were 7.8 ± 0.3 µg/g in carbohydrate-fed rats and 5.6 ± 0.3 µg/g in control rats, $P < 0.001$. N = number of samples assayed. Each striatum and hypothalamus sample refers to pooled sections from 4 animals. Data are presented as means ± standard errors of the means. (From Colmenares et al., 1975.)

Region	5-HT (nmole/g)	N	5-HIAA (nmole/g)	N	5-HT + 5-HIAA (nmole/g)
Cerebral cortex					
Control	1.08 ± 0.03	(27)	1.15 ± 0.05	(27)	2.23 ± 0.06
Carbohydrate-fat	1.18 ± 0.03*	(27)	1.53 ± 0.07***	(27)	2.70 ± 0.08***
Striatum					
Control	2.14 ± 0.08	(6)	2.58 ± 0.17	(6)	4.71 ± 0.19
Carbohydrate-fat	2.24 ± 0.02	(5)	2.61 ± 0.43	(5)	4.84 ± 0.20
Hypothalamus					
Control	5.04 ± 0.14	(6)	3.09 ± 0.20	(6)	8.13 ± 0.27
Carbohydrate-fat	5.22 ± 0.16	(5)	3.70 ± 0.06	(5)	8.91 ± 0.21
Rest of brain					
Control	3.17 ± 0.10	(26)	2.82 ± 0.09	(26)	5.99 ± 0.15
Carbohydrate-fat	3.39 ± 0.10	(25)	3.10 ± 0.08*	(25)	6.50 ± 0.19**
Brain stem					
Control	3.57 ± 0.09	(26)	3.58 ± 0.11	(25)	7.15 ± 0.15
Carbohydrate-fat	3.90 ± 0.08**	(27)	4.20 ± 0.10***	(27)	8.10 ± 0.17***
Spinal cord					
Control	2.65 ± 0.10	(27)	1.45 ± 0.07	(28)	4.10 ± 0.11
Carbohydrate-fat	2.94 ± 0.07*	(26)	1.80 ± 0.08**	(27)	4.79 ± 0.09***

* $P < 0.05$ compared with control values.
** $P < 0.01$ compared with control values.
*** $P < 0.001$ compared with control values.

This effect was independent of the route by which the insulin was administered; moreover, it was associated with a 55% fall in plasma glucose and also with major reductions in the plasma concentrations of most other amino acids, including the neutral ones that compete with tryptophan for uptake into the brain (Guroff and Udenfriend, 1962; Blasberg and Lajtha, 1965). Two hours after rats received the insulin, brain tryptophan levels were elevated by 36% ($P < 0.01$) and brain serotonin by 28% ($P < 0.01$).

The increase in brain serotonin observed in rats receiving insulin might have been artifactual, resulting not from increased availability of substrate, but from central reflexes activated by hypoglycemia. To determine whether the *physiological* secretion of insulin in *normoglycemic* rats also increased plasma and brain tryptophan and brain serotonin, we measured these indoles in rats that were given free access to a carbohydrate diet after a 15-hr fast. Plasma tryptophan levels were found to be significantly elevated 1–3 hr after food presentation; brain tryptophan and serotonin were also significantly elevated after 2 and 3 hr (Fernstrom and Wurtman, 1971b). In subsequent studies we noted that the carbohydrate-induced increases in brain serotonin and 5-HIAA were present both in regions containing the cell bodies of serotoninergic neurons (i.e., the brain stem) and in regions containing only axons and terminals (spinal cord, cerebral cortex) (Table I; Colmenares et al., 1975). Some variation among the responses in particular regions has been noted: the least effects on tryptophan and the hydroxyindoles occur in the striatum and hypothalamus. Using 5-HTP accumulation after administration of a decarboxylase inhibitor as an index of 5-hydroxyindole synthesis, we have further shown that the elevation in brain serotonin levels that follows carbohydrate consumption actually does result from accelerated synthesis of the amine (Table II; Colmenares et al., 1975).

TABLE II

Effect of carbohydrate-fat consumption on the accumulation of 5-hydroxytryptophan in various brain regions and spinal cord following Ro 4-4602 administration

Rats fasted overnight were given access to a carbohydrate-fat diet the next morning; control animals continued to fast. One hour later, all animals were injected with Ro 4-4602 (80 mg/kg, i.p.), and killed 1 hr thereafter. N = number of samples assayed. Each striatum, hypothalamus and septum sample refers to pooled sections from 4 animals. Data are presented as means ± standard errors of the means. (From Colmenares et al., 1975.)

Region	5-HTP (nmole/g)			
	Control	N	Carbohydrate-fat	N
Cerebral cortex	0.75 ± 0.04	(17)	0.97 ± 0.04***	(19)
Striatum	1.40 ± 0.06	(4)	1.75 ± 0.07**	(4)
Hypothalamus	2.65 ± 0.32	(4)	3.72 ± 0.32	(4)
Septum	1.58 ± 0.06	(4)	2.51 ± 0.23*	(3)
Rest of brain	1.52 ± 0.07	(6)	2.11 ± 0.09***	(6)
Brain stem	2.15 ± 0.12	(18)	2.85 ± 0.11***	(19)
Spinal cord	0.79 ± 0.08	(11)	0.98 ± 0.05*	(12)

* $P < 0.05$ compared with control levels.
** $P < 0.01$ compared with control levels.
*** $P < 0.001$ compared with control levels.

II.3. Suppression of brain serotonin synthesis by dietary protein

Because the eliciting of insulin secretion by carbohydrate consumption raised plasma tryptophan levels and, ultimately, the concentrations of tryptophan and serotonin in the brain, we anticipated that the consumption of a "balanced" diet would cause an even greater rise in brain serotonin: the dietary carbohydrate would elevate plasma tryptophan by causing insulin secretion, and the dietary protein would contribute directly to plasma tryptophan; brain tryptophan and serotonin would presumably increase accordingly. However, when we gave fasted rats access to diets containing either natural protein or complete mixtures of amino acids, we found that, despite a major increase in plasma tryptophan (about 60%, $P < 0.001$), *no* increases in brain tryptophan or serotonin occurred (Fig. 4; Fernstrom and Wurtman, 1972b; Fernstrom et al., 1973).

It seemed possible that brain tryptophan had failed to increase after protein ingestion because the plasma concentrations of competing neutral amino acids had increased even more than that of tryptophan. To test this hypothesis, we allowed groups of animals to eat either a synthetic diet containing carbohydrates plus all of the amino acids (in the same proportions as are present in an 18% casein diet), or this diet minus the 5 amino acids thought to share a common transport system with tryptophan (tyrosine, phenylalanine, leucine, isoleucine, and valine) (Blasberg and Lajtha, 1965). Both diets significantly increased plasma tryptophan levels above those found in fasted controls. However, large increases in brain tryptophan, serotonin, or 5-HIAA occurred only when the competing amino acids were deleted from the diet (Fernstrom and Wurtman, 1972b). In the repeat experiments, we omitted aspartate and glutamate — two acidic amino acids — from the diet, instead of the 5 neutral amino acids; plasma tryptophan concentrations again increased 70–80% above those of fasted controls ($P < 0.001$); however, brain tryptophan, serotonin, and 5-HIAA remained unaffected.

On the basis of these observations, we postulated that brain tryptophan and 5-hydroxyindole levels do not simply reflect plasma tryptophan levels: they must also depend on the plasma concentrations of the other neutral amino acids. This relationship was confirmed by an analysis correlating brain tryptophan concentration with the ratio of plasma tryptophan to the 5 competing amino acids, in individual rats given diets containing various amounts of each amino acid (Fernstrom and Wurtman, 1972b). This analysis yielded a correlation of 0.95 ($P < 0.001$ that $r = 0$), whereas the correlation between brain tryptophan and plasma tryptophan alone was less striking ($r = 0.66$; $P < 0.001$ that $r = 0$). Similarly, the correlation coefficient for brain 5-hydroxyindoles (serotonin plus 5-HIAA) versus the plasma amino acid ratio was 0.89 ($P < 0.001$), and that of 5-hydroxyindoles versus tryptophan alone was only 0.58 ($P < 0.001$). The reason that brain tryptophan and serotonin had appeared, in our earlier studies, to depend upon plasma tryptophan alone was that all of the psysiological manipulations that had been examined (tryptophan injections, insulin injections, carbohydrate consumption) raised the numerator (i.e., plasma tryptophan) in the fraction, while either lowering the denominator (competing amino acids) or leaving it unaltered. Only when rats consumed protein did *both* the numerator and the denominator increase. The effect of food consumption on 5-hydroxyindole synthesis in rat brain may now be modeled as in Fig. 5.

Serum tryptophan in humans does not increase after glucose consumption or insulin

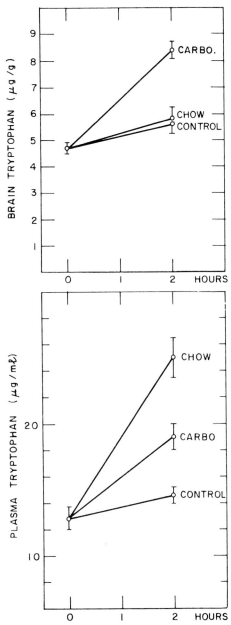

Fig. 4. Changes in brain and plasma tryptophan concentrations following the consumption of different foods. Rats were killed 2 hr after diet presentation. Vertical bars represent standard errors of the mean. Plasma tryptophan concentrations were significantly higher in rats consuming either of the two diets than in the fasting controls (chow: $P < 0.001$; carbohydrate: $P < 0.01$). Brain tryptophan levels were significantly elevated in rats consuming the protein-free (carbohydrate) diet ($P < 0.001$). (From: Fernstrom et al., 1973.)

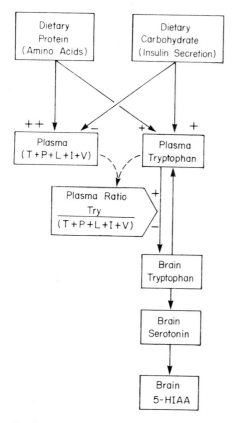

Fig. 5. Proposed sequence describing diet-induced changes in brain serotonin concentration in the rat. The ratio of tryptophan to the combined levels of tyrosine, phenylalanine, leucine, isoleucine, and valine in the plasma is thought to control the tryptophan level in the brain. The three-branched amino acids may be more significant in suppressing brain tryptophan uptake than the two aromatic amino acids. (From: Fernstrom and Wurtman, 1972b.)

injection; it also does not substantially decrease (Lipsett et al., 1973). In contrast, the plasma concentrations of competing neutral amino acids fall markedly (Wurtman, 1970). Hence the plasma ratio of total tryptophan to the sum of the competing amino acids increases, just as it does in rats. We might then anticipate that brain tryptophan and serotonin levels would also rise in humans following insulin injection or carbohydrate consumption. (It has not been possible, of course, to test this assumption directly. We have recently found that the amounts of 5-HIAA excreted into the urine do vary as a direct function of dietary protein content in both rats and humans (Nomura et al., 1975). However, virtually all of this increment reflects serotonin synthesized outside the central nervous system inasmuch as it can be suppressed by administration of a peripherally active decarboxylase inhibitor.)

II.4. Lack of effect of tryptophan binding to albumin on brain serotonin synthesis

Tryptophan in plasma is distributed between two pools: about 10–20% circulates as the free amino acid, and the remainder is bound to serum albumin (McMenamy and Oncley, 1958). No other circulating amino acid binds appreciably to plasma proteins. Because binding generally implies storage, several investigators have suggested that the plasma-free tryptophan pool determines the availability of circulating tryptophan to brain and other tissues (e.g., Knott and Curzon, 1972).

A variety of lipid-soluble compounds that bind to albumin in the blood (e.g., hormones, drugs, non-esterified fatty acids (NEFA)) may displace each other from binding sites. Thus, for example, increasing the serum concentration of NEFA in vitro causes the concentration of albumin-bound tryptophan to fall, and that of free tryptophan to rise. In collaboration with Drs. Bertha Madras and Hamish Munro, we examined this relationship in vivo by feeding rats diets that were expected to alter serum NEFA levels, and then measuring the changes in serum-free and albumin-bound tryptophan. We then examined the correlations between diet-induced changes in brain tryptophan or 5-hydroxyindoles and (1) serum-free tryptophan, (2) total serum tryptophan, and (3) the ratio of serum tryptophan to the sum of the competing neutral amino acids.

These studies showed that diet-induced changes in brain tryptophan levels exhibit no consistent correlations with the changes in serum-free tryptophan (Madras et al., 1973, 1974). For example, the consumption by rats of a carbohydrate diet *decreases* serum-free tryptophan (presumably via the insulin-mediated decline in serum NEFA, which causes an increase in the affinity of circulating albumin for tryptophan); at the same time, the carbohydrate diet *increases* brain tryptophan and 5-hydroxyindoles (probably by raising the serum ratio of tryptophan to competing neutral amino acids). Thus, the binding of serum tryptophan to albumin does not appear to limit the availability of the amino acid to the brain. Indeed, the physiological significance of tryptophan binding appears to be just the opposite: it allows serum tryptophan levels to remain elevated after insulin is secreted, at a time when the serum concentrations of the amino acids that compete with tryptophan for brain uptake are declining. Thus albumin binding makes it possible for carbohydrate consumption to elevate brain tryptophan and, consequently, to increase the synthesis of 5-hydroxyindoles.

II.5. Physiological significance of diet-induced changes in brain serotonin levels

An increase in brain serotonin content, whether induced by diet or by any other treatment, is physiologically significant only if it is associated with a corresponding change in the amount of the neurotransmitter that is secreted into synaptic clefts. This effect, in turn, probably depends upon two factors: the number of action potentials traversing serotoninergic neurons and the number of neurotransmitter molecules released per action potential. Of course, no direct methods are available for measuring the number of serotonin molecules released into brain synapses; hence, it is also impossible to test directly whether a relationship exists between the increased brain serotonin levels in rats consuming carbohydrates and the amount of the released neurotransmitter. One *indirect* estimate of serotonin release into brain synapses can be obtained by examining a particular behavior in rats that has previously been shown to depend on serotoninergic neurotransmission (i.e., flinching and jumping when an electric current is applied to the grid on which they stand). (Of course, this behavior — and all behaviors — are mediated by neurons utilizing many additional transmitters.) Processes that decrease the number of serotonin molecules interacting with postsynaptic receptors make animals more sensitive to this painful stimulus (i.e., they lower the flinch/jump threshold); treatments that enhance the activation of serotonin receptors tend to be analgesic. My associate Dr. Loy Lytle finds that the lowering of brain serotonin by a nutritional treatment (chronic consumption of a diet containing corn — which is poor in tryptophan — as the sole protein source) makes rats more sensitive to shock; this effect can rapidly be reversed either by an injection of tryptophan or by consumption of a meal that raises brain serotonin (Lytle et al., 1975). Numerous other "brain outputs" thought to involve serotoninergic neurons (including the secretion of several pituitary hormones) are now being examined to see if they are altered following ingestion of meals that affect brain serotonin metabolism.

II.6. Interactions between diet-induced changes in brain serotonin and drug treatments

Brain 5-HIAA levels fall when rats are treated with chlorimipramine, a drug that blocks the presynaptic uptake of serotonin and thus probably potentiates its postsynaptic effects. This action probably reflects a decrease in the rate of firing among serotoninergic neurons, which is mediated by a multisynaptic reflex arc or by presynaptic receptors responding to increased amounts of serotonin within synaptic clefts. My former associate Dr. Jacob Jacoby has shown that the consumption of a carbohydrate meal by chlorimipramine-treated rats causes increments in brain 5-hydroxyindole levels that are comparable to those observed in untreated animals (Jacoby et al., 1975). Because chlorimipramine, in the doses used in this study, virtually shuts off action potentials along serotonin neurons, we interpret these results as showing that the "external" feedback control of the physiological activity of serotonin neurons does not override the "internal" regulation of serotonin synthesis by diet and brain tryptophan. We also found that carbohydrate consumption elevates brain serotonin levels even among animals pretreated with a monoamine oxidase inhibitor (which already gave them markedly increased brain serotonin levels (Jacoby et al., 1975)). This finding suggests that intraneuronal serotonin concentrations do not normally exert feedback control over serotonin synthesis. In contrast, both of the above factors — the rate at which the neuron fires and the intraneuronal concentration of the monoamine — seem to affect greatly brain catecholamine synthesis.

It might be anticipated that the potencies of drugs that act by releasing brain serotonin (such as fenfluramine) would be considerably altered by diet-induced changes in brain serotonin. The same could be said for drugs that interact with serotonin receptors.

III. CONTROL OF BRAIN CATECHOLAMINE SYNTHESIS BY BRAIN TYROSINE LEVELS

The administration of tryptophan to rats causes rapid elevations in brain 5-hydroxyindole concentrations (Fig. 3); the ease with which this phenomenon could be demonstrated greatly facilitated acceptance of the hypothesis that brain tryptophan levels *normally* control brain serotonin synthesis. In contrast, little if any published evidence is available showing that tyrosine administration raises brain catecholamine levels. The lack of such evidence, plus the fact that whole-brain tyrosine levels are approximately equal to the K_m for brain tyrosine hydroxylase (0.14 mM for whole rat brain (Coyle, 1972); 0.1 mM for sheep caudate nuclei (Poillon, 1971)), have led most investigators to believe that brain tyrosine levels are not a significant factor controlling brain catecholamine synthesis.

We suspected that brain tyrosine hydroxylase might not be saturated in vivo and that the failure of brain catecholamine *levels* to rise after tyrosine administration might not be indicative of possible changes in catecholamine *synthesis*. We have explored this relationship by estimating brain catechol (i.e., DOPA) synthesis rates in animals given, first, amino acids that raise and lower brain tyrosine and, then, a centrally acting decarboxylase inhibitor (Ro 4-4602; 800 mg/kg, i.p.). In animals given a dose of tyrosine that raised brain tyrosine by 80% (50 mg/kg), DOPA accumulated 13% faster ($P < 0.05$) in the hour after administration of the decarboxylase inhibitor than it did in rats not given the tyrosine (Wurtman et al., 1974). In contrast, treatment of rats with another neutral amino acid that lowered brain tyrosine by 18% (tryptophan, 50 mg/kg) caused DOPA accumulation to decline by 32% ($P < 0.001$). When other amino acids were tested, we observed that only those that decreased brain tyrosine also suppressed brain DOPA synthesis (Table III). The high degree of correlation between brain tyrosine levels and brain catechol synthesis was shown in experiments in which rats received the synthetic neutral amino acid p-chlorophenylalanine (PCPA) (300 mg/kg, i.p.) at various intervals before they received Ro 4-4602, and were then killed an hour after the decarboxylase inhibitor (Fig. 6). The longer the interval between the injection of PCPA and autopsy, the greater were the declines in both brain tyrosine level and brain catechol synthesis.

TABLE III

Effects of various amino acids on accumulation of DOPA in rat brain

Groups of 7–9 rats received the amino acid intraperitoneally 15 min after Ro 4-4602 (800 mg/kg, i.p.) and 45 min before they were killed. Control rats received saline instead of the amino acid, and then the decarboxylase inhibitor. Data are given as percentages of controls ± the standard errors. (From Wurtman et al., 1974.)

Treatment	Dose (mg/kg)	Tyrosine (%)	DOPA (%)
Leucine	100	78 ± 7*	75 ± 5**
Histidine	100	94 ± 3	99 ± 4
Alanine	100	104 ± 3	101 ± 5
Lysine	100	96 ± 2	99 ± 5
Phenylalanine	50	128 ± 7*	114 ± 8
Phenylalanine	100	103 ± 3	80 ± 5*
PCPA	300	53 ± 1**	48 ± 3**

* Differs from control means, $P < 0.01$.
** Differs from control means, $P < 0.001$.

By examining brain catechol synthesis with centrally active decarboxylase inhibitors, one assures that any differences in DOPA formation among various experimental groups will not cause parallel differences in catechol*amine* synthesis or release (i.e., because DOPA is not being transformed to amines). Hence, no feedback mechanisms dependent upon receptor stimulation by neurotransmitters will act differentially to suppress (or activate) tyrosine hydroxylase activity. On the other hand, when catechol synthesis is estimated by measuring brain catecholamine levels after treatment with monoamine oxidase inhibitors — or brain homovanillic acid levels after probenecid — changes in catechol synthesis are able, at least theoretically, to modify neurotransmitter release and, thereby, to activate feedback mechanisms that compensate for the effects of altered precursor levels. Preliminary experiments using these latter paradigms suggest that, in the intact, untreated organism, naturally occurring changes in brain tyrosine level (e.g., those occurring after consumption of a high-protein meal) may be so well compensated by feedback changes in tyrosine hydroxylase activity as to have relatively little prolonged effect on catecholamine synthesis. If subsequent studies support this speculation, they will imply a fundamental difference in the biologies of serotoninergic and catecholaminergic neurons.

IV. CONTROL OF BRAIN ACETYLCHOLINE SYNTHESIS BY BRAIN CHOLINE LEVELS AND DIETARY CHOLINE CONTENT

Acetylcholine is synthesized from choline and acetylcoenzyme A in a reaction catalyzed by choline acetyltransferase (Hebb, 1972). Unlike liver, mammalian brain may not be capable of de novo choline synthesis by methylation of phosphatidylethanolamines (Ansell and Spanner, 1971); thus, central cholinergic neurons may largely depend on the blood for their supply of the acetylcholine precursor (Schuberth and Jenden, 1975). We have recently found that changes in brain choline level, produced either by injecting choline (60 mg/kg, i.p., of the chloride) or by feeding rats diets containing various amounts of choline, cause major changes in brain acetylcholine levels.

Twenty minutes after choline injection, brain choline rises significantly to $233 \pm 7\%$ of control values ($P < 0.001$); it subsequently falls, reaching control values 60 min after injection (Fig. 6). Brain acetylcholine concentration rises significantly ($P < 0.001$) to $122 \pm 2\%$ of control values after 40 min, and returns to basal levels 40 min later. The time course of the changes in brain choline and acetylcholine concentrations thus exhibits a precursor—product relationship. The increase in brain acetylcholine concentration 40 min after i.p. choline chloride is dose dependent between doses of 15 and 60 mg/kg; greater doses of choline do not cause greater increases in acetylcholine concentration.

Compared to rats consuming no choline, those consuming 20 or 129 mg of choline daily for 7 days exhibit caudate acetylcholine levels that are 28% and 45% higher, respectively (Cohen and Wurtman, 1975; Cohen et al., 1975). This range of choline ingestion is not dissimilar from the range chosen by humans in a normal day's diet.

The observation that choline injection or ingestion raises brain acetylcholine levels (probably by stimulating acetylcholine biosynthesis) may lead to a useful approach for modifying the functional activity of cholinergic neurons — provided that intraneuronal

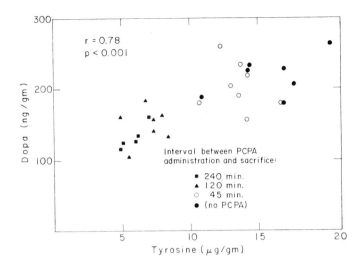

Fig. 6. Relationship between brain DOPA concentration and brain tyrosine level after PCPA administration. Rats received PCPA (300 mg/kg, i.p.) and Ro 4-4602 (800 mg/kg, i.p.) and were killed 1 hr after the Ro 4-4602.

acetylcholine content is indeed coupled to the release of the neurotransmitter. A recent case report suggests that this relationship does hold true in vivo: a patient with tardive dyskinesia, a disorder possibly resulting from deficient striatal cholinergic activity, was treated successfully with large doses of choline (Davis et al., 1975). The physiological and evolutionary significance of having the mammalian brain couple acetylcholine synthesis to the vagaries of food choice awaits clarification.

V. SUMMARY

The rate at which serotonin-containing neurons in rat brain synthesize their neurotransmitter varies with the concentration of tryptophan, the amino acid precursor, as measured in homogenates of whole brain. Brain tryptophan levels rise after the administration of tryptophan, the injection of insulin, or the consumption of a single protein-free, high-carbohydrate meal; soon thereafter, the levels of serotonin and its major metabolite, 5-hydroxyindole acetic acid (5-HIAA), also rise. These elevations are observed throughout the brain, i.e., both in regions containing the cell bodies of serotoninergic neurons and in regions containing only the axons and terminals. The addition of protein to the meal suppresses the increases in brain tryptophan, serotonin, and 5-HIAA, because protein contributes the other neutral amino acids (e.g., leucine, phenylalanine) to plasma in considerably greater quantities than it does tryptophan; these other amino acids interfere with the uptake of tryptophan into the brain.

The rate at which the brain accumulates 5-hydroxytryptophan in animals treated with an inhibitor of aromatic L-amino acid decarboxylase is also proportional to the brain's tryptophan concentration. This observation indicates that the precursor-induced changes in brain 5-hydroxyindole levels result from alterations in serotonin *synthesis*. Serotoninergic neurons continue to synthesize serotonin according to the level of tryptophan even after treatments which raise their serotonin content (e.g., monoamine oxidase inhibition) or which slow their firing rate (e.g., the administration of drugs that block serotonin reuptake); this finding suggests that these neurons are subject to "open-loop" control. Nutritionally induced changes in brain serotonin level cause inverse changes in pain sensitivity similar to those reported previously in animals treated with drugs affecting serotoninergic transmission; hence, neurotransmitter *levels* appear to affect neurotransmitter *release*, at least for central serotoninergic neurons.

The elevation of brain tyrosine (by its injection or by consumption of a single 40%-protein meal) accelerates brain catechol synthesis, as shown by the increased accumulation of brain DOPA in animals treated with a decarboxylase inhibitor. Under normal circumstances (i.e., in animals not treated with decarboxylase inhibitors), brain tyrosine concentrations are probably less important in determining catecholamine synthesis rates than is tryptophan in controlling serotonin synthesis: as soon as changes in catecholamine synthesis begin to affect the release of neurotransmitter molecules into synapses, feedback mechanisms cause compensatory changes in the rate of impulse flow along catecholaminergic neurons. This altered impulse flow causes, in turn, parallel alterations in tyrosine hydroxylase activity.

Choline availability also controls brain acetylcholine synthesis; hence, normally occurring fluctuations in dietary choline intake can cause significant variations in brain acetylcholine levels.

REFERENCES

Albrecht, P., Visscher, M.B., Bittner, J.J. and Halberg, F. (1956) Daily changes in 5-hydroxytryptamine concentration in mouse brain. *Proc. Soc. exp. Biol. (N.Y.)*, 92: 703–706.

Andén, N.-E. and Modigh, K. (1972) Effects of *p*-chlorophenylalanine and an M.A.O. inhibitor on the serotonin in the spinal cord after transection. *J. Neural Transm.*, 33: 211–222.

Ansell, G.B. and Spanner, S. (1971) Studies on the origin of choline in the brain of the rat. *Biochem. J.*, 122: 741–750.

Blasberg, R. and Lajtha, A. (1965) Substrate specificity of steady-state amino acid transport in mouse brain slices. *Arch. Biochem. Biophys.*, 112: 361–377.

Cohen, E.L. and Wurtman, R.J. (1975) Brain acetylcholine: increase after choline administration. *Life Sci.*, 16: 1095–1102.

Cohen, E.L., Unger, M., Verbiese, N. and Wurtman, R.J. (1975) Abstract presented at the Meeting of Soc. of Neurosci., New York, November, 1975.

Colmenares, J.L., Wurtman, R.J. and Fernstrom, J.D. (1975) Effects of ingesting a carbohydrate-fat meal on the levels and synthesis of 5-hydroxyindoles in various regions of the rat central nervous system. *J. Neurochem.*, 25: 825–829.

Coyle, J.T. (1972) Tyrosine hydroxylase in rat brain: cofactor requirements, regional and subcellular distribution. *Biochem. Pharmacol.*, 21: 1935–1944.

Davis, K.L., Berger, P.A. and Hollister, L.E. (1975) Choline for tardive dyskinesia. *New Engl. J. Med.*, 293: 152.

Fernstrom, J.D. and Wurtman, R.J. (1971a) Brain serotonin content: physiological dependence on plasma tryptophan levels. *Science*, 173: 149–152.
Fernstrom, J.D. and Wurtman, R.J. (1971b) Brain serotonin content: increase following ingestion of carbohydrate diet. *Science*, 174: 1023–1025.
Fernstrom, J.D. and Wurtman, R.J. (1972a) Elevation of plasma tryptophan by insulin in the rat. *Metabolism*, 21: 337–342.
Fernstrom, J.D. and Wurtman, R.J. (1972b) Brain serotonin content: physiological regulation by plasma neutral amino acids. *Science*, 178: 414–416.
Fernstrom, J.D. and Wurtman, R.J. (1974) Nutrition and the brain. *Scient. Amer.*, 230: 84–91.
Fernstrom, J.D., Larin, F. and Wurtman, R.J. (1971) Daily variations in the concentrations of individual amino acids in rat plasma. *Life Sci.*, 10: 813–819.
Fernstrom, J.D., Larin, F. and Wurtman, R.J. (1973) Correlations between brain tryptophan and plasma neutral amino acids following food consumption in rats. *Life Sci.*, 13: 517–524.
Fishman, B., Wurtman, R.J. and Munro, H.N. (1969) Daily rhythms in hepatic polysome profiles and tyrosine transaminase activity: role of dietary protein. *Proc. nat. Acad. Sci. (Wash.)*, 64: 677–682.
Guroff, G. and Udenfriend, S. (1962) Studies on aromatic amino acid uptake by rat brain in vivo. *J. biol. Chem.*, 237: 803–806.
Hebb, C. (1972) Biosynthesis of acetylcholine in nervous tissue. *Physiol. Rev.*, 52: 918–957.
Hökfelt, T., Fuxe, K., Goldstein, M. and Johansson, O. (1974) Immunohistochemical evidence for the existence of adrenaline neurons in the rat brain. *Brain Res.*, 66: 235–251.
Ikeda, M., Spector, S., Sjoerdsma, A. and Udenfriend, S. (1966) A kinetic study of bovine adrenal tyrosine hydroxylase. *J. biol. Chem.*, 241: 4452–4456.
Jacoby, J., Colmenares, J.L. and Wurtman, R.J. (1975) Failure of decreased 5-hydroxytryptamine uptake or monoamine oxidase inhibition to block the acceleration in brain 5-hydroxyindole synthesis that follows food consumption. *J. Neural Transm.*, 37: 25–32.
Joh, T.H., Geghman, C. and Reis, D. (1973) Immunochemical demonstration of increased accumulation of tyrosine hydroxylase protein in sympathetic ganglia and adrenal medulla elicited by reserpine. *Proc. nat. Acad. Sci. (Wash.)*, 70: 2767–2771.
Knott, P.J. and Curzon, G. (1972) Free tryptophan in plasma and brain tryptophan metabolism. *Nature (Lond.)*, 239: 452–453.
Lindqvist, M., Kehr, W. and Carlsson, A. (1975) Attempts to measure endogenous levels of DOPA and 5-HTP in rat brain. *J. Neural Transm.*, 36: 161–176.
Lipsett, D., Madras, B., Wurtman, R.J. and Munro, H.N. (1973) Serum tryptophan levels after carbohydrate ingestion: selective decline in non-albumin bound tryptophan coincident with reduction in serum free fatty acids. *Life Sci.*, 12: 57–64.
Lovenberg, W., Weissbach, H. and Udenfriend, S. (1962) Aromatic L-amino acid decarboxylase. *J. biol. Chem.*, 237: 89–93.
Lovenberg, W., Jequier, E. and Sjoerdsma, A. (1968) Tryptophan hydroxylase in mammalian systems. *Advanc. Pharmacol.*, 6A: 21–36.
Lytle, L.D., Hurko, O., Romero, J.A., Cottman, K., Leehey, D. and Wurtman, R.J. (1972) The effects of 6-hydroxydopamine pretreatment on the accumulation of DOPA, and dopamine in brain and peripheral organs following L-DOPA administration. *J. Neural Transm.*, 33: 63–71.
Lytle, L.D., Messing, R.B., Fischer, L. and Phebus, L. (1975) The effects of long term corn consumption on brain serotonin and the response to electric shock. *Science*, 190: 692–694.
Madras, B.K., Cohen, E.L., Fernstrom, J.D., Larin, F., Munro, H.N. and Wurtman, R.J. (1973) Dietary carbohydrate increases brain tryptophan and decreases serum free tryptophan. *Nature (Lond.)*, 244: 34–35.
Madras, B.K., Cohen, E.L., Messing, R., Munro, H.N. and Wurtman, R.J. (1974) Relevance of serum free tryptophan to tissue tryptophan concentration. *Metabolism*, 23: 1107–1116.
Marliss, E.B., Aoki, T.T., Unger, R.H., Soeldner, J.S. and Cahill, G.F. (1970) Glucagon levels and metabolic effects in fasting man. *J. clin. Invest.*, 49: 2256–2270.

McMenamy, R.H. and Oncley, J.L. (1957) Specific binding of tryptophan to serum albumin. *J. biol. Chem.*, 233: 1436–1447.

Moir, A.T.B. and Eccleston, D. (1968) The effects of precursor loading on the cerebral metabolism of 5-hydroxyindoles. *J. Neurochem.*, 15: 1093–1108.

Mueller, R.A., Thoenen, H. and Axelrod, J. (1969) Adrenal tyrosine hydroxylase: compensatory increase in activity after chemical sympathectomy. *Science*, 163: 468–469.

Nagatsu, T., Levitt, M. and Udenfriend, S. (1964) Tyrosine hydroxylase. The initial step in norepinephrine biosynthesis. *J. biol. Chem.*, 239: 2910–2917.

Nomura, M., Fernstrom, J.D., Hammarstrom, B., Munro, H.N., Rand, W. and Wurtman, R.J. (1975) Abstract presented at the Meeting of Soc. for Neurosci., New York, November, 1975.

Pohorecky, L.A., Zigmond, M.J., Karten, H. and Wurtman, R.J. (1969) Enzymatic conversion of norepinephrine to epinephrine by the brain. *J. Pharmacol. exp. Ther.*, 165: 190–195.

Poillon, W.N. (1971) Kinetic properties of brain tyrosine hydroxylase and its partial purification by affinity chromatography. *Biochem. biophys. Res. Commun.*, 44: 64–70.

Romero, J.A., Lytle, L.D., Ordonez, L.A. and Wurtman, R.J. (1973) Effects of L-DOPA administration on the concentrations of DOPA, dopamine, and norepinephrine in various rat tissues. *J. Pharmacol. exp. Ther.*, 184: 67–72.

Roth, R.H., Salzman, P.M. and Morgenroth, V.H., III (1974) Noradrenergic neurons: allosteric activation of hippocampal tyrosine hydroxylase by stimulation of the locus coeruleus. *Biochem. Pharmacol.*, 23: 2779–2784.

Schuberth, J. and Jenden, D.J. (1975) Transport of choline from plasma to cerebrospinal fluid in the rabbit with reference to the origin of choline and to acetylcholine metabolism in brain. *Brain Res.*, 84: 245–256.

Shiman, R., Akino, M. and Kaufman, S. (1971) Solubilization and partial purification of tyrosine hydroxylase from bovine adrenal medulla. *J. biol. Chem.*, 246: 1330–1340.

Weiner, N. (1970) Regulation of norepinephrine biosynthesis. *Ann. Rev. Pharmacol.*, 10: 273–290.

Wolstenholme, G.E.W. (Ed.) (1974) *CIBA Foundation Symposium on Aromatic Amino Acids in the Brain*, Churchill, London.

Wurtman, R.J. (1970) Diurnal rhythms in mammalian protein metabolism. In *Mammalian Protein Metabolism, Vol. 4*, H.N. Munro (Ed.), Academic Press, New York, Ch. 36.

Wurtman, R.J. and Fernstrom, J.D. (1972) L-Tryptophan, L-tyrosine, and the control of brain monoamine biosynthesis. In *Perspectives in Neuropharmacology*, S.H. Snyder (Ed.), Oxford University Press, New York, pp. 143–193.

Wurtman, R.J. and Fernstrom, J.D. (1975) Control of brain monoamine synthesis by diet and plasma amino acids. *Amer. J. clin. Nutr.*, 28: 638–647.

Wurtman, R.J., Rose, C.M., Chou, C. and Larin, F. (1968) Daily rhythms in the concentrations of various amino acids in human plasma. *New Engl. J. Med.*, 270: 171–175.

Wurtman, R.J., Larin, F., Mostafapour, S. and Fernstrom, J.D. (1974) Brain catechol synthesis: control by brain tyrosine concentration. *Science*, 185: 183–184.

Zivkovic, B., Guidotti, A. and Costa, E. (1973) Increase of tryptophan hydroxylase activity elicited by reserpine. *Brain Res.*, 57: 522–526.

Chapter 9

Neurotransmitter interactions with neuroendocrine tissue

CLAUDE KORDON, JACQUES EPELBAUM, ALAIN ENJALBERT and JEFFREY McKELVY

Unité de Neurobiologie de l'INSERM, 2 ter rue d'Alésia, 75014 Paris (France) and Department of Anatomy, University of Connecticut Health Center, Farmington, Conn. 06032 (U.S.A.)

I. INTRODUCTION

The regulation of gonadotropin-releasing neurohormones depends upon a great number of neural inputs, which carry humoral, sensory, somesthetic and behavioral information. These inputs traverse a highly hierarchized neuronal circuitry, and are integrated at the level of neurosecretory effector neurons, which represent the "final common pathway" of neuroendocrine regulating structures.

This circuitry involves several distinct regulatory levels, which are represented schematically in Fig. 1. *Effector* neurons, like luteinizing hormone-releasing hormone (LH-RH) producing cells, receive inputs from *integrator* structures, i.e., groups of neurons which respond to a variety of signals originating outside the hypothalamus, check their relevancy to a given neuroendocrine situation, and make the final decision of activating or not activating effector neurons. In the case of cyclic regulation of gonadotropic secretion, most data based on lesion (Barraclough, 1963; Flerko, 1963; Kordon, 1967), stimulation (Everett and Radford, 1961) as well as electrophysiological experiments (Terasawa and Sawyer, 1969) suggest that such an integrator structure lies within the preoptic area of the hypothalamus.

The inputs to integrator structures can be categorized in different classes: *humoral inputs*, as for instance those involved in steroid feedback processes, and interacting with specific hormone receptors located at various hypothalamic levels (Stumpf, 1968; Warembourg, 1970). The next category comprises *neural inputs*, which also can be subdivided in at least 3 classes:

(a) Specific somesthetic and sensory afferent pathways, like those transmitting information relative to the visual or the olfactory environment of the animal. Such information involves, for instance, retinohypothalamic fibers terminating in the hypothalamic supraoptic nucleus; limbicohypothalamic connections traveling through the stria terminalis, as well as the bundle of Schutz, which relays neuroendocrine reflexes such as the suckling reflex.

(b) Fibers projecting to the anteromedial part of the preoptic area, and which establish connections with postsynaptic elements according to a sexually differentiated pattern (Raisman and Field, 1971). These fibers may have a role in modulating rhythmic activation of gonadotropic secretion in the female, although direct evidence to characterize their precise function is still lacking.

(c) Diffuse projections, mainly originating in the midbrain. These are represented by specialized ascending projections and contain a particular variety of neurotransmitters referred to as monoamines.

Monoaminergic projections can affect neuroendocrine regulating structures in a variety of ways, which seem to depend upon the site of their interaction in this complex neuronal circuitry.

Monoamine-containing nerve endings appear highly concentrated in most hypothalamic structures important for processing neuroendocrine information. The anterior hypothalamus, which includes the sexually differentiated area, is rich in noradrenaline (NA) terminals; the suprachiasmatic nucleus, which plays a very important role in regulating autonomous and endocrine (Barraclough, 1963; Kordon, 1967) cyclic processes, has very high concentrations of 5-hydroxytryptamine (5-HT) (Brownstein et al., 1974); the arcuate nucleus, which is likely to contain a large number of hypophysiotropic neurosecretory effector neurons, has large amounts of dopamine (DA) and NA (Palkovits et al., 1974a). Finally, the median eminence, where peptidergic neurons are articulated onto the hypothalamo-hypophyseal portal system, is extremely rich in DA terminals (Fuxe and Hökfelt, 1966; Palkovits et al., 1974a), and the distribution of LH-RH and DA-containing endings shows extensive areas of overlap.

It follows that monoaminergic regulation has a different physiological significance, according to whether it interacts directly with integrator or effector structures, or merely with the transmission of external inputs to these structures (Fig. 1).

We will present here some examples of these interactions, describing separately aminergic effects on the preoptic integrator structures and on the inputs which modulate their activity. We will then review arguments supporting a possible involvement of neurotransmitters in the area of the anterior hypothalamus connected with sexual differentiation of gonadotropic regulation. Finally we will envisage those aminergic inputs which affect directly peptidergic neurons, and review recent developments which may make it possible in the near future to study the mechanisms of amine—hormone interactions at the cellular level, and possibly distinguish separate effects on the synthesis or the release of hypothalamic neurohormones.

II. AMINE METABOLISM AND ACTIVITY OF INTEGRATOR STRUCTURES

As stated before, several experiments suggest that the preoptic area is the main structure integrating neural information relevant for gonadotropic control. Interestingly enough, endogenous levels of NA in this region (Stefano and Donoso, 1967; Donoso and Cukier, 1968) as well as the rate of incorporation of [^{3}H]tyrosine into catecholamines (Fig. 2), more precisely into NA (Donoso et al., 1969), are influenced by the endocrine condition of the animal. These effects are limited to the anterior hypothalamus and seem to involve a modulation of the activity of enzymes involved in the biosynthesis of catecholamines (Kizer et al., 1974).

The precise role of NA in the function of anterior hypothalamic regulating structures

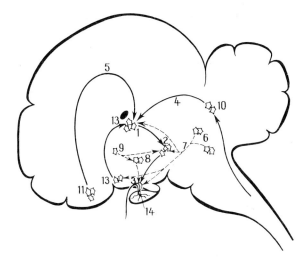

Fig. 1. Schematic representation of the various regulatory structures and afferents involved in neuroendocrine control. *Integrator structures*: 1, preoptic area of the anterior hypothalamus. *Effector structures*: 2, releasing hormone producing neurons; 3, neurovascular junctions of the median eminence zona externa. *Afferent pathways of neuroendocrine information*: 4, bundle of Schütz; 5, stria terminalis. *Monoaminergic projections involved in neuroendocrine control*: 6, mesencephalic nuclei with noradrenergic and serotoninergic cell bodies; 7, medial forebrain bundle; 8, tuberoinfundibular dopaminergic tract; 9, anterior hypothalamic catecholaminergic cell bodies. *Other relevant structures:* 10, mesencephalic projections of spinal ascending pathways involved in neuroendocrine reflexes; 11, amygdalar nuclei; 13, suprachiasmatic nucleus and its proposed projections to the preoptic area: anterior commissure in black; 14, hypophysial portal system.

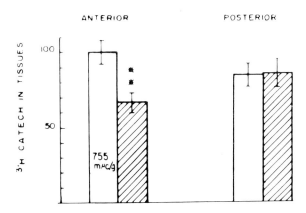

Fig. 2. In vitro incorporation of [^{3}H]tyrosine into catecholamines in slices of anterior or posterior hypothalamic tissue. The hypothalamus was collected in either proestrus (open bars) or estrus (hatched bars); ** $P < 0.01$.

is still unclear. Changes in the metabolism of the transmitter could theoretically translate changes in the neuronal activity which triggers LH-RH producing neurons, like that involved in the progesterone-induced facilitation of LH release (Kalra et al., 1972); however, they also could reflect overall changes in peripheral endocrine balance, as for instance variations in circulating steroid levels.

III. AMINE INTERACTIONS WITH SPECIFIC INPUTS AFFECTING NEUROENDOCRINE CONTROL

III.1. Cholinergic influences upon afferent information to the anterior and mediobasal hypothalamus

A positive, cholinergic regulation of the phasic release of LH and of ovulation has first been suspected by Sawyer et al. (1955), who had shown that atropine did block ovulation when administered at an appropriate stage of the estrous cycle. More recently, Lichtensteiger (1973) observed that atropine was also able to block the effect of electrical stimulation of the preoptic area on the release of both tuberoinfundibular dopamine and LH. This experiment suggests that a cholinergic component may be involved in the transfer of gonadotropic releasing information to dopaminergic and/or LH-RH-containing neurons.

A participation of acetylcholine in the transmission to the hypothalamus of inputs originating in amygdalar nuclei (Velasco and Taleisnik, 1969) has also been postulated. This type of mechanism could also account for the data on cholinergic dependent release of LH (Fiorindo et al., 1975).

III.2. Effects of serotonin in the phasic release of FSH and LH

The occurrence of well timed discharge of FSH and LH during the estrous cycle of the rat (Everett and Sawyer, 1950) seems to result from interaction of two distinct levels of regulation. The progressive increase in the amount of circulating steroids, particularly estradiol, which occurs between diestrous and proestrous (Schwartz, 1969) sensitizes the pituitary to LH-RH, and, thus, amplifies the release of FSH and LH to a given amount of endogenous neurohormone. In addition, a circadian trigger synchronizes the actual release of the hormone within a well timed interval of the photoperiod, as originally demonstrated by Everett and Sawyer (1950). An experimental preparation permitting elimination of one of these levels of fluctuation, in order to study more accurately the other, has recently been devised; it consists of stabilizing circulating concentrations of estradiol by implanting solid sources of the hormone to castrated animals (Chazal et al., 1974; Legan et al., 1975). Under these conditions, pituitary sensitivity no longer varies as a function of time (Chazal et al., 1974; Héry et al., 1975a), it remains elevated and, consequently, the release of FSH and LH exhibits a very clear-cut circadian cyclicity (Héry et al., 1975a).

This cyclicity can be assumed to reflect activation of a neural trigger activating LH release and synchronized by the light–dark cycle (Chazal et al., 1974).

Inhibition of seroninergic synaptic transmission blocks this circadian gonadotropic release (Héry et al., 1976) as it inhibits cyclic ovulation (Héry et al., 1975b). This effect of the amine has been shown to be specific. It can be induced by either pharmacological inhibition of synthesis of the amine (Héry et al., 1976) or by destruction of the cell bodies of origin of 5-HT ascending projections (Kordon et al., 1976). Reestablishment of normal endogenous levels of the amine after synthesis inhibition induces a parallel restoration of the phasic release of FSH and LH. Under all these circumstances, the amplitude of the circadian hormone fluctuation shows a good correlation with the hypothalamic concentration of the main metabolite of 5-HT, 5-hydroxyindole acetic acid (5-HIAA), which has been shown to be a good index of serotonin turnover (Héry et al., 1972).

It thus seems that 5-HT plays a role in modulating the transmission of a phasic neural information to LH-RH neurons. The precise site of this interaction in the brain is still unknown; however, indirect data suggest that this effect could involve the suprachiasmatic nucleus. Discrete lesions of this nucleus inhibit cyclic ovulation (Barraclough, 1963; Flerko, 1963; Kordon, 1967) as well as circadian LH release (Héry, unpublished observation); in the latter case, the effect of the lesion is very similar to that of 5-HT depletion.

III.3. Effects of serotonin on the suckling-induced release of prolactin in lactating animals

Neural inputs generated by suckling reach the hypothalamus via two distinct pathways, the bundle of Schutz and the medial forebrain bundle (Averill and Purves, 1966). In lactating animals, they induce a very rapid rise in plasma prolactin levels (Terkel et al., 1972; Grosvenor and Withworth, 1974), associated with a concomitant depletion in pituitary prolactin store (Grosvenor and Withworth, 1974).

Serotonin synthesis inhibition blocks the prolactin response to suckling (Kordon et al., 1973). Subsequent administration of the precursor of the amine to animals treated with the synthesis inhibitor induces a parallel restoration of endogenous concentrations of the neurotransmitter and of the ability of the animal to release prolactin as a response to suckling (Kordon et al., 1973). Activation of serotoninergic neurons for instance, by administration of the precursor, does not per se induce a significant release of prolactin, if the suckling stimulus is not concomitantly applied. This suggests that serotonin-containing neurons do not represent the main trigger of prolactin release, but that they merely act by sensitizing hypothalamic effector neurons to inputs generated by suckling.

This modulatory effect of 5-HT is likely to be of physiological importance, since suckling effectively elicits an increased release of the amine within the hypothalamus (Mena et al., 1976). This increased release appears correlated with the prolactin response, because, under weaning conditions, the same stimulus is no longer able to induce release of either hypothalamic serotonin or pituitary prolactin (Mena et al., 1976).

It is thus tempting to speculate that, under these conditions, serotoninergic fibers traveling through the medial forebrain bundle, which represents the main pathway of aminergic innervation of the hypothalamus (Fuxe et al., 1970), play a role in the sensitization of the hypothalamus towards the suckling reflex. Progressive deactivation of these fibers during weaning could thus be the cause of the non-responsive state which terminates lactation. In addition, such lactation-induced changes in the aminergic regulation of the mediobasal hypothalamus could also account for the well known ovulation blocking effect of lactation, in view of the effects of 5-HT on the release of LH-RH which we will review in Section V below.

IV. AMINERGIC CORRELATES OF SEX DIFFERENTIATION

The sexual difference in gonadotropic secretory pattern has been shown not to depend upon pituitary mechanisms (Harris, 1955) but upon an irreversible imprinting of the hypothalamus. Such imprinting appears correlated with the differentiation of a few synapses located in a discrete anterior hypothalamic area (Raisman and Field, 1971). The fact that differentiation in the male or the female sense can be elicited in either sex, provided the animal is exposed at the critical time to the appropriate hormonal environment (Raisman and Field, 1971), indicates that this ontogenetical process is not under direct genetic control; it rather seems that at a given stage, hormones can interfere with the genetically programmed pattern of synaptic articulation.

This sex dependent pattern of synaptic connections seems to have *biochemical* and *physiological* implications (Fig. 3). Depending upon the presence of synapses differentiated according to the female or the male mode, neurons of this area do or do not acquire the capacity to produce specific estradiol receptor proteins; this results in a relatively lower capacity to accumulate and to store [^{3}H]estradiol in the male than in the female (Vertes and King, 1971). This receptor deficit in the male seems limited to the anterior hypothalamus, since estrogen receptors in other areas, including the posterior

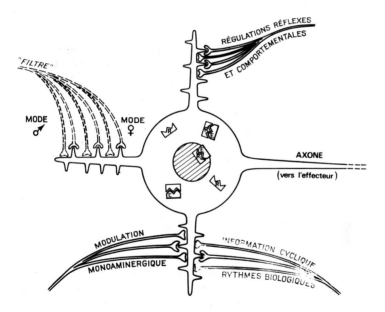

Fig. 3. Schematical representation of biochemical and psysiological correlates to the pattern of synaptic articulation found in the "sexually differentiated zone" of male or female rats. Depending upon the type of articulation (preferentially on dendritic spines in the female and on dendritic shafts in the male), postsynaptic neurons exhibit a varying degree of specific estradiol receptor synthesis. Parallely, 5-HT concentrations in this area are affected during development, and the ability of the anterior hypothalamus to activate rhythmically LH-RH release (a 5-HT dependent process) does or does not appear. See text for explanations and references.

hypothalamus, are of comparable magnitude in both sexes.

The *physiological* implication of the sexual dimorphism exhibited by preoptic synapses has been recognized long ago (Pfeiffer, 1936). It is represented by the inability of the male differentiated pattern to activate gonadotropic release in a rhythmic way (Barraclough, 1963), as compared with the phasic pattern of the female type.

Interestingly, the differentiation of preoptic synapses according to the male or the female type induces correlative variations in endogenous 5-HT concentrations in the anterior hypothalamus. In particular, 5-HT levels are higher between day 10 and day 20 in the female than in the male (Ladosky and Wandscheer, 1975); females adrogenized after birth, in which differentiation of the "female" morphological and physiological pattern is inhibited, show a male-like evolution of 5-HT concentrations (Ladosky and Wandscheer, 1975).

The physiological significance of such differences is still unclear. It should be noted, however, that they parallel the different pattern of gonadotropin secretion which is observed in male and female rats at the same time of postnatal development (McCann and Ramirez, 1964). In addition, 5-HT has been shown to play a role in the transfer of cyclic neural information to gonadotropin release regulating structures, as will be described in the next section. A different regulation of anterior hypothalamic content of the transmitter could thus be involved in the ontogenesis of 5-HT dependent cyclic regulation. This example of amine–hormone interaction, summarized in Fig. 3, may prove to be a great prospective interest for studying the morphogenetical induction of neuroendocrine regulating mechanisms but such analysis has not yet been undertaken at the cellular level.

V. EFFECTS OF MONOAMINES ON NEUROHORMONE-CONTAINING NEURONS

Dopamine has been shown to facilitate the release of LH (Kordon and Glowinski, 1969; Kamberi et al., 1970; Schneider and McCann, 1970). Micropharmacological studies have shown that the site of this effect lies within the arcuate-median eminence region (Kordon, 1969). This conclusion is in good agreement with previous findings that activity of dopaminergic tuberoinfundibular neurons (Fuxe and Hökfelt, 1966) is well correlated with changes in gonadotropic secretion (Lichtensteiger, 1969). However, this effect is still challenged by some authors (see the chapter by Fuxe in the present volume) on the basis of pharmacological experiments and histofluorescence data. These contradictions may be explained by possible differential effects of DA on LH-RH release under different endocrine conditions; in particular, it is possible that DA does not affect basal control and induced release of gonadotropins in the same way. Conclusive demonstration of the precise effect of the amine will only be made possible by cellular approaches, such as those described below in this chapter.

Dopamine also exhibits a clear inhibitory effect on the release of prolactin (Lu and Meites, 1971). This involves at least partially a direct action on the pituitary itself (MacLeod and Lehmeyer, 1974), where specific receptors of the amine have been demonstrated (Swalstig et al., 1974). In addition, the turnover of tuberoinfundibular DA is affected by prolactin (Hökfelt and Fuxe, 1972); this represents a "short" feedback loop.

However, DA is most likely not the only prolactin inhibiting factor present in the hypothalamus, since in vitro and in vivo inhibition of prolactin release can still be elicited from catecholamine free extracts of median eminence (see below).

The inhibitory action of 5-HT on gonadotropic release (Kordon et al., 1968; Kamberi et al., 1971) is also likely to affect directly the activity or the release of LH-RH-containing neurons; it seems to require changes in 5-HT synaptic levels within the arcuate-median eminence only (Kordon, 1969).

VI. APPROACHES TO THE CELLULAR MECHANISMS UNDERLYING AMINE—HORMONE INTERACTIONS

As stated before, only separate evaluations of neurotransmitter effects on synthesis, axonal transport or release of neurohormones into the portal system will ultimately permit an understanding of their mode of action, and the settling of remaining contradictions (such as the conflicting reports of a stimulatory vs. an inhibitory role of DA in LH secretion). Such studies require a better knowledge of the cellular and subcellular distribution of neurohormones, and of the capacity of isolated neurosecretory endings to retain functional properties. Some preliminary studies related to this will be described now.

VI.1. Subcellular distribution of neurohormones in neuroendocrine tissue

Regional analysis of LH-RH content in the hypothalamus (McCann, 1962; Palkovits et al., 1974), as well as cytoimmunochemical evidence (see review in Zimmerman, this volume) make it clear that most, if not all, of the active peptide is concentrated within the mediobasal hypothalamus. Cytoimmunochemistry also suggests, in the rat, that most of the material is present at the level of peptidergic nerve endings (Kordon et al., 1974; Baker et al., 1975).

Experiments of complete deafferentation of the mediobasal hypothalamus have not revealed a clear-cut picture of the distribution of LH-RH cell bodies, although they are strongly suggestive of the existence of two separate neuron systems contributing fibers to the median eminence and to the lamina terminalis (Weiner et al., 1975).

It was tempting to take advantage of the fractionation techniques originally devised by Whittaker (1969), which permit the separation of nerve endings on the basis of physical properties of presynaptic membranes. Under appropriate homogenization conditions, endings form spherical, artefactual entities which trap the content of nerve terminals and can be separated by differential and density gradient centrifugation.

This technique was first applied to neuroendocrine tissue by Clemente et al. (1971) and has been used since by a number of authors (Shin et al., 1974; Ramirez et al., 1975; Taber and Karavolas, 1975). We will describe here the data concerning the distribution of both LH-RH and prolactin inhibiting activities in the fractions which can be separated on sucrose gradients at the subcellular level.

VI.1.1. Distribution of LH-RH. The combination of two methods, a biological assay highly sensitive (Ramirez and McCann, 1963) and a specific radioimmunoassay of the decapeptide published by Kerdelhué et al. (1973), were used to test LH-RH distribution. Well correlated data were only obtained after proper extraction of the fractions (Ramirez et al., 1975) (Table I); electron micrographs and the cytoplasmic marker enzyme lactate dehydrogenase (LDH) were used to check the adequacy of each fraction; this proved important since satisfactory fractionation of medial basal hypothalamic (MBH) tissue necessitated slight changes in the centrifugation parameters originally described for other brain structures.

After low speed centrifugation in 0.32 M sucrose at $0°C$, to sediment nuclei and large membrane debris, almost all activity is recovered from the supernatant fraction (S1); only 8% of the total is measured in the nuclear pellet P1 (Table II). This is most likely due to contamination of the nuclear pellet, since the contamination of this fraction by the cytoplasmic marker enzyme amounts to the same proportion (Table II).

When S1 is further fractionated at high speed, 69% of its radioimmunoassayable activity and 60% of its bioassayable activity is recovered in the crude mitochondrial pellet (P2) (Table II). The microsomal supernatant (S2) has no assayable activity; in a radioimmunoassay system, it contains only 2% of the original activity of S1.

Recovery studies with exogenous LH-RH have shown that the absence of the peptide in the S2 fraction could not be accounted for by enzymatic degradation, which does not affect significantly the concentration of the peptide under the conditions and the duration of the fractionation procedure (which is carried out at $0°C$).

When the crude mitochondrial fraction is further purified on a sucrose gradient, LH-RH and LDH activities show a well correlated distribution in each fraction (myelinic fraction A, synaptosomal fraction B and mitochondrial fraction C) (Fig. 4). About two-thirds of both activities are recovered in the B interface. Electron microscopic examination of the fractions shows that most nerve endings are located in the B fraction; however, the C pellet shows a certain degree of synaptosomal contamination, which is likely to account for the amounts of LDH and LH-RH recovered from it.

PIF activity was assayed in vitro and in vivo (inhibition of prolactin release in lactating, urethane anesthetized female rats under conditions of electrical stimulation of mammary efferent nerves). A PIF activity was distributed evenly in both the mitochondrial pellet and the microsomal supernatant. However, in view of the presence of DA in both fractions, and of the reports quoted above that the amine has direct prolactin inhibiting

TABLE I

Influence of extraction conditions on LH-RH activity in the crude mitochondrial fraction

Extraction medium	Fractions	LH-RH activity (ng/hypothalamus)
Sucrose (0.32 M)	P2	0.58 ± 0.05 (3)
Sucrose (0.32 M) after osmotic shock	P2	2.7 ± 0.5 (2)
Sucrose (0.32 M + 0.1 N HCl)	P2	2.58 ± 0.15 (2)

effects on the pituitary itself, crude hypothalamic extracts, as well as mitochondrial and microsomal fractions, were submitted to extraction on alumina in order to eliminate catecholamines. The alumina eluted fractions were then assayed for PIF activity. Under these conditions, prolactin inhibiting activity was lost from the supernatant fraction in both assay systems. In the total homogenate and the crude mitochondrial fractions, however, a significant PIF activity was retained. This is strongly suggestive that a prolactin inhibitory substance different from DA is present in the mediobasal hypothalamus (Enjalbert, unpublished observations). This conclusion is consistent with recent data demonstrating PIF activity in non-aminergic fractions obtained after chromatographic purification of median eminence extracts (Takahara et al., 1974).

The distribution of thyrotropin-releasing hormone (TRH) (Barnea et al., 1975; McKelvy et al., 1975) and corticotropin-releasing hormone (CRH) (Epelbaum et al., 1975) is also preferentially synaptosomal. It thus seems that, with the exception of very low contaminating values, hypophysiotropic factors are only present in fractions derived from nerve endings under subcellular fractionation conditions. Moreover, this conclusion is consistent with cytoimmunochemical data obtained at the electron microscopic level (Pelletier et al., 1974; Goldsmith and Ganong, 1975); these authors reported that immunoreactive material is associated with electron dense material in nerve endings.

VI.2. In vitro release of neurohormones from nerve endings

Joint incubation of median eminence fragments with pituitary tissue has been widely used in the past as an indirect approach to study neurohormone release (Guillemin and Rosemberg, 1955; Moszkowska, 1959). The existence of sensitive radioimmunoassay methods for these hormones has made direct approaches now possible. In particular, it is of interest to study amine effects on the in vitro release from synaptosomal preparations,

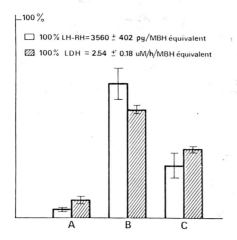

Fig. 4. Distribution of LH-RH and LDH after fractionation of the crude mitochondrial pellet (P2) on a discontinuous sucrose gradient. (LDH in μmoles $NADH_2$/hr/mg medial basal hypothalamic (MBH) equivalent ± S.E.M.; LH-RH in pg/MBH equivalents ± S.E.M.)

TABLE II

Subcellular distribution of LH-RH in the mediobasal hypothalamus

Fraction	Protein (µg/mg hypothalamic equiv.)	LDH activity (µmoles NADH$_2$/hr/hypothalamic equiv.)	LH-RH (ng/hypothalamic equiv.)		
			Radioimmunoassay HCl extraction		Biological activity sucrose (0.32 M)
Homogenate (H)	89.74 ± 7.29	115.00 ± 10.00	5.10 ± 0.30 (6)	5.50 ± 0.9 (3)	0.54 ± 0.30 (2)
S1	66.44 ± 9.40	97.35 ± 10.55	4.80 ± 0.70 (4)	5.00 ± 0.57 (3)	0.41 ± 0.30 (6)
P1	16.20 ± 7.20	7.27 ± 1.52	0.40 ± 0.02 (4)	—	0.63 ± 0.40 (2)
S2	23.30 ± 2.70	50.15 ± 7.61	0.10 ± 0.01 (4)	N.D. (6)	N.D. (6)
P2	42.40 ± 6.10	40.35 ± 9.70	3.30 ± 0.60 (10)	2.90 ± 0.62 (4)	—

TABLE III

Effect of K^+ on the liberation of LH-RH

Locke	LH-RH (pg/hypothalamic equivalent)					
	MBH			P2		
	Tissue	% of control	Medium	Tissue	% of control	Medium % of control
K^+ (5 mM)	4710 ± 316 (4)	100	—	2881 ± 336 (4)	100	675 ± 50 (4) 100
K^+ (55 mM)	2696 ± 638 (4)	57	—	1974 ± 76 (4)	68	860 ± 35 (4) 128

since this will permit distinguishing between perikaryal and presynaptic effects of the substances to be tested.

Membrane depolarization by high K^+ concentrations induces a 30% depletion of tissue LH-RH. The extent of this depletion is identical when the whole median eminence, the low speed supernatant (S1) or the crude mitochondrial fraction containing the synaptosomes (P2) are placed in the incubation medium (Fig. 5). However, as shown in Table III, only a very small portion of the released hormone is recovered from the incubation medium. This is very likely due to contamination of or active release into the medium of peptidases shown to be very potent inactivators of LH-RH (Griffiths et al., 1974; Koch et al., 1974; Kochman et al., 1975).

Bennett et al. (1975) have recently shown that not only field depolarization, but also addition of DA, to incubates of median eminence synaptosomes could release LH-RH. This observation supports the hypothesis of the LH-RH releasing effect of the amine, which was discussed in an earlier section of this chapter. However, the exact role of such in vitro interactions still has to be further substantiated by demonstrating that specific DA receptors are involved at the synaptosomal level, and by testing LH-RH release under conditions in which most of its concentration in the medium is not degraded by proteolytic enzymes.

VI.3. Biochemical implications of the synaptosomal distribution of neurohormones

As mentioned above, immunocytochemical and direct assay studies on releasing factor localization indicate that the peptides are present in highest concentrations in nerve terminals, predominantly in the median eminence. A hypothesis arising from these findings is that biologically active releasing factors may be generated at their release site by processing of biologically inactive precursor molecules. Such prohormonal forms could be made on perikaryal ribosomes and transported to nerve terminals. A corollary of this

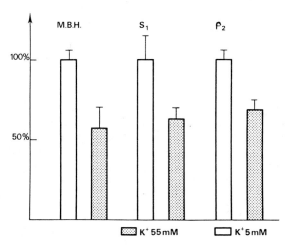

Fig. 5. Effect of potassium ion concentration on the tissue content of LH-RH of MBH and subcellular fractions prepared therefrom. (100% MBH = 4710 ± 316 pg/MBH equivalent; 100% S1 = 3794 ± 601 pg/MBH equivalent; 100% P2 = 2881 ± 336 pg/MBH equivalent, all ± S.E.M.)

hypothesis, of course, is that important regulatory events, mediating the extent of precursor processing, and therefore levels biologically active hormone, could take place in nerve terminals of the peptidergic neurons. Such a process would allow for rapid modulation of releasing factor production at the median eminence level in response to tuberoinfundibular dopaminergic and ascending monoaminergic influences, independent over short time periods of axonal transport. Interactions of these monoamines with peptidergic nerve terminals could occur via axo-axonic synapses (Fuxe and Hökfelt, 1966) and/or by the diffusion of monoamines released in the vicinity of peptidergic neurons to receptive sites on or in these cells. The above hypothesis is not inconsistent with our current knowledge regarding hypothalamic releasing factor biosynthesis. While a claim has been made that TRH biosynthesis in the rat occurs via a soluble enzymatic mechanism uninfluenced by antibiotic inhibitors of cytoribosomal protein synthesis or by ribonuclease (Mitnick and Reichlin, 1972), it has recently been demonstrated that TRH biosynthesis in the newt is completely abolished by ribonuclease treatment (McKelvy and Grimm-Jørgensen, 1975; Grimm-Jørgensen and McKelvy, 1976). While these latter studies also showed that antibiotic inhibitors such as cycloheximide, chloramphenicol and diphtheria toxin had little effect on TRH synthesis, it should be noted that the possibility has not been excluded that relatively small peptides could be made on ribosomes in the presence of such inhibitors, especially when initiation has not been blocked. Studies on the biosynthesis of other releasing factors are less extensive and at present offer no information relative to the mechanism of synthesis. At the present time, therefore, it must be concluded that the mode by which releasing factors are synthesized is an open question.

As a means of testing such a prohormone hypothesis, we have carried out experiments in which we attempted to demonstrate, by labeling techniques, the appearance of radioactive LH-RH after appropriate sequences of in vivo and in vitro isotope exposure. Brief-

TABLE IV

Incorporation of L-[^{3}H] leucine and L-[^{3}H] tyrosine into LH-RH in hypothalami of normal and castrated male rats

Treatment	Counts/min/mg methanol insoluble protein*
Normal males, perfused and incubated in vitro	305
Castrated males, perfused and incubated in vitro	1742
Castrated males perfused and *not* incubated in vitro	676
Normal males, in vitro incubation only	N.D.**
Castrated males, in vitro incubation only	N.D.

* Values corrected for recovery (see text).
** Not detectable.

ly, we infused high specific activity tritium labeled amino acid precursors of LH-RH intraventricularly into normal and castrated male rats, and allowed incorporation of the amino acids to occur in vivo. After sacrifice of the animals, hypothalamus and other tissues were either extracted immediately and subjected to sequential purification, or incubated in vitro prior to extraction. After addition of carrier LH-RH, extracted tissues were subjected to acid alumina chromatography, and cellulose acetate electrophoresis first at pH 4.2 then at pH 9.0, with the radioactivity associated with Pauli positive synthetic LH-RH being collected at each step. An extraction standard consisting of synthetic [^{3}H]LH-RH (kindly provided by Dr. Pierre Fromageot) was carried through the purification sequence, and used to correct for recoveries for each sample. As can be seen in Table IV, in vitro incubation results in an increased yield of biosynthetically labeled LH-RH relative to the condition of no incubation, and the amount of LH-RH associated radioactivity is increased in animals in which a functional demand has been placed on gonadotropin secretion. These results are consistent with the hypothesis of prohormone conversion as part of LRH biosynthesis, but they must be interpreted with caution, since we have not yet unequivocally demonstrated the radiochemical purity of the radioactivity associated with LH-RH obtained with our purification scheme. These results do not provide evidence for prohormone conversion in nerve terminals, since blocks of tissue were used, but future studies, in which synaptosomal fractions are isolated after an in vivo pulse of radioactivity and then subjected to in vitro incubation, may shed light on this problem. This strategy offers hope that a subcellular approach to the metabolism of hypothalamic releasing factors will add useful information relative to an understanding of integrated neuroendocrine activity in reproductive tissues.

VII. CONCLUSIONS

Despite considerable research in recent years, the precise role of given transmitters in the regulation of adenohypophyseal hormones is still largely controversial. Most of our present knowledge is based upon pharmacological or histophysiological experiments. In the former case, administration of drugs affects synaptic release of monoamines in several brain structures at the same time. Under these conditions, only overall effects, resulting from simultaneous modulation of a number of inputs to neurosecretory effector neurons, are likely to be detected. On the other hand, histophysiological experiments involving fluorescence histochemistry do not permit to decide which, from variations in amine content or in pituitary secretion, is the cause or the consequence of the other.

At this stage, it thus seems that the possibility of investigating cellular mechanisms underlying amine—hormone interactions should provide for a better knowledge of the level at which they occur. One can anticipate that most subsisting contradictions on amine effects will be settled by this approach. In addition, a clear distinction of differential amine interactions at various structural levels should allow for the proving or disproving of our hypothesis of an "aminergic coding" of neuroendocrine information. According to this hypothesis, neural impulses triggering the release of a pituitary hormone have to cross a certain number of aminergic "gates". For instance, everything happens as if impulses triggering the phasic release of LH-RH can only reach their target provided (a) the proper noradrenergic tuning occurs within the preoptic area at the appropriate time of the cycle (see Section II above); (b) a serotonin dependent mechanism facilitates circadian synchronization of LH stimulation (Section III.2.); (c) a modulation of the turnover of tuberoinfundibular DA interacts directly with the release of LH-RH (Section V); and, finally, (d) activation of median eminence 5-HT projections does not antagonize LH-RH release. Closing of either of these "gates" by appropriate aminergic tuning prevents the neural triggering of the pituitary. Some aminergic inputs are common to several pituitary hormones (as for example the facilitatory effect of 5-HT which can trigger both gonadotropins and prolactin, as shown in Section III, and may thus explain the simultaneous release of these hormones in the afternoon of proestrus), but others are not; only a given aminergic "message" resulting from synchronized gate openings, could thus control the secretion of one pituitary hormone.

REFERENCES

Averill, R.W.L. and Purves, H.D. (1966) Differential effects of permanent hypothalamic lesions on reproduction and lactation in rats. *J. Endocr.*, 26: 463.

Baker, B.L., Dermody, W.C. and Reel, J.R. (1975) Distribution of gonadotropin releasing hormone in the rat brain as observed with immunocytochemistry. *Endocrinology*, 97: 125–135.

Barnea, A., Ben-Jonathon, N., Colston, C., Johnston, J.M. and Porter, J.C. (1975) Differential sub-cellular compartmentalization of thyrotropin-releasing hormone (TRH) and gonadotropin-releasing hormone (LRH) in hypothalamic tissue. *Proc. nat. Acad. Sci. (Wash.)*, 72: 3163.

Barraclough, C.A. (1963) Secretion and release of FSH and LH: discussion. In *Advances in Neuroendocrinology*, A.V. Nalbandov (Ed.), University of Illinois Press, Urbana, p. 224.

Bennett, G.W., Edwardson, J.A., Holland, D., Jeffcoate, S.L. and White, N. (1975) Release of immunoreactive luteinizing hormone-releasing hormone and thyrotrophin-releasing hormone from hypothalamic synaptosomes. *Nature (Lond.)*, 257: 323.

Brownstein, M.J., Saavedra, J.M., Palkovits, M. and Axelrod, J. (1974) Histamine content of hypothalamic nuclei of the rat. *Brain Res.*, 77: 151–156.

Chazal, G., Faudon, M., Gogan, F. and La Plante, E. (1974) Negative and positive effects of estradiol upon luteinizing hormone secretion in the female rat. *J. Endocr.*, 61: 511.

Clemente, F., Ceccarelli, B., Ceratie, M., De Monte, M., Felici, M., Motta, M. and Pecile, A. (1970) Subcellular localization of neurotransmitters and releasing factors in the rat median eminence. *J. Endocr.*, 48: 205.

Donoso, A. and Cukier, J.O. (1968) Oestrogen as a depressor of noradrenaline concentration in the anterior hypothalamus. *Nature (Lond.)*, 218: 198.

Donoso, A.O., de Gutierrez-Moyano, M.B. and Santolaya, R.C. (1969) Metabolism of noradrenaline in the hypothalamus of castrated rats. *Neuroendocrinology*, 4: 12–19.

Epelbaum, J., Gautron, J.P., Enjalbert, A., McKelvy, J., Pattou, E., Priam, M., Carbonell, L. and Kordon, C. (1975) Subcellular distribution of neurohormones in the mediobasal hypothalamus of the rat. *Abstracts, Europ. Neurosci. Meeting, Munich*, September, 1975.

Everett, J.W. and Radford, H.M. (1961) Irritative deposits from stainless steel electrodes in the preoptic rat brain causing release of pituitary gonadotropin. *Proc. Soc. exp. Biol. (N.Y.)*, 108: 604.

Everett, J.W. and Sawyer, C.H. (1950) A 24 hour periodicity in the LH release apparatus of the female rat, disclosed by barbiturate sedation. *Endocrinology*, 47: 198.

Fiorindo, R., Justo, G., Motta, M., Simonovic, I. and Martini, L. (1975) Acetylcholine and the secretion of pituitary gonadotropins. In *Hypothalamic Hormones,* M. Motta, P.G. Crosignani and L. Martini (Eds.), Academic Press, London, p. 195.

Flerko, B. (1963) Secretion and release of FSH and LH. In *Advances in Neuroendocrinology*, A.V. Nalbandov (Ed.), University of Illinois Press, Urbana, p. 211.

Fuxe, K. and Hökfelt, T. (1966) Further evidence for the existence of tubero-infundibular dopamine neurons. *Acta physiol. scand.*, 66: 243.

Fuxe, K., Hökfelt, T. and Jonsson, G. (1970) Participation of central monoaminergic neurons in the regulation of anterior pituitary secretion. In *Neurochemical Aspects of Hypothalamic Function*, L. Martini and J. Meites (Eds.), Academic Press, New York, pp. 61–88.

Goldsmith, P.C. and Ganong, W.F. (1975) Ultrastructural localization of luteinizing hormone-releasing hormone in the median eminence of the rat. *Brain Res.*, 97: 181–193.

Griffiths, E.C., Hooper, K.C., Jeffcoate, S.L. and Holland, D.T. (1974) Presence of peptidases in the rat hypothalamus inactivating luteinizing hormone releasing hormone (LH-RH). *Acta endocr. (Kbh.)*, 77: 435–441.

Grimm-Jørgensen, Y. and McKelvy, J.F. (1976) Biosynthesis of TRF in a cell free system from newt brain. *Brain Res. Bull.*, in press.

Grosvenor, C.E. and Withworth, N. (1974) Evidence for a steady rate of secretion of prolactin following suckling in the rat. *J. Dairy Sci.*, 57: 900.

Guillemin, R. and Rosemberg, B. (1955) Humoral hypothalamic control of anterior pituitary: a study with combined tissue cultures. *Endocrinology*, 57: 599.

Harris, G.W. (1965) *Neural Control of the Pituitary Gland*, E. Arnold, London.

Héry, F., Rouer, E. and Glowinski, J. (1972) Daily variations of serotonin metabolism in the rat brain. *Brain Res.*, 43: 445–466.

Héry, M., LaPlante, E. and Kordon, C. (1975a) Role of pituitary sensitivity and of adrenal secretion in the effect of serotonin depletion on luteinizing hormone regulation. *J. Endocr.*, 67: 3.

Héry, M., LaPlante, E. et Kordon, C. (1975b) Interactions de la sérotonine cérébrale avec la libération cyclique de LH chez le rat. *Ann. Endocr. (Paris)*, 36: 123.

Héry, M., LaPlante, E. and Kordon, C. (1976) Participation of serotonin in the phasic release of LH: evidence from pharmacological studies. *Endocrinology*, in press.

Höfkelt, T. and Fuxe, K. (1972) Effects of prolactin and ergot alkaloids on the tuberoinfundibular dopamine neurons. *Neuroendocrinology*, 9: 100.

Kalra, P.S., Kalra, S.P., Krulich, L., Fawcett, C.P. and McCann, S.M. (1972) The involvement of norepinephrine in transmission of the stimulatory influence of progesterone on gonadotropin release. *Endocrinology*, 90: 1168–1176.

Kamberi, I.A., Mical, R.S. and Porter, J.C. (1970) Effect of anterior pituitary perfusion and intraventricular injection of catecholamines and indoleamines on LH release. *Endocrinology*, 87: 1.

Kamberi, I.A., Mical, R.S. and Porter, J.C. (1971) Effects of melatonin and serotonin on the release of FSH and prolactin. *Endocrinology*, 88: 1288.

Kerdelhué, B., Jutisz, M., Gillessen, D. and Studer, R.O. (1973) Obtention of antisera against a hypothalamic decapeptide (luteinizing hormone/follicle stimulating hormone releasing hormone) which stimulates the release of pituitary gonadotropins and development of its radioimmunoassay. *Biochim. biophys. Acta (Amst.)*, 297: 540.

Kizer, J.S., Palkovits, M., Zivin, J., Brownstein, M., Saavedra, J.M. and Kopin, I.J. (1974) The effect of endocrinological manipulations on tyrosine hydroxylase and dopamine-β-hydroxylase activities in individual hypothalamic nuclei of the adult male rat. *Endocrinology*, 95: 799–812.

Koch, Y., Baram, T., Chobsieng, P. and Fridkin, M. (1974) Enzymic degradation of LH-RH by hypothalamic tissue. *Biochem. biophys. Res. Commun.*, 61: 95.

Kochman, K., Kerdelhué, B., Zor, U. and Jutisz, M. (1975) Studies of enzymatic degradation of luteinizing hormone-releasing hormone by different tissues. *FEBS Lett.*, 50: 190.

Kordon, C. (1967) Contrôle nerveux du cycle ovarien. *Arch. Anat. micr. Morph. exp.*, 56, Suppl. 34: 458.

Kordon, C. (1969) Effect of selective experimental changes in regional hypothalamic monoamine levels on superovulation in the immature rat. *Neuroendocrinology*, 4: 129.

Kordon, C. and Glowinski, J. (1969) Selective inhibition of superovulation by blockade of dopamine synthesis during the critical period in the immature rat. *Endocrinology*, 85: 924.

Kordon, C., Javoy, F., Vassent, G. and Glowinski, J. (1968) Blockade of superovulation in the immature rat by increased brain serotonin. *Europ. J. Pharmacol.*, 4: 169.

Kordon, C., Blake, C.A., Terkel, J. and Sawyer, C.H. (1973) Participation of serotonin-containing neurons in the suckling-induced rise in plasma prolactin levels in lactating rats. *Neuroendocrinology*, 13: 312.

Kordon, C., Kerdelhué, B., Pattou, E. and Jutisz, M. (1974) Immunocytochemical localization of LH-RH in axons and nerve terminals of the rat median eminence. *Proc. Soc. exp. Biol. (N.Y.)*, 147: 122—127.

Kordon, C., Enjalbert, A. and Héry, M. (1976) Neurotransmitter control of pituitary function. In *Proc. Int. Symp. Hyp. and Endocrine Function*, F. LaBrie (Ed.), Plenum Press, New York, in press.

Ladosky, W. and Wandscheer, D.E. (1975) Interaction between estrogen and biogenic amines in the control of LH secretion. *J. Steroid Biochem.*, 6: 1013.

Legan, S.J., Coon, G.A. and Karsch, F.J. (1975) Role of estrogen as initiator of daily LH surges in the ovariectomized rat. *Endocrinology*, 96: 50.

Lichtensteiger, W. (1969) Cyclic variation of catecholamine content in hypothalamic nerve cells during the estrus cycle of the rat with a concomitant study of the substantia nigra. *J. Pharmacol. exp. Ther.*, 165: 204.

Lichtensteiger, W. (1973) Changes in hypothalamic monoamines in relation to endocrine states: functional characteristics of tubero-infundibular dopamine neurons. In *Endocrinology*, Int. Congr. Ser. No. 273, R.O. Scow (Ed.), Excerpta Medica, Amsterdam, pp. 131—137.

Lu, K.H. and Meites, J.(1971) Inhibition by L-DOPA and monoamine oxidase inhibitors of pituitary prolactin release; stimulation by methyldopa and D-amphetamine. *Proc. Soc. exp. Biol. (N.Y.)*, 137: 480.

MacLeod, R.M. and Lehmeyer, J.E. (1974) Studies on the mechanism of dopamine-mediated inhibition of prolactin secretion. *Endocrinology*, 94: 1077—1085.

McCann, S.M. (1962) Hypothalamic luteinizing hormone releasing factor. *Amer. J. Physiol.*, 202: 395.

McCann, S.M. and Ramirez, V.D. (1964) The neuroendocrine regulation of hypophysial luteinizing hormone secretion. *Recent Progr. Hormone Res.*, 20: 131.

McKelvy, J.F. and Grimm-Jørgensen, Y. (1975) Studies on the biosynthesis of thyrotropin releasing hormone in vitro. In *Hypothalamic Hormones*, M. Motta, P.G. Crosignani and L. Martini (Eds.), Academic Press, London, p. 13.

McKelvy, J.F., Epelbaum, J., Grimm-Jørgensen, Y., Perrie, S. and Kordon, C. (1975) Subcellular distribution of TRF, LH-RH and their degradative enzymes. In preparation.

Mena, F., Enjalbert, A., Carbonel, L., Priam, M. and Kordon, C. (1976) Effect of suckling on plasma prolactin and hypothalamic monoamine levels in the rat. *Endocrinology*, in press.

Mitnick, M. and Reichlin, S. (1972) Enzymatic synthesis of thyrotropin-releasing hormone (TRH) by hypothalamic "TRH synthetase". *Endocrinology*, 91: 1145.

Moszkowska, A. (1959) Contribution à la recherche des relations du complexe hypothalamus-hypophysaire dans la fonction gonadotrope. Méthode in vivo et in vitro. *C. R. Soc. Biol. (Paris)*, 153: 1945.

Palkovits, M., Brownstein, M., Saavedra, J. and Axelrod, J. (1974a) Norepinephrine and dopamine content of hypothalamic nuclei of the rat. *Brain Res.*, 77: 137—149.

Palkovits, M., Arimura, A., Brownstein, M., Schally, A.V. and Saavedra, J.M. (1974b) Luteinizing hormone-releasing hormone (LH-RH) content of the hypothalamic nuclei in the rat. *Endocrinology*, 96: 554.

Pelletier, G., Labrie, F., Puviani, R., Arimura, A. and Schally, A.V. (1974) Electron microscope localization of luteinizing hormone-releasing hormone in the rat median eminence. *Endocrinology*, 95: 314—317.

Pfeiffer, C.A. (1936) Sexual differences of the hypophysis and their determination by the gonads. *Amer. J. Anat.*, 58: 195.

Raisman, G. and Field, D.M. (1971) Sexual dimorphism in the preoptic area of the rat. *Science*, 173: 731.

Ramirez, V.D. and McCann, S.M. (1963) A highly sensitive test for LH releasing activity: the ovariectomized estrogen-progesterone blocked rat. *Endocrinology*, 73: 193.

Ramirez, V.D., Epelbaum, J., Gautron, J.P., Pattou, E., Zamora, A. and Kordon, C. (1975) Distribution of LH-RH in subcellular fractions of mediobasal hypothalamus. *Molec. Cell. Endocr.*, 3: 339.

Sawyer, C.H., Critchlow, B.V. and Barraclough, C.A. (1955) Mechanism of blockade of pituitary activation in the rat by morphine, atropine and barbiturates. *Endocrinology*, 57: 345.

Schneider, H.P. and McCann, S.M. (1970) Estradiol and the neuroendocrine control of LH release in vitro, *Endocrinology*, 87: 330.

Schwartz, N.B. (1969) A model for the regulation of ovulation in the rat. *Recent Progr. Hormone Res.*, 25: 1.

Shin, S.H., Morris, A., Snyder, J., Hymes, W.C. and Milligan, J.V. (1974) Subcellular localization of LH-RH in the rat hypothalamus. *Neuroendocrinology*, 16: 191.

Stefano, F.J. and Donoso, A. (1967) Norepinephrine levels in the rat hypothalamus during the estrus cycle. *Endocrinology*, 81: 1405.

Stumpf, W.E. (1968) Estradiol concentrating neurons: topography in the hypothalamus by dry-mount autoradiography. *Science*, 162: 1001.

Swalstig, E.B., Sawyer, B.D. and Clemens, J.A. (1974) Inhibition of rat prolactin release by apomorphine in vivo and in vitro. *Endocrinology*, 95: 123.

Taber, C.A. and Karavolas, H.J. (1975) Subcellular localization of LH releasing activity in the rat hypothalamus. *Endocrinology*, 96: 446–452.

Takahara, J., Arimura, A. and Schally, A.V. (1974) Suppression of prolactin release by a purified porcine PIF preparation and catecholamines infused into rat hypophysial portal vessels. *Endocrinology*, 96: 2.

Terasawa, E. and Sawyer, C.H. (1969) Changes in electrical activity in the rat hypothalamus related to electrochemical stimulation of adenohypophysial function. *Endocrinology*, 85: 132.

Terkel, J., Blake, C.A. and Sawyer, C.H. (1972) Serum prolactin levels in lactating rats after suckling or exposure to ether. *Endocrinology*, 91: 49.

Velasco, M.E. and Taleisnik, S. (1969) Release of gonadotropin induced by amygdaloid stimulation in the rat. *Endocrinology*, 84: 132.

Vertes, M. and King, R.J. (1971) The mechanism of estradiol binding in rat hypothalamus: effect of androgenization. *J. Endocr.*, 271: 282.

Warembourg, M. (1970) Fixation de l'oestradiol-^{3}H au niveau des noyaux amygdaliens, septaux et du système hypothalamo-hypophysaire chez la souris femelle. *C. R. Acad. Sci. (Paris)*, 270: 152–154.

Weiner, R.I., Pattou, E., Kerdelhué, B. and Kordon, C. (1976) Differential effects of hypothalamic deafferentation upon luteinizing hormone releasing hormone in the median eminence and organum vasculosum of the lamina terminalis. *Endocrinology*, in press.

Whittaker, V.P. (1969) The synaptosome. In *Handbook of Neurochemistry, Vol. II*, A. Lajtha (Ed.), Plenum Press, New York, pp. 327–364.

Subcellular Mechanisms in Reproductive Neuroendocrinology, edited by
F. Naftolin, K.J. Ryan and J. Davies

Chapter 10

Hormonal regulation of the synthesis and metabolism of neurotransmitters

MICHAEL J. BROWNSTEIN

Laboratory of Clinical Science, National Institute of Mental Health, Bethesda, Md. 20014 (U.S.A.)

I. INTRODUCTION

Hormones must be able to exert some influence over the cells of the central nervous system which are involved in neuroendocrine regulation. In the last decade a great deal of information has been obtained about control of the synthesis and metabolism of neurotransmitters; and, based on this information, it is clear that hormones might interact with neurons in many ways.

This brief review is comprised of two parts. The first part is an outline of the biochemical events which underlie the process of neurotransmission. The second part contains a few examples of the effects of hormones on these biochemical events. Although there are many other sorts of compounds besides catecholamines which probably function as transmitters, most of the literature cited in this review deals with catecholaminergic neurons. Since certain features of the synthesis, storage, release, action, and metabolism of all neurotransmitters are similar, the suggestions presented below may be of some general value.

II. BIOSYNTHESIS OF MONOAMINES

The pathways which lead to the biosynthesis of the catecholamines (dopamine, norepinephrine, epinephrine) are depicted in Fig. 1. The enzymes in Fig. 1 are synthesized in the nerve cell body and transported down the axon to its terminals. This process, discovered in 1958 by Weiss and Hiscoe, is known as axoplasmic transport. Microtubules may serve as the tracks along which cellular ingredients travel from the cell body to more distal parts of the cell. Drugs such as colchicine and vinblastine, which probably interfere with the functional integrity of microtubules, block axoplasmic transport (Hökfelt and Dahlström, 1971).

Tyrosine is taken up from the blood stream into catecholaminergic nerve terminals where it is hydroxylated by tyrosine hydroxylase (Nagatsu et al., 1964) to form dihydroxyphenylalanine (DOPA). Under certain circumstances the precursors of neurotransmitters can be shown to limit their rates of synthesis. Dr. Wurtman will discuss the role of substrate availability and uptake in regulating the production of transmitters in a subsequent paper.

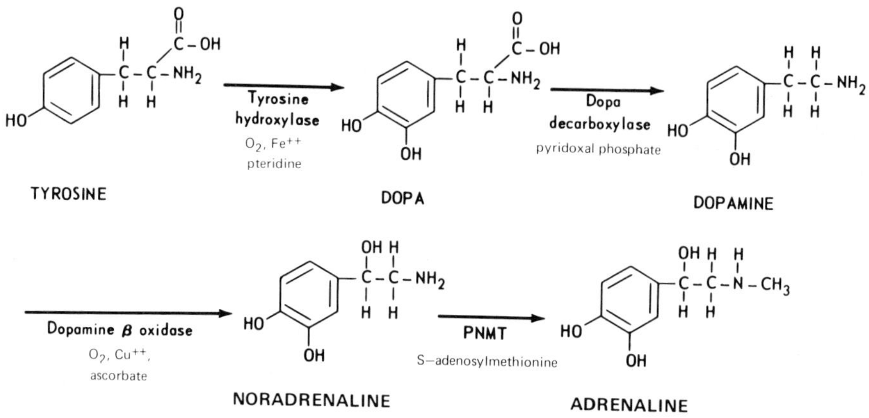

Fig. 1. Biosynthesis of catecholamines.

DOPA is decarboxylated by L-aromatic amino acid decarboxylase (Holtz et al., 1938) to form dopamine. The level of L-aromatic amino acid decarboxylase in catecholaminergic nerve endings is very high. Thus the synthesis of catecholamines is limited by the rate at which hydroxylation of tyrosine occurs (Spector et al., 1963; Levitt et al., 1965). Since there is such an excess of decarboxylase present in the nerve terminals, inhibition of its activity has not proven very effective in altering monoamine levels. Inhibition of tyrosine hydroxylase, on the other hand, is accompanied by marked reductions in amines (Spector et al., 1965).

Noradrenergic nerves contain an enzyme which dopaminergic nerves do not have — dopamine-β-hydroxylase (Friedman and Kaufman, 1965). Dopamine-β-hydroxylase is a copper-containing enzyme which requires ascorbic acid as a cofactor. Like tyrosine hydroxylase, it also requires molecular oxygen. It does not exhibit a high degree of substrate specificity; it can oxidize nearly any phenylethylamine to its corresponding phenylethanolamine.

Recently a population of neurons has been found in the brain which, like cells of the adrenal medulla, are capable of producing epinephrine (Hökfelt et al., 1973; Saavedra et al., 1974). These cells contain phenylethanolamine-N-methyltransferase (Axelrod, 1962).

II.1. Alterations in the levels and in the kinetic properties of biosynthetic enzymes

For many years it has been known that increasing the rate of release of catecholamines does not result in a decrease in their concentration in nervous tissue (von Euler and Hellner-Björkman, 1955). After a few days, increases in release seem to be followed by increases in the amounts of biosynthetic enzymes present in the nerve terminals (Mueller et al., 1969). Presumably new enzyme molecules must be synthesized at a greater rate and/or old enzyme molecules must be degraded more slowly.

Acute increases in the turnover of norepinephrine resulting from stimulation of noradrenergic nerves for an hour or less have been found recently to be accompanied by changes in the kinetic properties of tyrosine hydroxylase (Roth et al., 1974). The affinity of the enzyme for its substrate and cofactor increase and inhibition by its ultimate product, norepinephrine, decreases. Reversible changes in tyrosine hydroxylase which mimic those brought on by nerve stimulation can be produced in vitro by calcium (Morgenroth et al., 1974). Irreversible changes of the same sort can be produced by protein kinase in the presence of cyclic AMP (Harris et al., 1975; Morgenroth et al., 1975). This suggests that tyrosine hydroxylase can be phosphorylated and thereby activated in vivo.

II.2. Storage and release of monoamines

Having been synthesized, the monoamines are bound and stored by synaptic vesicles, small subcellular particles found in large numbers in the terminal swellings of nerves. In this way they are prevented from diffusing out of the neuron and are protected from destruction by monoamine oxidase, an intraneuronal degradative enzyme.

Little is yet known of the mechanism whereby axonally conducted impulses which invade the nerve terminal result in release of monoamines. Depolarization of the terminal results is an increase in its permeability to ions, among them Ca^{2+}. The increase in Ca^{2+} intracellularly appears to be primarily responsible for initiating the events which lead to the liberation of monoamines. Prostaglandins of the E series block the calcium-induced release of norepinephrine from nerves and may act as regulators of transmitter outflow (Hedqvist, 1970). Agents which inhibit phosphodiesterase increase the amount of norepinephrine released during nerve stimulation (Langer, 1973). Consequently, it is possible that cyclic AMP might also play some role in modulating the release of catecholamines.

Catecholamines seem to be released from chromaffin cells and from peripheral sympathetic nerve terminals mainly by a process known as exocytosis, the opposite of micropinocytosis. This process involves fusion of vesicles with the cell membrane and opening of the vesicles into the extracellular space. Thus, the contents of the vesicles are emptied entirely. Since exocytotic release is blocked by colchicine, vinblastine and cytochalasin-B, it appears that microtubules and microfilaments are involved in this sort of release (Thoa et al., 1972).

III. RECEPTORS

Monoamines exert their action by occupation of postsynaptic receptors. Major strides have been made in recent years in studying these receptors. Changes in the sensitivity of postsynaptic cells following denervation or chronic administration of agonists have been shown to be attributable to alterations in the number of receptors present on the surface of these cells (Hall, 1972). Denervation results in supersensitivity to agonists and is associated with an increase in receptor number; administration of agonists has the opposite effect (Kebabian et al., 1975). Isolation of receptors from membranes should allow their properties to be studied in detail. To date no interactions have been found between hormones and purified receptors for neurotransmitters (S. Snyder, personal communication).

Recently, presynaptic receptors have been shown to exist as well as postsynaptic ones

(Langer, 1974). These presynaptic receptors seem to monitor the amount of neurotransmitter in the synaptic cleft and alter the amount of transmitter released by the nerve terminal. Presynaptic receptors, if they are coupled to adenylate cyclase, might also be involved in activating enzymes involved in the biosynthesis of transmitters.

IV. TERMINATION OF THE ACTION OF NEUROTRANSMITTERS

There is more than one mechanism by which the effect of a neurotransmitter can be terminated. An important mode of inactivation is enzymatic degradation of the compound. The best example of this is the rapid cleavage of acetylcholine by acetylcholinesterase.

Another mechanism is reuptake into the presynaptic element (Hertting and Axelrod, 1961). In the case of norepinephrine this uptake and retention process is more important than degradation by either monoamine oxidase or catechol-O-methyltransferase (Crout, 1961).

A third means of terminating the action of transmitters involves their uptake by non-neuronal elements. This has been shown to occur, for example, in vascular smooth muscle and in cardiac muscle (Iversen, 1965). As opposed to reuptake by adrenergic nerves, which is not followed by catabolism of the amines, extraneuronal uptake is followed by their O-methylation. *Beta*-estradiol and corticosterone are fairly potent inhibitors of non-neuronal uptake and metabolism of catecholamines (Iversen and Salt, 1970); they are without effect on the neuronal uptake mechanism. Whether non-neuronal cells in the central nervous system are involved in inactivating any of the neurotransmitters remains to be demonstrated.

IV.1. The biochemical basis of the action of hormones on nerve cells

From the foregoing summary of the "life" of amines, it should be clear that a hormone or drug could have several actions. It could stimulate or interfere with the manufacture or axonal transport of synthesizing enzymes; it could activate or interfere with the activation of these enzymes; it could alter the availability of precursors or cofactors; it could interfere with the storage or release of neurotransmitters; it could antagonize the effect of the neurotransmitter on the receptor or potentiate it; it could block the process of reuptake or inhibit enzymes responsible for degrading the neurotransmitter.

To date, examples of these effects of hormones have not been sought extensively and only a few have been found.

IV.1.1. Synthesis and degradation of enzymes. The removal of the pituitary gland, the removal of the endocrine glands, and the administration of hormones have been shown to alter the levels of enzymes involved in the biosynthesis of neurotransmitters. The adrenal medulla has proven to be a good model for hormonal control of the enzymes which are responsible for synthesizing catecholamines. In the adrenal medulla the activities of tyrosine hydroxylase (TH) and phenylethanolamine-N-methyltransferase (PNMT) are under the control of both the cholinergic nerves which innervate the medulla and the adrenal corticosteroids (Axelrod, 1974). Conditions which cause an increase in adrenaline release, such as stress- or drug-induced hypotension, also lead to an increase in TH and PNMT activity. This increase can be blocked by cutting the splanchnic nerves. Thus the increase in TH and PNMT result from an increased neuronal input.

In addition to being under the influence of incoming nerves, the cells of the adrenal medulla also respond to ACTH and glucocorticoids. Marked falls in adrenal TH and PNMT occur after rats are hypophysectomized. The fall in PNMT can be prevented by treatment with ACTH or with dexamethasone. No other hormones — including FSH, prolactin, TSH, growth hormone, estrogen, testosterone, and mineralocorticoids — can substitute for ACTH and dexamethasone. The increase in PNMT which is produced by dexamethasone could either be due to a decrease in the rate of degradation of the enzyme (Ciaranello and Axelrod, 1973) or to an increase in its rate of synthesis. Unlike the case with PNMT, TH levels could be maintained by administration of ACTH but not dexamethasone.

In the brain, tyrosine hydroxylase activity can be altered in specific regions by endocrine manipulations (Kizer et al., 1974). For example, thyroidectomy causes an increase in the level of the enzyme in the hypothalamic median eminence. This increase is reversed by treating the rats with thyroxine. Similarly, the increase in TH which is found in the median eminence following gonadectomy is reversed by testosterone. Adrenalectomy results in a decrease in TH which is corrected by dexamethasone administration. At present it is not known whether the above changes occur because of an increased neuronal input to catecholaminergic nerves or because of a direct effect of altering hormonal levels on these nerves.

IV.1.2. Activation of enzymes. The fact that tyrosine hydroxylase appears to be activated by a cyclic AMP stimulated protein kinase suggests that hormones, many of which act by increasing intracellular cyclic AMP, could activate this and other neuronal enzymes. This hypothesis has yet to be tested.

IV.1.3. Inhibition of enzymes. A number of years ago it was reported that 2-hydroxyestradiol is a potent inhibitor of catechol-O-methyltransferase (COMT) (Kruppen et al., 1969). This compound is formed from estradiol by microsomal enzymes in the liver (Daly et al., 1965). Probably because of this, large doses of estradiol cause significant inhibition of COMT in the liver (Cohn and Axelrod, 1971).

Another hormonal metabolite, 3-iodotyrosine, is a potent inhibitor of tyrosine hydroxylase (Udenfriend et al., 1965). It is unlikely that iodotyrosine normally acts to inhibit this enzyme, however. Dehalogenases in tissue must inactivate it quite rapidly.

Thyroxine itself is a fairly strong inhibitor of dopamine-β-hydroxylase (Nagatsu et al., 1968).

IV.1.4. Transport, storage and release. No compounds, other than those which alter microtubules and microfilaments, have been shown to alter axoplasmic transport.

Dratman et al. (1973) have shown that thyroxine is taken up by adrenergic nerves and deiodinated. Whether the product of deiodination is stored and released is not known. If the metabolite (1) were able to displace catecholamines from their storage sites, and (2) were inactive, their effect might be to decrease neuronal stimulation of postsynaptic structures. In this case the metabolite would be called a "false transmitter". On the other hand, the metabolite could be as active as, or more active than, the normally occurring transmitter and, upon being released, could act in concert with it. The fact that thyroid hormones are sympathomimetic only in animals with an intact nervous system suggests that the latter possibility is the correct one.

A number of pharmacological and physiological manipulations have been shown to alter the turnover of neurotransmitters. The mechanisms by which hormones altering turnover rates remain to be determined.

IV.1.5. Neuronal uptake. A large number of hormones have been studied with regard to their action on the neuronal uptake of monoamines. Estrogen and progesterone have recently been found to alter the uptake of dopamine and serotonin respectively (Wirz-Justice et al., 1974). Prostaglandin E_2, angiotensin, vasopressin, oxytocin and various other peptides are without effect on amine uptake.

V. CONCLUSION

Hormones may produce many of their central feedback effects simply by altering the firing rates of their target cells. But hormones may also interact with the biochemical machinery of neurons and by doing so they may effect long-lasting changes in transmitter metabolism. The search for interactions between hormones and nerves at the biochemical level is just beginning.

REFERENCES

Axelrod, J. (1962) Purification and properties of phenylethanolamine-N-methyltransferase. *J. biol. Chem.*, 237: 1657–1660.

Axelrod, J. (1974) Regulation of the neurotransmitter norepinephrine. In *The Neurosciences. Third Study Program*, F.O. Schmitt and R.F. Worden (Eds.), MIT Press, Cambridge, Mass., pp. 863–876.

Ciaranello, R.D. and Axelrod, J. (1973) Genetically controlled alterations in the rate of degradation of phenylethanolamine-N-methyltransferase. *J. biol. Chem.*, 248: 5616–5623.

Cohn, C.K. and Axelrod, J. (1971) Effect of estradiol on catechol-O-methyltransferase activity in rat liver. *Life Sci.*, 10: 1351–1354.

Crout, J.R. (1961) Effect of inhibiting both catechol-O-methyltransferase and monoamine oxidase on cardiovascular responses to noradrenaline. *Proc. Soc. exp. Biol. (N.Y.)*, 108: 482–484.

Daly, J., Inscoe, J.K. and Axelrod, J. (1965) The formation of O-methylated catechols by microsomal hydroxylation of phenols and subsequent enzymatic catechol O-methylation. Substrate specificity. *J Med. Chem.*, 8: 153–157.

Dratman, M.B., Axelrod, J., Crutchfield, F.L., Coyle, T. and Utiger, R.D. (1973) Catecholamine enzymes and thyroid hormone metabolism: role of tyrosine hydroxylase. *Endocrinology*, 92: A-156.

Euler, U.S. von and Hellner-Björkman, S. (1955) Effect of increased adrenergic nerve activity on the content of noradrenaline and adrenaline in cat organs. *Acta physiol. scand.*, 33, Suppl. 118: 17–20.

Friedman, S. and Kaufman, S. (1965) 3,4-Dihydroxyphenylethylamine β-hydroxylase: physical properties, copper content, and role of copper in the catalytic activity. *J biol. Chem.*, 240: 4763–4773.

Hall, Z. (1972) Release of neurotransmitters and their interaction with receptors. *Ann. Rev. Biochem.*, 41: 925–952.

Harris, J.E., Baldessarini, R.J., Morgenroth, III, V.H. and Roth, R.H. (1975) Activation by cyclic 3':5'-adenosine monophosphate of tyrosine hydroxylase in the rat brain. *Proc. nat. Acad. Sci. (Wash.)*, 72: 789–793.

Hedqvist, P. (1970) Antagonism by calcium of the inhibitory action of prostaglandin E_2 on sympathetic neurotransmission in the cat spleen. *Acta physiol. scand.*, 80: 269–275.

Hertting, G. and Axelrod, J. (1961) The fate of tritiated noradrenaline at the sympathetic nerve-endings. *Nature (Lond.)*, 192: 172–173.

Hökfelt, T. and Dahlström, A. (1971) Effects of two mitosis inhibitors (colchicine and vinblastine) on the distribution and axonal transport of noradrenaline storage particles, studied by fluorescence and electron microscopy. *Z. Zellforsch.*, 119: 460–482.

Hökfelt, T., Fuxe, K., Goldstein, M. and Johansson, D. (1973) Evidence for adrenaline neurons in the rat brain. *Acta physiol. scand.*, 89: 286–288.

Holtz, P., Heise, R. und Lüdtke, K. (1938) Fermentativer Abbau von 1-Dioxyphenylalanin (Dopa) durch Niere. *Naunyn-Schmiedeberg's Arch. exp. Path. Pharmak.*, 191: 87–118.

Iversen, L.L. (1965) The uptake of catecholamines at high perfusion concentrations in the rat isolated heart: a novel catecholamine uptake process. *Brit. J. Pharmacol.*, 25: 18–33.

Iversen, L.L. and Salt, P.J. (1970) Inhibition of catecholamine uptake by steroids in the isolated rat heart. *Brit. J. Pharmacol.*, 40: 528–530.

Kebabian, J.W., Zatz, M., Romero, J.A. and Axelrod, J. (1975) Rapid changes in rat pineal β-adrenergic receptor: alterations in ^{3}H-(−)-alprenolol binding and adenylate cyclase. *Proc. nat. Acad. Sci. (Wash.)*, 72: 3735–3739.

Kizer, J.S., Palkovits, M., Zivin, J., Brownstein, M., Saavedra, J.M. and Kopin, I.J. (1974) The effect of endocrinological manipulations on tyrosine hydroxylase and dopamine-β-hydroxylase activities in individual hypothalamic nuclei of the adult male rat. *Endocrinology*, 95: 799–812.

Kruppen, R. von, Höller, M., Tilmann, D. und Breuer, H. (1969) Wirkung von Östrogenen auf den Abbau und die Methylierung von Adrenalin bei der Maus *in vivo*. *Hoppe-Seyler Z. Physiol. Chem.*, 350: 1301–1309.

Langer, S.Z. (1973) The regulation of transmitter release elicited by nerve stimulation through a presynaptic feed-back mechanism. In *Frontiers in Catecholamine Research*, E. Usdin and S.H. Snyder (Eds.), Pergamon Press, New York, pp. 543–549.

Langer, S.Z. (1974) Presynaptic regulation of catecholamine release. *Biochem. Pharmacol.*, 23: 1793–1800.

Levitt, M., Spector, S., Sjoerdsma, A. and Udenfriend, S. (1965) Elucidation of the rate-limiting step in norepinephrine biosynthesis in the perfused guinea-pig heart. *J. Pharmacol. exp. Ther.*, 148: 1–8.

Morgenroth, III, V.H., Boadle-Biber, M. and Roth, R.H. (1974) Tyrosine hydroxylase: activation by nerve stimulation. *Proc. nat. Acad. Sci. (Wash.)*, 71: 4283–4287.

Morgenroth, III, V.H., Hegstrand, L.R., Roth, R.H. and Greengard, P. (1975) Evidence for involvement of protein kinase in the activation by adenosine 3':5'-monophosphate of brain tyrosine 3-monooxygenase. *J. biol. Chem.*, 250: 1946–1948.

Mueller, R.A., Thoenen, H. and Axelrod, J. (1969) Increase in tyrosine hydroxylase activity after reserpine administration. *J. Pharmacol. exp. Ther.*, 169: 74–79.

Nagatsu, T., Levitt, M. and Udenfriend, S. (1964) Tyrosine hydroxylase: the initial step in norepinephrine biosynthesis. *J. biol. Chem.*, 239: 2910–2917.

Nagatsu, T., Van der Schoot, J.B., Levitt, M. and Udenfriend, S. (1968) Factors influencing dopamine β-hydroxylase activity and epinephrine levels in guinea pig adrenal gland. *J. Biochem.*, 64: 39–43.

Roth, R.H., Salzman, P.M. and Morgenroth, III, V.H. (1974) Noradrenergic neurons: allosteric activation of hippocampal tyrosine hydroxylase by stimulation of the locus coeruleus. *Biochem. Pharmacol.*, 23: 2779–2784.

Saavedra, J.M., Palkovits, M., Brownstein, M.J. and Axelrod, J. (1974) Localisation of phenylethanolamine N-methyltransferase in rat brain nuclei. *Nature (Lond.)*, 248: 695—696.

Spector, S., Zaltzman-Nirenberg, P., Levitt, M. and Udenfriend, S. (1963) Norepinephrine synthesis from tyrosine-C^{14} in isolated perfused guinea pig heart. *Science*, 139: 1299—1301.

Spector, S., Sjoerdsma, A. and Udenfriend, S. (1965) Blockade of endogenous norepinephrine synthesis by α-methyl-tyrosine, an inhibitor of tyrosine hydroxylase. *J. Pharmacol. exp. Ther.*, 147: 86—95.

Thoa, N.B., Wooten, G.F., Axelrod, J. and Kopin, I.J. (1972) Inhibition of release of dopamine-β-hydroxylase and norepinephrine from sympathetic nerves by colchicine, vinblastine or cytochalasin-B. *Proc. nat. Acad. Sci. (Wash.)*, 69: 520—522.

Udenfriend, S., Zaltzman-Nirenberg, P. and Nagatsu, T. (1965) Inhibitors of purified beef adrenal tyrosine hydroxylase. *Biochem. Pharmacol.*, 14: 837—845.

Wirz-Justice, A., Hackmann, F. and Lichtsteiner, M. (1974) The effect of oestradiol dipropionate and progesterone on monoamine uptake in the rat brain. *J. Neurochem.*, 22: 187—189.

Weiss, P. and Hiscoe, H. (1958) Experiments on the mechanism of nerve growth. *J. exp. Zool.*, 107: 315—396.

Subcellular Mechanisms in Reproductive Neuroendocrinology, edited by
F. Naftolin, K.J. Ryan and J. Davies
© 1976 Elsevier Scientific Publishing Company–Amsterdam, The Netherlands

Chapter 11

On the role of neurotransmitters and hypothalamic hormones and their interactions in hypothalamic and extrahypothalamic control of pituitary function and sexual behavior

K. FUXE, T. HÖKFELT, A. LÖFSTRÖM, O. JOHANSSON, L. AGNATI, B. EVERITT,
M. GOLDSTEIN, S. JEFFCOATE, N. WHITE, P. ENEROTH, J.-Å. GUSTAFSSON and P. SKETT

Departments of Histology and Chemistry, Karolinska Institute, Stockholm (Sweden), (M.G.) New York University Medical Center, New York, N.Y. (U.S.A.), (S.J. and N.W.) Department of Chemical Pathology, St. Thomas's Hospital Medical School, London (Great Britain) and (P.E.) Department of Obstetrics and Gynecology, Karolinska Hospital, Stockholm (Sweden)

I. INTRODUCTION

The research on the role of hypothalamus and related brain structures in the control of pituitary function has undergone an explosive development due to the biochemical characterization of many transmitters, such as dopamine (DA), noradrenaline (NA), adrenaline (A) and 5-hydroxytryptamine (5-HT) and hypothalamic hormones such as TRF, LHRF and GHIF (somatostatin).

In the present review article the cellular localization of the various neurotransmitters and hypothalamic hormones will be described as well as their possible functions.

II. MAPPING OF NEUROTRANSMITTER SYSTEMS

II.1. DA neuron systems

In the hypothalamus the most well known DA system is the tubero-infundibular DA system (Carlsson et al., 1962; Fuxe, 1964; Fuxe and Hökfelt, 1969; Hökfelt and Fuxe, 1972; Björklund et al., 1973). The cell bodies are mainly found in the nucleus arcuatus (Fig. 1) and in the anterior periventricular hypothalamic nucleus (Fig. 2). This system gives rise to a rich DA innervation of the lateral external layer (Fig. 4) of the median eminence and a substantial DA innervation is also present in the medial external layer (Fig. 5) and in the external layer of the infundibular stalk. The calculated DA distribution is shown in Table I (Löfström et al., 1975a). The most anterior part of the nucleus arcuatus sends DA axons down into the intermediate lobe and the posterior lobe of the hypophysis (Björklund et al., 1973).

Other DA cell groups are localized in the posterior periventricular diencephalon (group

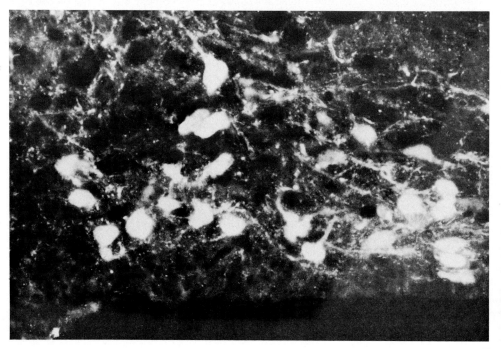

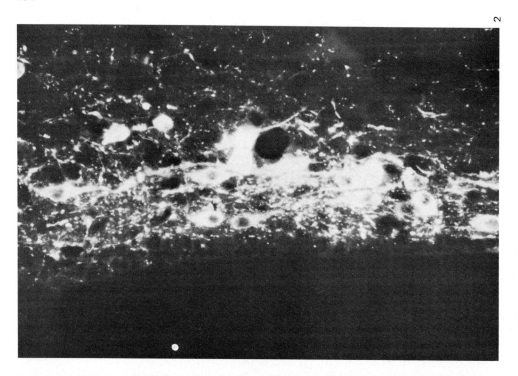

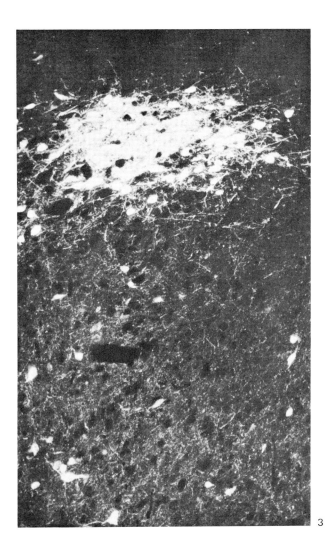

3

Fig. 1. Dopadecarboxylase immunofluorescence. Specific fluorescence is observed in DA cell bodies of the arcuate nucleus (A12). Positive nerve terminals are also seen surrounding the DA nerve cells. x 400.

Fig. 2. Tyrosine hydroxylase immunofluorescence. Specific immunofluorescence is observed in DA cell bodies of the anterior periventricular hypothalamic nucleus (A12). Positive nerve terminals and axons are also observed. x 400.

Fig. 3. Tyrosine hydroxylase immunofluorescence. Specific immunofluorescence is observed in DA cell bodies of the A13 group, which is a densely packed cell group located in the medial zona incerta and dorsal hypothalamus. A few cell bodies are also located in the dorsomedial hypothalamic nucleus present below the A13 group in the picture. Weakly fluorescent dots are also observed in this latter nucleus, probably representing networks of NA and A terminals. x 160.

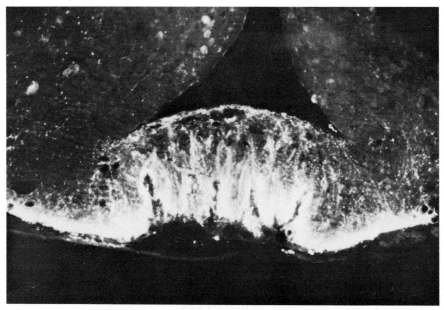

Fig. 4. CA fluorescence in the median eminence. A strong diffuse specific fluorescence is observed in the entire external layer and positive nerve terminals are also found in the subependymal layer. A few DA cell bodies are seen in the arcuate nucleus. x 160.

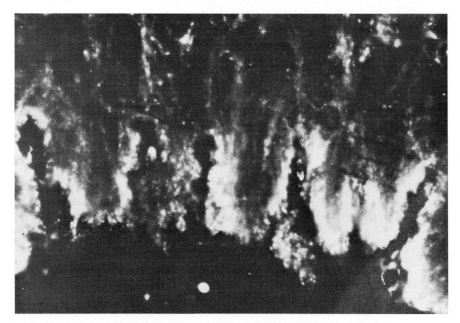

Fig. 5. CA fluorescence in the medial external layer of the median eminence. Specific CA fluorescence is seen outlining the short capillary loops. This area contains a mixture of DA and NA nerve terminals (Löfström et al., 1975a).

TABLE I

Calculated relative DA distribution in the median eminence

The figures represent the common means of the fluorescence intensity after BH deafferentation and VB lesions expressed as a percentage of the intensity in respective areas of untreated control (Löfström et al., 1975a). SEL = subependymal layer; PMZ = medial palisade zone; LPZ = lateral palisade zone.

Region	SEL (%)	MPZ (%)	LPZ (%)
Rostral	20	75	80
Central	5	60	95
Caudal	10	50	90

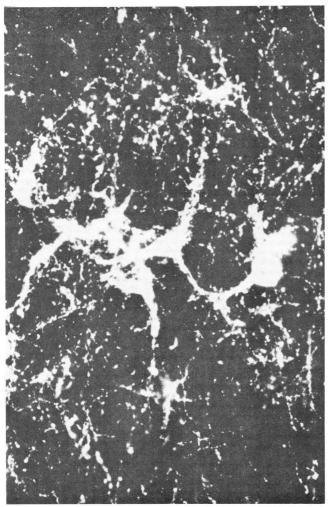

Fig. 6. CA fluorescence is observed in probable DA nerve terminals in the lateral septal nucleus which completely outline the dendrites and cell bodies of some septal cells, so that they almost appear as if they were fluorescent (Hökfelt et al., to be published).

A11, Dahlström and Fuxe, 1964) and in the dorsal hypothalamus and in the zona incerta (A13, Fuxe, 1965; Fig. 3). These DA cell groups have been claimed to innervate the periventricular hypothalamus and the anterior hypothalamic nucleus (Björklund et al., 1975). The DA cell bodies in the most anterior periventricular hypothalamic area and in the preoptic area have been called A14 (Björklund et al., 1973) and are supposed to innervate the periventricular preoptic area (Björklund et al., 1975). For biochemical mapping of DA neurons see Koslow et al. (1974) and Palkovits et al. (1974).

A new system of mesencephalic periaqueductal DA cell bodies projecting into the diencephalon has also recently been described (Hökfelt, Johansson, Fuxe and Goldstein, to be published).

There also exists an extensive DA innervation of various nuclei of the limbic system such as the nucleus septalis lateralis (Fig. 6), nucleus amygdaloideus centralis, the nucleus accumbens, tuberculum olfactorium and the limbic cortex (see Fuxe, 1965; Fuxe et al., 1974; Hökfelt et al., 1975a). This DA innervation has a mesencephalic origin (see Fuxe et al., 1970; Hökfelt et al., 1975a; see also Lindvall and Björklund, 1974).

II.2. NA neuron systems

The extensive NA nerve terminal systems of the hypothalamus have been described in detail by Fuxe (1965). The hypothalamic NA nerve terminals are fairly thick compared with the very fine to fine cortical NA networks (Fig. 7). In the median eminence the NA nerve terminals are mainly found in the subependymal layer and in the medial external

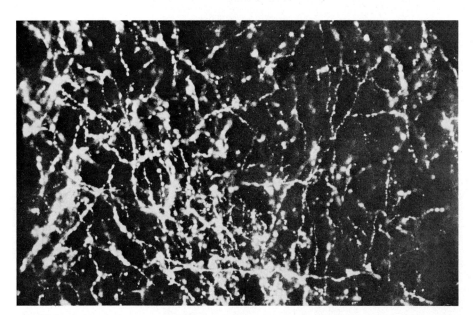

Fig. 7. CA fluorescence in the medial septal nucleus. Vibratom section. The fluorescence is located in probable NA nerve terminals. Two types of NA terminals are observed, one fairly thick (to the left) which mainly is observed in hypothalamus and subcortical limbic structures and one fine (to the right) which mainly is observed in cortical regions (Hökfelt et al., to be published). x 400.

layer (Fig. 8). The NA cell bodies giving rise to these systems are localized to the pons and medulla oblongata and the axons form the so-called ventral NA bundle in the mesencephalon. The locus coeruleus mainly innervates the cortical regions (see Fuxe et al., 1970a, b, 1971; Hartman et al., 1972; Goldstein et al., 1974; Hökfelt et al., 1973a, 1975a).

II.3. The adrenaline neuron systems

As seen in Fig. 8 the adrenaline cell bodies, as visualized in immunohistochemical studies on phenylethanolamine-N-methyltransferase (PNMT), are localized in the reticular formation of the medulla oblongata. These nerve cells send ascending axons, e.g., to the hypothalamus and descending axons to the sympathetic lateral column (Hökfelt et al., 1973b, 1974a). In Fig. 9 the adrenaline nerve terminals in the mediobasal hypothalamus are illustrated. The networks are mainly found in the ventral part of the arcuate nucleus and in the area immediately lateral of the median eminence but also in the subependymal and internal layer of the median eminence. Later studies on the distribution of PNMT activity in brain have fully confirmed the distribution of the adrenaline nerve terminals and cell bodies as shown by immunohistochemistry (Saavedra et al., 1974a).

II.4. 5-HT neuron systems

The 5-HT cell bodies innervating the hypothalamus are mainly found in the nucleus raphe dorsalis (B7) and nucleus raphe medianus (B8) (Dahlström and Fuxe, 1964) (Fig. 10). The axons ascend medially in the midbrain and innervate various nuclei of the hypothalamus (Fuxe and Jonsson, 1974). Due to technical difficulties it has only been possible to observe a dense network of 5-HT nerve terminals in the nucleus suprachiasmaticus (Fuxe, 1965). Recently, using radioenzymatical techniques the distribution of 5-HT in various nuclei of the hypothalamus has been mapped out (Saavedra et al., 1974b).

Interestingly, a high concentration of 5-HT was found in the arcuate nucleus, which thus probably is controlled by a 5-HT mechanism.

II.5. Possible histamine-containing neurons

Regional analysis has demonstrated that histamine is found in the highest concentrations in the hypothalamus (5.00 nmole/g). A number of findings on the subcellular localization of histamine indicate its presence in synaptosomes. Therefore, histamine may be a putative neurotransmitter, since also the synthesizing enzymes are present and it can be released by depolarization (see Dismukes and Snyder, 1974).

Recently, histamine contents have been determined in various hypothalamic nuclei. The highest concentrations were found in the median eminence, the ventral premammillary nucleus and the nucleus suprachiasmaticus (Brownstein et al., 1974a). It is therefore of great interest that the two latter areas are rich in nerve cell bodies which contain dopadecarboxylase but lack tyrosine hydroxylase (Fig. 11; Fuxe et al., 1974; Hökfelt et al., 1975a). Also these nerve cells have been shown to take up and decarboxylate DOPA (Lidbrink et al., 1974). It is therefore tempting to speculate that these nerve cells are histamine-containing neurons and that their demonstration depends upon the fact that histidine decarboxylase can cross-react with the antibodies against dopadecarboxylase. This problem is now being investigated.

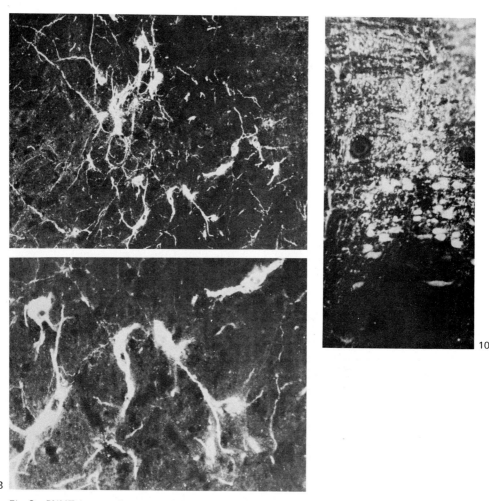

Fig. 8. PNMT immunofluorescence. Nerve cell bodies and processes with a strong immunofluorescence are observed in the ventrolateral reticular formation of the medulla oblongata. x 160 (above) and x 400 (below). (From Hökfelt et al., 1974a.)

Fig. 9. PNMT immunofluorescence in the median eminence and arcuate nucleus. Positive nerve terminals are found mainly in the ventral part of the arcuate nucleus close to the ventral surface of the brain. A few are also present in the inner layers and in the medial external layer of the median eminence. x 160.

Fig. 10. Monoamine fluorescence in nucleus raphe medianus (B8). 5-HT cell bodies and 5-HT fibers of passage (transversely cut) show a strong fluorescence following the intracerebral administration of 6-HT (4 μg/4 μl) into the B8 region after pretreatment with nialamide (100 mg/kg, i.p.) 30 min earlier. 6-HT gives a strong fluorescence yield and is taken up and accumulated in the 5-HT nerve cells.

Fig. 11. Dopadecarboxylase (DDC) immunofluorescence. Premammillary area. Two groups of DDC positive nerve cells are observed. The group to the left is present in the border zone between the ventral and dorsal premammillary nuclei and it is built up of fairly large cell bodies. The one to the right consists of fairly small nerve cell bodies and is localized periventricularly. x 160.

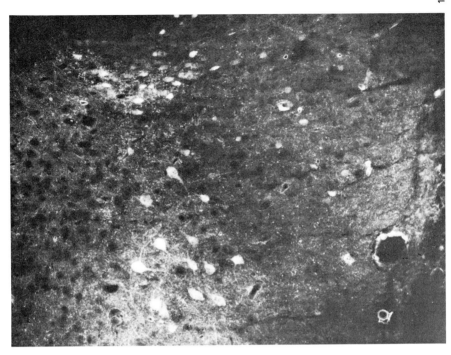

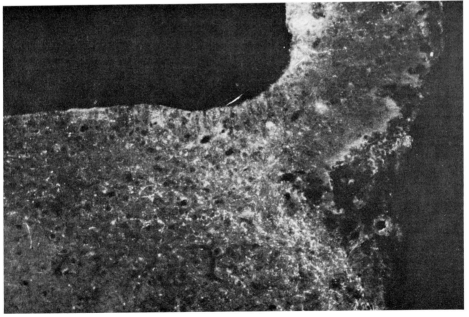

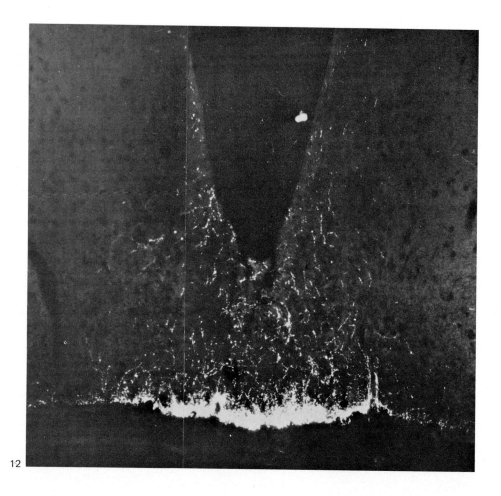

Figs. 12—14. LRF immunofluorescence of rat at the rostral (12), middle (13) and caudal level (14) of the median eminence. A very dense meshwork of LRF-containing terminals is observed in the entire external layer of the dorsal (Fig. 12) and ventral (Fig. 14) lip. In the median eminence proper the LRF terminals are concentrated to the lateral external layer with only few terminals in the medial external layer. The infundibular stalk is practically devoid of LRF-containing terminals. Note the existence of LRF-containing terminals periventricularly and in the subependymal layer. × 160.

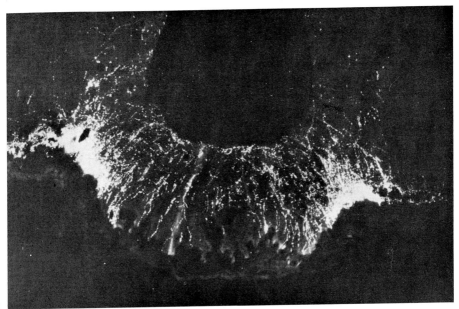

13

14

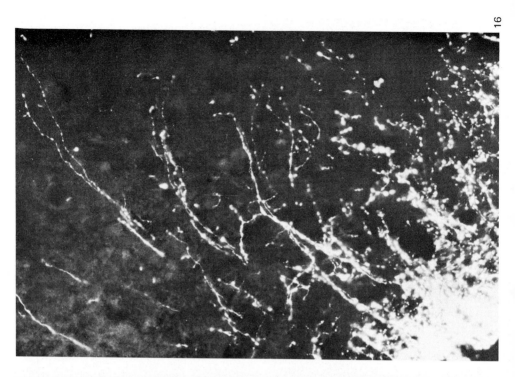

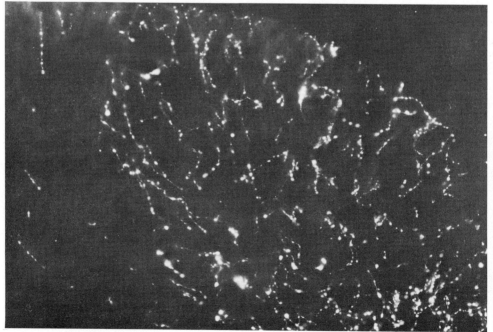

II.6. Cholinergic neuron systems

On the basis of acetylcholinesterase stainings, the distribution of cholinergic cell bodies and terminals in the hypothalamus has been tentatively described (Shute and Lewis, 1966). Recently, the distribution of choline acetylase in various hypothalamic nuclei has also been shown (Brownstein et al., 1975). Choline acetylase is a much better marker of cholinergic systems than acetylcholinesterase. It is of considerable interest to note that a fairly high choline acetylase activity was present in the median eminence, suggesting the existence of a cholinergic mechanism operating at this level. In general, the results agreed with those obtained with acetylcholinesterase stainings.

II.7. GABA-containing neuron systems

It is well known that gabergic mechanisms must operate in the hypothalamus in view of high GABA levels and a high GABA uptake (Hökfelt et al., 1970; Okada et al., 1971).

Recently, also in autoradiographical studies on GABA uptake it has been possible to show the existence of probable GABA-containing interneurons in the medial hypothalamus (Makara et al., 1975).

II.8. Substance P-containing neurons

Hökfelt and collaborators (Nilsson et al., 1974; Hökfelt et al., 1975b, c) have used antibodies against Substance P (SP) to localize Substance P in brain, spinal cord and the peripheral nervous system. The evidence suggests that Substance P is localized to certain populations of primary sensory neurons and to extensive terminal networks in the brain. Networks of moderate density are found throughout the hypothalamus, but no innervation exists of the median eminence except for a few fibers in the internal layer. It should be pointed out that there exists a dense network of Substance P terminals in the medial amygdaloid nucleus. For a review, see Leeman (1974).

II.9. LRF-containing neurons

Barry and coworkers (Barry et al., 1973, 1974) were the first to demonstrate the localization of LRF-containing cell bodies and nerve terminals in the hypothalamus and other parts of the tel-diencephalon. The LRF cell bodies were mainly located in the preoptic and septal area and the highest densities were found in the optic crest and in the median eminence. The localization of LRF-containing nerve terminals, as described by the French workers, has largely been confirmed by other groups (Pelletier et al., 1974; Hökfelt et al., 1975a, e; Sotelo et al., 1975). It should be underlined that DA is not present in the LRF neurons since, e.g., the LRF contents is not reduced by 6-OH-DA treatment (Kizer et al., 1975).

Fig. 15. LRF immunofluorescence. Nucleus arcuatus. LRF-containing nerve terminals are observed. Several are reaching the surface of the third ventricle (to the right). x 400.

Fig. 16. LRF immunofluorescence. Nucleus arcuatus and adjacent lateral median eminence area. LRF preterminal axons from the arcuate nucleus appear to descend to the lateral external layer (lower left). x 400.

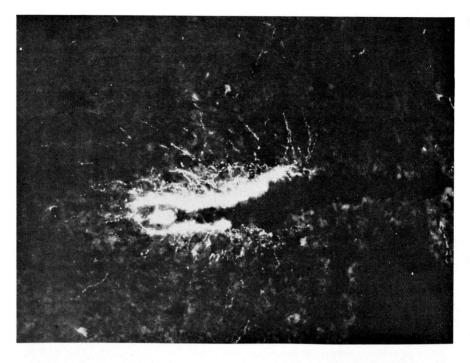

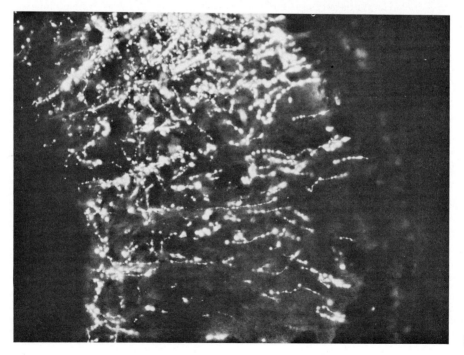

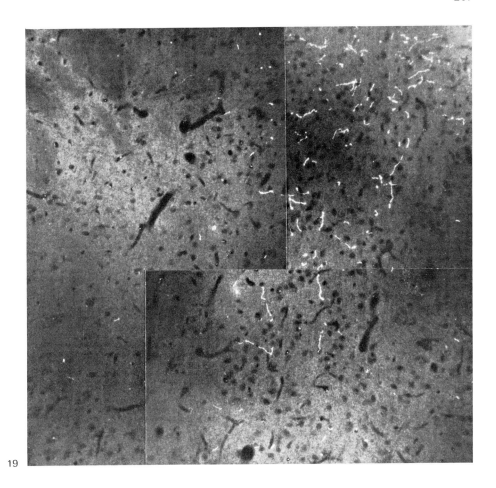

Fig. 17. LRF immunofluorescence. The lateral part of the medial external layer. A network of moderate density is seen in all layers of the median eminence. The networks become dense at the border towards the lateral part of the median eminence (to the right). x 400.

Fig. 18. LRF immunofluorescence. Optic crest. A zone of densely packed LRF-containing nerve terminals is observed facing the surface of the brain. x 400.

Fig. 19. LRF immunofluorescence. Supramammillary commissure. A network of LRF terminals is observed in the midline laterally of the tractus mammillothalamicus (to the upper left, dark structures). x 160.

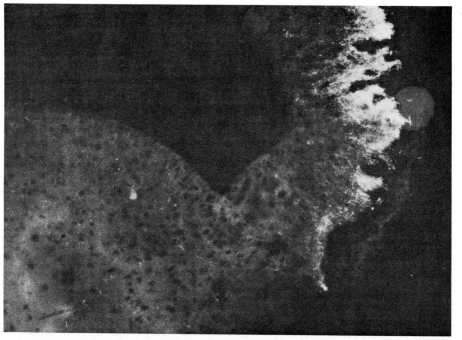

Figs. 20—22. TRF immunofluorescence at the rostral (Fig. 20), middle (Fig. 21) and caudal (Fig. 22) level of the median eminence. In the dorsal lip (Fig. 20) only a sparse plexus of TRF terminals is observed in the external layer. At the other two levels a dense innervation of TRF nerve terminals is observed in the medial external layer surrounding the short capillary loops. x 160.

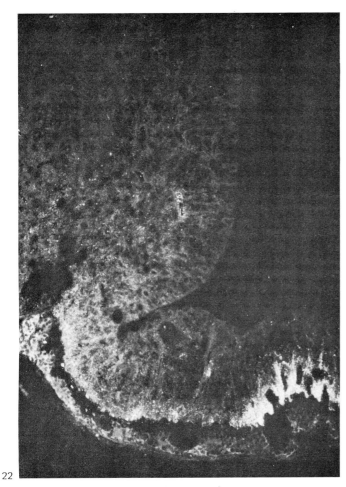

22

In Figs. 12—14 the distribution of LRF nerve terminals is shown at various levels of the median eminence. It should be noted that there are only few LRF nerve terminals in the medial external layer and in the infundibular stalk, whereas a high density is found in the lateral external layer. In the latter area also the highest density of DA nerve terminals is found, giving a morphological basis for an axo-axonic interaction between DA- and LRF-containing nerve terminals. From above it also follows that the DA and NA nerve terminals in the medial external layer and in the infundibular stalk are mainly concerned in the control of other adenohypophyseal hormones, perhaps mainly prolactin (see below). The existence of LRF nerve terminals in the arcuate nucleus reaching the ependyma of the third ventricle is shown in Fig. 15. In Fig. 16 LRF axons can be traced into the arcuate nucleus. The existence of LRF nerve terminal plexa in the subependymal layer, internal layer and certain parts of the medial external layer is illustrated in Fig. 17. In view of these findings it has to be considered that interactions between LRF and NA and A terminals can occur in these regions. The LRF innervation of the optic crest is shown in Fig. 18. As pointed out by Hökfelt et al. (1975e) the LRF secretion in this region is probably not under a dopaminergic control, since only few CA nerve terminals are found in this region. The histological analysis suggests the existence of a primary capillary plexus, also in this region draining into the portal vessels. In Fig. 19 it is shown that LRF terminals can innervate also brain regions not directly secreting hypothalamic hormones exemplified in this case by the presence of LRF terminals in the supramammillary commissure. It may be, therefore, that LRF can act as a neurotransmitter in these terminals.

In view of the recent findings that LRF can facilitate sexual behavior in the ovariectomized estrogen-primed female rat (see McCann and Moss, 1975), it may be speculated that LRF terminals in this region can be involved in the control of sexual behavior. With regard to the location of the LRF-containing cell bodies, other workers have not been able to confirm the presence of LRF cell bodies in the septal region and preoptic area. In our own work on rat single LRF cell bodies have been observed in the most anterior part of the preoptic region at the same level as the optic crest. No LRF-containing cell bodies have been observed in the arcuate nucleus.

II.10. TRF-containing neurons

Recently Hökfelt and collaborators (Hökfelt et al., 1975f, g) have in immunohistochemical studies been able to demonstrate TRF nerve terminal plexuses of moderate to high density in the median eminence (Figs. 20—23), in other hypothalamic regions such as the dorsomedial hypothalamic nucleus (Figs. 24, 25), and the medial part of the ventromedial hypothalamic nucleus, in the septal area, nucleus accumbens, reticular formation (Fig. 26) and several motor nuclei of the brain stem and spinal cord (Figs. 27, 28). In the median eminence the TRF nerve terminals are mainly found in the medial external layer (Figs. 21—23) and in the infundibular stalk. Thus, there exists the possibility for DA and NA nerve terminals to influence TRF secretion via an axo-axonic interaction in these parts of the median eminence. The fact that many regions of the brain and spinal cord receive an TRF innervation suggests that TRF can subserve a transmitter function in large parts of the CNS, particularly in the motor systems (see Plotnikoff et al., 1972) in view of its distribution in the brain stem and spinal cord. Also TRF has in this laboratory been shown to induce a special reflex syndrome in the acutely spinalized rat involving crossing and flexion of the hindlimbs and an increase in muscle tone (Fuxe, unpublished data).

Also the distribution of TRF nerve terminals is such that they could be involved in the mediation of the euphoriant and antidepressant effects of TRF (Prange et al., 1972) as well as its effect on sleep—wakefulness. Certainly the results obtained underline the importance of peptides in behavioral control (see De Wied et al., 1972).

The present results on the distribution of TRF terminals are in good agreement with results obtained in radioimmunological studies (Brownstein et al., 1974b; Jackson and Reichlin, 1974; Krulich et al., 1974).

II.11. Somatostatin-containing neurons

Hökfelt and collaborators (Hökfelt et al., 1974b; Hökfelt et al., 1975d) have recently demonstrated somatostatin-containing nerve terminals in the external layer of the median eminence and of the infundibular stalk, in the internal layer of the median eminence and in certain hypothalamic nuclei, such as ventromedial nucleus and the ventral premammil-

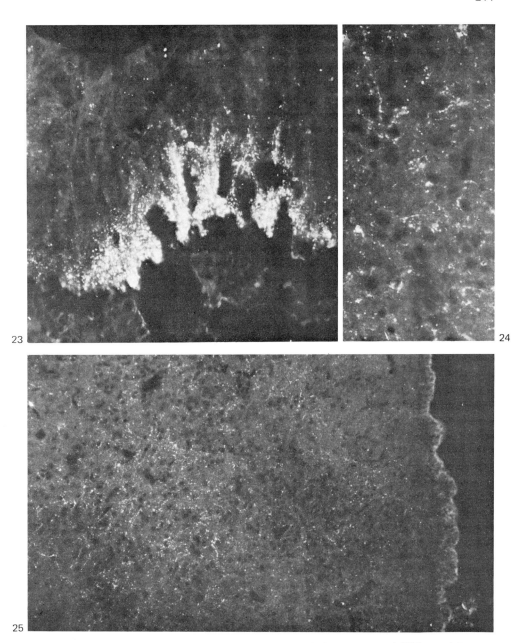

Fig. 23. TRF immunofluorescence. Medial external layer. A dense plexus of TRF-containing terminals is exclusively localized to this area surrounding the short capillary loops. x 400.

Figs. 24–25. TRF immunofluorescence. Dorsomedial hypothalamic nucleus. There exists a network of TRF-containing nerve terminals, which is sparse in the medial part close to the third ventricle but which is of moderate density in the lateral part. The terminals appear to have a varicose appearance. Fig. 25, x 160; Fig. 24, x 400.

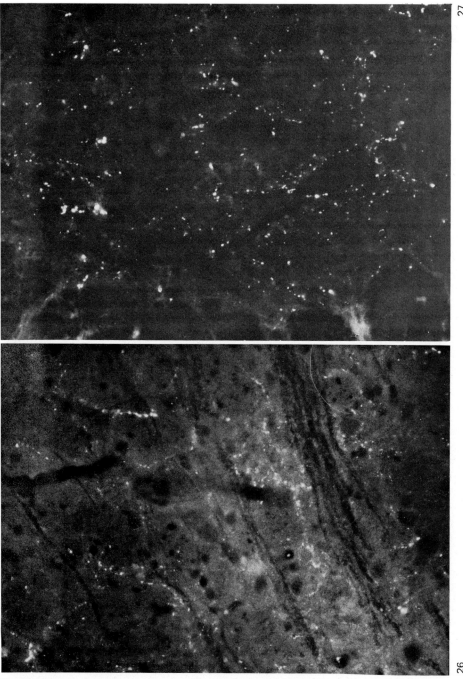

Fig. 26. TRF immunofluorescence. A sparse network of varicose TRF-containing nerve terminals is observed in the reticular formation of the medulla oblongata. x 400.

Fig. 27. TRF immunofluorescence. Ventral horn of the lumbar spinal cord. A meshwork of varicose TRF-containing terminals is observed in the most anterior part of the ventral horn. x 400.

Fig. 28. Same treatment and area as in Fig. 27 after adsorption of the TRF antiserum with TRF. Specific immunofluorescence is no longer present. x 400.

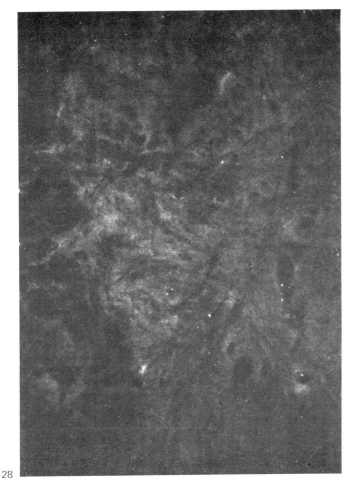

28

lary nucleus. Similar results have been obtained by other groups (Dubois et al., 1974; Pelletier et al., 1975; Setalo et al., 1975).

More recently Hökfelt and collaborators have found numerous somatostatin-containing cell bodies in the anterior periventricular area and networks have now also been found in the suprachiasmatic nucleus, in the neostriatum, and in the tuberculum olfactorium and in many other brain areas and spinal cord (Hökfelt et al., 1975e; Johansson et al., to be published).

These findings indicate that this hypothalamic hormone can subserve a transmitter role in large parts of the brain. In this connection it should be pointed out that Renaud et al. (1975) have suggested that somatostatin neurons project both to the median eminence and to hypothalamic nuclei such as the ventromedial hypothalamic nucleus, where they may inhibit GRF neuron activity. The existence of recurrent inhibitory collaterals of the supraoptic system has also been described (Nicoll and Barker, 1971).

The findings of Hökfelt et al. (1974b) and of other groups clearly suggest the possibility of a dopaminergic control of the somatostatin release from the median eminence.

TABLE II

CA turnover in the various areas and regions of the median eminence

Rate constants (k/hr ± 95% conf. interval) of CA fluorescence decline after tyrosine hydroxylase inhibition (H44/68, 250 mg/kg, i.p.) for the various areas and regions of the median eminence. NM = normal male; AM = adrenalectomized male; CM = castrated male; CF = castrated female; PM = pregnant female; Ro = rostral region; Ce = central region; Ca = caudal region. (From Löfström et al., 1975b.)

	NM	AM	CM	CF	PF
Subependymal layer					
Ro	0.084 ± 0.102	0.228 ± 0.438	0.162 ± 0.084	0.084 ± 0.162	0.468 ± 0.474
Ce	0.162 ± 0.108	0.420 ± 0.474	0.102 ± 0.156	0.132 ± 0.168	0.390 ± 0.354
Ca	0.186 ± 0.102	0.384 ± 0.420	0.438 ± 0.210	0.204 ± 0.186	0.438 ± 1.056
Medial palisade zone					
Ro	0.240 ± 0.060	0.444 ± 0.270	0.402 ± 0.108	0.276 ± 0.126	0.654 ± 0.342
Ce	0.318 ± 0.060	0.372 ± 0.342	0.330 ± 0.150	0.342 ± 0.144	0.468 ± 0.180
Ca	0.246 ± 0.078	0.564 ± 0.336	0.330 ± 0.210	0.168 ± 0.156	0.504 ± 0.570
Lateral palisade zone					
Ro	0.168 ± 0.054	0.210 ± 0.258	0.264 ± 0.114	0.276 ± 0.126	0.504 ± 0.336
Ce	0.228 ± 0.054	0.264 ± 0.240	0.258 ± 0.114	0.300 ± 0.0078	0.308 ± 0.204
Ca	0.216 ± 0.078	0.288 ± 0.216	0.306 ± 0.132	0.168 ± 0.0078	0.570 ± 0.474

III. THE FUNCTIONAL ROLE OF HYPOTHALAMIC AND EXTRAHYPOTHALAMIC NEUROTRANSMITTER SYSTEMS

III.1. DA neuron systems

For many years this laboratory has studied the changes that occur in DA and NA turnover in the median eminence and in other hypothalamic nuclei in various endocrine states (see Fuxe and Hökfelt, 1969; Hökfelt and Fuxe, 1972; Fuxe et al., 1975a, b, c; Löfström et al., 1975a, b). With the use of quantitative microspectrofluorimetry it has been possible to show that the DA turnover in the medial external layer (MPZ) is higher than that in the lateral external layer (LPZ) (see Tables II and III; Löfström et al., 1975b). Thus, there may exist at least two types of DA pathways, one lateral controlling the LRF secretion (see above) and one medial controlling, e.g., prolactin secretion. As seen in Tables II and III there occur no certain changes in DA turnover of the median eminence after castration of female and male rats or after adrenalectomy. However, in agreement with previous findings (see Fuxe and Hökfelt, 1969; Hökfelt and Fuxe, 1972) in pregnancy the DA turnover was increased in the median eminence both in the medial palisade zone (MPZ) and lateral palisade zone (LPZ) (Löfström et al., 1975b) as shown in the correlation matrix of preferential mediation (Table IV). Thus, the tuberoinfundibular DA neurons may be involved in the control of gonadotrophin secretion. In support of this view, there are variations in DA turnover of the median eminence in relation to the estrous cycle with a low DA turnover in relation to proestrous and early estrous (see Fuxe and Hökfelt, 1969; Hökfelt and Fuxe, 1972). Löfström (1975, to be published) has recently shown a half-life of approximately 3 hr in relation to proestrous of 5 day cycling

TABLE III

Half-lives $(t_{1/2})$ of CA fluorescence decline after tyrosine hydroxylase inhibition

$t_{1/2} = \dfrac{0.6931}{k}$, where k is obtained on the basis of analysis of variance (Table II). Ro = rostral region; Ce = central region; Ca = caudal region; NM = normal male; AM = adrenalectomized male; CM = castrated male; CF = castrated female; PF = pregnant female.

	NM	AM	CM	CF	PF
Subependymal layer (SEL)					
Ro	8.3	3.0	4.3	8.3	1.5
Ce	4.3	1.7	6.8	5.3	1.8
Ca	3.7	1.8	1.6	3.4	1.6
Medial palisade zone (MPZ)					
Ro	2.9	1.6	1.7	2.5	1.1
Ce	2.2	1.9	2.1	2.0	1.5
Ca	2.8	1.2	2.1	4.1	1.4
Lateral palisade zone (LPZ)					
Ro	4.1	3.3	2.6	2.5	1.4
Ce	3.0	2.6	2.7	2.3	2.3
Ca	3.2	2.4	2.3	4.1	1.2

Sprague—Dawley rats (Fig. 29), whereas in estrous the half-life is approximately 1.8 hr. Also recently a detailed quantitative analysis has been made on the changes that occur in DA and NA turnover of the median eminence after PMS treatment of prepubertal female rats on day 30 (Löfström, Agnati, Fuxe and Hökfelt, to be published). In relation to the endogenous LH peak reduction on day 32 (see Fig. 36) there occurs a slowing of DA turnover, which is maximal at 7—9 p.m. (Fig. 30). Taking together all the findings, it may be stated that a low DA turnover in the LPZ is correlated with a high secretion of LH, and vice versa. We have therefore suggested that the lateral tuberoinfundibular DA pathway exerts an inhibitory action on the peak LRF secretion (Fuxe and Hökfelt, 1969; Hökfelt and Fuxe 1972; Fuxe et al., 1975a, b, c).

It may be mentioned that our k values for DA fluorescence disappearance in endocrine states after tyrosine hydroxylase inhibition agree well with recent findings from other groups on DA turnover in the median eminence (Bacopoulus et al., 1975; Versteeg et al., 1975).

In our studies, evidence has been obtained that the inhibitory feedback action of estrogen on LH secretion (Smith and Davidson, 1974; Sawyer, 1975) may be mediated via increased activity in the tuberoinfundibular DA neurons (see Fuxe and Hökfelt, 1969; Hökfelt and Fuxe, 1972; Fuxe et al., 1975c; Löfström et al., to be published). After

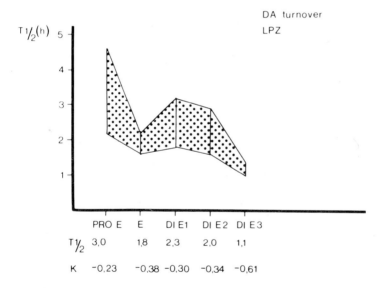

Fig. 29. DA turnover changes during the ovarian cycle in the lateral external layer (LPZ) of the median eminence of the 5 day cycling rat. On the x-axis the various stages of the ovarian cycle are given. ProE = proestrus; E = estrus; DiE1 = diestrus day 1; DiE2 = diestrus day 2; DiE3 = diestrus day 3. Under each stage of the ovarian cycle the half-life ($T_{1/2}$) is given, which has been calculated from the rate constant ($k \times hr^{-1}$). The values are based on studies on 16—23 rats and obtained by measuring the disappearance of DA fluorescence after injection of the tyrosine hydroxylase inhibitor α-methyl-tyrosine methylester (H44/68), 30, 60 and 120 min after injection using quantitative microfluorometry. On the y-axis the half-lives with their 95% confidence intervals (dotted area) are shown. (From Löfström, to be published.)

repeated daily injections of estradiol benzoate (20 μg, s.c.) to ovariectomized rats there occurs a 5-fold increase in the DA turnover of MPZ and LPZ (Fig. 31) which is correlated with a reduction of serum LH levels. As seen in Fig. 32, estrogen (20 μg, 72 hr) given to ovariectomized rats causes a dose-dependent increase in DA turnover of the MPZ and LPZ. This increase is seen to be well correlated with a dose-dependent reduction of serum LH levels (Löfström, Eneroth, Gustafsson and Skett, to be published).

Intraindividual correlations between the lowering of LH levels and the increase of DA turnover after estrogen have been made using treatment with the tyrosine hydroxylase inhibitor α-methyl-tyrosine methylester (H44/68) to study DA turnover. For LPZ the

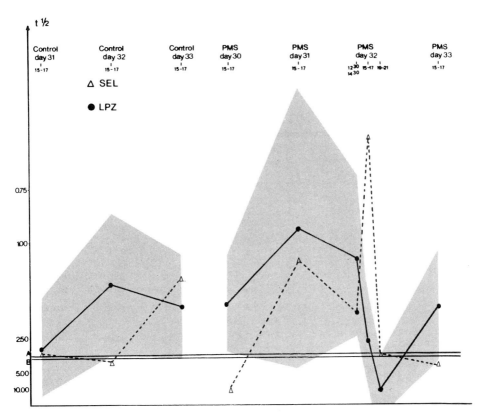

Fig. 30. CA turnover changes occurring in the lateral external layer (LPZ, mainly DA, ●) and in the subependymal layer (SEL, △, mainly NA) of the median eminence after PMS treatment (10 I.U., s.c.) or saline treatment on day 30 of female prepubertal rats. On the x-axis the various time-intervals studied after PMS or saline treatment on day 30 are shown. On the y-axis the half-lives ($T_{1/2}$) in hours are indicated. The half-lives were calculated from the rate constants ($k \times hr^{-1}$). The rate constants were obtained as described in text to Fig. 29. The shaded area indicates the 95% confidence interval for LPZ (●). Line A indicates the higher confidence interval for the saline treated group on day 32, 3—5 p.m. Line B indicates the lower confidence interval for the PMS treated group on day 32, 7—9 p.m. Note that the confidence intervals do not overlap (from Löfström, Agnati, Fuxe and Hökfelt, to be published).

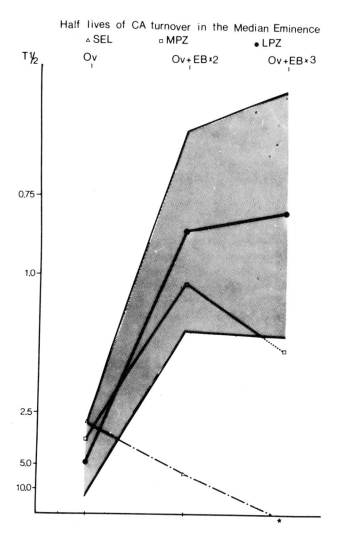

Fig. 31. The effect of repeated estrogen treatment on CA turnover in the median eminence of ovariectomized rats. On the x-axis the various treatments are shown. Ov = ovariectomy (one month) plus treatment with oil (solvent for 17β-estradiolbenzoate). Ov + EB x 2 = 17β-estradiolbenzoate (20 μg, s.c.) given twice with 24 intervals, the last injection given 2–4 hr before killing. Ov + EB x 3 = 17β-estradiolbenzoate (20 μg, s.c.) given three times with 24 hr intervals, the last injection given 2–4 hr before killing. On the y-axis the half-lives are shown for the various regions (△ = SEL; □ = MPZ; ● = LPZ) calculated from their respective rate-constants (k x hr^{-1}). The k-values are based on studies on 20–23 rats and obtained by studying the disappearance of CA fluorescence after H44/68 as described in text to Fig. 29. The shaded area represents the 95% confidence interval for the LPZ (from Löfström, Eneroth, Gustafsson, Skett, to be published).

correlation coefficient is + 0.870*** (Pearson's product moment correlation coefficient ***: $P < 0.001$) and for MPZ + 0.509* (*: $P < 0.05$) clearly indicating that the median eminence DA terminals and especially those in LPZ may mediate at least partly the inhibitory estrogen feedback on LRF secretion (see Table V). In agreement with this view Grant and Stumpf (1973) have found that estradiol accumulates in some of the DA cell bodies in the median eminence. Also the time curve for the estrogen-induced increase in DA turnover is relatively well correlated with the lowering of LH and FSH secretion. In Fig. 33, the time curve for the effects of estrogen (20 µg, s.c.) on DA turnover, serum LH and FSH levels is shown in the ovariectomized rat. The FSH and LH levels were determined without H44/68 treatment in a separate experiment. The percent change from the control (± 0) seen after estrogen at each time interval is plotted. The confidence intervals around the respective control areas (± 0) have also been given. The results show that the prolonged increase in DA turnover in LPZ till day 4 is correlated with a prolonged reduction of serum LH and FSH levels. Also the initial reduction of LH and FSH levels seen after estrogen is accompanied by an increase of DA turnover in LPZ.

TABLE IV

Correlation matrix of preferential mediation

The calculations are based on the k-values obtained from Table V. The highest k-value is given the rank 1 and the lowest k-value the rank 3. Likewise, the area with the highest proportion of DA nerve terminals (Löfström et al., 1975a) is assigned the rank 1. According to the Spearman's rank correlation, a positive correlation signifies a correlation of CA turnover with DA and a negative value a correlation of CA turnover with NA. Regarding some assumptions underlying this correlation matrix, see Löfström et al., 1975a. NM = normal male; AM = adrenalectomized male; CM = castrated male; CF = castrated female; PF = pregnant female; Ro = rostral region; Ce= central region; Ca = caudal region.

	NM	AM	CM	CF	PF
Ro	+0.5	−0.5	+0.5	+0.1	+0.5
Ce	+0.5	−1.0	+0.5	+0.5	−0.5
Ca	+0.5	−0.5	−1.0	−0.5	+1.0

TABLE V

Intraindividual correlations between CA fluorescence and plasma LH concentrations in estrogen-primed ovariectomized rats

For details on treatment see Fig. 32. SEL: subependymal layer; MPZ: medial palisade zone; LPZ: lateral palisade zone; MPOA: medial preoptic area; Caud: nucleus caudatus.

Areas	N	Without H44/68	N	With H44/68
SEL	15	+ 0.105	18	− 0.777***
MPZ	15	+ 0.361	18	+ 0.509*
LPZ	15	− 0.230	18	+ 0.870***
MPOA	13	+ 0.473§	26	− 0.438*
Caud	15	+ 0.094	29	− 0.097

Pearson's product moment correlation coefficients are shown: ***: $P < 0.001$. **: $P < 0.01$. *: $P < 0.05$. §: $0.05 < P < 0.1$.

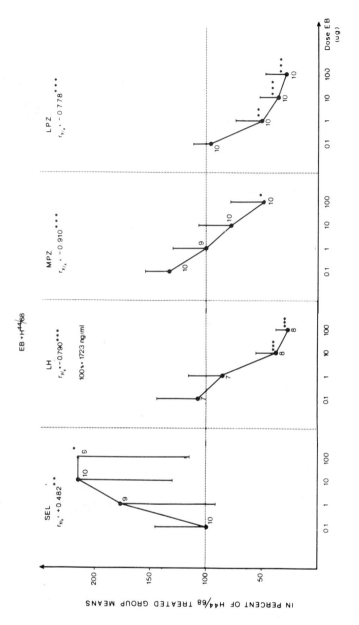

Fig. 32. The effect of various doses of estrogen on the CA turnover of the median eminence and on the serum LH levels of the ovariectomized rat. All animals received H44/68 (250 mg/kg, i.p., 2 hr before killing). On the x-axis the doses of 17β-estradiolbenzoate (72 hr before killing, s.c.) are given on a logarithmic scale. On the y-axis the fluorescence values and serum LH levels are given in percent of oil and H44/68 treated group means ± S.D. The sample size is shown at each respective point. Statistical significance was performed according to Student's t-test. * $P < 0.05$; ** $P < 0.01$; *** $P < 0.001$. The dose—effect relationship was studied by calculating Pearson's product moment correlation coefficient ($r_{y/x}$). (From Löfström, Eneroth, Gustafsson and Skett, to be published.)

All these and previous findings from our group speak against the view of McCann and coworkers (McCann et al., 1972) that estrogen blocks LRF secretion by blocking a stimulating action of DA on the LRF neurons.

The mechanism for the marked change in median eminence DA turnover caused by estrogen is not known. It may be mentioned that no corresponding changes in DA turnover are found in the striatum and the limbic forebrain (Fuxe et al., 1975a; Löfström et al., to be published). Since an increase of DA turnover was observed, the change cannot be explained on the basis of a catecholamine-O-methyltransferase inhibition (Breuer and Köster, 1974) due to the possible formation of catechol estrogens (Fishman and Norton, 1975). Direct effects on DA uptake and release by estrogens can certainly be involved, but no clear effects on DA uptake by estrogen have so far been reported (Endersby and Wilson, 1973; Wirz-Justin et al., 1974). In view of the work of Grant and Stumpf (1973) it may be that at least some of the DA cell bodies contain estradiol receptors. The activation of these receptors by estrogen may subsequently lead to, e.g., a change in the sensitivity of the receptors for transmitters located on the surface of the DA cell bodies, which could result in an increase in nervous impulse flow and thus in an increase of DA turnover.

In agreement with our view a DA receptor agonist, ET 495 (see Fuxe et al., 1975d) has been found to lower the LH and to some extent also FSH secretion in the ovariectomized female rat (Fuxe et al., 1975c; Agnati, Wuttke, Fuxe and Hökfelt, to be published). This reduction is counteracted to a large extent by pretreatment with pimozide, a DA receptor blocking agent. Pimozide itself and also H44/68 were not influencing the serum LH and FSH levels of the ovariectomized rat when given acutely (see Fig. 34b) in agreement with previous results (Carr et al., 1975; Cocchi et al., 1975). After repeated injections of pimozide a certain reduction of FSH and LH levels, however, does occur in the ovariectomized rat, which may be related to reduction of NA turnover that develops in the hypothalamus under these conditions. A trend for a reduction of LH secretion by pimozide has also been observed by Ojeda et al. (1974).

It is known from the work of Sawyer and coworkers (see Blake et al., 1974; Sawyer, 1975) that nicotine can reduce FSH, LH and prolactin secretion in various endocrine states and cause a delay in the appearance of the LH and FSH peak secretion in proestrous. Also the pulsatile discharge of LH in the ovariectomized rat is blocked by nicotine as is seen after estrogen (Blake et al., 1974). In view of these findings it has been tested whether nicotine can exert these inhibitory actions on gonadotrophin secretion via activation of the inhibitory dopaminergic mechanism in the median eminence. As seen in Fig. 34a nicotine causes a certain depletion of the DA stores and an enhancement of the H44/68-induced DA fluorescence disappearance in the median eminence of the ovariectomized rat, suggesting that nicotine may be a granular releaser of DA stores in the median eminence. Similar effects were observed in both LPZ and MPZ but other DA systems did not show signs of being activated. These changes could be in the same group of rats correlated with a reduction of LH, FSH and prolactin secretion. Furthermore, pimozide counteracted the effects of nicotine on DA turnover and this effect appeared to be correlated with a reduction of the effects of nicotine on LH and FSH serum levels. Also nicotine, when injected prior to the critical period in PMS treated prepubertal rats, was

222

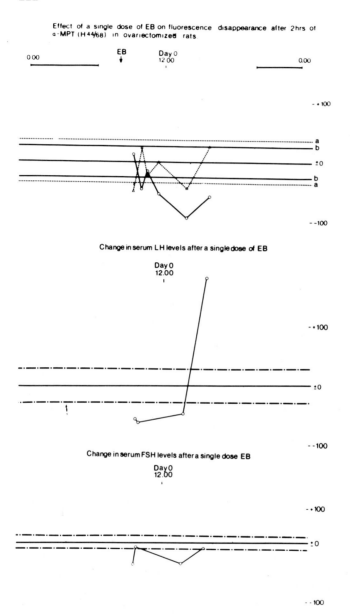

Fig. 33. The effect of an acute injection of 17β-estradiolbenzoate on the CA turnover and serum FSH and LH levels in the ovariectomized rat at various time intervals following the injection. The changes in CA turnover were studied in a separate experiment using H44/68 treatment (250 mg/kg, i.p., 2 hr before killing). In a second experiment the effects of estrogen on serum FSH and LH levels were studied without any concomitant treatment with H44/68. On the x-axis the time intervals after estrogen are shown. At each time interval a pair of rats have been studied, i.e., a control and an

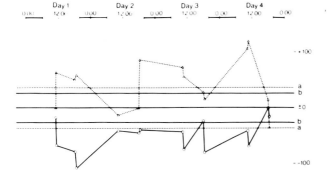

Effect of a single dose of EB on fluorescence disappearance after 2 hrs of α-MPT (H44/68) in ovariectomized rats

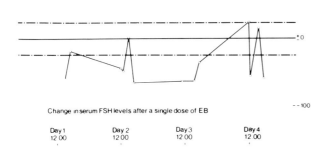

Change in serum LH levels after a single dose of EB

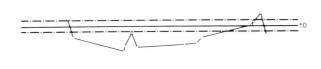

Change in serum FSH levels after a single dose of EB

estrogen treated animal (20 μg, s.c.). On the y-axis the difference between the control and estrogen treated animals are shown in percent of the respective control animals. △ = SEL; 0 = LPZ. The control animals are set as 0 and around the zero line the 99.9% confidence intervals for the SEL fluorescence (line a) and LPZ fluorescence (line b) is shown for the overall control group means and the 95% confidence interval for the FSH and LH levels (hatched line) of the overall control group means (from Löfström, Eneroth, Gustafsson and Skett, to be published).

found to increase DA release in the median eminence and this was associated with a delay in the appearance of the LH peak secretion (Figs. 35 and 36). In view of this, studies on nicotine have given further support for the hypothesis that median eminence DA nerve terminals inhibit LRF and prolactin secretion. Pharmacological studies on ovulation in prepubertal rats treated with PMS have also given support for our view that the lateral tuberoinfundibular DA pathway inhibits LRF secretion (see Fuxe et al., 1975a–d). After i.p. injections of PMS (10 I.U.) on day 30 there occurs ovulation on day 33 in 90% of the rats as seen in Fig. 37. The median number of eggs is 10 with a 95% confidence interval of

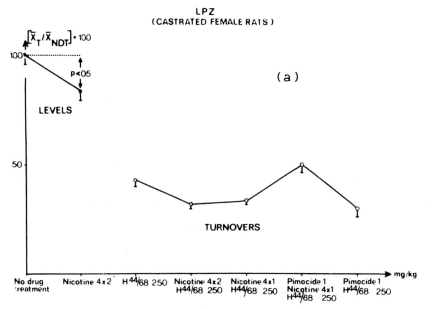

Fig. 34a. The effects of nicotine and pimozide alone or together on the H44/68-induced CA fluorescence disappearance in the median eminence of the ovariectomized female rat. On the x-axis the different treatments are given. Nicotine was given 4 times with 30 min intervals, the first injection being given immediately before the H44/68 injection at 2 p.m. (250 mg/kg, 2 hr before killing, i.p.). Pimozide (1 mg/kg, i.p.) was given 2 hr before the H44/68 injection. On the y-axis the mean values ± S.E.M. ($\bar{X}_T$) for each treatment are given in percent of the no drug treatment group mean value ($\bar{X}_{NDT}$). The mean sample size is 7 rats. The statistical analysis was made according to one-way analysis of variance and subsequently multiple comparisons were made: sum of squares simultaneous test procedure (experimentalwise error rate $\alpha = 0.05$). The means not significantly different are embraced by the same line in the tables of the figure. Student's t-test was used to compare CA fluorescence found in the no drug treatment group and the nicotine treated group. Lateral palisade zone (LPZ). (From Eneroth et al., 1975.)

Pimozide 1 mg/kg + H44/68
Nicotine 4x2 mg/kg + H44/68
Nicotine 4x1 mg/kg+ H44/68
H44/68 mg/kg
Pimozide 1 mg/kg + nicotine 4x1 mg/kg + H44/68

± 1. After treatment with DA receptor agonists such as ergocornine and CB 154 the proportion of non-ovulating rats is increased from 10 to 80% (hatched bar in Fig. 37). In Fig. 38 it can be seen that pimozide counteracts significantly the inhibitory effects of ET 495 on the number of eggs. Similar results have been obtained with the other types of DA receptor agonists mentioned above (Agnati et al., 1975; Fuxe et al., 1975a–d). Again these studies strongly speak against the view of McCann and collaborators (McCann et al., 1972; McCann and Moss, 1975) that DA increases LRF secretion. This hypothesis was a matter of fact first suggested by Kordon (see review from 1973) and supported also by

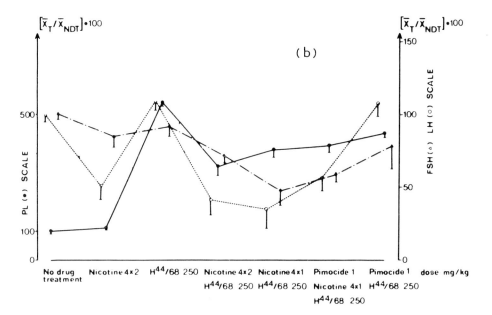

Fig. 34b. The effects of nicotine, H44/68, and pimozide alone or together on the serum FSH, LH and prolactin levels in the ovariectomized female rat. The different treatments are shown on the x-axis (for the complete schedule, see Fig. 34a). On the y-axis the percentage ratio between the mean values after each treatment ($\bar{X}_T$) and the mean value of the control group ($\bar{X}_{NDT}$, no drug treatment) is given. The serum hormone levels were as follows: (means ± S.E.) 1107 ± 190 ng/ml (7 rats, LH), 2974 ± 279 ng/ml (7 rats, FSH) and 26.7 ± 1.9 ng/ml (7 rats, prolactin). It should be noted that for FSH and LH a different scale was used from the one used for prolactin. For each treatment the mean ± S.E. (N = 7) is given. The statistical analysis was performed according to one-way analysis of variance and subsequently multiple comparisons were made: sum of squares simultaneous test procedure (experimentalwise error rate $\alpha = 0.05$). The means not significantly different are embraced by the same line in the tables of the figure. (From Eneroth et al., 1975.)

PL	LH	FSH
No drug treatment	Nicotine 4 × 1 mg/kg + H44/68	Nicotine 4 × 1 mg/kg + H44/68
Nicotine 4 × 2 mg/kg	Nicotine 4 × 2 mg/kg + H44/68	Pimozide 1 mg/kg + nicotine 4 × 1 mg/kg + H44/68
Nicotine 4 × 2 mg/kg + H44/68	Nicotine 4 × 2 mg/kg	Nicotine 4 × 2 mg/kg + H44/68
Nicotine 4 × 1 mg/kg + H44/68	Pimozide 1 mg/kg + nicotine 4 × 1 mg/kg + H44/68	
Pimozide 1 mg/kg + nicotine 4 × 1 mg/kg + H44/68	No drug treatment	Pimozide 1 mg/kg + H44/68
Pimozide 1 mg/kg + H44/68	Pimozide 1 mg/kg + H44/68	Nicotine 4 × 2 mg/kg
H44/68 250 mg/kg	H44/68 250 mg/kg	H44/68 250 mg/kg
		No drug treatment

Lichtensteiger (1973). Recently, however, support for our hypothesis has been obtained in several laboratories in studies on the action of DA on LH secretion (see Miyachi, 1973; Sawyer, 1975). In our opinion, selective DA receptor agonists for the median eminence DA receptor should have a great potential in control of fertility both in males and females. A large number of studies have suggested that median eminence DA nerve terminals participate in the inhibitory control of prolactin secretion (Maanen and Smelik, 1968; Hökfelt and Fuxe, 1972; McCann et al., 1972; Carr et al., 1975; Lawson and Gala, 1975). Recent findings suggest that DA may be one of the prolactin inhibitory factors (Macleod and Lehmeyer, 1974). It has been found by our group that ovine prolactin increases DA turnover in the MPZ and LPZ, particularly in the hypophysectomized rat (see Fig. 39), when given i.v. or i.p. twice daily for 2–3 days. The incraese in DA turnover still remains after prolactin has long disappeared from the circulation since the DA

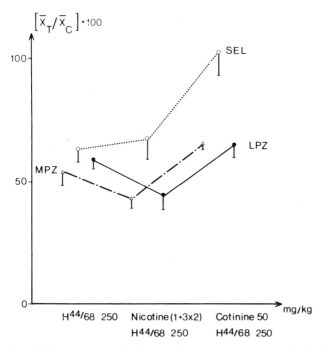

Fig. 35. The effect of nicotine, and cotinine on the H44/68-induced CA fluorescence disappearance in the median eminence of PMS treated prepubertal rats. On the x-axis the different treatments are shown. Cotinine was given i.p. 15 min before H44/68. The first nicotine dose was 1 mg/kg, i.p. (2 p.m., day 32), for further details, see Fig. 34a. On the y-axis the mean values ($\bar{X}_T$) ± S.E. are given in percent of the PMS alone treated groups mean value ($\bar{X}_C$). The sample size is 7 rats. The statistical analysis was performed according to one-way analysis of variance and subsequently multiple comparisons were made (experimentalwise error rate $\alpha = 0.05$). The means not significantly different are embraced by the same line in the tables of the figure. (From Eneroth et al., 1975.)

LPZ	MPZ	SEL
Nicotine (1 + 3 × 2) mg/kg + H44/68	Nicotine (1 + 3 × 2) mg/kg + H44/68 ⎫	H44/68 mg/kg
H44/68 250 mg/kg ⎫	H44/68 250 mg/kg ⎬	Nicotine (1 + 3 × 2) mg/kg + H44/68 ⎫
Cotinine 50 mg/kg ⎭	Cotinine 50 mg/kg + H44/68 ⎭	Cotinine 50 mg/kg + H44/68 ⎭

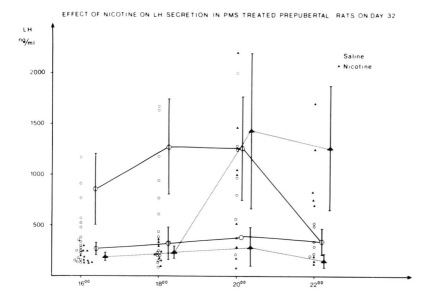

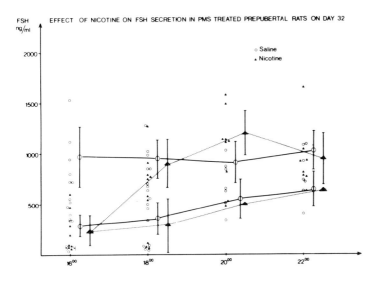

Fig. 36. The effect of nicotine treatment on the LH and FSH peak secretion of PMS treated prepubertal rats on day 32. For details on nicotine treatment, see Figs. 34—35. The first injection was made at 2 p.m. On the x-axis the time intervals studied are shown. On the y-axis the individual LH and FSH values are shown; o = saline treated group; ▲ = nicotine treated group. Also the group means of the basal and peak LH and FSH levels are shown (larger symbols) with 95% confidence intervals. (From Eneroth et al., 1975.)

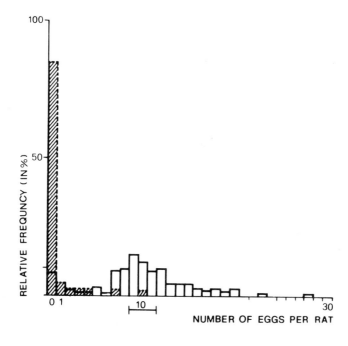

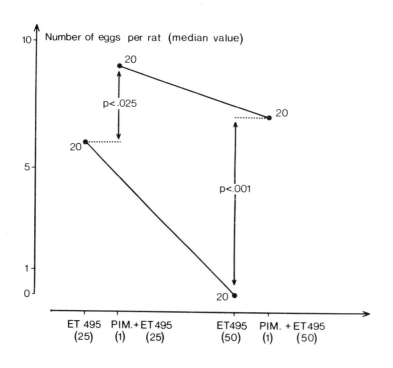

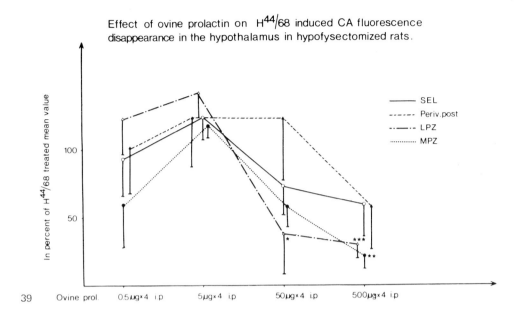

Fig. 37. Relative frequency distribution of eggs in control (N = 107) and animals treated with DA receptor stimulating agents (striped bars; N = 44). The median value with its 95% confidence interval is shown for the control distribution (quantile test). (From Löfström, Agnati, Fuxe and Hökfelt, to be published; see also Fuxe et al., 1975d.)

Fig. 38. The effect of pimozide on the ET 495-induced inhibition of ovulation. ET 495 is a well known DA receptor agonist (see Fuxe et al., 1975d). On the x-axis the treatments are given. ET 495 was given s.c. in two doses at 13.30 and 14.30. Total dose is shown in figure. Pimozide (1 mg/kg) was given i.p. at 12.30. On the y-axis the number of eggs are given. Pimozide counteracts the inhibitory effects of ET495 on number of eggs (slippage test). Sample size is also shown. (From Löfström, Agnati, Fuxe and Hökfelt, to be published.)

Fig. 39. The effect of different doses of ovine prolactin on CA turnover in the hypothalamus of hypophysectomized male rats. Hypophysectomy was performed one month earlier. On the x-axis the doses of prolactin are given on a logarithmic scale. The doses of prolactin were given day and morning twice daily for 2 days and the rats were killed the following morning after 2 hr of H44/68 treatment (250 mg/kg, i.p.). On the y-axis the fluorescence values are given in percent of H44/68 group mean value ± S.D. (3–5 rats per group). Statistical analysis according to Student's t-test: * $P < 0.05$; ** $P < 0.01$; *** $P < 0.001$. Notice also the trend for acceleration in MPZ for the lowest dose of prolactin used (0.5 μg × 4) and the trend for acceleration in the SEL and posterior periventricular hypothalamus (Periv. post.) in the highest dose (500 μg × 4). All comparisons are made with the control group.

turnover is studied 12—14 hr after the last injection. These results suggest that prolactin may mediate its inhibitory feedback on its own secretion via increasing DA release in the median eminence. Since DA turnover is increased both in MPZ and LPZ, also the inhibitory effects of prolactin on LH secretion could partly involve inhibition of LRF release in view of the increased DA release in LPZ (see Fuxe et al., 1975a—d).

Interestingly it was found that ovine prolactin also increases DA turnover in the nucleus caudatus but not in the limbic forebrain (Fig. 40). These results may offer a possible partial neurochemical basis for the effects of prolactin on nesting and maternal behavior.

DA neurons in the median eminence may not only be involved in the control of LRF and prolactin secretion but also in control of GH and TRF secretion as already suggested previously in this article. Thus, pharmacological studies in the rat have indicated that DA neurons may inhibit TRF secretion (Tuomisto et al., 1975) and GH secretion (Müller, 1973). DA neurons are also involved in the control of sexual behavior. Recent pharmacological studies suggest that DA pathways inhibit sexual behavior (Everitt et al., 1975a, b). As seen in Fig. 41, however, it is found in the lower dose range that DA receptor agonists such as ET 495 and apomorphine can enhance sexual behavior in the ovariectomized

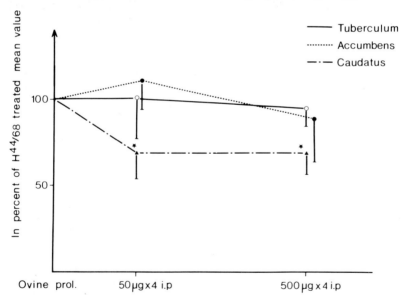

Fig. 40. The effect of ovine prolactin on the DA turnover in the forebrain hypophysectomized male rats. On the x-axis the doses of prolactin are given. The values are from the same experiment as shown in Fig. 39. On the y-axis the group means are given in percent of the H44/68 treated group means ± S.D. (3—5 rats per group). Student's t-test. * $P < 0.05$. All comparisons are made with the control group.

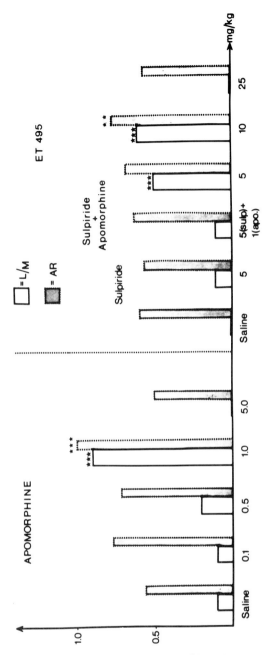

Fig. 41. The effect of various doses of apomorphine and ET 495 on the sexual receptivity of estrogen primed ovariectomized rats. On the x-axis the different treatments are shown. All rats received a daily 0.5 µg injection of 17β-estradiolbenzoate. Apomorphine was given 10 min before testing. Sulpiride was given i.p. 30 min before apomorphine. Sulpiride may preferentially block pre-DA synaptic receptors. ET 495 was given i.p. 30 min before testing. On the y-axis the L/M (= number of lordosis responses ÷ number of adequate mounts by male) and AR (= acceptance ratio = number of mounts ÷ number of mounts + number of refused mounts). Mann–Whitney U-test. All comparisons are made with the saline treated control group. ** $P < 0.01$; *** $P < 0.001$. (Everitt and Fuxe, to be published.)

estrogen primed rat. The explanation is probably that in the lower doses ET 495 and apomorphine preferentially stimulate presynaptic DA receptors on the DA cell bodies (Everitt and Fuxe, to be published). It is of considerable interest to note that whereas LRF causes ovulation and favors sexual behavior, DA inhibits ovulation and sexual behavior.

III.2. NA neuron systems

Sawyer and collaborates (Sawyer et al., 1949; Sawyer, 1975) were the first to suggest on the basis of studies on adrenergic blocking agents that NA in the hypothalamus may be essential for LH secretion and ovulation. In agreement, after castration, there also occurs an increase of whole brain NA turnover of female rats (Anton-Tay et al., 1969)

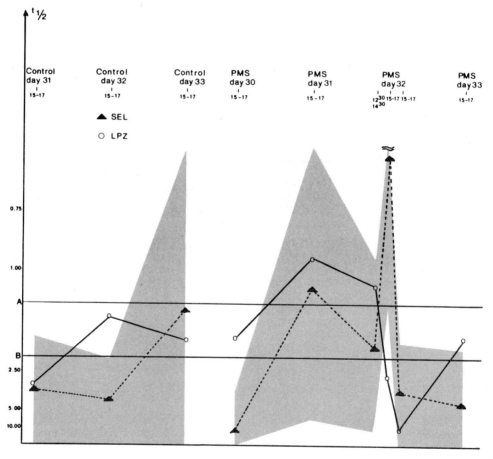

Fig. 42. Same experiment as in Fig. 41. In this figure, however, the 95% confidence interval (shaded area) has been given for the SEL (▲) instead of for LPZ (○). Line A indicates the higher confidence interval for the PMS treated group, day 32, 3–5 p.m. Line B shows the lower confidence interval for the saline treated group, day 32, 3–5 p.m. Notice that no overlap occurs.

and in hypothalamic NA turnover (Donoso et al., 1969). In whole brain of castrated male rats, however, no change in NA turnover has been found (Fuxe and Hökfelt, 1969; Bernard and Paolino, 1974). In Tables II and III it can be seen that there is a trend for an increased NA turnover in the subependymal layer (SEL) of castrated male and female rats (Löfström et al., 1975b), in agreement with the work of Bernard and Paolino (1974) also showing an increase of NA turnover in the hypothalamus of castrated rats. In our laboratory it has recently been possible to demonstrate an increase of NA turnover in SEL in relation to the critical period of proestrous (Löfström, to be published) and on day 32 in prepubertal rats treated with PMS 2 days earlier (Fig. 42). Note the sharp increase of NA turnover as studied between 3–5 p.m. (from Löfström, Agnati, Fuxe, Hökfelt, to be published). As seen in Figs. 31–33 the inhibitory feedback action of estrogen on LRF secretion when given to ovariectomized rats may involve a reduction of NA neuron activity. Thus, both after single and repeated doses of estrogen there is a significant trend for reduction of NA turnover in the SEL and medial preoptic area. Intraindividual correlations between fluorescence in these regions after estrogen and H44/68 treatment and serum LH levels also show a significant inverse relationship (Löfström, Everitt, Gustafsson and Skett, to be published). Also biochemical analyses of NA in brain stem and cortex

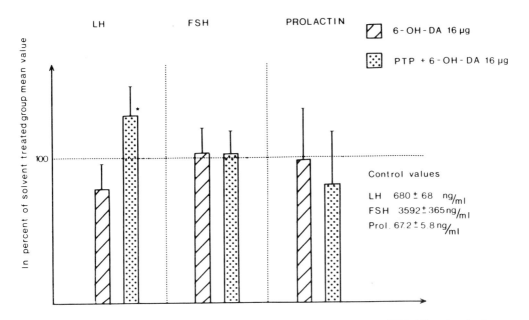

Fig. 43. Effect of intrategmental injection of 6-OH-DA (16 μg/8 μl) on serum FSH, LH and prolactin levels of ovariectomized female rats (3 months after ovariectomy) one week after the 6-OH-DA injection. Means ± S.E.M. (5–7 rats per group) are shown in percent of ovariectomized solvent injected group means. Student's t-test. * $P < 0.05$. The comparisons have been made between the two 6-OH-DA groups one being pretreated with protriptyline, the other not. The absolute levels of the hormones in the solvent treated group were as follows. FSH: 3592 ± 365 ng/ml; LH = 680 ± 68 ng/ml; prolactin = 67.2 ± 5.8 ng/ml. (From Fuxe et al., 1975.)

cerebri after estrogen treatment of ovariectomized rats reveal a significant slowing of NA turnover (Everitt et al., 1975a).

Recent studies on 6-OH-DA-induced lesion of the ventral NA pathway in ovariectomized rats have also given evidence that this pathway, possibly together with the A pathway, facilitates LH secretion. Thus, the protection of the hypothalamic NA nerve terminals from degeneration with protriptyline is associated with a selective increase of LH secretion (Figs. 43, 44; Fuxe et al., 1974e). In agreement are the findings that after 6-OH-DA treatment the NA induces an enhanced release of LH (Bacha and Donoso, 1974). Also Tima and Flerkó (1974) have been able to induce ovulation by NA in anovulatory rats (see also Sawyer, 1975). Pharmacological studies with dopamine-β-hydroxylase inhibitors have also suggested the importance of NA for the postcastration rise of LH (Cocchi et al., 1975; see also McCann and Moss, 1975). At this point it should, however, be pointed out that the large number of studies performed with DOPA is of no or little value when trying to differentiate between DA and NA involvement in various neuroendocrine processes. Thus, DOPA penetrates poorly into the brain and is decarboxylated in both DA, NA and 5-HT neurons.

Several years ago (Fuxe et al., 1970d) we postulated that NA terminals may inhibit the activity of the tuberoinfundibular DA neurons, possibly partly via increasing prolactin secretion, since α-adrenergic blocking agents and dopamine-β-hydroxylase inhibitors were found to increase DA turnover in the median eminence and clonidine was found to reduce it. Thus, the triggering of the NA neurons in relation to the critical period will then also ensure the turning off of the inhibitory DA terminals in the median eminence.

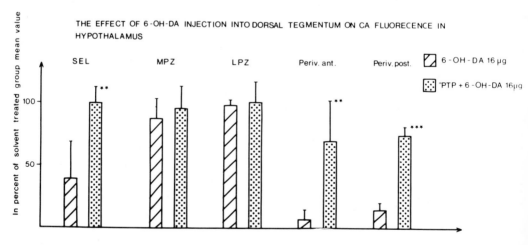

Fig. 44. The effect of intrategmental injections of 6-OH-DA (16 μg/8 μl) with or without prior treatment with protriptyline on amine fluorescence in various CA nerve terminal plexa of the median eminence and other hypothalamic regions of ovariectomized female rats. The rats are the same as those described in Fig. 43. Means ± S.D. (5–7 rats per group) are shown in percent of ovariectomized solvent treated group means. The explanation for the abbreviation used for the various regions is found in the text. The 6-OH-DA group and the protriptyline + 6-OH-DA group have been compared using Student's t-test. ** $P < 0.01$; *** $P < 0.001$. (From Fuxe et al. 1975.)

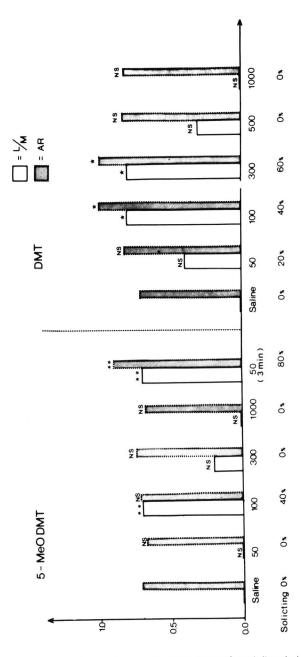

Fig. 45. The effect of 5-methoxy-dimethyltryptamine (5-MeO-DMT) and dimethyltryptamine (DMT) on the sexual behavior of the estrogen-primed ovariectomized rat. Estrogen priming was performed as described in text to Fig. 41. On the x-axis the different doses are given in µg/kg. All injections were made i.p. and 10 min before testing except when indicated otherwise in the figure. Below each treatment also the % of animals showing soliciting behavior (hopping and darting) is shown. On the y-axis the L/M and AR are shown (see text to Fig. 41). 5—10 rats were used in each group. Mann—Whitney U-test. * $P < 0.05$; ** $P < 0.01$. (Everitt and Fuxe, to be published.)

This interesting interaction is now being further investigated.

Many pharmacological studies suggest that the NA neurons also are involved in control of ACTH, TSH and GH secretion. According to the view of Ganong, van Loon, Scapagnini and collaborators (see Ganong, 1973), NA neurons may inhibit ACTH secretion. Our work (Fuxe et al., 1973) on 6-OH-DA-induced lesions of the ventral NA pathway to the hypothalamus would suggest that the hypothalamic NA terminals mainly inhibit stress-induced increases in ACTH secretion. It is important that the hypothalamic NA terminals degenerate, since otherwise no effects are observed (Fuxe et al., 1973; cf. Kaplanski et al. 1974). Also it should be pointed out that effects are observed only on the day following the lesion, since compensatory effects probably occur (see also Kaplanski et al., 1974). In agreement with an inhibitory effect of NA on CRF secretion, it has been found that NA inhibits CRF release caused by acetylcholine (Hillhouse et al., 1975).

It is also well known from previous work that stress or adrenalectomy increases NA turnover in large parts of the brain (see Fuxe et al., 1970c) including the SEL (Löfström et al., 1975b). The latter finding is of particular interest, since Réthelyi (1975) recently has found nerve cells in SEL that may produce CRF. The increases in NA turnover that occur after adrenalectomy all over the brain can be counteracted by treatment with glyco- and mineralocorticoids (see Fuxe et al., 1973). These widespread changes can probably be best related to changes in acquisition and extinction of conditional avoidance responses that occur in states with different degrees of activity in the pituitary-adrenal axis.

GH secretion appears to be controlled in the same way as LH secretion. Thus, NA appears to increase GH secretion whereas DA may inhibit GH secretion (Müller, 1973; Chihara et al., 1975). This view is mainly based on pharmacological evidence. It may be mentioned that high doses of GH have been found to increase DA turnover in LPZ (see Hökfelt and Fuxe, 1972). Again it should be pointed out that in man DA receptor agonists such as apomorphine increase GH secretion.

Also TSH secretion may be controlled by a stimulating NA mechanism and an inhibitory DA mechanism (Grimm and Reichlin, 1973; Tuomisto et al., 1975), the evidence being mainly pharmacological in nature.

In view of the above it seems as if a principal pattern develops with hypothalamic NA terminals increasing adenohypophyseal secretion and hypothalamic DA terminals decreasing it. The DA neurons in the medial basal hypothalamus probably represent an important braking mechanism inhibiting the discharge of the releasing factors such as LRF or directly inhibiting the secretion of adenohypophyseal hormones such as prolactin by releasing DA into the primary capillary plexus. The hypothalamic NA neurons obviously favor certain patterns of hormonal secretion from the pituitary gland increasing LRF, TRF and GH secretion and reducing CRF secretion. An important site of action is probably also here the median eminence since, e.g., after gonadectomy, increases in tyrosine hydroxylase activity have only been observed in the median eminence (Kizer et al., 1974). The endocrine function of the adrenaline and possible histamine pathways is still unknown. It is interesting to note, however, that adrenaline may be more potent in causing ovulation than NA (see Sawyer, 1975).

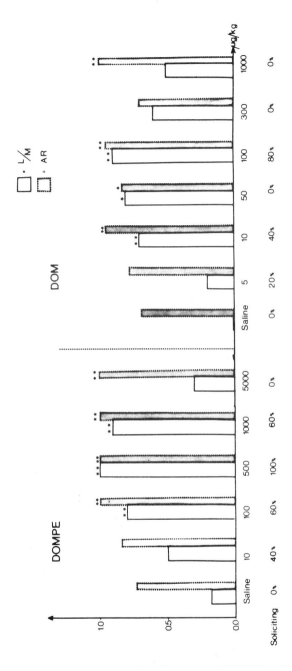

Fig. 46. The effect of 3,5-dimethoxy-4-methyl-phenylethylamine (DOMPE) and 3,5-dimethoxy-4-methylamphetamine (DOM) on sexual behavior in the ovariectomized female rat. For estrogen priming, see text to Fig. 41. On the y-axis the various doses are given in μg/kg. All injections were made i.p. 10 min before testing soliciting behavior, L/M and AR are shown on the y-axis as described in text to Figs. 41 and 45. 5–10 rats were used per group. Mann–Whitney U-test. * $P < 0.05$; ** $P < 0.01$. (Everitt and Fuxe, to be published.)

III.3. The 5-HT neuron systems

In this laboratory it has not been possible to obtain an inhibition of ovulation in PMS treated prepubertal rats after treatment with 5-HT releasing agents such as fenfluramine or 5-HT receptor stimulating drugs such as D-LSD either in low doses (10 µg/kg), which preferentially acts on presynaptic 5-HT receptors, or in high doses, which activates also the postsynaptic 5-HT receptors (Fuxe et al., 1975d). These results therefore do not agree with the view that 5-HT systems may inhibit LRF secretion (see Kordon, 1973). However, estrogen has been shown to change 5-HT turnover in the ovariectomized rat, and this change is counteracted by subsequent treatment with progesterone (Fuxe et al., 1974b; Everitt et al., 1975a). Therefore, 5-HT neurons may still be involved in the control of LRF secretion. Interestingly, the effects of estrogen on 5-HT turnover are dependent on the sleep—wakefulness cycle. Thus, at night estrogen increases 5-HT turnover, whereas it reduces at day time (Everitt et al., 1975a). Recently, Kordon (1973) has obtained evidence that 5-HT pathways mediate the suckling-induced rise in prolactin secretion and other workers (Caligaris and Taleisnik, 1974) have found that the estrogen-induced increase in prolactin secretion involves the activation of 5-HT pathways. The main function of 5-HT neurons may therefore be the control of prolactin and not LRF secretion.

Studies on the effect of ACTH and corticosterone on 5-HT turnover and on the effects of 5,6-HT-induced degeneration of ascending 5-HT pathways (Fuxe et al., 1973; Telegdy and Vermes, 1975) have indicated that 5-HT pathways may be involved in the control of diurnal rhythm and stress-induced changes of ACTH secretion, having an inhibitory action on ACTH secretion. Also the inhibitory adrenocortical steroidal feedback on ACTH secretion may to some extent involve activation of 5-HT neurons, since corticosterone in low doses (1 mg/kg) acutely increases 5-HT and 5-HIAA levels in limbic structures and hypothalamus (Telegdy and Vermes, 1975).

It should be mentioned, however, that there also exists evidence for the existence of one type of 5-HT system, which has a stimulatory influence on ACTH secretion (Fuller et al., 1975).

Findings are also available that 5-HT neurons may inhibit TRF secretion based on in vitro effects of 5-HT on TRF release from hypothalamic slices (Grimm and Reichlin, 1973) and on the effect of 5-HTP on TSH secretion (Tuomisto et al., 1975). In view of the above hypothalamic 5-HT neurons like the NA neurons may favor certain patterns of hormonal secretions from the hypothalamus increasing, e.g., prolactin secretion and reducing, e.g., ACTH and TRF secretion.

It should be underlined that the 5-HT pathways have a powerful inhibitory action on sexual behavior (see Everitt et al., 1975a) and therefore the changes in 5-HT turnover induced by gonadal steroids can also be related to their actions on sexual behavior. In Figs. 45 and 46 it can be seen that 5-HT receptor agonists such as 5-methoxy-dimethyltryptamine (5-MeO-DMT) and 3,5-dimethoxy-4-methylamphetamine (DOM) in low doses enhance sexual behavior and in high doses reduce it. The explanation for this is that in low doses they preferentially act on presynaptic 5-HT receptors hereby reducing 5-HT neurotransmission, whereas in higher doses they act also on the postsynaptic 5-HT receptor, hereby increasing 5-HT neurotransmission.

In performing a pharmacological analysis of the effects of DA and 5-HT receptor agonists on hormonal secretion it is very important that dose-effect curves are performed, so that possible effects on pre- and postsynaptic monoamine receptors can be distinguished. Otherwise, interpretation will be difficult.

III.4. Acetylcholine neuron systems

Sawyer and coworkers were able to show already in 1949 (Sawyer et al., 1949) that atropine could block ovulation when given prior to the critical period, suggesting the existence of a cholinergic mechanism in the hypothalamus stimulating the LRF secretion. These findings have been confirmed and extended by McCann's group (see McCann and Moss, 1975), who obtained evidence that the effects of atropine are centrally induced. In an in vitro system Martini's group (Simonovic et al., 1974) has also obtained evidence that acetylcholine can release LRF from the median eminence. Therefore, it seems possible that cholinergic mechanisms in the median eminence can increase LRF secretion.

Acetylcholine-containing systems may also increase ACTH secretion, since atropine can prevent increases in ACTH secretion and very low amounts (10–100 pg/me) of acetylcholine can cause release of CRF (see review by Wilson, 1974).

III.5. GABA-containing neurons

Recent evidence suggests that intraventricular injections of GABA result in a stimulation of LH secretion (Ondo, 1974) and in inhibition of ACTH secretion (Makara and

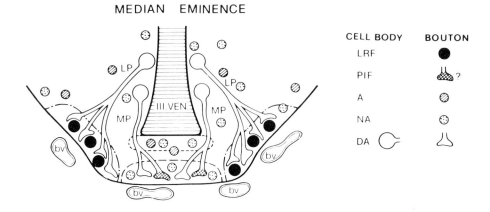

LPZ   LATERAL PALISADE ZONE ——·——·——
MPZ   MEDIAL PALISADE ZONE ——————-
SEL   SUBEPENDYMAL LAYER ------------
LP   LATERAL PATHWAY
MP   MEDIAL PATHWAY
bv   Blood vessels of primary capillary network

Fig. 47. Schematic diagram over the distribution of DA, NA, A and LRF in the rat median eminence. It is indicated that some of the medial DA terminals may be one of the PIFs. It is also indicated that the lateral DA pathways may mainly be concerned with inhibition of LRF secretion.

Stark, 1974). GABA has also been shown to inhibit CRF release from hypothalamic slices (Burden et al., 1974). These results suggest that probable GABA interneurons are connected to the circuitry controlling LRF and CRF secretion from the median eminence.

In conclusion, there is no doubt that various neurotransmitter systems in the hypothalamus play a crucial role in regulating the activity of the peptidergic releasing factor and inhibitory factor systems. In Fig. 47 the distribution and relation of the DA, NA and A systems to the LRF system in the median eminence is shown. It is also indicated that DA may be one of the PIFs. It should also be pointed out that somatostatin and TRF not only act as hypothalamic hormones but also as central neurotransmitters. Furthermore, Substance P should be considered as a transmitter which can participate in the control of neuroendocrine functions.

ACKNOWLEDGEMENTS

This work was supported by grants from the Swedish Medical Research Council (04X-715, 04X-2887, 19X-3411), M. Bergwall's Stiftelse, Knut and Alice Wallenbergs' Stiftelse and from Harald and Greta Jeanssons' Stiftelse.

REFERENCES

Anton-Tay, F., Pelham, R.W. and Wurtman, R.J. (1969) Increased turnover of ^{3}H-norepinephrine in rat brain following castration or treatment with ovine follicle stimulating hormone. *Endocrinology*, 84: 1489–1492.

Bacha, J.C. and Donoso, A.O. (1974) Enhanced luteinizing hormone release after noradrenaline treatment in 6-hydroxydopamine-treated rats. *J. Endocr.*, 62: 169–170.

Bacopoulus, N.G., Bhatnagar, R.K., Schnute, W.J. and Van Orden, III, L.S. (1975) On the use of the fluorescence histochemical method to estimate catecholamine content in brain. *Neuropharmacology*, 14: 291–299.

Barry, J., Dubois, M.P. and Poulain, P. (1973) LRF producing cells of the mammalian hypothalamus. *Z. Zellforsch.*, 146: 351–366.

Barry, J., Dubois, M.P. and Carette, B. (1974) Immunofluorescence study of the preoptico-infundibular LRF neurosecretory pathway in the normal, castrated or testosterone-treated male guinea pig. *Endocrinology*, 95: 1416–1423.

Bernard, B.K. and Paolino, R.M. (1974) Time-depending changes in brain biogenic amine dynamics following castration in male rats. *J. Neurochem.*, 22: 951–956.

Björklund, A., Moore, R.Y., Nobin, A. and Stenevi, U. (1973) The organization of tubero-hypophyseal and reticulo-infundibular catecholamine neuron systems in the rat brain. *Brain Res.*, 51: 171–191.

Björklund, A., Lindvall, O. and Nobin, A. (1975) Evidence of an incerto-hypothalamic dopamine neuron system in the rat. *Brain Res.*, 89: 29–42.

Blake, C.A., Norman, R.L. and Sawyer, C.H. (1974) Localization of the inhibitory actions of estrogen and nicotine on release of luteinizing hormone in rats. *Neuroendocrinology*, 16: 22–35.

Breuer, H. and Köster, G. (1974) Interaction between estrogens and neurotransmitter: biochemical mechanism. *Advanc. Biosci.*, 15: 287–300.

Brownstein, M., Saavedra, J.M., Palkovits, M. and Axelrod, J. (1974a) Histamine content of hypothalamic nuclei of the rat. *Brain Res.*, 77: 151–156.

Brownstein, M.J., Palkovits, M., Saavedra, J.M., Bassiri, R.M. and Utiger, R.D. (1974b) Thyrotropin-releasing hormone in specific nuclei of rat brain. *Science*, 185: 267–269.
Brownstein, M., Kobayashi, R., Palkovits, M. and Saavedra, J.M. (1975) Choline acetyltransferase levels in diencephalic nuclei of the rat. *J. Neurochem.*, 24: 35–38.
Burden, J., Hillhouse, E.W. and Jones, M.T. (1974) The inhibitory action of GABA and melatonin on the release of corticotrophin-releasing hormone from the rat hypophysiotrophic area *in vitro*. *J. Physiol. (Lond.)*, 239: 116P.
Caligaris, L. and Taleisnik, S. (1974) Involvement of neurones containing 5-hydroxytryptamine in the mechanism of prolactin release induced by oestrogen. *J. Endocr.*, 62: 25–33.
Carlsson, A., Falck, B. and Hillarp, N.Å. (1962) Cellular localization of brain monoamines. *Acta physiol. scand.*, 56, Suppl. 196: 196.
Carr, L.A., Conway, P.M. and Voogt, J.L. (1975) Inhibition of brain catecholamine synthesis and release of prolactin and luteinizing hormone in the ovariectomized rat. *J. Pharmacol. exp. Ther.*, 192: 15–21.
Chihara, K., Kato, Y., Ohgo, S. and Imura, H. (1975) Effects of drugs influencing brain catecholamines on GH release in rats with hypothalamic surgery. *Neuroendocrinology*, 18: 192–203.
Cocchi, D., Fraschini, F., Jalanbo, H. and Müller, E.E. (1975) Role of brain catecholamines in the postcastration rise in plasma LH of prepubertal rats. *Endocrinology*, 95: 1649–1657.
Constantinidis, J., Geissbuhler, F., Gaillard, J.M., Aubert, C., Hovaguimian, Th. et Tissot, R. (1975) Formation de noradrénaline dans le cerveau du rat après administration de (+)-érythro-3,4-dihydroxyphénylsérine. *Psychopharmacologia (Berl.)*, 41: 201–209.
Dahlström, A. and Fuxe, K. (1964) Evidence for the existence of monoamine containing neurons in the central nervous system. I. Demonstration of monoamine in the cell bodies of brain stem neurons. *Acta physiol. scand.*, 62, Suppl. 232: 1–55.
De Wied, D., Delft, A.M.L. van, Gispen, W.H., Weijnen, J.A.W.M. and Van Wimersma Greidanus, Tj.B. (1972) The role of pituitary–adrenal system hormones in active-avoidance conditioning. In *Hormones and Behaviour*, S. Levine (Ed.), Academic Press, New York, pp. 135–171.
Dismukes, K. and Snyder, S.H. (1974) Dynamics of brain histamine In *Advances in Neurology*, F. McDowell and A. Barbeau (Eds.), Raven Press, New York, pp. 101–109.
Donoso, A.O., De Gutierrez Moyano, M.B. and Santolaya, R.C. (1969) Metabolism of noradrenaline in the hypothalamus of castrated rats. *Neuroendocrinology*, 4: 12–20.
Endersby, C.A. and Wilson, C.A. (1973) The effect of ovarian steroids on the accumulation of ^{3}H-labelled monoamines by hypothalamic tissue *in vitro*. *Brain Res.*, 73: 321–331.
Eneroth, P., Fuxe, K., Gustafsson, J.Å., Hökfelt, T., Löfström, A. and Skett, P. (1975) Studies on the nicotine induced reduction of gonadotrophin secretion. Evidence for selective activation of an inhibitory dopaminergic mechanism in the median eminence and for a differential control of LH and FSH secretion. *Neuroendocrinology*, in press.
Everitt, B.J., Fuxe, K., Hökfelt, T. and Jonsson, G. (1975a) Studies on the role of monoamines in the hormonal regulation of sexual receptivity in the female rat. In *Sexual Behaviour: Pharmacology and Biochemistry*, M. Sandler and G.L. Gessa (Eds.), Raven Press, New York, pp. 147–159.
Everitt, B.J., Fuxe, K. and Hökfelt, T. (1975b) Serotonin, catecholamines and sexual receptivity of female rats. Pharmacological findings. *J. Pharmacol. (Paris)*, 6: 269–276.
Fishman, J. and Norton, B. (1975) Catechol estrogen formation in the central nervous system of the rat. *Endocrinology*, 96: 1054–1059.
Fuller, R.W., Snoddy, H.D. and Molloy, B.B. (1975) Potentiation of the L-5-hydroxytryptophan-induced elevation of plasma corticosterone levels in rats by a specific inhibitor of serotonin uptake. *Res. Commun. Chem. Path. Pharmacol.*, 10: 193–196.
Fuxe, K. (1964) Cellular localization of monoamines in the median eminence and infundibular system of some mammals. *Z. Zellforsch.*, 61: 710–724.
Fuxe, K. (1965) Evidence for the existence of monoamine neurons in the central nervous system. IV. The distribution of monoamine nerve terminals in the central nervous system. *Acta physiol. scand.*, 64, Suppl. 247: 39–85.
Fuxe, K. and Hökfelt, T. (1969) Catecholamines in the hypothalamus and the pituitary gland. In

Frontiers in Neuroendocrinology, W.F. Ganong and L. Martini (Eds.), Oxford University Press, New York, pp. 47–96.

Fuxe, K. and Jonsson, G. (1974) Further mapping of central 5-hydroxytryptamine neurons: studies with the neurotoxic dihydroxytryptamines. *Advanc. Biochem. Psychopharmacol.*, 10: 10–12.

Fuxe, K., Hökfelt, T. and Ungerstedt, U. (1970a) Morphological and functional aspects of central monoamine neurons. *Int. Rev. Neurobiol.*, 13: 93–126.

Fuxe, K., Goldstein, M., Hökfelt, T. and Joh, T.H. (1970b) Immunohistochemical localization of dopamine-β-hydroxylase in the peripheral and central nervous system. *Res. Commun. Chem. Path. Pharmacol.*, 1: 627–636.

Fuxe, K., Corrodi, H., Hökfelt, T. and Jonsson, G. (1970c) Central monoamine neurons and pituitary-adrenal activity. In *Pituitary, Adrenal and the Brain, Progr. Brain Res., Vol. 32*, D. De Wied and J.A.W.M. Weijnen (Eds.), Elsevier, Amsterdam, pp. 42–56.

Fuxe, K., Hökfelt, T. and Jonsson, G. (1970d) Participation of central monoaminergic neurons in the regulation of anterior pituitary secretion. In *Neurochemical Aspects of Hypothalamic Function*, L. Martini and J. Meites (Eds.), Academic Press, New York, pp. 61–88.

Fuxe, K., Hökfelt, T., Goldstein, M. and Joh, T.H. (1971) Cellular localization of dopamine-β-hydroxylase and phenylethanolamine-N-methyltransferase as revealed by immunohistochemistry. In *Histochemistry of Nervous Transmission, Progr. in Brain Res., Vol. 34*, O. Eränkö (Ed.), Elsevier, Amsterdam, pp. 127–138.

Fuxe, K., Hökfelt, T., Jonsson, G., Levine, S., Lidbrink, P. and Löfström, A. (1973) Brain and pituitary–adrenal interactions. Studies on central monoamine neurons. In *Brain–Pituitary–Adrenal Interrelationships*, A. Brodish and E.S. Redgate (Eds.), Karger, Basel, pp. 239–269.

Fuxe, K., Goldstein, M., Hökfelt, T., Jonsson, G. and Lidbrink, P. (1974a) Dopaminergic involvement in hypothalamic function: extrahypothalamic and hypothalamic control. A neuroanatomical analysis. *Advanc. Neurol.*, 5: 405–419.

Fuxe, K., Schubert, J., Hökfelt, T. and Jonsson, G. (1974b) Some aspects of the interrelationship between central 5-hydroxytryptamine neurons and hormones. *Advanc. Biochem. Psychopharmacol.*, 10: 67–74.

Fuxe, K., Löfström, A., Agnati, L., Everitt, B.J., Hökfelt, T., Jonsson, G. and Wiesel, F.-A. (1975a) On the role of central catecholamine and 5-hydroxytryptamine neurons in neuroendocrine regulation. In *Anatomical Neuroendocrin. int. Conf., Chapel Hill, 1974*, W. Stumpf and L. Grant (Eds.), Karger, Basel, in press.

Fuxe, K., Löfström, A., Eneroth, P., Gustafsson, J.Å., Hökfelt, T., Scett, P., Wuttke, W., Fraser, H. and Jeffcoate, S. (1975b) Interactions between hypothalamic CA nerve terminals and LRF containing neurons. Further evidence for an inhibitory dopaminergic and a facilitory noradrenergic influence. Excerpta Medica, Amsterdam, In press.

Fuxe, K., Hökfelt, T., Agnati, L., Everitt, B., Johansson, O., Jonsson, G., Wuttke, W. and Goldstein, M. (1975c) Role of monoamines in the control of gonadotrophin secretion. In *Neuroendocrine Regulation of Fertility*, Karger, Basel, in press.

Fuxe, K., Agnati, L., Corrodi, H., Everitt, B., Hökfelt, T., Löfström, A. and Ungerstedt, U. (1975d) Action of dopamine receptor agonists in forebrain and hypothalamus: rotational behaviour, ovulation and dopamine turnover. In *Advances in Neurology, Vol. 9*, D.B. Calne, T.N. Chase and A. Barbeau (Eds.), Raven Press, New York, pp. 223–242.

Fuxe, K., Eneroth, P., Gustafsson, J.Å., Hökfelt, T., Jonsson, G., Löfström, A. and Skett, P. (1975e) Effects of 6-OH-DA induced lesions of the ascending noradrenaline and adrenaline pathways to the tel- and diencephalon on FSH, LH and prolactin secretion in the ovariectomized female rat. Paper presented at the Int. Meeting on Chemical Tools in Catecholamine Research, Göteborg, Sweden.

Ganong, W.F. (1973) Catecholamines and the secretion of renin, ACTH and growth hormone. In *Frontiers in Catecholamine Research*, E. Usdin and S. Snyder (Eds.), Pergamon Press, New York, pp. 819–824.

Goldstein, M., Anagnoste, B., Freedman, L.S., Roffman, M., Ebstein, R.P., Park, D.H., Fuxe, K. and Hökfelt, T. (1973) Characterization, localization and regulation of catecholamine synthesizing enzymes. In *Frontiers in Catecholamine Research*, E. Usdin and S. Snyder (Eds.), Pergamon Press, Oxford, pp. 69–78.

Grant, L.D. and Stumpf, W.E. (1973) Localization of ^{3}H-estradiol and catecholamines in identical neurons in the hypothalamus. *J. Histochem. Cytochem.*, 21: 404p.

Grimm, Y. and Reichlin, S. (1973) Thyrotropin releasing hormone (TRH): neurotransmitter regulation of secretion by mouse hypothalamic tissue *in vitro*. *Endocrinology*, 93: 626–631.

Hartman, B.L., Zide, D. and Udenfriend, S. (1972) The use of dopamine-β-hydroxylase as a marker for the noradrenergic pathways of the central nervous system in the rat. *Proc. nat. Acad. Sci. (Wash.)*, 69: 2722–2726.

Hillhouse, E.W., Burden, J. and Jones, M.T. (1975) The effect of various putative neurotransmitters on the release of corticotrophin releasing hormone from the hypothalamus of the rat *in vitro*. I. The effect of acetylcholine and noradrenaline. *Neuroendocrinology*, 17: 1–11.

Hökfelt, T. and Fuxe, K. (1972) On the morphology and the neuroendocrine role of the hypothalamic catecholamine neurons. In *Brain–Endocrine Interaction. Median Eminence: Structure and Function, Int. Symp. Munich,* K.M. Knigge, D.E. Scott and A. Weindl (Eds.), Karger, Basel, pp. 181–223.

Hökfelt, T., Jonsson, G. and Ljungdahl, Å. (1970) Regional uptake and subcellular localization of [^{3}H]gamma-aminobutyric acid (GABA) in rat brain slices. *Life Sci.*, 9: 203–212.

Hökfelt, T., Fuxe, K. and Goldstein, M. (1973a) Immunohistochemical studies on monoamine-containing cell systems. *Brain Res.*, 62: 461–470.

Hökfelt, T., Fuxe, K., Goldstein, M. and Johansson, O. (1973b) Evidence for adrenaline neurons in the rat brain. *Acta physiol. scand.*, 89: 286–288.

Hökfelt, T., Fuxe, K., Goldstein, M. and Johansson, O. (1974) Immunohistochemical evidence for the existence of adrenaline neurons in the rat brain. *Brain Res.*, 66: 235–251.

Hökfelt, T., Fuxe, K., Goldstein, M., Johansson, O., Fraser, H. and Jeffcoate, S. (1975a) Immunofluorescence mapping of central monoamine and releasing hormone (LHR) systems. In *Anatomical Neuroendocrinology*, W. Stumpf and L. Grant (Eds.), Karger, Basel, in press.

Hökfelt, T., Kellerth, J.-O., Nilsson, G. and Pernow, B. (1975b) Experimental immunohistochemical studies on the localization and distribution of Substance P in cat primary sensory neurons. *Brain Res.*, 100: 235–252.

Hökfelt, T., Kellerth, J.O., Nilsson, G. and Pernow, B. (1975c) Morphological support for a transmitter or modulator role of Substance P: immunohistochemical localization in the central nervous system and in primary sensory neurons. *Science,* in press.

Hökfelt, T., Efendic, S., Hellerström, C., Johansson, O., Luft, R. and Arimura, A. (1975d) Cellular localization of somatostatin in endocrine-like cells and neurons of the rat with special reference to the A$_1$-cells of the pancreatic islets and the hypothalamus. *Acta endocr. (Kbh.)*, 80, Suppl. 200: 5–41.

Hökfelt, T., Johansson, O., Fuxe, K., Goldstein, M., Park, D., Ebstein, R., Fraser, H., Jeffcoate, S., Efendic, S., Luft, R. and Arimura, A. (1975e) The mapping and relationship of hypothalamic neurotransmitters and hypothalamic hormones. In *Sixth Int. Congr. of Pharmacology, Helsinki*, Pergamon, Oxford, in press.

Hökfelt, T., Fuxe, K., Johansson, O., Jeffcoate, S. and White, N. (1975f) Distribution of thyrotropin releasing hormone (TRH) in the central nervous system as revealed with immunohistochemistry. *Europ. J. Pharmacol.,* in press.

Hökfelt, T., Fuxe, K., Johansson, O., Jeffcoate, S. and White, N. (1975g) Thyrotropin releasing hormone (TRH)-containing nerve terminals in certain brain stem nuclei and in the spinal cord. *Neurosci. Lett.*, 1: 133–139.

Hökfelt, T., Efendic, S., Johansson, O., Luft, R. and Arimura, A. (1975h) Immunohistochemical localization of somatostatin (growth hormone release-inhibiting factor) in the guinea pig brain. *Brain Res.*, 80: 165–169.

Jackson, I.M.D. and Reichlin, S. (1974) Thyrotropin-releasing hormone (TRH): distribution in hypothalamic and extrahypothalamic brain tissues of mammalian and submammalian chordates. *Endocrinology*, 95: 854–862.

Javoy, F., Glowinski, J. and Kordon, C. (1968) Effects of adrenalectomy on the turnover of norepinephrine in the rat brain. *Europ. J. Pharmacol.,* 4: 103—104.

Kaplanski, J., Delft, A.M.L. van, Nyakas, C., Stoof, J.C. and Smelik, P.G. (1974) Circadian periodicity and stress responsiveness of the pituitary-adrenal system of rats after central administration of 6-hydroxydopamine. *J. Endocr.,* 63: 299—310.

Kizer, J.S., Palkovits, M., Zivin, J., Brownstein, M., Saavedra, J.M. and Kopin, I.J. (1974) The effect on endocrinological manipulations on tyrosine hydroxylase and dopamine-β-hydroxylase activities in individual hypothalamic nuclei of the adult male rat. *Endocrinology,* 95: 799—812.

Kizer, J.S., Arimura, A., Schally, A.V. and Brownstein, M.J. (1975) Absence of luteinizing hormonereleasing hormone (LH-RH) from catecholaminergic neurons. *Endocrinology,* 96: 523—525.

Kordon, C. (1973) Effect of drugs acting on brain monoamines and the control of gonadotropic secretion. In *Endocrinology, Int. Congr. Ser. No. 273,* R.O. Scow (Ed.), Excerpta Medica, Amsterdam, pp. 120—124.

Koslow, S.H., Racagni, G. and Costa, E. (1974) Mass fragmentographic measurement of norepinephrine, dopamine, serotonin and acetylcholine in seven discrete nuclei of the rat tel-diencephalon. *Neuropharmacology,* 13: 1123—1130.

Krulich, L., Quijada, M., Hefco, E. and Sundberg, D.K. (1974) Localization of thyrotropin-releasing factor (TRF) in the hypothalamus of the rat. *Endocrinology,* 95: 9—17.

Lawson, D.M. and Gala, R.R. (1975) The influence of adrenergic, dopaminergic, cholinergic and serotoninergic drugs on plasma prolactin levels in ovariectomized estrogen-treated rats. *Endocrinology,* 96: 313—318.

Leeman, S.E. (1974) Substance P. *Life Sci.,* 15: 2033—2044.

Lichtensteiger, W. (1973) Changes in hypothalamic monoamines in relation to endocrine states: functional characteristics of tubero-infundibular dopamine neurons. In *Endocrinology, Int. Congr. Ser. No. 273,* R.O. Scow (Ed.), Excerpta Medica, Amsterdam, pp. 131—137.

Lidbrink, P., Jonsson, G. and Fuxe, K. (1974) Selective reserpine-resistant accumulation of catecholamines in central dopamine neurons after DOPA administration. *Brain Res.,* 67: 439—456.

Lindvall, O. and Björklund, A. (1974) The organization of the ascending catecholamine neuron systems in the rat brain as revealed by the glyoxylic acid fluorescence method. *Acta physiol. scand.,* Suppl. 412: 1.

Löfström, A., Jonsson, G. and Fuxe, K. (1975a) Microfluorimetric quantitation of catecholamine fluorescence in rat median eminence. I. Aspects on the distribution of dopamine and noradrenaline nerve terminals. *J. Histochem. Cytochem.,* in press.

Löfström, A., Jonsson, G., Wiesel, F.-A. and Fuxe, K. (1975b) Microfluorimetric quantitation of catecholamine fluorescence in rat median eminence. II. Turnover changes in hormonal states. *J. Histochem. Cytochem.,* in press.

Maanen, J.N. van and Smelik, P.G. (1968) Induction of pseudopregnancy in rats following local depletion of monoamines in the median eminence of the hypothalamus. *Neuroendocrinology,* 3: 177—186.

Makara, G.B. and Stark, E. (1974) Effect of gamma-aminobutyric acid (GABA) and GABA antagonist drugs on ACTH release. *Neuroendocrinology,* 16: 178—190.

Makara, G.B., Rappay, Gy. and Stark, E. (1975) Autoradiographic localization of ^{3}H-gamma-aminobutyric acid in the medial hypothalamus. *Exp. Brain Res.,* 22: 449—455.

McCann, S.M. and Moss, R.L. (1975) Putative neurotransmitters involved in discharging gonadotropin-releasing neurohormones and the action of LH-releasing hormone on the CNS. *Life Sci.,* 16: 833—852.

McCann, S.M., Kalra, P.S., Donoso, A.O., Bishop, W., Schneidler, H.P.G., Fawcett, C.P. and Krulich, L. (1972) The role of monoamines in the control of gonadotropin and prolactin secretion. In *Brain—Endocrine Interaction. Median Eminence: Structure and Function,* K.M. Knigge D.E. Scott and W. Weindl (Eds.), Karger, Basel, pp. 224—235.

Macleod, R.M. and Lehmeyer, J.E. (1974) Studies on the mechanism of the dopamine-mediated inhibition of prolactin secretion. *Endocrinology,* 94: 1077—1085.

Miyachi, Y. (1973) *In vitro* studies of pituitary-median eminence unit. *Endocrinology,* 93: 492—496.

Müller, E.E. (1973) Brain monoamines participation in the control of growth hormone secretion in different animal species. In *Frontiers in Catecholamine Research*, E. Usdin and S. Snyder (Eds.), Pergamon Press, New York, pp. 835–841.

Nicoll, R.A. and Barker, J.L. (1971) The pharmacology of recurrent inhibition of the supraoptic neurosecretory system. *Brain Res.*, 35: 501–511.

Nilsson, G., Hökfelt, T. and Pernow, B. (1974) Distribution of Substance P-like immunoreactivity in the rat central nervous system as revealed by immunohistochemistry. *Med. Biol.*, 52: 424–427.

Ojeda, S.R., Harms, P.G. and McCann, S.M. (1974) Effect of blockade of dopaminergic receptors on prolactin and LH release: median eminence and pituitary sites of action. *Endocrinology*, 6: 1650–1657.

Ondo, J.G. (1974) Gamma-aminobutyric acid effects on pituitary gonadotropin secretion. *Science*, 186: 738–739.

Palkovits, M., Brownstein, M., Saavedra, J.M. and Axelrod, J. (1974) Norepinephrine and dopamine content of hypothalamic nuclei of the rat. *Brain Res.*, 77: 137–149.

Pelletier, G., Labrie, F., Puviani, R., Arimura, A. and Schally, A.V. (1974) Immunohistochemical localization of luteinizing hormone-releasing hormone in the rat median eminence. *Endocrinology*, 95: 314–317.

Pelletier, G., Leclerc, R., Dube, D., Labrie, F., Puviani, R., Arimura, A. and Schally, A.V. (1975) Localization of growth hormone-releasing inhibiting hormone (somatostatin) in the rat brain. *Amer. J. Anat.*, 142: 397–401.

Pickel, V.M., Joh, T.H., Field, P.M., Becker, C.G. and Reis, D. (1975) Cellular localization of tyrosine hydroxylase by immunohistochemistry. *J. Histochem. Cytochem.*, 23: 1–12.

Plotnikoff, N.P., Prange, A.J., Jr., Breese, G.R., Anderson, M.S. and Wilson, I.C. (1972) Thyrotropin releasing hormone: enhancement of DOPA activity by a hypothalamic hormone. *Science*, 178: 417–418.

Prange, A.J., Wilson, I.C., Lara, P.P., Alltorp, L.B. and Breese, G.R. (1972) Effects of thyrotropin-releasing hormone in depression. *Lancet*, 2: 999–1002.

Renaud, L.P., Åartin, J.B. and Brazeau, P. (1975) Depressant action of TRH, LH-RH and somatostatin on activity of central neurones. *Nature (Lond.)*, 255: 233–235.

Réthelyi, M. (1975) Neurons in the subependymal layer of the rat median eminence. *Neuroendocrinology*, 17: 330–339.

Saavedra, J.M., Palkovits, M., Brownstein, M.J. and Axelrod, J. (1974a) Localization of phenylethanolamine N-methyl transferase in the rat brain nuclei. *Nature (Lond.)*, 248: 695–696.

Saavedra, J.M., Palkovits, M., Brownstein, M.J. and Axelrod, J. (1974b) Serotonin distribution in the nuclei of the rat hypothalamus and preoptic region. *Brain Res.*, 77: 157–165.

Sawyer, C.H. (1975) First Geoffrey Harris Memorial Lecture. Some recent developments in brain-pituitary-ovarian physiology. *Neuroendocrinology*, 17: 97–124.

Sawyer, C.H., Markee, J.E. and Townsend, B.F. (1949) Cholinergic and adrenergic components in the neurohumoral control of the release of LH in the rabbit. *Endocrinology*, 44: 18–37.

Setalo, G., Vigh, S., Schally, A.V., Arimura, A. and Flerkó, B. (1975a) LH-RH-containing neural elements in the rat hypothalamus. *Endocrinology*, 96: 135–142.

Setalo, G., Vigh, S., Schally, A.V., Arimura, A. and Flerkó, B. (1975b) GH-RIH containing neural elements in the hypothalamus. *Brain Res.*, 90: 352–356.

Shute, C.C.D. and Lewis, P.R. (1966) Cholinergic and monoaminergic pathways in the hypothalamus. *Brit. med. Bull.*, 22: 222–226.

Simonovic, I., Motta, M. and Martini, L. (1974). *Endocrinology*, 95: 1373–1379.

Smith, E.R. and Davidson, J.M. (1974) Feedback suppression of LH with maintained pituitary sensitivity: evidence for a cerebral action of estrogen. *Neuroendocrinology*, 14: 369–373.

Takahashi, T. and Otsuka, M. (1975) Regional distribution of Substance P in the spinal cord and nerve roots of the cat and the effect of dorsal root section. *Brain Res.*, 87: 1–11.

Telegdy, G. and Vermes, I. (1975) Effect of adrenocortical hormones on activity of the serotoninergic system in limbic structures in rats. *Neuroendocrinology*, 18: 16–26.

Tima, L. and Flerkó, B. (1974) Ovulation induced by norepinephrine in rats made anovulatory by various experimental procedures. *Neuroendocrinology*, 15: 346–354.

Tuomisto, J., Ranta, T., Männistö, P., Saarinen, A. and Leppäluoto, J. (1975) Neurotransmitter control of thyrotropin secretion in the rat. *Europ. J. Pharmacol.*, 30: 221–229.

Versteeg, D.H.G., Gugten, J. van der and Ree, J.M. van (1975) Regional turnover and synthesis of catecholamines in rat hypothalamus. *Nature (Lond.)*, 256: 502–503.

Wirz-Justin, A., Hackmann, E. and Lichsteiner, M. (1974) The effect of oestradioldipropionate and progesterone on monoamine uptake in the brain. *J. Neurochem.*, 22: 187–189.

Chapter 12

Endogenous steroids in neuroendocrine tissues

J.R.G. CHALLIS[*], F. NAFTOLIN[*], I.J. DAVIES, K.J. RYAN and T. LANMAN

Department of Obstetrics and Gynecology and Laboratory for Human Reproduction and Reproductive Biology, Harvard Medical School, Boston, Mass. 02115 (U.S.A.) and (J.R.G.C.) Nuffield Department of Obstetrics and Gynecology, University of Oxford, Headington, Oxford (Great Britain)

I. INTRODUCTION

The relationship between a hormone and its target organ may be considered from the standpoint of the biological response evoked in the target tissue by the hormone, or from an examination of the effects of the tissue on the hormone in question (Axelrod, 1971). The latter approach includes the study of the metabolism of hormones by target tissues, and measurements of the affinity of the tissue for the hormone. More specifically, for steroids, this involves the demonstration of specific soluble and nuclear receptor proteins, and a study of the modulation of the concentrations and the affinity of such receptors under a variety of natural and experimentally defined conditions.

The product of steroid uptake and metabolism resolves itself into a somewhat static measurement of the concentration of steroid in a given tissue at the particular moment in time under examination. It will already be apparent that there are severe limitations in the investigation of any one of the above factors in isolation, since all contribute to a multi-complex interaction, and ideally all should be studied in parallel. However, detailed experiments concerned with steroid uptake and metabolism in neuroendocrine tissues (NET) are presented elsewhere in this volume, and it is the purpose of the present chapter to consider the tissue concentrations of steroids that may result from these processes. One approach to this problem, as discussed later, is to determine a range of different steroids within a single tissue sample in order to expedite a dynamic interpretation from such static measurements.

The underlying theme to this chapter will be the relationship between hormone concentrations in blood, and their concentration and distribution within target organs. This fundamental relationship depends on a multiplicity of factors, including not only the activity of steroid receptors, but also the physicochemical form of the steroid in the vascular compartment, the presence and extent of globulin binding therein, and the metabolic clearance rate (MCR) of the steroid and the permeability of the "blood–brain" barrier. Sadly, there is a paucity of information correlating these different parameters.

---

[*] Present address: Department of Obstetrics and Gynecology, Women's Pavilion, Royal Victoria Hospital, Montreal, P.Q., Canada.

Initially therefore, some consideration will be given to the concept derived from in vivo experiments that cranial tissue may be a major contributor to MCR. Measurements of steroids in NET and other target tissues will then be presented for two models that have been studied recently by using radioimmunoassay, and the significance of these measurements will be reviewed in conjunction with the limited information in the literature on steroid levels in NET determined previously by alternative procedures.

II. THE HEAD AS A MAJOR COMPONENT OF METABOLIC CLEARANCE RATE

Extensive studies have been carried out in conscious sheep utilizing tracer isotope kinetics techniques to measure at steady state the MCR and production rate of progesterone and estrogens. These studies have provided unequivocal evidence that in the conscious unstressed animal the cranial tissue is a major site of steroid hormone uptake and metabolism. Progesterone may be taken as an example (Bedford et al., 1973). During the continuous infusion of tracer amounts of [^{3}H]progesterone into a jugular vein, samples were taken at steady state from a carotid artery, the hepatic portal vein and hepatic vein, the uterine vein and, in animals with ovarian and adrenal transplants, from the venous effluent of these glands. In one anesthetized animal samples were taken from a renal vein. Radioactive progesterone was isolated specifically from these samples, and the organ uptake and the contribution of individual organ clearances to the total metabolic clearance rate were calculated. Table I shows that in non-pregnant sheep the percentage extraction ($(A-V)/(A) \times 100\%$) of [^{3}H]progesterone across the gut and liver (total splanchnic) exceeds 90%. There is also appreciable extraction across the kidney (66%), adrenal

TABLE I

Organ extraction and organ clearance rates of progesterone in non-pregnant sheep***

Organ	Percentage extraction*	Clearance rate** (l/min)
Splanchnic	92	1.5
Kidney	66	0.6
Head	26	0.35
Uterus	20	0.1
Adrenal	30	< 0.02
Ovary	16	< 0.02

Total metabolic clearance rate ≃ 2.75 l/min
Cardiac output ≃ 5 l/min

* $\frac{A-V}{A} \times 100$, where A is [^{3}H]progesterone in affluent vessel, and V is [^{3}H]progesterone in effluent vessel.
** Organ clearance rate = percentage extraction x blood flow.
*** Data taken and calculated from Bedford et al. (1973).

(30%), uterus (20%) and of particular relevance to present discussion, the head (26%). When corrected for blood flow, approximately 50% of the total MCR (2.75 l/min) can be accounted for by splanchnic clearance (1—1.5 l/min). The kidney also makes a major contribution to MCR, but the third major area of metabolism after the kidney is the head, which accounts for approximately 13% of MCR. Baldwin and Bell (1963) have equated blood flow to the cranial tissue in sheep with the total carotid blood flow, so that this metabolism is due, in large part, to brain tissue. Evidence for hormone production in this region was also found, since the concentrations of radioactivity isolated as 20α-dihydro-progesterone in the jugular vein was significantly greater than in the carotid artery, indicating the presence of an active 20α-hydroxysteroid dehydrogenase enzyme system.

Similar experiments showed that there was a significant uptake (10—15%) of tritiated estrone and estradiol across cranial tissue, and demonstrated the presence of a 17β-oxidoreductase system; the radioactivity in $E_2\beta$ in the jugular vein being consistently higher than in the carotid artery during the continuous infusion of tritiated estrone (Challis et al., 1973b). These results have since been confirmed by Kazama and Longcope (1974). Recently, in vitro studies have been used to demonstrate 17β-oxidoreductase activity in anterior pituitary, hypothalamic. and cortical tissue from the sheep (Jenkin et al., 1975) as well as the rabbit (Reddy et al., 1974). The work on sheep confirmed the earlier finding of steroid sulfatase activities in these neural tissues (Payne et al., 1973). In pituitary tissue from fetal lambs the steroid sulfatase activity was appreciably higher than in the mother (mean conversion $E_1S \rightarrow E_1$, approximately 12 nmole/g pituitary/2 hr incubation), raising the question of whether this enzyme has any role in vivo in view of the high levels of circulating estrogen sulfates in the ovine fetus (Currie et al., 1973).

In summary these in vivo and in vitro studies provide clear evidence for a metabolic and biosynthetic function of NET in sheep , and lead naturally to the question of the concentrations of hormones in NET during various reproductive states.

III. HORMONE CONCENTRATIONS IN NET IN SHEEP FOLLOWING EXOGENOUS STEROID ADMINISTRATION

Ideally one would like to follow changes in hormone levels in NET throughout the estrous cycle and to correlate these with ovarian steroid production rates, behavioral responses, and gonadotrophin release. Our initial approach to this problem has been to create a model estrous cycle in ovariectomized sheep treated with physiological amounts of progesterone and/or estradiol (Challis et al., 1975). At slaughter the concentrations of these steroids have been measured at target sites in the uterus and brain.

Sixteen sexually mature Clun Forest or Border Leicester ewes were ovariectomized through a mid-ventral incision during the period of seasonal anestrum. The animals were divided into 4 groups of 4 animals each; group I, control group receiving 0.5 ml ethanol (vehicle for steroid injections) i.m. b.i.d. for 10 days; group II, receiving 15 μg $E_2\beta$ i.m. b.i.d. for 10 days; group III, receiving 10 mg progesterone i.m. b.i.d. for 10 days; group IV, receiving 10 mg progesterone i.m. b.i.d. for 10 days + 15 μg $E_2\beta$ i.m. b.i.d. on days

8 through 10. The amounts of progesterone and estradiol were calculated as approximately twice the production rate of these steroids during the luteal phase and at estrus respectively, to allow for differences between i.m. injections and endogenous secretion directly into the vascular compartment. On day 10, 12 hr after the final hormone injection the sheep were killed with an overdose of pentobarbitone, and brain and uterine tissues were dissected out and immediately frozen in liquid nitrogen. Subsequently the frozen tissues were thawed at 4 °C and homogenized at 4 °C in a saturated solution of indomethacin (Merck, Sharp and Dohme). Progesterone and estradiol-17β were measured by radioimmunoassay (see below).

Table II shows the concentration of progesterone (mean ± S.E.M.) in the 4 groups of animals. The following tissues were harvested; the uterine caruncles, a residual section of the uterus (endometrium + myometrium), pituitary, hypothalamus and cortex. In the control animals (group I), low levels of progesterone (0.21–0.82 ng/g) were found in all tissues, which could reflect uptake of the adrenal contribution to progesterone production (Baird et al., 1973). Treatment with progesterone provoked an elevation in the mean progesterone concentration in the uterine tissues, and in pituitary, hypothalamus and cortex, although the range of values achieved (e.g., 2.7–15.8 ng/g for pituitary tissue in group III), for the relatively few animals meant that statistical significance was not always accomplished. The pituitary and hypothalamus accumulated progesterone to a similar extent as the uterine tissues, and the mean values tended to be higher than for cortex (Table II; groups III and IV). Addition of estradiol had no significant effect on the concentration of progesterone in any one of the 5 tissues studied (group IV).

The pattern of estradiol concentration was clearly different to that of progesterone (Table III). Although the uterine caruncles showed a highly significant increase in their concentration of $E_2\beta$ (see groups II and IV) and there was a significant increase in $E_2\beta$ in the pituitary in group II, neither the hypothalamus nor cortex showed any significant elevation in $E_2\beta$ over the levels found in the control group I. In contrast to progesterone, the elevation in pituitary $E_2\beta$ levels was approximately 10–15 fold greater than the expected elevation in plasma E_2 during administration of the steroid.

From these studies it may be concluded that the maximum concentration of progesterone and $E_2\beta$ in sheep pituitary and hypothalamus during the luteal phase of the cycle is unlikely to exceed 6 ng/g and 150 pg/g, respectively. After regression of the corpus luteum, and during the onset of estrus, progesterone concentrations in both neural tissues are likely to be < 0.5 ng/g, whilst the concentration of $E_2\beta$ in the pituitary may be expected to exceed 250 pg/g. The failure to find the concomitant increase in $E_2\beta$ in other NET could be due to differences in estrogen receptor activities, or to an increased catabolism of the steroid. Even so, the increase in $E_2\beta$ seen in the pituitary is significantly less than that observed in another target tissue, the uterine caruncles.

IV. ENDOGENOUS STEROID LEVELS IN THE MALE RAT

Whilst the foregoing study provides information about estradiol and progesterone concentrations in NET it does little to indicate the dynamic interaction that may exist

TABLE II

Progesterone concentration in uterus and NET of ovariectomized sheep treated with progesterone and/or estradiol (4 animals each group)

For further explanation of treatment groups see text. (From Challis, Louis and Robinson, 1975.)

Group	Progesterone concentration (ng/g); mean ± S.E.M.				
	Uterine caruncles	Endometrium + myometrium	Pituitary	Hypothalamus	Cortex
I, control	0.82 ± 0.55	0.64 ± 0.28	0.49 ± 0.28	0.61 ± 0.33	0.21 ± 0.10
II, estradiol	1.93 ± 0.70	1.34 ± 0.56	0.48 ± 0.23	0.41 ± 0.18	0.48 ± 0.30
III, progesterone	6.58 ± 2.19*	3.82 ± 1.11*	6.42 ± 3.13	3.47 ± 2.16	1.75 ± 0.49*
IV, progesterone + estradiol	4.82 ± 1.14*	4.66 ± 1.33*	2.32 ± 0.54*	3.73 ± 1.76	2.93 ± 0.89*

* $P < 0.05$ vs. corresponding value in group I.

TABLE III

Estradiol concentration in uterus and NET of ovariectomized sheep treated with progesterone and/or estradiol (4 animals each group)

For further explanation of treatment groups see text. (From Challis, Louis and Robinson, 1975.)

Group	Estradiol concentration (pg/g); mean ± S.E.M.				
	Uterine caruncles	Endometrium + myometrium	Pituitary	Hypothalamus	Cortex
I, control	99 ± 42	195 ± 65	57 ± 17	129 ± 54	74 ± 21
II, estradiol	1631 ± 111**	648 ± 258	348 ± 103*	< 38	97 ± 29
III, progesterone	64 ± 14	102 ± 53	141 ± 52	106 ± 66	58 ± 12
IV, progesterone + estradiol	954 ± 125**	415 ± 365	264 ± 89	184 ± 55	97 ± 35

* $P < 0.05$ vs. corresponding value in group I.
** $P < 0.001$ vs. corresponding value in group I.

between different steroids. In the following example we have attempted to perform a study which would allow correlation with other indices of steroid metabolic activities of NET, in the male rat. A central question is whether C_{19} androgens are metabolized preferentially in this species through a 5α reduction pathway, or are aromatized to estrogen. One approach to this problem is to determine the concentrations of the appropriate C_{19} and C_{18} steroids in the same sample of NET in order to elucidate the relative amounts of these steroids that are present in vivo. In this study progesterone has been measured as a representative C_{21} steroid; androstenedione and testosterone as 3-keto, Δ^4, C_{19} steroids; 5α-DHT as a 5α ring A reduced C_{19} steroid; and E_1 and $E_2\beta$ as potential C_{18} products of aromatization.

Sexually mature (greater than 300 g) male rats (Charles River Co., Mass. U.S.A.) were used in this study. One group of 3 animals was castrated under diethyl ether anesthesia, and the rats were killed 3 weeks later. All the animals were killed by decapitation, and plasma and weighed organ samples were immediately frozen. Immediately before assay, the tissue was homogenized in a known volume of distilled water at 4 °C, and tracer amounts of radioactive progesterone (P), androstenedione (A), 5α-dihydro testosterone (DHT), testosterone (T), E_1 and $E_2\beta$ in phosphate buffer were added to replicate samples. The steroids were extracted with diethyl ether, and separated from each other on microcolumns of celite coated with ethylene glycol and operated under nitrogen pressure (Abraham et al., 1970; Challis et al., 1973a). Briefly, the dried ether extract was applied to the columns with 1.0 + 0.5 ml iso-octane. Elution of steroids from the column was carried out as follows (Fig. 1). Fraction 1, containing P, was eluted with 1.5 ml iso-octane. Fraction 2 in which P and A overlap was discarded. Fraction 3, containing A, was then eluted with 2.5 ml iso-octane. Fraction 4, which contained DHT, was eluted with 3.5 ml iso-octane, followed by fraction 5, containing T, with 6 ml 30% benzene in iso-octane. E_1 was eluted in fraction 6 with 3.5. ml of 15% ethyl acetate in iso-octane, and $E_2\beta$ was then eluted in fraction 7 with 3.5 ml of 35% ethyl acetate in iso-octane. The different fractions were collected into glass vials (Packard Co.), and evaporated to dryness under nitrogen.

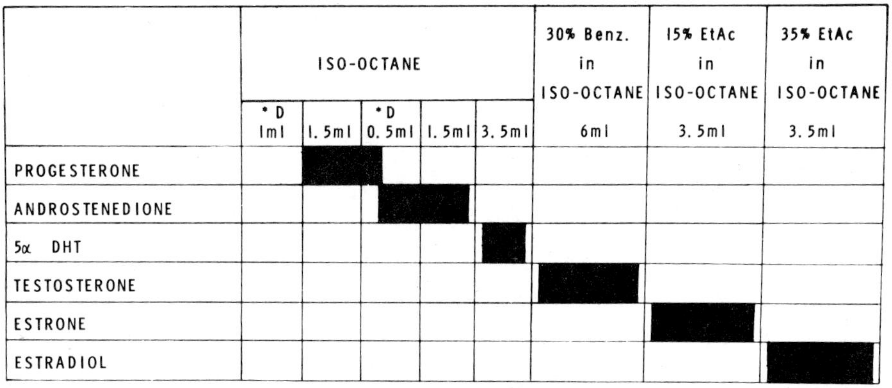

	ISO-OCTANE					30% Benz. in ISO-OCTANE	15% EtAc in ISO-OCTANE	35% EtAc in ISO-OCTANE
	*D 1ml	1.5ml	*D 0.5ml	1.5ml	3.5ml	6ml	3.5ml	3.5ml
PROGESTERONE		■						
ANDROSTENEDIONE				■				
5α DHT					■			
TESTOSTERONE						■		
ESTRONE							■	
ESTRADIOL								■

Fig. 1. Steroid elution profile on celite microcolumns. *D, fraction discarded; Benz., benzene; EtAc, ethyl acetate.

The steroids were redissolved in 1 ml 0.1 M phosphate buffer, pH 7.2 containing 0.1% gelatin (w/v) and aliquots were taken for liquid scintillation counting to determine recovery, and for radioimmunoassay. The latter was performed using previously described conditions and antibodies. In the present study, A was assayed using an antiserum prepared against A-6β-ol-hemisuccinate-BSA (Steraloids, Pawling N.Y.) which at an initial dilution of 1:4000 showed the following cross reactivities; P (0.24%), pregnenolone (0.07%), cortisol (0.02%), DHEA (4.2%), DHT, 17α-hydroxy-progesterone and 20α-dihydro-progesterone (< 0.01%).

Radioimmunoassay was performed using techniques which have been described and evaluated previously (Challis et al., 1973a, 1975). The incubations were carried out at 4°C for between 2.5 to 18 hr. The bound and free steroids were separated by exposure at 4°C for 15 min to 0.2 ml of 0.625% charcoal coated with 0.0625% dextran T70 (w/v) and suspended in 0.1 M phosphate buffer.

The concentrations of steroids found in the serum and NET from the intact and castrate rats are shown in Table IV, and may be compared with values presented by Robel et al. (1973). In intact males testosterone was present in serum at a concentration of about 2 ng/ml, which is similar to the value reported by other workers (Lloyd, 1972; Robel et al., 1973). Serum testosterone became virtually undetectable after orchidectomy. The steroid was present in the pituitary, hypothalamus and cortex, at concentrations 2–4 times greater than those in serum, with the pituitary containing significantly more testosterone than the hypothalamus ($P < 0.05$) or cortex ($P < 0.05$). In castrate animals the concentration of testosterone in NET was too low for accurate determination. These values for intact rats show a similar pattern to that reported by Robel et al. (1973), using gas liquid chromatography with electron capture detection, although in a latter study, concentrations of 60 ng/g and 13 ng/g in the pituitary and hypothalamus respectively were found. No immediate explanation can be offered for this discrepancy; it could be related to differences between the two experimental groups of animals, or to methodological differences. Androstenedione was present in neuroendocrine tissues at concentrations about one-half those of T. The concentration of A in the pituitary was significantly greater ($P < 0.01$) than in the cortex, but all the NET appeared to concentrate the steroid to some extent. A clear decrease in the concentration of A in all tissues was seen after castration. The concentration of DHT in serum was much lower (0.22 ng/ml) than T. Increased concentrations of DHT relative to serum were found in the brain and pituitary, and although the pattern was similar to that seen for T, the absolute values were significantly less ($P < 0.01$). After castration, the values for DHT became undetectable. Robel et al. (1973) reported concentrations of DHT in the hypothalamus of about 2 ng/g, and in the pituitary and the cortex of less than 6 and 0.4 ng/g respectively, which in general agree well with the present results.

In the present study both estradiol and estrone were detected in NET, particularly in the pituitary, at concentrations which appeared to be much greater than those previously reported for plasma (Labhsetwar, 1972; de Jong et al., 1973). However, the amounts of steroid measured in the RIA were low, and with the small quantity of tissue available, more precise quantitation was not felt justified. Estradiol is thought to arise in male rats partly from testicular secretion, but largely from the peripheral conversion of testos-

TABLE IV

Steroid concentrations in NET of male rats

Number of animals in each group in parentheses.

Steroid	Tissue	Concentration (ng/g or ng/ml; mean ± S.E.M.)			
		Intact		Castrate	
Testosterone	Serum	⩾ 1.8		⩽ 0.004**	(3)
	Pituitary	8.2 ± 1.2	(3)	⩽ 0.12	(3)
	Hypothalamus	4.1 ± 0.9	(6)	⩽ 0.13	(3)
	Cortex	4.0 ± 1.1	(6)	⩽ 0.08	(3)
Androstenedione	Serum	0.26	(2)	0.014 ± 0.006	(3)
	Pituitary	4.28 ± 0.8	(4)	⩽ 0.7	(3)
	Hypothalamus	2.2 ± 0.7	(6)	⩽ 1.4	(3)
	Cortex	1.5 ± 0.9	(6)	⩽ 0.3	(3)
5α-DHT	Serum	0.22	(2)	⩽ 0.005	(3)
	Pituitary	3.12 ± 0.4	(6)	⩽ 0.05	(3)
	Hypothalamus	1.80 ± 0.3	(6)	⩽ 0.05	(3)
	Cortex	2.10 ± 0.6	(7)	⩽ 0.05	(3)
Estradiol	Serum	0.002 –0.02*		—	
	Pituitary	⩽ 2.4	(3)	⩽ 1.2	(3)
	Hypothalamus	⩽ 0.6	(3)	⩽ 0.4	(3)
	Cortex	⩽ 0.3	(3)	⩽ 0.09	(3)
Estrone	Serum	⩽ 0.02	(3)	⩽ 0.02	(3)
	Pituitary	⩽ 1.6	(3)	⩽ 0.8	(3)
	Hypothalamus	⩽ 0 5	(3)	⩽ 0.2	(3)
	Cortex	⩽ 0.4	(3)	⩽ 0.3	(3)
Progesterone	Serum	0.59 ± 0.18	(3)	0.24 ± 0.05	(3)
	Pituitary	3.10 ± 0.5	(7)	⩽ 2.2	(3)
	Hypothalamus	2.40 ± 0.6	(6)	⩽ 1.7	(3)
	Cortex	2.20 ± 0.7	(7)	1.40 ± 0.4	(3)

* Range of published values (see text).
** Values too low to be measured accurately; their upper limit is given.

TABLE V

Steroid concentrations in serum and NET of male rats relative to that of testosterone

Steroid	Percentage of testosterone concentrations			
	Serum	Pituitary	Hypothalamus	Cortex
Testosterone	100	100	100	100
Androstenedione	14	52	54	38
5α-DHT	12	38	44	53
Estradiol	< 1	29*	15*	7*
Progesterone	33	38	59	55

* Calculated from the upper limit value (see Table IV).

terone (de Jong et al., 1973), which could occur in a number of sites including NET. It was, therefore, of interest that after castration an apparent decline in the NET content of estradiol and estrone was noted.

P was also measured in the present study. This steroid was circulating at quite high concentrations in serum, and the values were not rendered undetectable by orchidectomy, thereby indicating a possible adrenal source. In the intact animals, P was found in NET at concentrations 4–6 times those of serum, and the increase in the pituitary concentration was statistically significant ($P < 0.01$). The concentration of P in NET fell after castration, although the steroid could still be detected, and the concentration in the cortex was significantly greater than serum ($P < 0.05$).

When the concentrations of individual steroids in different tissues were related to the concentration of testosterone in that tissue (Table V) it could be seen that there was a significant concentration of A, DHT and E_2 relative to T in all the NET studied, but particularly in the pituitary and hypothalamus. The data could be interpreted as indicating either 5β reduction or aromatization of T in NET, or ascribed to variations in the receptor activity for these different steroids. Detailed discussions of the activities of these different enzyme and receptor systems are to be found elsewhere in this volume.

V. DISCUSSION

The measurements reported in this paper, whilst still leaving many questions unanswered, extend considerably the information available concerning the endogenous levels of steroid hormones in the brain and pituitary.

Previous investigations into the nature of steroids naturally present in NET are, at best, sparse, and stem from the early report of Touchstone et al. (1966), who identified by a number of different criteria the presence of cortisol in human brain and peripheral nerves at concentrations considerably higher than plasma. Later Henkin et al. (1968) measured cortisol (F) and corticosterone (B) in the brain, cord, sciatic nerve and pituitary of intact and adrenalectomized cats. All these tissues, but especially the pituitary, had significantly greater concentrations of both corticoids than plasma. The higher levels of F than B in plasma were reflected in most of the tissues studied. Adrenalectomy reduced the corticoid levels in the tissues to 5–10% of those in the intact animal.

The presence of endogenous gonadal steroids in the brain and pituitary has perhaps always been inferred, both from the evidence showing uptake of exogenous radioactive steroids, and from the ability to reduce the uptake of labeled hormone by pretreatment with unlabeled steroid or by endogenous hormone. Such studies have indicated the presence of corticosterone in the hippocampus of the rat (McEwen et al., 1968), estradiol in the rat pituitary and hypothalamus (Kato and Villee, 1967b; Whalen et al., 1973), progesterone (Luttge and Wallis, 1973) and testosterone (Naess and Attramadal, 1974) in the rat pituitary and posterior hypothalamus. Further presumptive evidence for the presence of steroids in NET is derived from studies such as those of Whalen and Luttge (1971) in which they showed that the uptake of radioactivity by diencephalic and mesencephalic areas of male and female rats following the administration of tritiated progesterone was significantly greater in adrenalectomized than intact animals. The present studies have shown that progesterone is present in the circulation of the male rat after castration, and is presumably adrenal in origin. This would be consistent with Whalen and Luttge's (1971) interpretation of their data, although in that study it was suggested that adrenal progesterone was released in response to an ACTH stimulus from the stress of sampling. However, despite the above data and despite the many sophisticated studies of steroid uptake into brain and pituitary tissues, it is surprising that so little information has appeared on endogenous steroid levels in NET. From a purely experimental viewpoint, the size of the endogenous steroid pool in areas of brain and pituitary may have profound influences on the quantitative interpretation of both uptake and metabolism studies and may account for some of the discrepancies that have been reported.

A further potential source of steroids is the cerebrospinal fluid (CSF) and Anand Kumar and co-workers (Anand Kumar and Thomas, 1968; David and Anand Kumar, 1974) have argued that steroids in CSF could be of importance in the regulation of gonadotrophin release. In 1967, Lurie and Weiss published the first data showing the presence of progesterone in the CSF of 2 out of 4 women at term pregnancy. The levels were 0–5% those of peripheral plasma. Later Anand Kumar and Thomas (1968) isolated [^{3}H]E_1 and [^{3}H]$E_2\beta$ from CSF of ovariectomized monkeys, following the injection of [^{3}H]$E_2\beta$ into the blood. CSF radioactivity was at maximum by 1–2 hr after injection, at which time the activity in E_1 and E_2 was 5% of the radioactivity in plasma. In a later more extensive study, David and Anand Kumar (1974) showed that testosterone, progesterone, 17α-hydroxy-progesterone, norethynodrel and mestranol injected into the saphenous vein of intact or castrate monkeys passed into the CSF. Interestingly, while the levels of testosterone and progesterone remained higher in plasma, the other steroids were higher in CSF than in plasma 1 hr after injection. The reason for this difference is not clear, although may relate to differences in the metabolism and binding in the vascular compartment.

In the present studies, exogenous progesterone was shown to be taken up by the brain and pituitary of the ovariectomized ewe, and was measured at concentrations higher than serum in the NET of the male rat. In the male rat, the precise function of the progesterone is unclear, although its presence raises the question of whether it has in both the male and female a modulating effect on gonadotrophin release. Whalen and Luttge (1971) showed that following injection of [^{3}H]progesterone there was uptake of radioactivity by the anterior and posterior hypothalamus, cerebral peduncle, cortex and pituitary of both male and female castrate rats, although, after pretreatment with 1 μg estradiol benzoate the uptake in the female was greater. They consistently found that mesencephalic structures retained more progesterone than diencephalic structures, consistent with the hypothesis that mid-brain structures are sensitive to progesterone and may be involved in mediating steroid effects on sex receptivity or be related to the hypnotic actions of the steroid. The distribution of endogenous progesterone in the male rat suggested that it was preferentially accumulated in the pituitary, but it was still present in the hypothalamus and cortex, and would be consistent with this steroid having a rather wider area of distribution.

It is now established that the progesterone is metabolized in neural tissues and pituitary to 5α-pregnane-3,20-dione and 3α-hydroxy-5α-pregnan-20-one (Cheng and Karavolas, 1973). Tabei et al. (1974) have also demonstrated 20α reduction of progesterone, and shown that the activity of both the 20α-hydroxy steroid dehydrogenase and the 5α reductase is greater in the pituitary than in the hypothalamus or cortex. Neither the 5α-pregnane compounds nor 20α-dihydro-progesterone would have been measured in the present radioimmunoassay study. It is now recognized that the 5α reductase is present in brain and pituitary tissues from a number of different species (see Tabei et al., 1974) raising the question of whether progesterone or one of its metabolites is actually the active progestogen at the target site.

In the present study on male rats, we have found that the concentration of androstenedione is higher in the pituitary, hypothalamus and cortex than in the plasma, and that the concentration of androstenedione in the pituitary is significantly greater than that in the cortex, thereby implying some specific concentrating mechanism. At the present time it is not known whether androstenedione itself has any modulating effect on gonadotrophin release. However, the presence of a 17β-oxidoreductase system and the fact that androstenedione can be converted to estrogen in NET from a number of species (Reddy et al., 1973; Naftolin et al., 1975) clearly indicates the potential for further metabolism of this steroid. Recently Perez et al. (1974) have shown that androstenedione is metabolized to a number of other C_{19} steroids, of which 5α-androstenedione is the major product, by the pituitary, hypothalamus and hippocampus of intact and castrate male rats. Weisz and Gibbs (1974) have isolated labeled androstenedione from the hypothalamus, cortex and pituitary of newborn rats after the i.p. injection of tritiated testosterone.

The binding of testosterone and estradiol by neural and pituitary tissues has been extensively investigated, and there is evidence that uptake of both steroids may occur. Early studies show that estrogen was not metabolized extensively within the pituitary and hypothalamus (Michael, 1965; Kato and Villee, 1967a), but it is now recognized that testosterone can be metabolized both to 5α-DHT and other ring A reduced androgens (Massa et al., 1974) or be aromatized to estrogen (Naftolin et al., 1975). The present study has not resolved the issue of which of these processes predominates since both testosterone and DHT were present in brain and pituitary tissue at concentrations greater than plasma. Although estrone and estradiol were present in NET at concentrations far in excess of those in plasma, the small weights of tissue involved made accurate quantitation of the estrogen difficult. Nevertheless, their presence would indicate either an avid receptor system, or appreciable metabolism from C_{19} steroids. It was, therefore, of interest that in the experiments on the ovariectomized sheep, significant uptake of estradiol was found in the pituitary, although the retention of the steroid was very much less than in a second target system, the uterine caruncles. The present results in the rat are consistent with the early studies showing that the uptake of estradiol by female rat hypothalamic and pituitary tissues occurred rapidly, and that the steroid was retained in an unmetabolized form in the tissue (Kato and Villee, 1967b). Furthermore, the uptake of E_2 was sex-independent (Green et al., 1969; Korach and Muldoon, 1974). Pretreatment of the animals with unlabeled estradiol-17β, but not estradiol-17α, progesterone or testosterone prevented the pituitary and hypothalamic uptake, implying a specific receptor system of limited capacity (Kato and Villee, 1967b). It is of interest therefore that in the present experiments, the pituitary uptake of $E_2\beta$ was not significantly different in either the presence or absence of progesterone. Presumably one might not expect to find changes in the uptake of tritiated $E_2\beta$ administered at different stages of the estrous cycle, since endogenous estrogen levels are below those required for receptor saturation (Kato and Villee, 1967b). However, the published data on this point remain conflicting (see discussion by Korach and Muldoon, 1974).

Several workers have demonstrated uptake of radioactivity by brain and pituitary tissues after the administration of radioactive testosterone to male rats (Sar and Stumpf, 1972, 1973; Naess and Attramadal, 1974; Weisz and Gibbs, 1974). The presence of cytosol binding of testosterone and DHT has been reported (Korach and Muldoon, 1974; Naess et al., 1975), although in the latter study displacement with unlabeled steroid was not achieved. In NET, testosterone is metabolized to DHT and 3α-androstenedione (Denef et al., 1974; Massa et al., 1974) and the conversion of testosterone to DHT is greatest in the mid-brain, hypothalamic and pituitary areas. The high concentration of DHT found in the hypothalamus and pituitary of the male rat would be consistent with its formation from testosterone, since the pituitary to serum and hypothalamus to serum ratios for DHT were significantly greater than for testosterone, although the absolute concentrations of hormone were less. In Korach and Muldoon's (1974) study the apparent binding of DHT to hypothalamic cytosol was greater than testosterone, but because of the large plasma concentration difference in these two hormones would be inadequate, in itself, to explain the high concentration relative to plasma of DHT over testosterone.

In the present study, the level of DHT in the pituitary was greater than in the hypothalamus, which would again be consistent with the observation that the conversion of testosterone to DHT is greater in the pituitary (Massa et al., 1974). A similar distribution of testosterone and DHT between brain and pituitary tissues was recorded by Robel et al. (1973). These workers also showed that although testosterone was found in a variety of other androgen-dependent tissues, including the prostate, and levator ani muscle, DHT could not be detected in the latter. The significance of this distribution was considered in the context of testosterone as a true hormone in some tissues, but a pre-hormone in others, such as the ventral prostate. In view of the compulsive evidence that C_{19} steroids may be aromatized in NET, and the arguments that have been advanced to attribute their biological effects to this metabolic transformation (Naftolin et al., 1975), it is apparent that testosterone must be considered as a pre-hormone not only for DHT, but also for estrogen.

Whilst the present study has provided new information on hormone levels in the brain and pituitary, it has indicated clearly the importance of combining the various aspects of steroid metabolism, binding, and concentration, in order to provide meaningful information. The developmental and experimental aspects of this work are still largely uncharted areas, so too is the general applicability to other animals of concepts developed and tested in the rat. If an integrated approach to these challenging questions is made, it must surely reveal many of the answers.

ACKNOWLEDGEMENTS

We are indebted to Doctors B.V. Caldwell (Yale University) and D. Tulchinsky (Boston Hospital for Women) for providing antisera used in these studies. We thank Julie Siu for excellent technical assistance, and Doctors T.M. Louis and J.S. Robinson for allowing us to use unpublished information.

This work was supported in part by United Cerebral Palsy Grant R-226-72, and by a grant from the British M.R.C.

REFERENCES

Abraham, G.E., Tulchinsky, D. and Korenman, S.G. (1970) Chromatographic purification of estradiol-17β for use in radio-ligand assay. *Biochem. Med.*, 3: 365–368.

Anand Kumar, T.C. and Thomas, G.H. (1968) Metabolites of ^{3}H-estradiol-17β in the cerebrospinal fluid of the rhesus monkey. *Nature (Lond.)*, 219: 628–629.

Axelrod, L.R. (1971) The metabolism of corticosteroids by incubated and perfused brain tissues. In *Influence of Hormones on the Nervous System, Proc. int. Soc. Psychoneuroendocrinol., Brooklyn, N.Y., 1970*, Karger, Basel, pp. 74–84.

Baird, D.T., McCracken, J.A. and Goding, J.R. (1973) Studies in steroid synthesis and secretion with the autotransplanted sheep ovary and adrenal. In *The Endocrinology of Pregnancy and Parturition. Experimental Studies in the Sheep*, C.G. Pierrepoint (Ed.), Alpha Omega Alpha, Cardiff, pp. 5–16.

Baldwin, B.A. and Bell, F.R. (1973) The anatomy of the cerebral circulation of the sheep and ox. The dynamic distribution of the blood supplied by the carotid and vertebral arteries to cranial regions. *J. Anat. (Lond.)*, 97: 203–215.

Bedford, C.A., Harrison, F.A. and Heap, R.B. (1973) The kinetics of progesterone metabolism in the pregnant sheep. In *The Endocrinology of Pregnancy and Parturition. Experimental Studies in the Sheep*, C.G. Pierrepoint (Ed.), Alpha Omega Alpha, Cardiff, pp. 83–93.

Challis, J.R.G., Davies, I.J. and Ryan, K.J. (1973a) The concentrations of progesterone, estrone and estradiol-17β in the plasma of pregnant rabbits. *Endocrinology*, 93: 971–976.

Challis, J.R.G., Harrison, F.A. and Heap, R.B. (1973b) The kinetics of oestrogen metabolism in the pregnant sheep. In *The Endocrinology of Pregnancy and Parturition. Experimental Studies in the Sheep*, C.G. Pierrepoint (Ed.), Alpha Omega Alpha, Cardiff, pp. 73–82.

Challis, J.R.G., Davies, I.J., Benirschke, K., Hendrickx, A.G. and Ryan, K.J. (1975) The effects of dexamethasone on the peripheral plasma concentrations of adrostenedione, testosterone and cortisol in the pregnant rhesus monkey. *Endocrinology*, 96: 185–192.

Cheng, Y.J. and Karavolas, H.J. (1973) Conversion of progesterone to 5α-pregnane-3,20-dione and 3α-hydroxy-5α-pregnan-20-one by rat medial basal hypothalami and the effects of estradiol and stage of estrous cycle on the conversion. *Endocrinology*, 93: 1157–1162.

Currie, W.B., Wong, M.S.F., Cox, R.I. and Thorburn, G.D. (1973) Spontaneous or dexamethasone-induced parturition in the sheep and goat: changes in plasma concentrations of maternal prostaglandin F and foetal oestrogen sulphate. *Mem. Soc. Endocr.*, 20: 95–118.

David, G.F.X. and Anand Kumar, T.C. (1974) Transfer of steroidal hormones from blood to the cerebrospinal fluid in the rhesus monkey. *Neuroendocrinology*, 14: 114–120.

Denef, C., Magnus, C. and McEwen, B.S. (1974) Sex dependent changes in pituitary 5α-dihydrotestosterone and 3α-androstanediol formation during postnatal development and puberty in the rat. *Endocrinology*, 94: 1265–1274.

Green, R., Luttge, W.G. and Whalen, R.E. (1969) Uptake and retention of tritiated estradiol in brain and peripheral tissue of male, female and neonatally androgenized female rats. *Endocrinology*, 85: 373–378.

Henkin, R.I., Casper, A.G.T., Brown, R., Harlan, A.B. and Bartter, F.C. (1968) Presence of corticosterone and cortisol in the central and peripheral nervous system of the cat. *Endocrinology*, 82: 1058–1061.

Jenkin, G., Henville, A. and Heap, R.B. (1975) Metabolism of oestrone sulphate and binding of oestrogens by the brain and pituitary of foetal and adult sheep. *J. Endocr.*, 64: 22P–23P.

de Jong, F.H., Hey, A.H. and van der Molen, H.J. (1973) Effect of gonadotrophins on the secretion of oestradiol-17β and testosterone by the rat testis. *J. Endocr.*, 57: 277–284.

Kato, J. and Villee, C.A. (1967a) Preferential uptake of estradiol by the anterior hypothalamus of the rat. *Endocrinology*, 80: 567–575.

Kato, J. and Villee, C.A. (1967b) Factors affecting uptake of estradiol-6,7-^{3}H by the hypophysis and hypothalamus. *Endocrinology*, 80: 1133–1138.

Kazama, N. and Longcope, C. (1974) In vivo studies on the metabolism of estrone and estradiol-17β by the brain. *Steroids*, 23: 469–481.

Korach, K.S. and Muldoon, T.G. (1974) Studies on the nature of the hypothalamic estradiol-concentrating mechanism in the male and female rat. *Endocrinology*, 94: 785–793.

Labhsetwar, A.P. (1972) Peripheral serum levels of immunoreactive "oestradiol" in rats during various reproductive states: adrenal contribution of immunoreactive material. *J. Endocr.*, 52: 399–400.

Lloyd, B.J. (1972) Plasma testosterone and accessory sex glands in normal and cryptorchid rats. *J. Endocr.*, 54: 285–296.

Lurie, A.O. and Weiss, J.B. (1967) Progesterone in cerebrospinal fluid during human pregnancy. *Nature (Lond.)*, 215: 1178.

Luttge, W.G. and Wallis, C.J. (1973) In vitro accumulation and saturation of ^{3}H-progestins in selected brain regions and in the adenohypophysis, uterus and pineal of the female rat. *Steroids*, 22: 493–502.

Massa, P., Justo, S. and Martini, L. (1974) Conversion of testosterone into 5α-reduced metabolites in the anterior pituitary and in the brain of maturing rats. In *Sexual Endocrinology of the Perinatal Period*, M.G. Forest and J. Bertrand (Eds.), INSERM, Paris, pp. 219–232.

McEwen, B.S., Weiss, J.M. and Schwartz, L.S. (1968) Selective retention of corticosterone by limbic structures of rat brain. *Nature (Lond.)*, 220: 911–912.

Michael, R.P. (1965) Oestrogens in the central nervous system. *Brit. med. Bull.*, 21: 87–90.

Naess, O. and Attramadal, A. (1974) Uptake and binding of androgens by the anterior pituitary gland, hypothalamus, preoptic area and brain cortex of rats. *Acta endocr. (Kbh.)*, 76: 417–430.

Naess, O., Attramadal, A. and Aakvaag, A. (1975) Androgen binding proteins in the anterior pituitary, hypothalamus, preoptic area and brain cortex of the rat. *Endocrinology*, 96: 1–9.

Naftolin, F., Ryan, K.J., Davies, I.J., Reddy, V.V., Flores, F., Petro, Z. and Kuhn, M. (1975) The formation of estrogens by central neuro-endocrine tissues. *Recent Progr. Hormone Res.*, 31: 295.

Payne, A.H., Lawrence, C.C., Foster, D.L. and Jaffe, R.B. (1973) Intranuclear binding of 17β-estradiol and estrone in female ovine pituitaries following incubation with estrone sulfate. *J. biol. Chem.*, 248: 1598–1602.

Perez, A.E., Ortiz, A., Cabeza, M., Beyer, C. and Perez-Palacios, G. (1974) In vitro metabolism of ^{3}H-androstenedione by the male rat pituitary, hypothalamus, and hippocampus. *J. Steroid Biochem.*, 5: 391 (abstr.).

Reddy, V.V., Naftolin, F. and Ryan, K.J. (1973) Aromatization in the central nervous system of rabbits: effects of castration and hormone treatment. *Endocrinology*, 92: 589–594.

Reddy, V.V., Naftolin, F. and Ryan, K.J. (1974) Steroid 17β-oxidoreductase activity in the rabbit CNS and adenohypophysis. *J. Endocr.*, 62: 401–402.

Robel, P., Corpechot, C. and Baulieu, E.E. (1973) Testosterone and androstanolone in rat plasma and tissues. *FEBS Lett.*, 33: 218–220.

Sar, M. and Stumpf, W.E. (1972) Cellular localization of androgen in the brain and pituitary after the injection of tritiated testosterone. *Experientia (Basel)*, 28: 1364–1366.

Sar, M. and Stumpf, W.E. (1973) Autoradiographic localization of radioactivity in the rat brain after the injection of 1,2-^{3}H-testosterone. *Endocrinology*, 92: 251–256.

Tabei, T., Haga, H., Heinrichs, W.L. and Herrmann, W.L. (1974) Metabolism of progesterone by rat brain, pituitary gland and other tissues. *Steroids*, 23: 651–666.

Touchstone, J.C., Kasparow, M., Hughes, P.A. and Horwitz, M.A. (1966) Corticosteroids in human brain. *Steroids*, 4: 205–211.

Weisz, J. and Gibbs, C. (1974) Metabolism of testosterone in the brain of the newborn female rat after an injection of tritiated testosterone. *Neuroendocrinology,* 14: 72–86.

Whalen, R.E. and Luttge, W.G. (1971) Differential localization of progesterone uptake in brain. Role of sex, estrogen pretreatment and adrenalectomy. *Brain Res.*, 33: 147–155.

Whalen, R.E., Gorzalka, B.B. and Luttge, W.G. (1973) Steroid extraction and tissue digestion in the assay of radioactivity in rat brain following tritiated estradiol-17β. *Steroids,* 21: 219–231.

Subcellular Mechanisms in Reproductive Neuroendocrinology, edited by
F. Naftolin, K.J. Ryan and J. Davies
© 1976 Elsevier Scientific Publishing Company—Amsterdam, The Netherlands

Chapter 13

Specific binding of steroids by neuroendocrine tissues

I.J. DAVIES, F. NAFTOLIN, K.J. RYAN and J. SIU

Department of Obstetrics and Gynecology, Harvard Medical School, Boston, Mass. 02115 (U.S.A.) and (F.N.) Department of Obstetrics and Gynecology, McGill University Faculty of Medicine and Royal Victoria Hospital, Montreal, Que. H3A 1A1 (Canada)

I. INTRODUCTION

According to what is currently known, cytoplasmic receptors are a prerequisite for most, if not all, biologic responses to steroid hormones and the specificity and the nature of the response is determined in large part by the interaction of the steroid with a receptor and the resulting association of the complex with nuclear chromatin. This review will consider briefly some aspects of steroid—receptor interactions in the brain and pituitary which may contribute to our understanding of the action and interaction of gonadal steroids in these tissues. Attention will be directed primarily to the identity of the steroid receptors, their specificity for compounds of biologic interest, and interactions which may modulate the action of gonadal steroids in the brain and pituitary at the receptor level.

Most of our information pertaining to steroid—receptor interactions is derived from studies of non-nervous tissues. Therefore, it is important to know to what extent information gained from the study of other tissues is applicable to the brain and pituitary. For estrogens, the evidence indicates that receptors in different tissues and in different species are similar, and that information derived from models such as the rat uterus are to a large extent applicable to the brain and pituitary. For other steroids this has not been clearly established at the present time.

II. METHODS OF DEMONSTRATION OF STEROID RECEPTORS

A basic approach to the demonstration of steroid receptors is the injection of radiolabeled hormone and subsequent observation of "uptake" or concentration of radioactivity in a specific tissue. However, brain tissues take up steroids in relatively large amounts in a non-specific unsaturable manner. This tends to obscure, or might mimic, receptor-related specific retention. Demonstration of retention which is specifically inhibited by pretreatment with the unlabeled steroid or a specific steroid antagonist is important, and the association of the retained radioactivity with the nuclear fraction provides additional evidence for the presence of a specific receptor. However, both the possibility of inhibition of uptake by a non-competitive mechanism and the conversion of the injected steroid to metabolites must be considered.

Direct evidence for the presence of receptors is provided by tissue fractionation and the physicochemical demonstration of a receptor-like component in the supernatant. The dependence of nuclear binding of the steroid on the presence of the cytoplasmic fraction provides strong evidence for the identity of the receptor. Steroid receptors in the pituitary and brain have been studied mostly in the rat and the information which will be reviewed here pertains to the rat except where otherwise indicated.

III. ESTROGENS

III.1. Uptake studies

Uptake of [^{3}H]estradiol by the pituitary and hypothalamus-preoptic area was demonstrated by Eisenfeld and Axelrod in 1965. Pretreatment with unlabeled estradiol diminished the uptake of [^{3}H]estradiol and the radioactivity in these tissues was identified as being primarily unchanged [^{3}H]estradiol. These observations have subsequently been confirmed and extended by many investigators. The specific nature of the hormone retention is evidenced by inhibition of uptake by pretreatment with estradiol, diethylstilbestrol, clomiphene, and other antiestrogens (Eisenfeld and Axelrod, 1967; Zigmond and McEwen, 1970; Chazal et al., 1975). Comparable doses of estrone, 17α-estradiol, and testosterone did not inhibit the uptake of [^{3}H]estradiol (Zigmond and McEwen, 1970; Luttge and Whalen, 1972). Injection of [^{3}H]estrone did not reveal any limited-capacity uptake in pituitary or brain tissues (Luttge and Whalen, 1972).

Retained [^{3}H]estradiol was shown to be associated with the nuclear fraction (Zigmond and McEwen, 1970; Kato et al., 1970; Chazal et al., 1975), and administration of unlabeled estradiol in vivo, followed by tissue fractionation, showed that this treatment caused an increase in exchangeable bound estradiol in the nuclear fraction (Anderson et al., 1973). Following [^{3}H]estradiol injection, the concentration of radioactivity in the cytoplasmic fraction was observed to rise relatively rapidly. The concentration of radioactivity in the nuclear fraction continued to rise while that in the cytosol was declining (Mowles et al., 1971). Similar observations were made in vitro with pituitary cell suspensions (Leavitt et al., 1973). The uptake of [^{3}H]estradiol in vitro by the nuclear fraction of rabbit hypothalamus was shown to require the presence of the soluble fraction (Chader and Villee, 1970). These studies and others clearly establish the presence of an estrogen receptor and a nuclear translocation process in the pituitary and hypothalamus comparable to that which has been described in the uterus.

An interesting facet of Eisenfeld and Axelrod's (1965) original uptake experiments was the observation that pretreatment with the progestin, norethynodrel, reduced [^{3}H]estradiol uptake. These authors subsequently reported reduction in the uptake of [^{3}H]estradiol by pretreatment with a similar progestin, norethindrone, but not with two other progestins, chlormadinone and medroxyprogesterone (Eisenfeld and Axelrod, 1967). Similar results with these compounds and with norgestrel were reported by others (Banerjee et al., 1973). Studies with a larger group of synthetic progestins indicated that the inhibition of [^{3}H]estradiol uptake was characteristic of progestins of the 19-nor configuration but not with other progestins (Van Kordelaar et al., 1975a). Ciacco and Lisk (1972), employing implantation technics, reported that priming with progesterone itself diminished [^{3}H]estradiol uptake. The implications of these observations will be considered below.

III.2. Characterization of the receptor

Physicochemical characterization of the estrogen receptor has not revealed any substantial difference from that in uterine tissue. In our laboratory the estradiol–receptor complexes from both pituitary and anterior hypothalamus were observed to sediment in sucrose density gradients as 8.3 S ± 0.06 (n = 7) relative to human gamma-globulin (S = 7.1) (Davies et al., 1975a and unpublished). This is in agreement with other investigators, both for the rat (Kato, 1970; Notides, 1970; Korach and Muldoon, 1973, 1974a), and the mouse (Fox and Johnston, 1974). Receptors extracted from nuclear fractions have been reported to be 6 S (Kato, 1970) or 7 S (Mowles et al., 1971).

The equilibrium association constant determined by saturation analysis is not measurably different between the pituitary and anterior hypothalamus, the mean value for both tissues being 0.9 ± 0.2 (S.E.) $\times 10^{10}$ M^{-1} (n = 8) (Davies, et al., 1975a and unpublished). Values of similar magnitude have been reported by others (Notides, 1970; Leavitt et al., 1973; Ginsburg et al., 1974a; Korach and Muldoon, 1974a, b; Vreeburg et al., 1974). The latter authors also determined the rate constants of association and dissociation. Therefore, both in sedimentation characteristics and in binding kinetics, the estrogen receptors in pituitary and brain structures do not appear to differ substantially from uterine receptors.

The high-affinity binding to the estrogen receptors in brain and pituitary cytosol in vitro is highly specific for estrogens. Employing concentrations of competitors up to 1000-fold, there was no substantial reduction in binding at equilibrium by testosterone, progesterone or dihydrotestosterone (DHT) (Notides, 1970; Korach and Muldoon, 1973, 1974a; Davies et al., 1975a). The radiolabeled forms of these compounds, and also norethindrone, were tested directly for binding to the estrogen receptor and no high-affinity binding was observed (Korach and Muldoon, 1973, 1974a; Davies et al., 1975a). At high concentrations of competing DHT, some small inhibition of [^{3}H]estradiol binding at equilibrium may be observed which cannot be ascribed to high-affinity interaction (Davies, 1975a and unpublished). Korach and Muldoon (1975) have recently reported that, with pituitary cytosol in vitro at 0–4 °C, DHT inhibits the initial association rate of [^{3}H]estradiol with the receptor. They attribute this effect to low-affinity competitive interaction of DHT with the estradiol binding sites. This will be discussed further below.

As previously indicated, uptake studies in vivo indicated an inhibition of [^{3}H]estradiol retention in neural tissues by 19-nor progestins which suggested a competitive interaction. In subsequent competition experiments with uterine cytosol *in vitro*, these compounds were reported to inhibit [^{3}H]estradiol binding. However, very high concentrations of the competitors were required (1500-fold) and the degree of inhibition did not correlate well with the relative effectiveness of different compounds in vivo (Van Kordelaar et al., 1975b). When [^{3}H]norethindrone was incubated with pituitary and hypothalamic cytosol no high-affinity binding was observed (Davies et al., 1975a). While it seems likely that these 19-nor progestins may inhibit [^{3}H]estradiol uptake by a competitive effect, it should be empasized that progesterone itself is not a competitor. Furthermore, progestins, including progesterone itself, have been found to have an inhibitory effect on estrogen replenishment which seems physiologically more important. This will be discussed further below.

The relative affinities of different estrogens and estrogen antagonists have been determined from competition curves (Table I) (Davies et al., unpublished). As might be expected, the relatively potent synthetic estrogens are very good competitors, the classical naturally occurring estrogens other than 17β-estradiol being much less effective. The anti-estrogens compete in vitro only at rather high concentrations. These results are consistent with previous reports with both pituitary (Notides, 1970) and uterus (Korenman, 1969).

Of current interest are the 2-hydroxyestrogens, or catechol estrogens. Both 2-hydroxyestrone and 2-hydroxyestradiol have been found to have affinities for the estrogen receptor which are within one order of magnitude of their parent compounds (Davies et al., 1975b). These catechol estrogens are major metabolites of estrogen in both animals and man, may be synthesized in brain tissues (Fishman and Norton, 1975), and have been demonstrated to have biologic activity which may be anti-estrogenic (Naftolin et al., 1975). The physiologic significance of these observations has not been established.

III.3. Control of receptor activity

Injection of estradiol results in a rapid diminution in receptor concentration in the pituitary and brain, reflecting translocation to the nucleus. This is followed in a period of

TABLE I

Relative affinities* of some estrogens and anti-estrogens

	Pituitary	Anterior hypothalamus
17β-Estradiol	100	100
Diethylstilbestrol	49	—
17α-Ethinyl-estradiol	48	—
2-Hydroxyestradiol	37	22
Estriol	11	5
17α-Estradiol	8	2
Estrone	7	2
2-Hydroxyestrone	3	2
CI-628	3	2
Cis-clomiphene	0.9	0.3
Trans-clomiphene	< 0.1	< 0.1

* The relative affinity was calculated as the concentration of 17β-estradiol required to reduce saturable [^{3}H]estradiol binding by 50%, divided by the concentration of competitor required to cause the same reduction, multiplied by 100% (Davies et al., 1975b).

hours by restoration of the cytoplasmic receptors to a concentration which may exceed initial control values. The replenishment process is inhibited by cyclohexamide, and it is believed that synthesis or activation of receptor is one of the responses to estradiol. The maintenance of relatively normal concentrations of receptor in castrate or hypophysectomized animals suggests that the synthesis of receptor is not entirely estrogen-dependent but is estrogen stimulated (Cidlowski and Muldoon, 1974).

With the uterus, it has been shown that progesterone, while not interfering with estrogen binding to the receptors or with estrogen-receptor translocation, inhibits receptor replenishment. This inhibition of replenishment results in lower receptor concentrations and a diminution in the uterotrophic response to estradiol (Hsueh et al., 1975).

In adult cycling female rats, the quantity of available estrogen receptors in the uterus is reported to vary during the estrus cycle and to correlate with the estrogen responsiveness of the tissue (Lee, 1974). Similarly, in human endometrium the quantity of estrogen receptor is higher during the proliferative phase than during the secretory phase. Furthermore, administration of medroxyprogesterone during the proliferative phase lowers the quantity of receptors to that which is found in the secretory phase (Tseng and Gurpide, 1975). While we are aware of no direct evidence that these observations pertain to the brain and pituitary, it seems likely that estrogen receptor levels are modulated positively by estradiol and negatively by progesterone, thereby varying the sensitivity of these tissues to estradiol.

While no physiologic effect of androgens on estrogen receptors has been established, some possibilities should be noted. As indicated above, DHT was shown to decrease the initial association rate of estradiol with the receptor in cytosol in vitro at $0-4\,°C$. There is other evidence which indicates that DHT influences the estradiol receptor. Rochefort et al. (1972), on the basis of in vitro experiments with uterine tissues, reported that, although DHT and testosterone do not bind to estrogen receptor sites, these androgens do induce translocation of the estrogen receptors to the nucleus. They suggested that this might be related to the uterotrophic and anti-estrogenic response observed when non-physiologic doses of testosterone are administered in vivo. Ruh et al. (1975) did experiments utilizing whole uteri incubated at $37\,°C$. Incubation with either DHT or testosterone induced nuclear accumulation of estradiol receptors. This was associated with a subsequent diminution in the ability of the tissue to take up [^{3}H]estradiol and concentrate it in the nuclear fraction. This effect was not inhibited by cyproterone, indicating non-involvement of an androgen receptor. It was inhibited by anti-estrogens, suggesting interaction of the androgens with the estrogen receptors. More recently it has been reported that injection of pharmacological doses of DHT into immature rats resulted in binding of DHT to uterine estrogen receptors, nuclear translocation, and a uterotrophic response. No uterotrophic response was observed with a smaller dose of DHT which was sufficient to saturate the androgen receptors. These data suggest a pharmacological role for this interaction of DHT with estrogen receptors (Rochefort and Garcia, 1975).

A possible role for other androgen metabolites has been suggested. Poortman et al. (1974) reported that some C_{19} compounds can compete weakly with estradiol for receptor sites. The strongest competitor was androstenediol, which competed with a relative affinity of 2%.

III.4. Ontogeny

The quantity of high-affinity estrogen receptors in the brain of the newborn rat is either very limited or difficult to demonstrate (Barley et al., 1974). By 7 days of age, an 8 S receptor, which is physicochemically like the adult receptor, becomes demonstrable in hypothalamic cytosol by sucrose density gradient analysis. The concentration increases gradually at first, more rapidly between 14 and 21 days, and reaches adult levels at about 28 days (Kato et al., 1971, 1974). Another report indicates that the receptor is demonstrable by day 5 of life (Barley et al., 1974). Nuclear binding following in vivo injection of [^{3}H]estradiol was low at 7 days, the major rise to achieve adult levels occurring around day 25 (Plapinger and McEwen, 1973). In vitro tissue incubations indicated the appearance of saturable binding of [^{3}H]estradiol between days 3 and 5 (Kulin and Reiter, 1972).

Consideration of estrogen dynamics in the newborn rat is complicated by the presence in plasma and brain tissues of an estrogen binding protein which is believed to be identical with α-fetoprotein. This extracellular estrogen binding protein disappears during the first 4 weeks of life while the number of estrogen receptors appears to be rising (Plapinger et al., 1973). Raynaud (1973) has shown that the uterotrophic activity of 17β-estradiol, which binds to this plasma protein, is markedly impeded compared to a synthetic estrogen, R 2858, which is not bound to the plasma protein. The relative effectiveness of estradiol increases during the first 4 weeks of life as the plasma estradiol binding protein disappears. One implication of these observations is that in the newborn rat circulating estradiol might be relatively ineffective as compared to estradiol which might be synthesized intracellularly from androgen precursors.

III.5. Males vs. females

In experiments with 28-day-old rats the number of estrogen binding sites in female anterior hypothalami was found to be slightly greater than in males ($\simeq 15\%$). The pituitaries showed no sex difference (Davies et al., 1975c). Nuclear concentration of injected [^{3}H]estradiol in the hypothalamus of castrate adult animals has also been observed to be somewhat greater in females (McEwen and Pfaff, 1970; Whalen and Massicci, 1975). While it is clear that males and females are more alike than different in this regard, these observations are of interest in connection with the hormone-induced constant estrus syndrome. While not all reports have been consistent (Whalen and Massicci, 1975), a number of investigators have observed diminished hypothalamic [^{3}H]estradiol uptake in these masculinized animals as compared to normal females (for references, see Davies et al., 1975c).

IV. ANDROGENS

The information currently available on androgen receptors is complex and potentially confusing. In a recent review Liao (1974) lists a variety of androgen-binding proteins which have been found in different androgen-sensitive tissues which have varying characteristics and uncertain functions. The most thoroughly studied androgen receptor is that in ventral prostate cytosol. This component sediments at 3–4 S in hypertonic solutions. In hypotonic extracts an 8–9 S component is observed which dissociates to 3–4 S in high-salt solutions. The 3–4 S binding component can be separated into an α and β component by ammonium sulfate precipitation. The β component binds DHT very specifically and it is this species which has clearly been shown to be an androgen receptor. However, there is evidence which suggests that DHT does not account for all of the actions of androgens. Testosterone itself or other metabolites of testosterone may be responsible for some androgenic actions of testosterone in the prostate and in other tissues (Liao, 1974). A recent report describes separation of 4 different androgen-binding components in prostatic cytosol, all of which can bind with DHT to prostatic chromatin but which differ in their steroid specificity (Goldman and Katsumata, 1975).

When [^{3}H]testosterone is injected into rats, the differential retention of radioactivity in specific areas of the brain as compared to other areas of the brain or blood is quite limited. Most of the uptake is of an unsaturable nature and the interpretation of the findings is complicated by the metabolism of the steroid (McEwen et al., 1970a, 1970b). The characterization of the cytoplasmic receptors and the identification of steroid specifically bound to nuclei is critical to the elucidation of androgen effects on brain function.

Jouan et al. (1971) identified a testosterone-binding macromolecule in pituitary and anterior hypothalamic cytosol employing both tissue-slice incubations and in vivo injections. Kato and Onouchi (1973a, b), employing sucrose density gradient analysis, demonstrated an 8 S DHT-binding component in cytosol from these tissues. Macromolecular binding of androgens has subsequently been reported to be present not only in pituitary and hypothalamus, but also in the amygdala, the preoptic area, and the cerebral cortex (Ginsburg et al., 1974b; Naess and Attramadal, 1974). While most of the studies described utilize castrate male rats, the binding component is also demonstrable in females (Ginsburg et al., 1974b).

Naess et al. (1975a) have characterized this androgen-binding component to some extent. It is reported to be a heat-labile protein which is sensitive to sulfhydryl reagents and binds testosterone, DHT, and cyproterone with high affinity. While Jouan et al. (1971) have focused primarily on testosterone binding, Kato and Onouchi (1973a) observed a greater affinity of the 8 S component for DHT than for testosterone. Naess et al. (1975b) reported that the affinity for DHT and testosterone is essentially the same. The relative affinity, as compared to testosterone, was 16% for progesterone and 8% for estradiol. Cortisol did not compete (Naess et al., 1975b). Fifteen minutes after in vivo administration of [^{3}H]testosterone, most of the radioactivity bound in the cytoplasm was testosterone with lesser amounts of DHT. Estradiol could not be identified. In similar experiments, Jouan et al. (1971) also found testosterone to exceed DHT in the bound fraction in the cytosol. The latter investigators subsequently examined the nature of the radioactivity in the nuclear fraction of the pituitary 1 hr after [^{3}H]testosterone injection (Thieulant et al., 1973). The concentration of radioactivity in the nuclei was 7-fold that in the cytosol and 63% of the bound fraction was testosterone while 21% was DHT. Lieberburg and McEwen (1975) carried out experiments specifically looking for nuclear-bound radiolabeled estrogen following administration of [^{3}H]testosterone to 5-day-old male and female rats. While the proportion of radioactivity identified as estradiol in the whole homogenates of limbic tissues was very small, as much as 50% of the radioactivity associated with the nuclear fraction was identified as estradiol. In autoradiographic studies, Sheridan et al. (1974) found that the concentration of radioactivity by nuclei in brain structures following [^{3}H]testosterone injection was completely inhibited by pretreatment with either testosterone or estradiol, but not by DHT.

While the presence of a receptor-like protein or proteins for androgens in brain and pituitary is established, the relationship of this component with the translocation process has not been adequately explored and the relative importance of testosterone and its metabolites remains open to further study.

V. PROGESTERONE

A receptor-translocation mechanism for progesterone, well studied in the reproductive tract, has not been fully established in pituitary and brain tissues at this time. The uptake of [^{3}H]progesterone in vivo has been reported to be higher in some areas of the brain than in others. Uptake in mesencephalic structures was observed to be greater than in diencephalic structures, and uptake in the anterior hypothalamus and pituitary was higher than that in the cortex. However, the observed differences were not large and, contrary to what might be expected from data derived from uterine tissue, neither estrogen priming nor progesterone pretreatment appeared to alter the uptake (Whalen and Luttge, 1971a, b; Seiko and Hattori, 1971, 1973; Whalen and Gorzalka, 1974). Furthermore, extraction and chromatography of the radioactivity in the areas of apparent concentration indicated that a substantial amount of the radioactivity represented metabolites of progesterone (Whalen and Gorzalka, 1974; Seiko and Hattori, 1973). The failure of progesterone pretreatment to alter the pattern of [^{3}H]progesterone uptake by brain structures in the rat has also been reported for the guinea pig (Wade and Feder, 1972). In the experiments of Seiko and Hattori (1973), in which [^{3}H]progesterone appeared to be preferentially retained in the median eminence and pituitary, almost all of the radioactivity was found to be in the cytosol rather than the nuclear fraction. To our knowledge, there is no evidence from tissue fractionation experiments to indicate nuclear retention of progesterone or its metabolites.

Autoradiographic studies have given variable results. With the guinea pig, Sar and Stumpf (1973) observed nuclear labeling of neurons in the hypothalamus following injection of [^{3}H]progesterone. Furthermore, nuclear labeling required estrogen priming and was specifically inhibited by progesterone pretreatment. In similar experiments with the rat, these investigators were unable to observe localization of [^{3}H]progesterone (Stumpf, 1971).

There are two reports of the identification of progesterone receptor-like components in the cytosol of hypothalamus and pituitary. With rat tissues, incubation of cytosol with [^{3}H]progesterone followed by Sephadex chromatography demonstrated a macromolecular component which bound progesterone. Specificity and limited capacity were indicated by the ability of competing unlabeled progesterone, but not corticosterone or estradiol, to eliminate the binding. This component was demonstrable with median eminence and pituitary but not with the remainder of the hypothalamus. However, it was in these experiments that the radioactivity retained in these tissues following in vivo injection of [^{3}H]progesterone was found to be associated almost entirely with the cytoplasmic fraction (Seiko and Hattori, 1973). Another group of investigators (Iramain et al., 1973), utilizing rabbits and guinea pigs, demonstrated a high-affinity, limited capacity, stereospecific progesterone-binding protein in both pituitary and anterior hypothalamus. The binding of [^{3}H]progesterone could be inhibited by competing progesterone, 5α-dihydroprogesterone, synthetic progestins, and to a lesser extent by testosterone, but not by estradiol or corticosterone.

Atger et al. (1974) were unable to demonstrate high-affinity binding of progesterone in pituitary or hypothalamic cytosol of guinea pigs in vitro. Furthermore, no binding was observed either in the cytosol or in the nuclear 0.3 M KCl extracts following injection of [^{3}H]progesterone in vivo. The concentration of radioactivity in the crude nuclear pellet exceeded that of the cytosol in the hypothalamus but the radioactivity was not extractable with 0.3 M KCl, implying that it was not bound to a receptor. If there is specific nuclear retention of radioactivity in the hypothalamus following injection of [^{3}H]progesterone, the identity of the radioactivity has not been established and evidence for a translocation process has not been reported.

VI. SUMMARY

In summary, our interpretation of the current information concerning steroid receptors in the brain and pituitary is as follows. The estrogen receptor in these tissues is the same or very similar to that in uterine tissue. The presence of the receptor is not entirely estrogen-dependent but the synthesis or activation of the receptor is stimulated by estradiol. By analogy with uterine tissue, receptor replenishment is probably inhibited by progesterone, and estradiol and progesterone together may modulate the sensitivity of these tissues to estradiol. DHT and testosterone may have the capacity to diminish estradiol activity in vivo by means of a low-affinity interaction with estrogen receptors, but a physiologic role for this interaction has not been established. Most investigators agree that the number of estrogen receptors in the newborn rat is small and rises to adult levels by 28 days. In addition, the estrogen-binding protein in the plasma of the newborn rat appears to markedly impede the effectiveness of circulating estradiol. If estradiol is synthesized intracellularly, it might be relatively more potent than that in the plasma. While immature normal males and hormone-masculinized females appear to have some diminution in the quantity of estrogen receptors in the hypothalamus, they are very similar in this regard and show no difference in the pituitary.

While an androgen receptor-like component has been identified in pituitary and brain tissues, reports are not entirely consistent as to its specificity. It appears to have the ability to bind both DHT and testosterone. Following [^{3}H]testosterone administration, both DHT and testosterone have been identified in the bound fraction in the cytosol, and testosterone, DHT, and estradiol have all been identified in the nuclear fraction. Autoradiographically, nuclear labeling of brain structures following [^{3}H]testosterone injection was prevented by pretreatment with testosterone or estradiol, but not by DHT. The relative importance of testosterone itself and metabolites of testosterone in these tissues has not been defined by these studies.

For progesterone, there are reports of receptor-like components in hypothalamic and pituitary cytosol of the rat, rabbit, and guinea pig. However, tissue fractionation has failed to reveal nuclear localization of progesterone or its metabolites. Autoradiographic data indicate specific retention of progesterone or its metabolites in the pituitary and hypothalamus of the guinea pig but similar experiments with the rat have given negative results. At the present time, we do not have complete evidence for a cytoplasmic receptor and nuclear translocation mechanism for progesterone in these tissues.

REFERENCES

Anderson, J.N., Peck, Jr., E.J. and Clark, J.H. (1973) Nuclear receptor estrogen complex: accumulation retention and localization in the hypothalamus and pituitary. *Endocrinology*, 93: 711–717.

Atger, M., Baulieu, E.E. and Milgrom, E. (1974) An investigation of progesterone receptors in guinea pig vagina, uterine cervix, mammary glands, pituitary and hypothalamus. *Endocrinology*, 94: 161–167.

Banerjee, R.C., Brazeau, Jr., P., Saucier, R. and Husain, S.M. (1973) Effect of norethindrone and norgestrel on the tissue distribution of [^{3}H]estradiol-17β in ovariectomized rats. *Steroids*, 21: 133–145.

Barley, J., Ginsburg, M., Greenstein, B.D., MacLusky, N.J. and Thomas, P.J. (1974) A receptor mediating sexual differentiation? *Nature (Lond.)*, 252: 259–260.

Chader, G.J. and Villee, C.A. (1970) Uptake of estradiol by the rabbit hypothalamus. *Biochem. J.*, 118: 93–97.

Chazal, G., Faudon, M., Gogan, F. and Rotsztejn, W. (1975) Effects of two estradiol antagonists upon the estradiol uptake in the rat brain and peripheral tissues. *Brain Res.*, 89: 245–254.

Ciacco, L.A. and Lisk, R.D. (1972) Effect of hormone priming on retention of [^{3}H]estradiol by males and females. *Nature New Biol.*, 236: 82–83.

Cidlowski, J.A. and Muldoon, T.G. (1974) Estrogenic regulation of cytoplasmic receptor populations in estrogen-responsive tissues of the rat. *Endocrinology*, 95: 1621–1629.

Davies, I.J., Siu, J., Naftolin, F. and Ryan, K.J. (1975a) In *Schering Workshop on Central Actions of Estrogenic Hormones. Advances in the Biosciences, Vol. 15*, G. Raspé (Ed.), Pergamon Press/Vieweg, Braunschweig, pp. 89–103.

Davies, I.J., Naftolin, F., Ryan, K.J., Fishman, J. and Siu, J. (1975b) The affinity of catechol estrogens for estrogen receptors in the pituitary and anterior hypothalamus of the rat. *Endocrinology*, 97: 554–557.

Davies, I.J., Naftolin, F., Ryan, K.J. and Siu, J. (1975c) Estradiol receptors in the pituitary and anterior hypothalamus of the rat: measurement by agar gel electrophoresis. *Steroids*, 25: 591–609.

Eisenfeld, A.J. and Axelrod, J. (1965) Selectivity of estrogen distribution in tissues. *J. Pharmacol. exp Ther.*, 150: 469–475.

Eisenfeld, A.J. and Axelrod, J. (1967) Evidence for estradiol binding sites in the hypothalamus. Effect of drugs. *Biochem. Pharmacol.*, 16: 1781–1785.

Fishman, J. and Norton, B. (1975) Catechol estrogen formation in the central nervous system of the rat. *Endocrinology*, 96: 1054–1059.

Fox, T.O. and Johnston, C.J. (1974) Estradiol receptors from mouse brain and uterus: binding to DNA. *Brain Res.*, 77: 330–336.

Ginsburg, M., Greenstein, B.D., MacLusky, N.J., Morris, I.D. and Thomas, P.J. (1974a) An improved method for the study of high-affinity steroid binding. Estradiol binding in brain and pituitary. *Steroids*, 23: 773–792.

Ginsburg, M., Greenstein, B.D., MacLusky, N.J., Morris, I.D. and Thomas, P.J. (1974b) Dihydrotestosterone binding in brain and pituitary cytosol of rats. *J. Endocr.*, 61: xxiv.

Goldman, A.S. and Katsumata, M. (1975) Multiple dihydrotestosterone cytosol receptors in rat ventral prostate by a novel micromethod of electrofocussing: blocking action by cyproterone acetate and uptake by nuclear chromatin. *J. Steroid Biochem.*, 5: 332.

Hsueh, A.J.W., Peck, E.J. and Clark, J.H. (1975) Progesterone antagonism of the estrogen receptor and estrogen-induced uterine growth. *Nature (Lond.)*, 254: 337–339.

Iramain, C.A., Danzo, B.J., Strott, C.A. and Toft, D.O. (1973) Progesterone binding in the hypothalamus and hypophysis of female guinea pigs and rabbits. In *IV Int. Congr. Int. Soc. Psychoneuroendocrinol.*, (Abstract).

Jouan, P., Samperez, S., Thieulant, M.L. et Mercier, L. (1971) Etude du récepteur cytoplasmique de la [1,2-^{3}H]testostérone dans l'hypophyse antérieure et l'hypothalamus du rat. *J. Steroid Biochem.*, 2: 223–236.

Kato, J. (1970) Estrogen receptors in the hypothalamus and hypophysis in relation to reproduction. In *Hormonal Steroids, Proceedings of the Third International Congress, Hamburg, Int. Congr. Ser. No. 219*, V.H.T. James and L. Martini (Eds.), Excerpta Medica, Amsterdam, pp. 764–773.

Kato, J. and Onouchi, T. (1973a) 5α-Dihydrotestosterone receptor in the rat hypothalamus. *Endocr. jap.*, 20: 429–432.

Kato, J. and Onouchi, T. (1973b) 5α-Dihydrotestosterone receptor in the rat hypophysis. *Endocr. jap.*, 20: 641–644.

Kato, J., Atsumi, Y. and Muramatsu, M. (1970) Nuclear estradiol receptor in rat anterior hypophysis. *J. Biochem.*, 67: 871–872.

Kato, J., Atsumi, Y. and Inabo, M. (1971) Development of estrogen receptors in the rat hypothalamus. *J. Biochem.*, 70: 1051–1053.

Kato, J., Atsumi, Y. and Inaba, M. (1974) Estradiol receptors in female rat hypothalamus in the developmental stages and during pubescence. *Endocrinology*, 94: 309.

Korach, K.S. and Muldoon, T.G. (1973) 17β-Estradiol-receptor interaction in the anterior pituitary of male and female rats. *Endocrinology*, 92: 322–326.

Korach, K.S. and Muldoon, T.G. (1974a) Studies on the nature of the hypothalamic estradiol-concentrating mechanism in the male and female rat. *Endocrinology*, 94: 785–793.

Korach, K.S. and Muldoon, T.G. (1974b) Characterization of the interaction between 17β-estradiol and its cytoplasmic receptor in the rat anterior pituitary gland. *Biochemistry*, 13: 1932–1938.

Korach, K.S. and Muldoon, T.G. (1975) Inhibition of anterior pituitary estrogen-receptor complex formation by low-affinity interaction with 5α-dihydrotestosterone. *Endocrinology*, 97: 231–236.

Korenman, S.G. (1969) Comparative binding affinity of estrogens and its relation to estrogenic potency. *Steroids*, 13: 163–177.

Kulin, H.E. and Reiter, E.O. (1972) Ontogeny of the in vitro uptake of tritiated estradiol by the hypothalamus of the female rat. *Endocrinology*, 90: 1371–1374.

Leavitt, W.W., Kimmel, G.L. and Friend, J.P. (1973) Steroid hormone uptake by anterior pituitary cell suspensions. *Endocrinology*, 92: 94–103.

Lee, C. (1974) Uterine responsiveness to estrogen in 4-day cycling and in estrogen-primed rats as assessed by uptake and incorporation of [^{3}H]leucine. *Endocrinology*, 95: 1754–1758.

Liao, S. (1974). In *Biochemistry of Hormones,* H.L. Kornberg and D.C. Phillips (Eds.), Butterworths, London, pp. 153–185.

Lieberburg, I. and McEwen, B.S. (1975) Estradiol-17β: a metabolite of testosterone recovered in cell nuclei from limbic areas of neonatal rat brains. *Brain Res.*, 85: 165–170.

Luttge, W.G. and Whalen, R.E. (1972) The accumulation, retention and interaction of oestradiol and oestrone in central, neural and peripheral tissues of gonadectomized female rats. *J. Endocr.*, 52: 379–395.

McEwen, B.S. and Pfaff, D.F. (1970) Factors influencing sex hormone uptake by rat brain regions. I. Effects of neonatal treatment, hypophysectomy, and competing steroid on estradiol uptake. *Brain Res.*, 21: 1—16.

McEwen, B.S., Pfaff, D.W. and Zigmond, R.E. (1970a) Factors influencing sex hormone uptake by rat brain regions. II. Effects of neonatal treatment and hypophysectomy on testosterone uptake. *Brain Res.*, 21: 17—28.

McEwen, B.S., Pfaff, D.W. and Zigmond, R.E. (1970b) Factors influencing sex hormone uptake by rat brain regions. III. Effects of competing steroids on testosterone uptake. *Brain Res.*, 21: 29—38.

Mowles, T.F., Ashkanazy, B., Mix, Jr., E. and Sheppard, H. (1971) Hypothalamic and hypophyseal estradiol-binding complexes. *Endocrinology*, 89: 484—491.

Naess, O. and Attramadal, A. (1974) Uptake and binding of androgens by the anterior pituitary gland, hypothalamus, preoptic area, and brain cortex of rats. *Acta endocr. (Kbh.)*, 76: 417—430.

Naess, O., Attramadal, A. and Aakvaag, A. (1975a) Androgen binding proteins in the anterior pituitary, hypothalamus, preoptic area and brain cortex of the rat. *Endocrinology*, 96: 1—9.

Naess, O., Attramadal, A., Hanson, V., Aakvaag, A. and Torgerson, O. (1975b) Androgen receptors in the cytosol fractions of the anterior pituitary gland, hypothalamus, preoptic area and brain cortex of the rat. *J. Steroid Biochem.*, 5: 391.

Naftolin, F., Morishita, H., Davies, I.J., Ryan, K.J. and Fishman, J. (1975) 2-Hydroxyestrone-induced rise in serum luteinizing hormone in the immature male rat. *Biochem. biophys. Res. Commun.*, 64: 905—910.

Notides, A.C. (1970) Binding affinity and specificity of the estrogen receptor of the rat uterus and anterior pituitary. *Endocrinology*, 87: 987—992.

Plapinger, L. and McEwen, B.S. (1973) Ontogeny of estradiol-binding sites in rat brain. I. Appearance of presumptive adult receptors in cytosol and nuclei. *Endocrinology*, 93: 1119—1128.

Plapinger, L., McEwen, B.S. and Clemens, L.E. (1973) Ontogeny of estradiol-binding sites in rat brain. II. Characteristics of a neonatal binding macromolecule. *Endocrinology*, 93: 1129—1139.

Poortman, J., Vroegindewey-Jie, D., Thijssen, J.H.H. and Schwarz, F. (1974) Inhibition of the binding of estradiol to its specific receptor by C_{19} steroids in human mammary tumour and myometrial tissue. *J. Endocr.*, 64: 25 P.

Raynaud, J.P. (1973) Influence of rat estradiol binding plasma protein (EBP) on uterotrophic activity. *Steroids*, 21: 249—258.

Rochefort, H. and Garcia, M. (1975) Pharmacological interaction of and action of androgens on the estrogen receptor site. *Tenth Meeting FEBS, Paris* (Abstract).

Rochefort, H., Lignon, F. and Capony, F. (1972) Formation of estrogen nuclear receptor in uterus: effect of androgens, estrone and nafoxidine. *Biochem. biophys. Res. Commun.*, 47: 662—670.

Ruh, T.S., Wassilak, S.G. and Ruh, M.F. (1975) Androgen-induced nuclear accumulation of the estrogen receptor. *Steroids*, 25: 257—273.

Sar, M. and Stumpf, W.E. (1973) Neurons of the hypothalamus concentrate [^{3}H]progesterone or its metabolites. *Science*, 182: 1266—1268.

Seiko, K. and Hattori, M. (1971) A more extensive study on the uptake of labelled progesterone by the hypothalamus and pituitary gland of rats. *J. Endocr.*, 51: 793—794.

Seiko, K. and Hattori, M. (1973) In vivo uptake of progesterone by the hypothalamus and pituitary of the female ovariectomized rat and its relationship to cytoplasmic progesterone-binding protein. *Endocr. jap.*, 20: 111—119.

Sheridan, P.J., Sar, M. and Stumpf, W.E. (1974) Interaction of exogenous steroids in the developing rat brain. *Endocrinology*, 95: 1749—1753.

Stumpf, W.E. (1971) Autoradiographic techniques and the localization of estrogen, androgen and glucocorticoid in the pituitary and brain. *Amer. Zool.*, 11: 725—739.

Thieulant, M.L., Samperez, S. and Jouan, P. (1973) Binding and metabolism of [^{3}H]testosterone in the nuclei of rat pituitary in vivo. *J. Steroid Biochem.*, 4: 677—685.

Tseng, L. and Gurpide, E. (1975) Effects of progestins on estradiol receptor levels in human endometrium. *J. clin. Endocr.*, 41: 402—404.

Van Kordelaar, J.M.G., Brockman, M.M.M. and Van Rossum, J.M. (1975a) Interaction of contraceptive progestins and related compounds with the estrogen receptor. I. Effect on [^{3}H]estradiol distribution pattern in the ovariectomized rat. *Acta endocr. (Kbh.)*, 78: 145–164.

Van Kordelaar, J.M.G., Vermorken, A.J.M., De Weerd, C.J.M. and Van Rossum, J.M. (1975b) Interaction of contraceptive progestins and related compounds with the estrogen receptor. II. Effect on [^{3}H]estradiol binding to the rat uterine receptor *in vitro*. *Acta endocr. (Kbh.)*, 78: 165–179.

Vreeburg, J.T.M., Schretlen, P. and Baum, M.J. (1975) Studies *in vitro* on estradiol high affinity binding in cytosols from different brain regions and pituitary of the adult male rat. *J. Endocr.*, 64: 24P.

Wade, G.N. and Feder, H.H. (1972) [1,2-^{3}H]Progesterone uptake by guinea pig brain and uterus: differential localization, time-course of uptake and metabolism, and effects of age, sex estrogen-priming and competing steroids. *Brain Res.*, 45: 525–543.

Whalen, R.E. and Gorzalka, B.B. (1974) Estrogen-progesterone interactions in uterus and brain of intact and adrenalectomized immature and adult rats. *Endocrinology*, 94: 214–223.

Whalen R.E. and Luttge, W.G. (1971a) Role of the adrenal in the preferential accumulation of progestin by mesencephalic structures. *Steroids*, 18: 141–146.

Whalen, R.E. and Luttge, W.G. (1971b) Differential localization of progesterone uptake in brain. Role of estrogen pretreatment and adrenalectomy. *Brain Res.*, 33: 147–155.

Whalen, R.E. and Massicci, J. (1975) Subcellular analysis of the accumulation of estrogen by the brain of male and female rats. *Brain Res.*, 89: 255–264.

Zigmond, R.E. and McEwen, B.S. (1970) Selective retention of oestradiol by cell nuclei in specific regions of the ovariectomized rat. *J. Neurochem.*, 17: 889–899.

Chapter 14

Steroid receptors in neuroendocrine tissues: topography, subcellular distribution, and functional implications

BRUCE S. McEWEN

The Rockefeller University, New York, N.Y. 10021 (U.S.A.)

I. INTRODUCTION

The discovery of intracellular receptor sites for steroid hormones in brain and non-neural target tissues was made possible by the synthesis in the late 1950's of high specific activity tritium-labeled steroids (Gupta, 1960; Jensen and Jacobson, 1962; Michael; 1965). The high specific activity of the steroid allowed the recognition of binding sites of limited capacity in certain tissues against a background of non-specific labeling in virtually all tissues (Jensen and Jacobson, 1962; Eisenfeld and Axelrod, 1965; Kato and Villee, 1967). The low decay energy of tritium made possible by means of autoradiography the resolution of cellular and especially of cell nuclear labeling (Attramadal, 1964; Michael, 1965; Stumpf and Roth, 1966; Pfaff, 1968a, b; Stumpf, 1968a, b). The proliferation and extension of information concerning receptors for estrogens, androgens, progestins, gluco- and mineralocorticoids has been traced and documented in a recent book on the subject (King and Mainwaring, 1974).

The purpose of this article is to briefly summarize the current status of information on pituitary and neural receptors for steroid hormones, their properties and their topography in the brain, and to consider the functional implications of the steroid–receptor interactions for neuroendocrine function and for behavior. Implicit in the cell nuclear localization of labeled steroid hormones is the notion that the initial event involves alteration of genomic function of the target cells, leading to alterations of RNA and protein formation. Other actions of steroid hormones independent of the cellular receptors described herein are by no means excluded, although they will not be discussed further.

II. STEROID HORMONE RECEPTORS IN NEUROENDOCRINE TISSUES

II.1. *Estrogen receptors in adult tissues*

Putative estradiol receptors in the rat brain and pituitary have been extensively studied. Their topography, as determined by autoradiography (Stumpf, 1968a, b, 1970; Anderson and Greenwald, 1969; Pfaff and Keiner, 1973) and cell fractionation experiments on dissected brain tissue (Zigmond and McEwen, 1970; Maurer and Woolley, 1974;

McEwen et al., 1975a; see Fig. 1), agrees well, where such information is available, with sites where implanted estradiol exerts activational effects on sexual behavior and on neuroendocrine function (see Lisk, 1967; McEwen and Pfaff, 1973). Similar patterns of estrophilic neurons, virtually identical in their distribution in the preoptic area and tuberal hypothalamus and differing somewhat in extrahypothalamic structures, have been seen among vertebrates, including fish, birds, amphibia, rodents, carnivores, and primates (see Morrell et al., 1975). This suggests an ancestral origin of the pattern of estrophilic neurons and a conservation of the basic pattern in the course of evolution.

The intensity of labeling of pituitary and of various brain regions is a function both of the density of estrophilic neurons and the intensity of labeling of individual cells (see Stumpf, 1968b, for pituitary; Warembourg, 1970, for brain). In the pituitary, where 60–80% of the cells of the anterior lobe are labeled, maximal cell nuclear binding capacity corresponds to 12,500 molecules/cell (McEwen et al., 1975a), which is similar to the average labeling of uterine cells (Notides, 1970). In the most intensely labeled parts of the hypothalamus and preoptic area, where more than 50% of the neurons may be labeled, maximal cell nuclear binding of estradiol is $\approx$ 4–5000 molecules/cell, while the corticomedial amygdala has labeling of $\approx$ 3000 molecules/cell. Other brain regions, including the rest of the hypothalamus and amygdala, are at least an order of magnitude lower (McEwen et al., 1975a). Neural and pituitary estrogen receptors appear to be at least partially occupied by peak endogenous estrogen levels during the estrous cycle (McGuire and Lisk, 1968; Kato, 1970).

Pituitary and estrophilic neural tissues contain soluble "estrogen receptors" which, like those in the uterus, have a sedimentation coefficient in sucrose density gradients of $\approx$ 8S and stereospecifically bind active estrogens and non-steroidal antiestrogens (see Luine and McEwen, 1976; and Davies, this volume, for references). That these molecules are precursors for cell nuclear binding of estrogens may be inferred from depletion studies in which injected estradiol reduces the available cytosol estradiol-binding capacity for up to 20 hr (Cidlowski and Muldoon, 1974; Maclusky, Lieberburg, McEwen, unpublished). Non-steroidal antiestrogens such as nitromifene citrate (CI628; Parke-Davis) prevent neural and pituitary cell nuclear labeling by [^{3}H]estradiol (Chazal et al., 1975) and may do so, at least in the uterus, by translocating with the cytosol receptor to the cell nuclei where they are estrogenic for a brief period and then maintain the depletion of cytosol receptor levels (Clark et al., 1974). Considerable interest now centers on the repletion of cytosol receptor levels that occurs after estrogen but not after antiestrogen administration, which process may well involve de novo synthesis of receptors (Clark et al., 1974; Cidlowski and Muldoon, 1974).

The cell nuclear estrogen–receptor complex from pituitary and neural tissue, which may be extracted by 0.3 M KCl, appears to exist as a molecule sedimenting around 5–7S (Kato et al., 1970; Mowles et al., 1971; Vertes and King, 1971). According to present views derived from work on the uterus, the 5S cell nuclear receptor may be a complex of a 4S estrogen-binding receptor subunit and an uncharacterized factor which does not bind estradiol (Notides and Nielsen, 1974; Yamamoto, 1974). The formation of this 5S complex appears to occur more slowly in hypothalamic cell nuclei than in cell nuclei of pituitary or uterus (Linkie, 1975). However, the time course of retention of total cell nuclear estradiol appears to be similar in pituitary and in estrophilic regions of brain (McEwen et al., 1975a).

Estrogen receptor sites are present in the male rat pituitary and brain (Eisenfeld and Axelrod, 1966; Pfaff, 1968c; Anderson and Greenwald, 1969; McEwen and Pfaff, 1970; Maurer and Woolley, 1974). Although neonatal androgenization of females has been reported to reduce estradiol binding in estrogen-sensitive tissues (Flerko and Mess, 1968; Vertes and King, 1971; Plapinger, 1973; Maurer and Woolley, 1974), quantitative differences in estrogen binding between gonadectomized males and females, indicative of a natural sex difference, have been difficult to find (Plapinger, 1973; Maurer and Woolley, 1974). In view of the ability of the adult rat hypothalamus and limbic brain to convert testosterone to estradiol (Naftolin et al., 1972; Weisz and Gibbs, 1974a; Lieberburg and McEwen, 1975b) and because of the efficacy of estradiol in restoring male sexual behav-

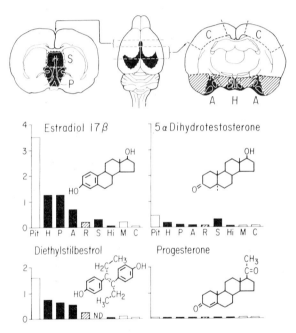

Fig. 1. Topography of cell nuclear binding of 4 ^{3}H-labeled, gonadal steroids in neuroendocrine tissues of ovariectomized-adrenalectomized (OVX-ADX) rats. Sampled brain regions, which were selected because they show by autoradiography high retention of [^{3}H]estradiol (Pfaff and Keiner, 1973), are depicted at the top of the figure. P, medial preoptic area; H, basomedial hypothalamus; A, corticomedial amygdala. The rest of hypothalamus and amygdala (R) contain fewer estrophilic neurons and have been pooled in these experiments. Septum (S) and hippocampus (Hi) are also shown at the top of the figure. Other sampled tissues are pituitary (Pit), midbrain central gray (M), and cerebral cortex (C). Details of the dissection scheme and data for [^{3}H]diethylstilbestrol ([^{3}H]DES) may be found in McEwen et al. (1975a). Dose of [^{3}H]DES was > 100 nmoles/kg body weight. Doses of [^{3}H]steroids were ≈ 10 nmoles/kg body weight. Survival time after tail vein infusion was 1–2 hr. Tissue was pooled from 3 to 4 rats in each experiment. For the [^{3}H]progesterone experiment, ADX-OVX rats received two daily priming injections of 15 μg estradiol benzoate in sesame oil (see McEwen et al., 1976). Data for other experiments are so far unpublished.

ior when injected systematically (see Pfaff, 1970) or implanted into the preoptic area of the castrated male rat (Christensen and Clemens, 1974), it is perhaps well to emphasize the similarities rather than the possible sex differences in neural estrogen receptor distribution and content.

II.2. Estrogen receptors in the neonatal brain

Estrogen receptor sites are present in the 2–3-day-old female rat brain and are found in both a soluble (Barley et al., 1974) and cell nuclear form (McEwen et al., 1975b) not only in the pituitary, hypothalamus, and amygdala, but also in the cerebral cortex. However, the neonatal hypothalamus, preoptic area and amygdala, and not the cerebral cortex, possess the ability to convert [^{3}H]testosterone to [^{3}H]estradiol in vivo (Weisz and Gibbs, 1974b; Lieberburg and McEwen, 1975a) and in vitro (Reddy et al., 1974; Weisz and Gibbs, 1974a). Considerable interest centers on this conversion and thus on hypothalamus, preoptic area and amygdala because of the demonstrated ability of both estrogens and aromatizable androgens (but not of ring A reduced androgens) to promote sexual differentiation of the brain during a neonatal critical period (see Plapinger and McEwen, 1976).

It was the remarkable concentration of testosterone-derived [^{3}H]estradiol in neonatal brain cell nuclei which first led us to recognize the neonatal estrogen receptors (Lieberburg and McEwen, 1975a). Our previous failure and the failure of others to find such receptors (see Plapinger and McEwen, 1973) may be traced to several factors: first, the presence of estrogen receptors in cerebral cortex up to the end of the third postnatal week (McEwen et al., 1975b and unpublished) since the cortex is used in the adult rat as a "non-binding" region for regional comparisons, and, second, to the presence in the neonatal blood and cerebrospinal fluid of a fetoneonatal estrogen-binding protein (fEBP) which sequesters estradiol-17β and prevents access of this steroid to tissue receptors (see Plapinger and McEwen, 1976, for literature).

Estrogens such as 11β-methoxy ethynylestradiol (RU-2858, Roussel-Uclaf) and, to a lesser extent, diethylstilbestrol (DES), which bind less well than 17β-estradiol to fEBP, bind well to neonatal cell nuclear estrogen receptor sites and provide the most convincing demonstration of these sites (McEwen et al., 1975b). RU-2858 and DES are potent agonists (the former 50 times better than estradiol) in inducing sexual differentiation of the brain (Ladosky, 1967; Clemens, 1974; Doughty et al., 1975). RU-2858 is more effective than estradiol in promoting uterine weight increases in neonatal female rats (Raynaud, 1973) and DES, given during human fetal development, is reported to induce masculinization of the reproductive tract of female children (Bongiovanni et al., 1959) and a predisposition to primary vaginal carcinoma in later life (Herbst et al., 1972).

Although it is tempting to conclude that neonatal estrogen receptors mediate the developmental effects of these hormones, and of androgen-derived estrogen, further studies are needed to conclusively prove this hypothesis (see Plapinger and McEwen, 1976, for discussion).

II.3. Androgen receptors in adult neuroendocrine tissues

The distribution of cells labeled with an injection of [^{3}H]testosterone (^{3}HT) in castrated adult male rats is similar but not identical to the distribution of cells labeled with [^{3}H]estradiol (Pfaff, 1968b, c; Toohimaa, 1971; Sar and Stumpf, 1972, 1973a; see also discussion by Zigmond, 1975), although the intensity of cellular labeling is generally lower for ^{3}HT than for [^{3}H]estradiol (^{3}HE$_2$). Highest concentrations of labeled cells are seen in the hypophysiotropic area, amygdala and pituitary, and also in lateral septum and hippocampus. ^{3}HT labels more cells than ^{3}HE$_2$ in lateral septum, hippocampus, dentate gyrus, subiculum, and ventromedial nucleus, while ^{3}HE$_2$ labels more cells than ^{3}HT in arcuate nucleus (Stumpf, 1970; Sar and Stumpf, 1973a). In the medial preoptic area a majority of the cells are labeled by both ^{3}HT and ^{3}HE$_2$ (see Zigmond, 1975, for discussion). This clearly indicates overlap of cells binding the two labels, but may well be explained by the known conversion of ^{3}HT to ^{3}HE$_2$. Between 35 and 50% of cell nuclear-bound radioactivity in adult male rat limbic structures has in fact been identified as ^{3}HE$_2$ following a ^{3}HT injection (see below and Lieberburg and McEwen, 1975b). Thus the conversion of testosterone to either estradiol or to 5α reduced metabolites makes difficult the interpretation of any labeling experiment with ^{3}HT. The pituitary is somewhat less complicated since no ^{3}HE$_2$ has been identified in cell nuclear extracts following ^{3}HT administration (Lieberburg and McEwen, 1975b). Autoradiographic studies reveal that ≈ 15% of rat pituitary cells, mostly basophils, are labeled after ^{3}HT, while 60—80% of the cells are labeled after ^{3}HE$_2$ (Stumpf, 1968b; Sar and Stumpf, 1973b, c).

In birds, cellular accumulation of radioactivity injected as ^{3}HT has been reported in periventricular areas of the hypophysiotropic area (Zigmond et al., 1972; Meyer, 1973). In addition, the accumulation of radioactivity after ^{3}HT administration has been demonstrated in the midbrain of the chaffinch, primarily in the nucleus intercollicularis, an area from which vocalizations can be stimulated in birds (Zigmond et al., 1973). Most of the radioactivity retained in cell nuclei in this last study appears to be testosterone or its 5α or 5β reduced metabolites.

Tissue uptake of radioactivity injected as ^{3}HT into castrated guinea pigs and rats is highest in prostate and seminal vesicles. Concentrations of radioactivity in pituitary and neural tissues, while lower than those in accessory sex glands, are nevertheless equal to or higher than those in serum or plasma. Pituitary generally is highest, followed by hypothalamus and cerebral cortex, although concentrations of radioactivity among CNS structures are similar, and regional differences which are reported by one group are not always found by another (Resko et al., 1967; Roy and Laumas, 1969; McEwen et al., 1970a, b; Phuong and Sauer, 1971; Stern and Eisenfeld, 1971; Perez-Palacios et al., 1973; Sar and Stumpf, 1973d; Dixit and Niemi, 1974). Perez-Palacios et al. (1973) compared the CNS tissue uptake of ^{3}HT, 5α-dihydrotestosterone (DHT) and androstenedione and found that [^{3}H]DHT was accumulated significantly more than the other two steroids relative to cerebral cortex by pituitary and by hippocampus and midbrain tegmentum. In this connection, it should be noted that [^{3}H]DHT is detected as a metabolite of ^{3}HT in pituitary and in brain regions (see below and Luine and McEwen, 1976). Estradiol has also been detected as a T metabolite in brain tissue (see above).

Uptake of radioactivity injected as ^{3}HT is significantly reduced by unlabeled testosterone in pituitary and in septum and less strongly inhibited in amygdala, preoptic area, hypothalamus and olfactory bulb (McEwen et al., 1970b; Stern and Eisenfeld, 1971). The anti-androgenic steroid, cyproterone, reduces tissue uptake of radioactivity injected as ^{3}HT in pituitary, septum, preoptic area, amygdala and hypothalamus (McEwen et al., 1970b; Stern and Eisenfeld, 1971; Sar and Stumpf, 1973d). Progesterone also reduces uptake of ^{3}HT radioactivity in pituitary, preoptic area, and central hypothalamus (Stern and Eisenfeld, 1971; Sar and Stumpf, 1973d), and this effect may be due to inhibition of Δ^4-3-keto steroid 5α-reductase by progesterone, a preferred substrate, and resulting reduction of the formation of [^{3}H]DHT (Stern and Eisenfeld, 1971; Massa and Martini, 1971/72).

It should be noted that cyproterone is an inhibitor of the aromatizing enzyme complex of placenta (Schwarzel et al., 1973) and thus its effects on ^{3}HT uptake noted above might be due in part to its ability to block ^{3}HE$_2$ formation. In this connection, it was noted by McEwen et al. (1970b) that unlabeled E$_2$ competed as well as, or better than, unlabeled T for the tissue uptake of ^{3}HT radioactivity by preoptic area, septum and olfactory bulb. In contrast, T competed better than E for uptake of ^{3}HT radioactivity by pituitary (McEwen et al., 1970b). These observations are consistent with the detection of ^{3}HE$_2$ as a metabolite of ^{3}HT in cell nuclei from limbic brain structures and failure to detect ^{3}HE$_2$ in pituitaries of the same animals (see below).

Besides the effects noted above on uptake of ^{3}HT radioactivity, several laboratories have reported that neonatal castration reduces the uptake of ^{3}HT radioactivity compared to adult castrate males in pituitary and in all brain structures (McEwen et al., 1970a), while androgen treatment of neonatal castrates selectively increases uptake of ^{3}HT radioactivity in anterior hypothalamus, pituitary, seminal vesicles and ventral prostate (Dixit and Niemi, 1974). Another treatment which reduces brainwide the uptake of ^{3}HT radioactivity is hypophysectomy (McEwen et al., 1970a).

Recent experiments in our laboratory have determined the regional distribution of tissue and cell nuclear retention of labeled metabolites of ^{3}HT (Lieberburg and McEwen, 1975c). In all tissue homogenates from both sexes total radioactivity was represented predominantly by ^{3}HT (11—25%), followed by [^{3}H]DHT (8—17%), and then by a variety of other [^{3}H]androgens and ^{3}HE$_2$ (all < 6%). However, radioactivity in purified cell nuclear fractions was represented almost entirely by ^{3}HE$_2$, DHT, and T. In nuclei from preoptic area, basomedial hypothalamus, corticomedial amygdala, and the rest of the hypothalamus, ^{3}HE$_2$ constituted 25—80% of total nuclear radioactivity, other regions being much lower. The percentage of nuclear ^{3}H radioactivity present as DHT was highest in pituitary (58—61%) and lower in brain regions (8—50%). Nuclear ^{3}HT was also highest in pituitary (33—34%) and lower in brain regions.

The pattern of cell nuclear retention of ^{3}HE$_2$ as a T metabolite differs somewhat from the pattern of cell nuclear retention of radioactivity infused as ^{3}HE$_2$. ^{3}HE$_2$ is not detected as a T metabolite in pituitary cell nuclei even though estrogen receptors exist in this tissue in high concentrations. ^{3}HE$_2$ as a T metabolite is particularly high in cell nuclei of corticomedial amygdala (A) of both sexes, higher than that in preoptic area (P) or in basomedial hypothalamus (H). In contrast, the cell nuclear labeling by radioactivity infused as ^{3}HE$_2$ is higher in both P and H than in A. These results imply that the aromatization system is absent from pituitary, a result confirmed by in vitro studies (Naftolin et al., 1972), and especially high in the amygdala, a result also consistent with in vitro experiments (Weisz and Gibbs, 1974a).

The pattern of cell nuclear retention of [^{3}H]DHT as a T metabolite is very similar to the pattern of cell nuclear retention of radioactivity infused as [^{3}H]DHT. A representative pattern is shown in Fig. 1 for cell nuclei from adrenalectomized-ovariectomized rats, and it can be seen that the labeling is highest in pituitary nuclei, followed by H, septum (S), P, A, rest of hypothalamus and amygdala (R), hippocampus (Hi), midbrain (M), and cerebral cortex (C).

Binding of ^{3}HT or [^{3}H]DHT to soluble macromolecules from brain regions and pituitary of castrated male rats has been described in a number of laboratories (Samperez et al., 1969a, b; Jouan et al., 1971a, b, 1973; Kato and Onouchi, 1973a, b; Monbon et al., 1973, 1974; Loras et al., 1974; Naess and Attramadal, 1974; Thieulant et al., 1974; Naess et al., 1975). According to one laboratory, the soluble form of the "receptor" from hypothalamus and pituitary has a sedimentation rate constant of 8.6S and a dissociation constant of 7×10^{-10} M and appears to bind DHT preferentially (Kato and Onouchi, 1973a, b). Another group reported that such molecules sediment at 6—7S and bind T and cyproterone as well as DHT but do not bind E_2 or cortisol (Naess et al., 1975). A third report indicates that cytosol DHT-binding sites are found in the following descending order of abundance in castrate male rats: ventral prostate, 100; pituitary, 32; hypothalamus, 14.5; amygdala, 7.4; cortex, 8.8 (Ginsburg et al., 1974). Dissociation constants in that study were on the order of $1-2 \times 10^{-9}$ M.

II.4. Glucocorticoid receptors in adult neuroendocrine tissues

The regional distribution of putative glucocorticoid receptors in rat brain differs markedly from the patterns described above for estrogen and 5α-dihydrotestosterone receptors. By infusing into adrenalectomized (ADX) rats a tritium-labeled form of the naturally occurring corticoid, [^{3}H]corticosterone (^{3}HB), one observes high concentrations of cell nuclear radioactivity in hippocampus, septum, and amygdala, and much lower concentrations in hypothalamus, midbrain and preoptic area (Fig. 2; see McEwen et al., 1972a, 1976). Autoradiographic studies substantiate this pattern of labeling and reveal other areas with labeled cells in the induseum griseum, anterior hippocampus, and scattered among neurons in the cerebral cortex (Gerlach and McEwen, 1972; Stumpf, 1972; Rhees et al., 1975; McEwen et al., 1975c; Warembourg, 1975). These studies show, moreover, that neurons are the principal cell type which is labeled by ^{3}HB, although the existence of glial cell receptor sites is not excluded.

Autoradiographic and cell fractionation experiments also demonstrate a predominance of septal-hippocampal labeling by ^{3}HB in the Pekin duck (Rhees et al., 1972), guinea pig (Warembourg, 1974), hamster (Kelley and McEwen, unpublished), and rhesus monkey (Gerlach et al., 1974). These results point to an ancestral origin in evolution for the brain glucocorticoid—receptor system. It is interesting to note that corticosterone, which may well be present in the adrenal secretion of all of the above species (Bush, 1953), is not the major glucocorticoid in some, e.g., the rhesus monkey. In this species, cortisol predominates, and we have observed that [^{3}H]cortisol labels hippocampal cell nuclear sites in ADX rhesus monkeys but only one-fourth as effectively as a comparable dose of ^{3}HB (Gerlach et al., 1974). In the rat, where endogenous cortisol is not detectable, cell nuclear labeling by [^{3}H]cortisol is negligible, i.e., less than one-tenth of that observed with comparable doses of ^{3}HB (McEwen et al., 1976). It is not known whether this species difference reflects intrinsic properties of the glucocorticoid receptor system or differences in systemic metabolism of these two glucocorticoids. It should be noted that transcortin of both rat and rhesus monkey displays preferential binding of corticosterone over cortisol (Murphy, 1967) and is thus not a factor in this difference.

The cellular process by which cell nuclear labeling by ^{3}HB takes place appears to be typical of steroid–target cell interactions. There are soluble (cytosol) binding proteins for the hormone (see below) which do not appear to be depleted in vivo when ^{3}HB is allowed to label cell nuclear receptor sites (Turner and McEwen, unpublished). Autoradiography reveals a progressive decrease of cytoplasmic labeling of hippocampal neurons and a progressive increase of cell nuclear labeling during the first 30 min after intraperitoneal injection of ^{3}HB (Rhees et al., 1975). Cell nuclear labeling in hippocampus and other brain regions has been observed in tissue slices incubated in vitro with ^{3}HB (McEwen and Wallach, 1973). Progesterone, which binds to soluble receptor sites but does not appear to label limited-capacity binding sites in cell nuclei, prevents the cell nuclear labeling in vitro by ^{3}HB (McEwen and Wallach, 1973). Such an inhibitory effect is reminiscent of the antiglucocorticoid action of progesterone in a variety of glucocorticoid–target cell interactions (see for references McEwen and Wallach, 1973; McEwen, 1974).

Soluble glucocorticoid receptor sites are found in all brain regions thus far studied. The sedimentation coefficient in sucrose density gradients has been reported as $\approx$ 7S for rat brain using [^{3}H]triamcinolone acetonide, a synthetic glucocorticoid (Chytil and Toft, 1972), and as $\approx$ 8–9S for chick optic tectum and chick neural retina using a number of [^{3}H]glucocorticoids (Wiggert and Chader, 1974, 1975). A high molecular weight ($>$ 200,000) is also indicated by gel exclusion chromatography (Grosser et al., 1971; De Kloet and McEwen, 1976a, b). The [^{3}H]corticosterone–cytosol receptor complex of the rat, labeled in vivo, is quantitatively precipitated by protamine sulfate, indicating that it possesses a net negative charge (McEwen et al., 1972b; Stevens et al., 1973). In this respect, and by reason of its sensitivity to sulfhydryl blocking agents, its presence after extensive perfusion of the brain, and its ability to bind synthetic glucocorticoids such as dexamethasone and triamcinolone acetonide, the brain cytosol receptor is different from transcortin found in blood (Chytil and Toft, 1972; McEwen et al., 1972b; Chader, 1973; McEwen and Wallach, 1973; Wiggert and Chader, 1975).

The regional distribution of [^{3}H]dexamethasone ([^{3}H]Dex) and ^{3}HB binding in rat cytosols from ADX rats perfused at sacrifice is summarized in Table I and it is evident that while hippocampus and septum have the highest in vitro binding capacity, regional differences in in vitro cytosol binding are less pronounced than in vivo regional differences in cell nuclear labeling (compare Table I with Fig. 2). For example, hypothalamus cytosol has a binding one-half of that of hippocampus (Table I) and yet hypothalamic cell nuclei show less than one-tenth the in vivo retention of ^{3}HB compared to hippocampus (Fig. 2; see also De Kloet et al., 1975). A closer parallel is seen between cell nuclear and cytosol binding when both are measured after in vivo labeling with ^{3}HB (Table II). Likewise a closer parallel is seen between cell nuclear and cytosol binding when both are measured by in vitro labeling with ^{3}HB (Table II). The regional pattern of cell nuclear

TABLE I

Glucocorticoid binding by brain cytosols from ADX rats perfused at sacrifice

Tissue	Cytosol binding in vitro*	
	[^{3}H]corticosterone	[^{3}H]dexamethasone
Hippocampus	306 ± 20	294 ± 12
Septum	N.D.**	271 ± 31
Amygdala	N.D.	177 ± 19
Cortex	N.D.	200 ± 29
Hypothalamus	149 ± 30	177 ± 15
Pituitary	352 ± 52	79 ± 11

* fmoles/mg protein; 2×10^{-8} M [^{3}H]steroid × 4 hr. Data from Olpe and McEwen (1976).
** N.D. = not determined.

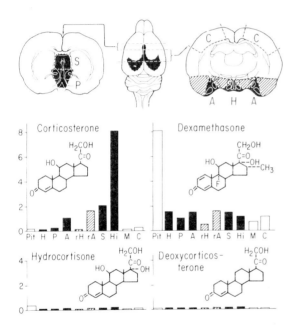

Fig. 2. Topography of cell nuclear binding of 4 ^{3}H-labeled, gluco- or mineralocorticoids in ADX-OVX rats. Identity of structures is same as in Fig. 1 except that rest of hypothalamus (rH) and rest of amygdala (rA) were analyzed separately. Steroid doses and survival times, which are comparable to those in the experiments summarized in Fig. 1, are indicated in McEwen et al. (1976).

labeling in tissue slices in vitro is therefore less highly differentiated than the in vivo labeling pattern (McEwen and Wallach, 1973).

An analogous difference between in vitro and in vivo labeling is observed when the binding of various ^{3}H-labeled steroids is compared for hippocampus from ADX rats

TABLE II

Comparison of in vivo and in vitro [^{3}H]corticosterone binding

Binding of [^{3}H]corticosterone* ± S.E.M.	Tissue		
	Hippocampus	Hypothalamus	Amygdala
In vivo			
Cytosol**	48 ± 11	5.5 ± 1.3	10.9 ± 2.7
Nuclei**	260 ± 70	20 ± 2	58 ± 13
In vitro			
Cytosol***	306 ± 20	149 ± 30	N.D. §§§
Nuclei§	300 ± 10	110 ± 10	129 §§

* fmoles/mg protein.
** Data from McEwen et al. (1972b).
*** Data from Olpe and McEwen (1976).
§ Data from De Kloet et al. (1975), using tissue slices.
§§ Data from McEwen and Wallach (1973), using tissue slices, and is the average of 2 determinations.
§§§ N.D. = not determined.

TABLE III

Binding of [^{3}H]steroids by hippocampus cytosol in vitro

Radioactive steroid at 2 × 10^{-8} M	Binding (fmoles/mg protein)	Per cent competition*
Corticosterone	546 ± 35	91
Dexamethasone	375 ± 29	88
Cortisol	250 ± 20	87
Deoxycorticosterone	422 ± 19	60
Progesterone	282 ± 21	75

* Competition by 2 × 10^{-6} M unlabeled corticosterone. Data from McEwen et al. (1976).

(McEwen et al., 1976). The in vivo pattern of cell nuclear labeling, shown in Figs. 1 and 2, reveals very little cell nuclear retention of radioactivity injected as [^{3}H]cortisol, [^{3}H]deoxycorticosterone or [^{3}H]progesterone, and a virtually uniform distribution of labeling by [^{3}H]Dex compared to the highly differentiated pattern of labeling by ^{3}HB. However, hippocampal cytosol from perfused ADX rats binds 40–60% as much [^{3}H]Dex, cortisol, progesterone, and deoxycorticosterone compared to ^{3}HB (Table III; McEwen et al., 1976).

Measurements of the cell nuclear binding of these same 5 [^{3}H]steroids in hippocampal tissue slices in vitro reveal a pattern more similar to the cytosol-binding pattern than the in vivo cell nuclear labeling pattern (McEwen et al., 1976). These observations, taken together with those on the regional pattern of binding described in the previous paragraph, suggest that factors operating in vivo, such as blood binding, systemic metabolism, and either blood flow or the kinetics of steroid penetration from the blood into the brain, selectively favor the entry and access to receptors of the naturally occurring glucocorticoid in the rat. At the same time, the in vitro experiments clearly point to higher concentrations of glucocorticoid receptors in the hippocampus and septum, compared to other brain regions.

The most surprising result of comparisons of the in vivo binding of various [^{3}H]corticosteroids is the unusually high labeling of pituitary cell nuclei by [^{3}H]Dex (Fig. 2; see De Kloet et al., 1975). This steroid is at least 5 or 6 times more effective in labeling pituitary cell nuclear sites than a comparable dose of ^{3}HB and is only one-sixth as good as ^{3}HB in labeling hippocampal cell nuclear sites in vivo (see Fig. 2; De Kloet et al., 1975; McEwen et al., 1976). It should also be noted that the in vivo cell nuclear labeling pattern for [^{3}H]Dex is fairly uniform across all brain regions measured (Fig. 2). The steroid specificity of labeling of hippocampus, hypothalamus and pituitary is also observed in cell nuclear binding experiments performed in vitro using tissue slices or fragments by the differences are less pronounced than those observed in vivo (De Kloet et al., 1975). The persistence of these differences in the in vitro system nevertheless encouraged us to look for the existence of more than one class of cytosol glucocorticoid receptors, as will be described below.

The anterior pituitary was the prime candidate for the study of multiple populations of binding proteins because of the unusual discrepancy between the higher cytosol binding of ^{3}HB than [^{3}H]Dex (Table I) and the higher in vivo and in vitro cell nuclear binding of [^{3}H]Dex than ^{3}HB. Anterior pituitary cytosol from perfused ADX rats yielded a macromolecule resembling transcortin with respect to its elution profile by both gel exclusion and DEAE-cellulose chromatography and by reason of its inability to bind [^{3}H]Dex (De Kloet and McEwen, 1976a). In addition, a receptor-like macromolecule of large molecular weight was found in pituitary cytosol. Unlike the transcortin-like component, this population of macromolecules binds both [^{3}H]Dex and ^{3}HB and the [^{3}H]steroid–receptor complex binds to a DNA-cellulose column (De Kloet and McEwen, 1976a) in a manner similar to glucocorticoid receptor complexes from liver (Kalimi et al., 1975) and lymphoma cells (Yamamoto et al., 1974). Other laboratories have recently reported similar high molecular weight glucocorticoid receptors in cytosols from mouse pituitary tumor cells (Watanabe et al., 1973) and from bovine anterior pituitary (Watanabe, 1975). The rat and mouse pituitary receptors have the additional property of degrading rapidly in cytosol in the absence of added [^{3}H]corticoid (Watanabe et al., 1973; De Kloet et al., 1975) whereas the bovine pituitary receptor appears to be more stable (Watanabe, 1975).

With respect to the question of multiple binding sites for glucocorticoids in brain tissue, our own studies suggest that hippocampus cytosol also contains two recognizably different populations of glucocorticoid receptors (De Kloet and McEwen, 1976b). These populations are distinguishable by the following properties. (1) The stability of [^{3}H]Dex-binding activity is less than ^{3}HB binding in the absence of added steroid. (2) In competition studies, ^{3}HB binding is preferentially suppressed by unlabeled B compared to Dex, whereas [^{3}H]Dex binding is equally suppressed by B and by Dex. A brief report by Anderson et al. (1974) indicates that they have obtained similar results in competition studies using whole rat brain cytosol. (3) ^{3}HB complexes with cytosol macromolecules display a different elution pattern in a gradient of NaCl from DEAE-cellulose columns than cytosol complexes with [^{3}H]Dex. The differences are such as to suggest that the

[^{3}H]Dex complex is less negatively charged than the ^{3}HB complex.

Attempts to demonstrate transcortin-like binding in hippocampal cytosol, always obtained from animals perfused at sacrifice, have been unsuccessful (De Kloet and McEwen, 1976b). It is thus tempting to attribute the different binding properties of the complexes with the two labeled steroids to two populations of receptor sites, and we have previously suggested that each may exist in a different cellular compartment, e.g., neurons and glial cells (De Kloet et al., 1975). Glial cells, which do not show noticeable in vivo labeling in autoradiography with ^{3}HB, are nevertheless responsive to glucocorticoids in the induction of glycerol phosphate dehydrogenase, and rat glial cell tumors in tissue culture contain glucocorticoid receptors (De Vellis and Inglish, 1968; De Vellis et al., 1971, 1974).

Endogenous corticosterone levels have been determined in brain tissue of intact rats and mice, and these levels rise in parallel with increased blood levels of the hormone produced by an ether stress (Butte et al., 1972). Diurnal elevations in plasma corticosterone in intact rats are accompanied by a decreased availability of brain cytosol receptor sites to labeling with ^{3}HB (Stevens et al., 1973). Thus the brain is "aware" of physiological fluctuations of plasma levels of corticosterone, and an important question remains as to how much of the available receptor capacity is occupied by endogenous corticosterone under resting and stress conditions. Using hippocampal tissue slices incubated in vitro with ^{3}HB we found that approximately 40% of the cell nuclear receptor sites were available to the radioactive hormone when the intact animals were sacrificed at 10 a.m. (McEwen et al., 1974). According to Stevens et al. (1973) brain cytosol receptors available to ^{3}HB at the diurnal peak fell to 50% or less of that than available at the morning trough. Our own work indicates a similar result, namely, that an injection of corticosterone into an ADX rat, sufficient to elevate plasma corticosterone levels to a value like that seen at the diurnal peak, leaves only 25–30% of the hippocampal cell nuclear sites available (McEwen et al., 1974).

Another approach to the problem of receptor occupancy has been to infuse ^{3}HB into ADX rats at a constant rate so as to achieve steady-state plasma levels of hormone for durations up to 45 min (Rotsztejn et al., 1975). Under these conditions hippocampal receptors show two in vivo saturation plateaus, the first at plasma levels less than 40 μg/100 ml, which is regarded as "specific", and the second at concentrations above 140 μg/100 ml, which is regarded as "non-specific". The pituitary receptors are saturated by concentrations between 40 and 115 μg/100 ml and it is in this range of blood corticosterone levels that ACTH blood levels are decreased, suggesting the operation of some form of relatively rapid negative feedback (Rotsztejn et al., 1975).

A complicating factor in the estimation of receptor occupancy is the occurrence of increased glucocorticoid receptor capacity in the absence of adrenal secretion (McEwen et al., 1974; Olpe and McEwen, 1976). Two postadrenalectomy increases in in vivo and in vitro ^{3}HB binding are observed, the first occurring as endogenous steroid is cleared and the second occurring between 12 hr and 5 days after ADX (McEwen et al., 1974). This second increase, which is similar in time course to the post-ADX increase in transcortin-binding capacity in the rat, may represent the synthesis of new receptor molecules (see McEwen et al., 1974). Fortunately, the second increase in receptor capacity does not begin until after 12 hr, and the 11 or 12 hr interval is a convenient one for the estimation of the "normal" receptor capacity. It is interesting to note that the second post-ADX increase occurs in all brain regions thus far examined with both ^{3}HB and [^{3}H]Dex as ligands, but the increase is greatest in hippocampus and septum and may not occur at all in pituitary when [^{3}H]Dex is used to selectively measure the receptor population as opposed to the transcortin-like macromolecules (Olpe and McEwen, 1976).

II.5. Progestin and mineralocorticoid receptors

Macromolecules capable of binding [^{3}H]deoxycorticosterone ([^{3}H]DOC) and [^{3}H]aldosterone have been described in rat brain (Swaneck et al., 1969; Lassman and Mulrow, 1974). Indeed, the brain appears to be capable of responding to mineralocorticoids with respect to regulation of specific salt hunger (see Lassman and Mulrow, 1974). A confounding factor in the study of mineralocorticoid receptors is the fact that glucocorticoid receptors have moderate ability to bind mineralocorticoids (Table III). Yet [^{3}H]DOC results in relatively little cell nuclear labeling in brains of adrenalectomized-ovariectomized rats (Fig. 2; see McEwen et al., 1976). It will be necessary in future work to devise means of distinguishing, both by hormonal specificity and by physical separation of binding proteins, between glucocorticoid receptors and bona fide mineralocorticoid binding sites.

Progesterone, like testosterone, is likely to undergo one of a number of metabolic transformations in the body (see Luine and McEwen, 1976). The progesterone story is thus, like that of testosterone, complicated by the question of metabolites which may mediate the hormonal effects and is additionally complicated at the level of the receptor sites by the difficulty so far encountered in demonstrating such sites which are specific for progesterone. As shown in Table III, progesterone is capable of interacting with cytosol glucocorticoid receptors in hippocampus. One report indicates the existence of progesterone-specific cytosol receptors in median eminence and pituitary tissue of estrogen-primed ovariectomized rats (Seiki and Hattori, 1973). Another report demonstrated cellular labeling by radioactivity injected as [^{3}H]progesterone in the hypophysiotropic region of estrogen-primed ovariectomized guinea pigs (Sar and Stumpf, 1973e). We have been unable to see significant cell nuclear retention of [^{3}H]progesterone radioactivity in ovariectomized-adrenalectomized rats primed with estradiol benzoate for two days (Fig. 2; see McEwen et al., 1976), and are at a loss to explain the discrepancies between our own and other negative results and the above mentioned findings (see Luine and McEwen, 1976, for fuller discussion).

III. APPROACHES TO UNDERSTANDING STEROID RECEPTOR FUNCTION IN NEUROENDOCRINE TISSUES

The implied action of cell nuclear steroid hormone receptors is the regulation of genomic activity, especially the production of RNA molecules along a DNA template. According to this scheme, secondary changes in cellular protein synthesis, directed by the altered population of messenger RNA molecules, would be responsible for hormone effects on cellular structure and function (see O'Malley and Means, 1974). We shall now consider evidence in support of this mechanism of action in neuroendocrine tissues, using selected examples from the literature on the actions of glucocorticoids and estrogens.

III.1. Glucocorticoid regulation of ACTH secretion and neural activity

There is now considerable evidence that the synthetic glucocorticoids like dexamethasone have their major blocking effect on stress-induced and resting ACTH secretion at the pituitary level (see De Kloet et al., 1975). A correlation has been shown in vitro between the potency of a steroid to suppress ACTH release from mouse pituitary cells grown in tissue culture and the ability of that steroid to compete for cytosol binding of [^{3}H]triamcinolone acetonide (Watanabe et al., 1974). Dexamethasone is known to suppress the ACTH release from pituitary cells in vitro by a process which requires macromolecular synthesis (Arimura et al., 1969). However, dexamethasone has also been shown to suppress ACTH release from pituitary glands under conditions in which de novo ACTH synthesis was not altered, although prolonged dexamethasone treatment did inhibit ACTH biosynthesis as well (Koch et al., 1974).

Corticosterone is far less potent than dexamethasone in vivo in suppressing ACTH release, and total suppression of ACTH release does not appear to be a normal physiological effect of the naturally occurring glucocorticoid in the rat (see McEwen et al., 1972b). The major exception to this is the rate-sensitive feedback of corticosterone, a rapid effect lasting on the order of minutes (see De Kloet and McEwen, 1976c, for references). It is also clear that corticosterone does exert a long-term controlling influence over the secretion of ACTH by the pituitary, as evidenced by the high rate of corticotropin-releasing factor (CRF)-induced ACTH secretion in pituitary cells from long-term ADX rats compared to intact animals (Sayers and Portanova, 1974). Portanova and Sayers (1974) have in fact shown that actinomycin D prevents corticosterone in vitro from suppressing ACTH secretion when pituitary cells are prepared from ADX rats, but not when cells are prepared from intact animals. As a working hypothesis, they propose that corticosterone induces the formation of a cellular factor with which the hormone subsequently interacts to suppress ACTH secretion. This hypothesis might account for the results of Koch et al. (1974) described in the previous paragraph. It should be pointed out that cell nuclear ^{3}HB labeling of excised pituitary glands in vitro is actually quite substantial (De Kloet et al., 1975), and thus consistent with the in vitro effects of this steroid on ACTH secretion by excised pituitary or pituitary cell suspensions (Kraicer and Milligan, 1970; Portanova and Sayers, 1974; Sayers and Portanova, 1974). It should also be noted that the smaller degree of labeling in vivo may be of physiological significance. Warembourg (1975) notes that many anterior pituitary cells are labeled generally by radioactivity injected as ^{3}HB, although only a few showed distinct cell nuclear localization of the radioactivity. She also points out that very few cells of pars intermedia are labeled, even though this zone is believed to be a major site of ACTH-containing cells.

A final note with respect to corticosterone action on pituitary adrenal function is that the release of CRF from hypothalamic synaptosomes, elicited by electrical stimulation, is strongly suppressed by 300 g of corticosterone acetate administered in the drinking water over 20 hr prior to sacrifice and by $7 \times 10^{-7} M$ corticosterone in the synaptosomal incubation medium (Edwardson and Bennett, 1974). It would thus appear that corticosterone modulation of pituitary adrenal secretion may involve actions on the central nervous system as well as the pituitary. Whether some of these effects may involve indirect actions on the hypothalamus via interaction of corticosterone with receptors in hippocampus, septum, and amygdala remains to be determined. (See McEwen et al., 1972b; and De Kloet and McEwen, 1976c, for references pertaining to feedback effects of corticoids in limbic structures.)

With respect to the functional significance of brain glucocorticoid receptors, we have already mentioned possible connections to the feedback modulation by corticosterone of ACTH secretion (see above). In addition, glucocorticoid receptors may mediate a number of neural and behavioral effects which have been observed following systemic glucocorticoid administration. These effects include suppression of hippocampal single unit electrical activity, suppression of rapid eye movement (REM) (i.e., paradoxical) sleep, restoration of normal thresholds of detection and recognition of sensory stimuli, affective disorders, and alterations of extinction rates of appetitively and aversively motivated behaviors (see McEwen et al., 1975c). Thus far these effects have not been studied with respect to possible genomic involvement in the hormone action.

III.2. Estrogen effects on behavior and neuroendocrine function

The best studied effects of a steroid hormone on neuroendocrine tissues are those of estrogens. Not only are the estrogen receptor sites extensively mapped in a variety of vertebrate species (see Morrell et al., 1975; as well as earlier discussion), but also the sites of estrogen action on lordosis behavior, locomotor activity, food intake, and the negative component of gonadotropin regulation are now being identified by local brain implantation of estrogens (see Lisk, 1967; McEwen and Pfaff, 1973). These sites of action lie within brain regions containing estrogen receptor sites. Moreover, other approaches have revealed a sequence of events beginning with estrogen interaction with the receptor sites and action at the genomic level and leading to alterations in cellular products which actually may mediate the neural and neuroendocrine responses.

III.2.1. Temporal aspects. A fundamental clue is timing. Peak estradiol levels in early proestrus precede by some hours both the luteinizing hormone (LH) surge and the onset of behavioral estrus. The time lag is also seen in experiments on ovariectomized rats which show lag periods of 20–30 hr for the facilitation by estradiol of lordosis responding (Green et al., 1970) and an LH surge (Jackson, 1972, 1973), respectively. The duration of [^{3}H]estradiol retention on brain and pituitary cell nuclear receptors following a single injection of 10 μg of the steroid appears to be less than 12 hr (McEwen et al., 1975a) and the intervening time up to the appearance of the physiological effect would appear to be due to a sequence of metabolic events initiated by the hormone. Thus the estrogen effect is not that of a "stimulus", i.e., required at the time the response occurs, but rather as a "permissive" agent, increasing the probability that appropriate stimuli occurring after an induction period will elicit the response. It is important to emphasize the essential role of the appropriate stimuli (e.g., the male palpating the female's flanks, eliciting lordosis; the day–night light cycle, leading to ovulation), for the hormone does not by itself induce the physiological response.

III.2.2. Pharmacological intervention. Another important piece of evidence for genomic involvement in estrogen action is the effectiveness of an RNA synthesis inhibitor, actinomycin D, in preventing estrogen induction of lordosis responding (Terkel et al., 1973; Whalen et al., 1974) or the LH surge (Jackson, 1972, 1973). Actinomycin D must be given before estradiol to block LH surge and is effective 6–12 hr after estradiol, but not later, in blocking the estrogen facilitation of behavioral estrus. In these latter experiments, actinomycin D was administered intracranially and led to reversible morphological alterations of nucleolar structure, the disappearance of which could be correlated with the reappearance some days later of estrogen sensitivity (Hough et al., 1974). Evidence for the participation of protein synthesis in the estrogen effects is provided by studies of intracranially applied cycloheximide, an inhibitor of protein formation, in which reversible blockade of lordosis facilitation was again observed (Quadagno and Ho, 1975). Cycloheximide was, like actinomycin D, effective only when applied from 6 hr before estradiol to 12 hr after it.

Another line of evidence relating directly to the cell nuclear estrogen "receptor" sites is the effectiveness of antiestrogenic compounds such as clomiphene, MER-25, and CI-628 to prevent both the LH surge (Shirley et al., 1968; Labhsetwar, 1970) and behavioral estrus (Arai and Gorski, 1968; Meyerson and Lindstrom, 1968; Komisaruk and Beyer, 1972; Whalen and Gorzalka, 1973; Södersten, 1974). These antiestrogens are known to make the receptor system unavailable to the natural estrogen (see earlier discussion). As in the case of actinomycin D, these antiestrogens must be given before or shortly after estrogen treatment to be effective on behavior or must be given on diestrus day 2 of the normal cycling female to block the LH surge.

III.2.3. Neurochemical evidence. Correlative studies of estrogen effects on pituitary and brain RNA and protein metabolism provide some evidence for altered metabolic states resulting from enhanced estrogen secretion or from estrogen administration (see Luine and McEwen, 1976). Measurements of brain and pituitary enzyme activities as a function of estrogen treatment also point to a variety of "inductive" effects on cellular metabolism. Where these effects have been examined in some detail they appear to be the direct result of estrogen action at the receptor level, although the definitive proof of this relationship is lacking. Thus an "inductive" neurochemical effect of estrogen may underlie a "permissive" action at the physiological level. Resolution of any paradox implicit in the juxtaposition of the terms "inductive" and "permissive" may rest on the fact that none of the neurochemical changes are increases from undetectable levels of a gene product. Rather they appear to be increases (or decreases) in level or activity of a constituent which is already present in substantial amount in the absence of the hormonal stimulus. Modulation in the amount of these constituents might be thought of as a means of "tuning" neuronal systems, i.e., of increasing (or decreasing) the functional efficiency of specific neural circuits. For a discussion of neural circuits which are involved in the mediation of the estrogen-dependent lordosis response, the reader is referred to the work of Pfaff (Pfaff et al., 1972; McEwen and Pfaff, 1973).

With respect to biochemical changes in the pituitary related to estradiol, hypophysial RNA levels were reported to be lowest during diestrus and to be highest in proestrus or estrus (Convey and Reece, 1969; Robinson and Leavitt, 1971). The changes in RNA concentrations were ascribed to the action of estrogen, since ovariectomy reduced concentrations of RNA in pituitary, and replacement therapy with estradiol restored normal RNA levels (Robinson and Leavitt, 1971). An excellent protein marker of estrogen action in pituitary is the enzyme glucose-6-phosphate dehydrogenase (G6PDH), which is the first enzyme of the pentose phosphate pathway of glucose metabolism, a pathway which is a source of reducing equivalents for reductive biosynthesis and of pentose sugars for nucleoside triphosphate and RNA synthesis (Luine et al., 1974, 1975a). This enzyme, which is also elevated by estrogen in the uterus and in the hypothalamus, is elevated in pituitary by estradiol-17β and by diethylstilbestrol and not by estradiol-17α or by testosterone (Luine et al., 1974). The "action spectrum" of these four steroids thus parallels the specificity of intracellular estrogen receptors. Estradiol elevation of G6PDH activity in pituitary is blocked by the antiestrogen MER-25, and this action is in keeping with and in fact is predicted by a cell nuclear receptor action of estradiol (Luine et al., 1975a). An important physiological effect of estrogen on the pituitary is the facilitation of its sensitivity to LH-RH. Increased sensitivity is seen in the course of the estrous cycle on the afternoon of proestrus, following elevation of circulating estrogen levels (Cooper et al., 1974; Gordon and Reichlin, 1974) and is observed 14 or more hours after estradiol administration to ovariectomized or intact rats (Libertun et al., 1974; Vilchez-Martinez et al., 1974). These effects of estradiol may be mediated by alterations in the amount or availability of the LH-RH receptor (Spona, 1974).

With respect to alterations in brain chemistry related to estradiol, there have been a number of reports of altered levels or labeling of RNA, which are somewhat conflicting so as not to produce a coherent picture (see Luine and McEwen, 1976). One reason for this may be the high ongoing level of RNA synthesis unrelated to hormonal stimulation. Another factor is the relatively low density of estrogen-responsive cells in neural tissue. Autoradiographic studies of amino acid incorporation into protein in brain cells as a function of estrogen stimulation have presented a more coherent picture (see Luine and McEwen, 1976). It appears that estrogen-dependent increases of incorporation into proteins are restricted, for the most part, to those brain regions which contain putative estradiol receptor sites when physiological fluctuations in estrogen levels are studied, but that if estradiol is given exogenously, particularly in large amounts, other brain regions respond with an increased or decreased incorporation rate. It is presently not known whether these more widespread changes in incorporation represent changed alterations in the amino acid pool size rather than altered rates of protein formation. Studies of brain enzyme activities as a function of estrogen treatment of ovariectomized rats have pointed to effects which are, so far as is presently known, specific to receptor-containing brain regions (Luine et al., 1974, 1975a, b). Among the enzymes which change are oxidative enzymes such as G6PDH (in basomedial hypothalamus) and isocitrate and malate dehydrogenases (in basomedial hypothalamus and corticomedial amygdala). The activity of monoamine oxidase using serotonin as a substrate is decreased by estrogen replacement therapy in basomedial hypothalamus and corticomedial amygdala, whereas the activity of choline acetyltransferase is increased in medial preoptic area and corticomedial amygdala by estrogen treatment (Luine et al., 1975b). The activities of hypothalamic peptidases have been shown to change during the estrous cycle and are increased by estrogen replacement therapy in ovariectomized rats (Heil et al., 1971; Griffiths and Hooper, 1973, 1974;

Kuhl et al., 1974). It remains to be seen whether these effects relate to the modulation of releasing hormone activities, since in one case the enzyme undergoing estrogen-dependent alteration inactivates LH-RH (Griffiths and Hooper, 1973, 1974).

Of primary interest in the ultimate understanding of estrogen action on sexual behavior and neuroendocrine function are the estrogen effects on neurotransmitter metabolism and action. For references concerning neurotransmitter systems and their physiological implications in reproductive function the reader should consult Crowley and Zemlan (1976). The actual evidence for hormonal effects at the neurochemical level is relatively scanty. First, with respect to brain catecholamines, estrogen suppression of dopamine-stimulated gonadotropin release in vitro in a combined median eminence-pituitary system was prevented by protein synthesis inhibitors, while stimulation by dopamine of gonadotropin release in vivo was shown to depend on prior estrogen priming (see McCann and Moss, 1975). In this regard, low doses of estradiol have been reported to increase the turnover of tuberoinfundibular dopamine (Fuxe et al., 1969) but dopamine turnover is also reported to be reduced during proestrus (see McCann and Moss, 1975) when estrogen titers have reached their peak. Resolution of this apparent discrepancy may well depend on elucidation of the time dependency of effects of a single estradiol injection, since as noted earlier in this section there is a large period between the presence of estradiol on its receptors and its physiological effects. Noradrenaline turnover is reported to be highest during states of high gonadotropin release, for example, during proestrus and as a result of gonadectomy (see McCann and Moss, 1975). A single estradiol injection, 56 hr before sacrifice, followed by a single progesterone injection, 6 hr before sacrifice, is reported to decrease noradrenaline turnover, especially in the preoptic-anterior hypothalamic region of the rat brain, and these effects parallel the decreased gonadotropin secretion which results (Bapna et al., 1971). Again, as in the case of dopamine turnover, the relationships to hormone amount and time are undoubtedly complex and beyond the scope of the present discussion. Nevertheless, it is worthwhile to note that current opinion favors a positive role for noradrenaline in the triggering of the LH surge and an inhibitory role for dopamine in prolactin secretion (see McCann and Moss, 1975). Before leaving the catecholamines it is worthwhile to note that two studies have reported increases after gonadectomy of hypothalamic activity of tyrosine hydroxylase (TH), a rate-limiting enzyme of catecholamine biosynthesis (Beattie et al., 1972; Kizer et al., 1974), but neither report was able to establish which gonadal steroids are involved in maintaining normal TH levels or whether this occurs by a direct steroid action or by an indirect effect, perhaps analogous to the reserpine or stress induction of this enzyme (Thoenen et al., 1969).

Second, with respect to serotonin, studies in the rat were unable to show changes in serotonin turnover in the hypothalamus following combined estradiol plus progesterone treatment in ovariectomized rats (Bapna et al., 1971). However, Gradwell et al. (1975) reported decreased serotonin turnover (which they did not localize as to brain region) in ovariectomized-adrenalectomized rhesus monkeys as a result of either estradiol or testosterone replacement therapy. Again, it must be pointed out that the time dependence of gonadal steroid effects has not been systematically investigated in either species, nor have the regional changes in serotonin metabolism been thoroughly investigated. However, it is quite apparent that there are important gonadal steroid influences on serotonin metabolism which await elucidation.

Third, acetylcholine metabolism may be influenced by estradiol, as suggested by the observations cited earlier that choline acetyltransferase activity is increased in medial preoptic area and corticomedial amygdala of female rats by estrogen replacement therapy (Luine et al., 1975b). These observations, however, must be extended to include other parameters of acetylcholine metabolism and to include a study of physiological levels of estradiol before a role for this transmitter system in the normal action of this hormone can be accepted.

The variety of possible neurochemical effects of estradiol serves to generate a working hypothesis or model of steroid hormone action on neurons (see Luine and McEwen, 1976). First, there are alterations in oxidative metabolism of neural tissue and of the pituitary as well which may provide increased amounts of energy for neuronal function as well as providing reducing equivalents for reductive biosynthesis of lipids and pentose sugars for RNA synthesis. Second, there may be hormone-induced alterations in biosynthetic and degradative enzymes for neurotransmitters and releasing hormones. In such cases, it might be expected that part of the "inductive lag" in the manifestation of hormonal effects would be occupied by the time required for the axonal or dendritic transport of the newly synthesized enzymes. Third, there may also be hormonally induced alterations in the amount of postsynaptic receptors for neurotransmitters. At the present time the only, albeit preliminary, evidence pointing in this direction deals with the sensitivity of the pituitary to LH-RH (see earlier discussion), but future work will undoubtedly yield interesting results as techniques for quantitatively measuring these receptors become available. Thus, genomic effects of steroid hormones like estradiol, which appear to account for the delayed "activation" of behavioral estrus and of ovulation, may well result in altered neuronal and synaptic efficiency within developmentally fixed neural pathways.

Genomic interactions also appear to be involved in the "organizational" actions of gonadal steroids during brain development and, as indicated earlier, the developing brain contains receptor sites for at least one steroid, estradiol, which are similar to those found in adult brain tissue. If indeed these receptors mediate the developmental effects of applied estrogens and also of testosterone via conversion to estradiol, then the difference between "activational" and "organizational" effects must involve the state of differentiation of the target neurons themselves. According to this notion, the immature cell genome is less permanently fixed in terms of genes which may be turned on or off, and the hormone—receptor interaction during the critical period can provide a signal for permanently altering the pattern of gene expression of those cells and for directing the pattern of neural connections formed by those cells.

ACKNOWLEDGEMENTS

This article summarizes work of a number of people who have recently been or are now in our laboratory: Dr. E.R. De Kloet, Mr. John Gerlach, Mr. Ivan Lieberburg, Dr. Victoria Luine, Dr. Neil Maclusky, Dr. H.-R. Olpe, Dr. Linda Plapinger and Dr. Barbara

Turner. I would also like to acknowledge the technical assistance of Ms. Claude Chaptal and Ms. Gislaine Wallach and the editorial assistance of Ms. Freddi Berg.

The research in this laboratory described in this article is supported by Grant NS 07080 from the NIH and by an institutional grant from The Rockefeller Foundation, RF 70095, for research in reproductive biology.

REFERENCES

Anderson, C.H. and Greenwald, S.S. (1969) Autoradiographic analysis of estradiol uptake in the brain and pituitary of the female rat. *Endocrinology*, 85: 1160—1165.
Anderson, N.S., III, Fanestil, D.D. and Ludens, J.H. (1974) Study of adrenal corticoid receptors in brain. *J. Steroid Biochem.*, 5: 335 (Abstract).
Arai, Y. and Gorski, R.A. (1968) Effect of anti-estrogen on steroid induced sexual receptivity in ovariectomized rats. *Physiol. Behav.*, 3: 351—353.
Arimura, A., Bowers, C.Y., Schally, A.V., Saito, M. and Miller, M.C. (1969) Effect of corticotropin-releasing factor, dexamethasone and actinomycin D on the release from rat pituitaries in vivo and in vitro. *Endocrinology*, 85: 300—311.
Attramadal, A. (1964) Distribution and site of action of oestradiol in the brain and pituitary gland of rat following intramuscular administration. In *Proc. 2nd int. Congr. of Endocrinology, Int. Congr. Ser. 83,* Excerpta Medica, Amsterdam, pp. 612—616.
Bapna, J., Neff, N.H. and Costa, E. (1971) A method for studying norepinephrine and serotonin metabolism in small regions of rat brain: effect of ovariectomy on amine metabolism in anterior and posterior hypothalamus. *Endocrinology,* 89: 1345—1349.
Barley, J., Ginsburg, M., Greenstein, B.D., Maclusky, N.J. and Thomas, P.J. (1974) A receptor mediating sexual differentiation? *Nature (Lond.),* 252: 259—260.
Beattie, C.W., Rodgers, C.H. and Soyka, L.F. (1972) Influence of ovariectomy and ovarian steroids on hypothalamic tyrosine hydroxylase activity in the rat. *Endocrinology*, 91: 276—279.
Bongiovanni, A.M., DiGeorge, A.M. and Grumbach, M.M. (1959) Masculinization of the female infant associated with estrogen therapy alone during gestation: four cases. *J. clin. Endocr.*, 19: 1004—1011.
Bush, I.E. (1953) Species differences in adrenocortical secretion. *J. Endocr.*, 9: 95—100.
Butte, J.C., Kakihana, R. and Noble, E.P. (1972) Rat and mouse brain corticosterone. *Endocrinology*, 90: 1091—1100.
Chader, G.J. (1973) Some factors affecting the uptake binding and retention of [^{3}H]cortisol by the chick embryo retina as related to enzyme induction. *J. Neurochem.*, 21: 1525—1532.
Chazal, G., Faudon, M., Gogan, F. and Rotsztejn, W. (1975) Effects of two estradiol antagonists upon the estradiol uptake in the rat and peripheral tissues. *Brain Res.*, 89: 245—254.
Christensen, L.W. and Clemens, L.G. (1974) Intrahypothalamic implants of testosterone or estradiol and resumption of masculine sexual behavior in long-term castrated male rats. *Endocrinology*, 95: 984—990.
Chytil, F. and Toft, D. (1972) Corticoid binding component in rat brain. *J. Neurochem.*, 19: 2877—2880.
Cidlowski, J.A. and Muldoon, T.G. (1974) Estrogenic regulation of cytoplasmic receptor populations in estrogen-responsive tissues of the rat. *Endocrinology*, 95: 1621—1629.
Clark, J.H., Peck, E.J. and Anderson, J.N. (1974) Oestrogen receptors and antagonism of steroid hormone action. *Nature (Lond.)*, 251: 446—448.
Clemens, L.G. (1974) Neurohormonal control of male sexual behavior. In *Reproductive Behavior,* W. Montagna and W.A. Sadler (Eds.), Plenum Press, New York, pp. 23—53.
Convey, E.M. and Reece, R.P. (1969) Influence of the estrous cycle on the nucleic acid content of the rat anterior pituitary. *Proc. Soc. exp. Biol. (N.Y.)*, 132: 878—880.

Cooper, K.J., Fawcett, C.P. and McCann, S.M. (1974) Variations in pituitary responsiveness to a luteinizing hormone/follicle stimulating hormone releasing factor (LH-RF/FSH-RF): preparation during the rat estrous cycle. *Endocrinology*, 95: 1293–1299.

Crowley, W.R. and Zemlan, F.P. (1976) The neurochemical basis of sexual behavior. In *Primer of Neuroendocrine Function and Behavior*, N.T. Adler (Ed.), Plenum Press, New York, in press.

De Kloet, E.R. and McEwen, B.S. (1976a) A putative glucocorticoid receptor and a transcortin-like macromolecule in pituitary cytosol. *Biochim. biophys. Acta (Amst.)*, 421: 115–123.

De Kloet, E.R. and McEwen, B.S. (1976b) Differences between cytosol receptor complexes with corticosterone and dexamethasone in hippocampal tissue from rat brain. *Biochim. biophys. Acta (Amst.)*, 421: 124–132.

De Kloet, E.R. and McEwen, B.S. (1976c) Glucocorticoid interactions with brain and pituitary. In *Molecular and Functional Neurobiology*, W.H. Gispen (Ed.), Elsevier, Amsterdam, pp. 257–307.

De Kloet, E.R., Wallach, G. and McEwen, B.S. (1975) Differences in corticosterone and dexamethasone binding to rat brain and pituitary. *Endocrinology*, 96: 598–609.

De Vellis, J. and Inglish, D. (1968) Hormonal control of glycerol phosphate dehydrogenase in the rat brain. *J. Neurochem.*, 15: 1061–1070.

De Vellis, J., Inglish, D., Cole, R. and Molson, J. (1971) Effects of hormones on the differentiation of cloned lines of neurons and glial cells. In *Influence of Hormones on the Nervous System*, D. Ford (Ed.), Karger, Basel, pp. 25–39.

De Vellis, J., McEwen, B.S., Cole, R. and Inglish, D. (1974) Relations between glucocorticoid binding and glycerolphosphate dehydrogenase induction in a rat glial cell line. *Trans. Amer. Soc. Neurochem.*, 5: 125.

Dixit, V.P. and Niemi, M. (1974) Uptake of exogenous ^{3}H-1,2-testosterone in the hypothalamus, endocrine glands and sex-accessory organs in neonatally castrated and androgenized male rats. *Endocr. exp.*, 8: 39–43.

Doughty, C., Booth, J.E., McDonald, P.G. and Parrott, R.F. (1975) Effects of oestradiol-17β, oestradiol benzoate and the synthetic oestrogen, RU2858 on sexual differentiation in the neonatal female rat. *J. Endocr.*, in press.

Edwardson, J.A. and Bennett, G.W. (1974) Modulation of corticotrophin-releasing factor release from hypothalamic synaptosomes. *Nature (Lond.)*, 251: 425–427.

Eisenfeld, A.J. and Axelrod, J. (1965) Selectivity of estrogen distribution in tissues. *J. Pharmacol. exp. Ther.*, 150: 469–475.

Eisenfeld, A.J. and Axelrod, J. (1966) Effect of steroid hormones, ovariectomy, estrogen pretreatment, sex and immaturity on the distribution ^{3}H-estradiol. *Endocrinology*, 79: 38–42.

Flerko, B. and Mess, B. (1968) Reduced oestradiol-binding capacity of androgen sterilized rats. *Acta physiol. Acad. Sci. hung.*, 33: 111–113.

Fuxe, K., Hökfelt, T. and Nilsson, O. (1969) Castration, sex hormones and tuberoinfundibular dopamine neurons. *Neuroendocrinology*, 5: 107–120.

Gerlach, J.L. and McEwen, B.S. (1972) Rat brain binds adrenal steroid hormone: radioautography of hippocampus with corticosterone. *Science*, 175: 1133–1136.

Gerlach, J.L., McEwen, B.S., Pfaff, D.W., Ferin, M. and Carmel, P.W. (1974) Rhesus monkey brain binds radioactivity from [^{3}H]estradiol and [^{3}H]corticosterone, demonstrated by nuclear isolation and autoradiography. *Proc. Endocr. Soc. 56th Ann. Meeting, Atlanta, Ga.*, Abstract No. 370.

Ginsburg, M., Greenstein, B.D., Maclusky, N.J., Morris, I.D. and Thomas, P.J. (1974) Dihydrotestosterone binding in brain and pituitary cytosol of rats. *J. Endocr.*, 61: XXIV.

Gordon, J.H. and Reichlin, S. (1974) Changes in pituitary responsiveness to luteinizing hormone-releasing factor during the rat estrous cycle. *Endocrinology*, 94: 974–977.

Gradwell, P.B., Everitt, B.J. and Herbert, J. (1975) 5-Hydroxytryptamine in the central nervous system and sexual receptivity of female rhesus monkeys. *Brain Res.*, 88: 281–293.

Green, R., Luttge, W.G. and Whalen, R.E. (1970) Induction of receptivity in ovariectomized rats by a single intravenous injection of estradiol-17β. *Physiol. Behav.,* 5: 137–141.

Griffiths, E.C. and Hooper, K.C. (1973) Changes in hypothalamic peptidase activity during the oestrous cycle in the adult female rat. *Acta endocr. (Kbh.),* 74: 41–48.

Griffiths, E.C. and Hooper, K.C. (1974) Peptidase activity in different areas of the rat hypothalamus. *Acta endocr. (Kbh.),* 77: 10–18.

Grosser, B.I., Stevens, W., Bruenger, F.W. and Reed, D.J. (1971) Corticosterone binding in rat brain cytosol. *J. Neurochem.,* 18: 1725–1732.

Gupta, G.N. (1960) *The Fate of Tritium-Labeled Estradiol-17β in Rat Tissues,* Ph.D. Thesis, University of Chicago.

Heil, H., Meltzer, V., Kuhl, H., Abraham, R. and Taubert, H.D. (1971) Stimulation of L-cystine-aminopeptidase activity by hormonal steroids and steroid analogs in the hypothalamus and other tissues of the female rat. *Fertil. and Steril.,* 22: 181–187.

Herbst, A.L., Kurman, R.J., Scully, R.E. and Poskanzen, D.C. (1972) Clear-cell adenocarcinoma of the genital tract in young females. *New Engl. J. Med.,* 287: 1259–1264.

Hough, J.C., Jr., Ho, K.-W.G., Cooke, P.H. and Quadagno, D.M. (1974) Actinomycin D: reversible inhibition of lordosis behavior and correlated changes in nucleolar morphology. *Horm. Behav.,* 5: 367–376.

Jackson, G.L. (1972) Effect of actinomycin D on estrogen-induced release of luteinizing hormone in ovariectomized rats. *Endocrinology,* 91: 1284–1287.

Jackson, G.L. (1973) Time interval between injection of estradiol benzoate and LH release in the rat and effect of actinomycin D or cycloheximide. *Endocrinology,* 93: 887–891.

Jensen, E.V. and Jacobson, H.I. (1962) Basic guides to the mechanism of estrogen action. *Recent Progr. Hormone Res.,* 18: 387–408.

Jouan, P., Samperez, S., Thieulant, M.L. et Mercier, L. (1971a) Neurobiologie moléculaire. Étude de la composition des stéroides liés aux "récepteurs" cytoplasmiques de l'hypophyse antérieure et de l'hypothalamus après incubation en présence de testostérone-³H. *C.R. Acad. Sci. (Paris),* 272: 2368–2371.

Jouan, P., Samperez, S., Thieulant, M.L. et Mercier, L. (1971b) Étude du récepteur cytoplasmique de la [1,2-³H]testostérone dans l'hypophyse antérieure et l'hypothalamus du rat. *J. Steroid Biochem.,* 2: 223–236.

Jouan, P., Samperez, S. and Thieulant, M.L. (1973) Testosterone "receptors" in purified nuclei of rat anterior hypophysis. *J. Steroid Biochem.,* 4: 65–74.

Kalimi, M., Colman, P. and Feigelson, P. (1975) The "activated" hepatic glucocorticoid receptor complex. *J. biol. Chem.,* 250: 1080–1086.

Kato, J. (1970) In vitro uptake of tritiated oestradiol by the rat anterior hypothalamus during the oestrous cycle. *Acta endocr. (Kbh.),* 63: 577–584.

Kato, J. and Onouchi, T. (1973a) 5-Alpha-dihydrotestosterone "receptor" in the rat hypothalamus. *Endocr. jap.,* 20: 429–432.

Kato, J. and Onouchi, T. (1973b) 5-Alpha-dihydrotestosterone "receptor" in the rat hypophysis. *Endocr. jap.,* 20: 641–644.

Kato, J. and Villee, C.A. (1967) Preferential uptake of estradiol by the anterior hypothalamus of the rat. *Endocrinology,* 80: 567–575.

Kato, J., Atsumi, Y. and Muramatsu, M. (1970) Nuclear estradiol receptor in rat anterior hypophysis. *J. Biochem. (Tokyo),* 67: 871–872.

King, R.J.B. and Mainwaring, W.I.P. (1974) *Steroid–Cell Interactions,* University Park Press, Baltimore, Md., 440 pp.

Kizer, J.S., Palkovits, M., Zivin, J., Brownstein, M., Saavedra, J.M. and Kopin, I.J. (1974) The effect of endocrinological manipulations on tyrosine hydroxylase and dopamine hydroxylase activities in individual hypothalamic nuclei of the adult male rat. *Endocrinology,* 95: 799–812.

Koch, B., Bucher, B. and Miahle, C. (1974) Pituitary nuclear retention of dexamethasone and ACTH biosynthesis. *Neuroendocrinology,* 15: 365–375.

Komisaruk, B.R. and Beyer, C. (1972) Differential antagonism, by MER-25, of behavioral and morphological effects of estradiol benzoate in rats. *Horm. Behav.*, 3: 63–70.

Kraicer, J. and Milligan, J.V. (1970) Suppression of ACTH release from adenohypophysis by corticosterone: an *in vitro* study. *Endocrinology*, 87: 371–376.

Kuhl, H., Rosniatowski, C., Oen, S. and Taubert, H. (1974) Sex steroids stimulate the activity of hypothalamic arylamidases in the rat. *Acta endocr. (Kbh.)*, 76: 1–14.

Labhsetwar, A.P. (1970) The role of oestrogens in spontaneous ovulation: evidence for positive oestrogen feedback in the 4-day oestrus cycle. *J. Endocr.*, 47: 481–493.

Ladosky, W. (1967) Anovulatory sterility in rats neonatally injected with stilbestrol. *Endokrinologie*, 52: 259–261.

Lassman, M.N. and Mulrow, P.J. (1974) Deficiency of deoxycorticosterone binding protein in the hypothalamus of rats resistant to deoxycorticosterone-induced hypertension. *Endocrinology*, 94: 1541–1546.

Libertun, C., Cooper, K.J., Fawcett, C.P. and McCann, S.M. (1974) Effects of ovariectomy and steroid treatment on hypophyseal sensitivity to purified LH-releasing factor (LRF). *Endocrinology*, 94: 518–525.

Lieberburg, I. and McEwen, B.S. (1975a) Estradiol 17β: a metabolite of testosterone recovered in cell nuclei from limbic areas of neonatal rat brains. *Brain Res.*, 85: 165–170.

Lieberburg, I. and McEwen, B.S. (1975b) Estradiol 17β: a metabolite of testosterone recovered in cell nuclei from limbic areas of adult male rat brains. *Brain Res.*, 91: 171–174.

Lieberburg, I. and McEwen, B.S. (1975c) Estradiol 17β and 5α-dihydrotestosterone: testosterone metabolites recovered in cell nuclei from adult rat brains. *Soc. Neurosci. Ann. Meeting, Nov. 1975*, p. 455.

Linkie, D.M. (1975) In vivo nuclear transformation of estrogen receptor complex (ER) in hypothalamus, pituitary and uterus. *Endocrine Soc., 57th Ann. Meeting, June 18–20, 1975*, Abstract No. 33, p. 67.

Lisk, R.D. (1967) Sexual behavior: hormonal control. In *Neuroendocrinology, Vol. 2*, L. Martini and W.F. Ganong (Eds.), Academic Press, New York, pp. 197–239.

Loras, B., Genott, A., Monbon, M., Beucher, F., Reboud, J.P. and Bertrand, J. (1974) Binding and metabolism of testosterone in the rat brain during sexual maturation. II. Testosterone metabolism. *J. Steroid Biochem.*, 5: 425–431.

Luine, V.N. and McEwen, B.S. (1976) Steroid hormone receptors in brain and pituitary: topography and possible functions. In *Sexual Behavior*, R.W. Goy and D.W. Pfaff (Eds.), Plenum Press, New York, in press.

Luine, V.N., Khylchevskaya, R.I. and McEwen, B.S. (1974) Oestrogen effects on brain and pituitary enzyme activities. *J. Neurochem.*, 23: 925–934.

Luine, V.N., Khylchevskaya, R.I., and McEwen, B.S. (1975a) Effect of gonadal hormones on enzyme activities in brain and pituitary of male and female rats. *Brain Res.*, 86: 283–292.

Luine, V.N., Khylchevska, R.I. and McEwen, B.S. (1975b) Effect of gonadal steroids on activities of monoamine oxidase and choline acetylase in rat brain. *Brain Res.*, 86: 293–306.

Massa, R. and Martini, L. (1971/72) Interference with the 5α-reductase system: a new approach for developing antiandrogens. *Gynec. Invest.*, 2: 253–270.

Maurer, R.A. and Woolley, D.E. (1974) Demonstration of nuclear ^{3}H-estradiol binding in hypothalamus and amygdala of female, androgenized-female and male rats. *Neuroendocrinology*, 16: 137–147.

McCann, S.M. and Moss, R.L. (1975) Putative neurotransmitters involved in discharging gonadotrophin-releasing neurohormones and the action of LH-releasing hormone on the CNS. *Life Sci.*, 16: 833–852.

McEwen, B.S. (1974) Adrenal steroid binding to presumptive receptors in the limbic brain of the rat. In *Neuroendocrinologie de l'Axe Corticotrope*, P. Dell (Ed.), INSERM (Colloque 1973, Vol. 22), Paris, pp. 79–94.

McEwen, B.S. and Pfaff, D.W. (1970) Factors influencing sex hormone uptake by rat brain regions. I. Effects of neonatal treatment, hypophysectomy, and competing steroid on estradiol uptake. *Brain Res.*, 21: 1–16.

McEwen, B.S. and Pfaff, D.W. (1973) Chemical and physiological approaches to neuroendocrine mechanism: attempts at integration. In *Frontiers in Neuroendocrinology*, W.F. Ganong and L. Martini (Eds.), Oxford Univ. Press, New York, pp. 267–335.

McEwen, B.S. and Wallach, G. (1973) Corticosterone binding to hippocampus: nuclear and cytosol binding *in vitro*. *Brain Res.*, 57: 373–386.

McEwen, B.S., Pfaff, D.W. and Zigmond, R.E. (1970a) Factors influencing sex hormone uptake by rat brain regions. II. Effects of neonatal treatment and hypophysectomy in testosterone uptake. *Brain Res.*, 21: 17–28.

McEwen, B.S., Pfaff, D.W. and Zigmond, R.E. (1970b) Factors influencing sex hormone uptake by rat brain regions. III. Effects of competing steroids on testosterone uptake. *Brain Res.*, 21: 29–38.

McEwen, B.S., Zigmond, R.E. and Gerlach, J.L. (1972a) Sites of steroid binding and action in the brain. In *Structure and Function of Nervous Tissue, Vol. 5,* G.H. Bourne (Ed.), Academic Press, New York, pp. 205–291.

McEwen, B.S., Magnus, C. and Wallach, G. (1972b) Soluble corticosterone-binding macromolecules extracted from rat brain. *Endocrinology*, 90: 217–226.

McEwen, B.S., Wallach, G. and Magnus, C. (1974) Corticosterone binding to hippocampus: immediate and delayed influences of the absence of adrenal secretion. *Brain Res.*, 70: 321–334.

McEwen, B.S., Pfaff, D.W., Chaptal, C. and Luine, V. (1975a) Brain cell nuclear retention of [^{3}H]estradiol doses able to promote lordosis: temporal and regional aspects. *Brain Res.*, 86: 155–161.

McEwen, B.S., Plapinger, L., Chaptal, C., Gerlach, J. and Wallach, G. (1975b) Role of fetoneonatal estrogen binding proteins in the associations of estrogen with neonatal brain cell nuclear receptors. *Brain Res.*, 96: 400–406.

McEwen, B.S., Gerlach, J.L. and Micco, D.J., Jr. (1975c) Putative glucocorticoid receptors in hippocampus and other regions of the rat brain. In *The Hippocampus: a Comprehensive Treatise*, R. Isaacson and K. Pribram (Eds.), Plenum Press, New York, pp. 285–322.

McEwen, B.S., De Kloet, E.R. and Wallach, G. (1976) Interactions *in vivo* and *in vitro* of corticoids and progesterone with cell nuclei and soluble macromolecules from rat brain regions and pituitary. *Brain Res.*, 105: 129–136.

McGuire, J.L. and Lisk, R.D. (1968) Estrogen receptors in the intact rat. *Proc. nat. Acad. Sci. (Wash.)*, 61: 497–503.

Meyer, C.C. (1973) Testosterone concentration in the male chick brain: an autoradiographic survey. *Science*, 180: 1381–1382.

Meyerson, B.J. and Lindstrom, L. (1968) Effects of an oestrogen antagonist ethamoxytriphentol (MER-25) on oestrous behaviour in rats. *Acta endocr. (Kbh.)*, 59: 41–48.

Michael, R.P. (1965) Oestrogens in the central nervous system. *Brit. med. Bull.*, 21: 87–90.

Monbon, M., Loras, B., Reboud, J.P. and Bertrand, J. (1973) Uptake, binding and metabolism of testosterone in rat brain tissue. *Brain Res.*, 53: 139–150.

Monbon, M., Loras, B., Reboud, J.P. and Bertrand, J. (1974) Binding and metabolism of testosterone in the rat brain during sexual maturation. I. Macromolecular binding of androgens. *J. Steroid Biochem.*, 5: 417–423.

Morrell, J.I., Kelley, D.B. and Pfaff, D.W. (1975) Sex steroid binding in the brains of vertebrates. In *The Ventricular System in Neuroendocrine Mechanisms, Proceedings of Second Brain-Endocrine Interaction Symposium,* K. Knigge, D.S. Scott and K. Kobayashi (Eds.), Karger, Basel, pp. 230–256.

Mowles, T.F., Ashkanazy, B., Mix, E., Jr. and Sheppard, H. (1971) Hypothalamic and hypophyseal estradiol-binding complexes. *Endocrinology*, 89: 484–491.

Murphy, B.E.P. (1967) Some studies of the protein-binding of steroids and their application to the routine micro- and ultramicro-measurement of various steroids in body fluids by competitive protein-binding radioassay. *J. clin. Endocr.*, 27: 973–990.

Naess, O. and Attramadal, A. (1974) Uptake and binding of androgens by the anterior pituitary gland, hypothalamus, preoptic area, and brain cortex of rats. *Acta endocr. (Kbh.)*, 76: 417–430.

Naess, O., Attramadal, A. and Aakvaag, A. (1975) Androgen binding proteins in the anterior pituitary, hypothalamus, preoptic area and brain cortex of the rat. *Endocrinology*, 96: 1–9.

Naftolin, F., Ryan, K.J. and Petro, Z. (1972) Aromatization of androstenedione by the anterior hypothalamus of adult male and female rats. *Endocrinology*, 90: 295–298.

Notides, A.C. (1970) Binding affinity and specificity of the estrogen receptor of the rat uterus and anterior pituitary. *Endocrinology*, 87: 987–992.

Notides, A.C. and Nielsen, S. (1974) The molecular mechanism of the *in vitro* 4S to 5S transformation of the uterine estrogen receptor. *J. biol. Chem.*, 249: 1866–1873.

Olpe, H.R. and McEwen, B.S. (1976) Glucocorticoid binding to receptor-like proteins in rat brain and pituitary: ontogenic and experimentally induced changes. *Brain Res.*, 105: 121–128.

O'Malley, B.W. and Means, A.R. (1974) Female steroid hormones and target cell nuclei. *Science*, 183: 610–620.

Perez-Palacios, B., Perez, A.E., Cruz, M.L. and Beyer, C. (1973) Comparative uptake of ^{3}H androgens by the brain and the pituitary of castrated male rats. *Biol. Reprod.*, 8: 395–399.

Pfaff, D.W. (1968a) Uptake of estradiol-17β-^{3}H in the female rat brain: an autoradiographic study. *Endocrinology*, 82: 1149–1155.

Pfaff, D.W. (1968b) Autoradiographic localization of radioactivity in rat brain after injection of tritiated sex hormones. *Science*, 161: 1355–1356.

Pfaff, D.W. (1968c) Autoradiographic localization of testosterone-^{3}H in the female rat brain and estradiol-^{3}H in the male rat brain. *Experientia (Basel)*, 24: 958–959.

Pfaff, D.W. (1970) Nature of sex hormone effects on rat sex behavior. *J. comp. physiol. Psychol.*, 73: 349–358.

Pfaff, D.W. and Keiner, M. (1973) Atlas of estradiol-concentrating cells in the central nervous system of the female rat. *J. comp. Neurol.*, 151: 121–158.

Pfaff, D., Lewis, C., Diakow, C. and Keiner, M. (1972) Neurophysiological analysis of mating behavior responses as hormone sensitive reflexes. In *Progress in Physiological Psychology, Vol. 5*, E. Stellar and J.M. Sprague (Eds.), Academic Press, New York, pp. 253–297.

Phuong, N.T. und Sauer, G. (1971) Die spezifische Aufnahme von markiertem Testeron im Hypothalamus. *Acad. Biol. Med. Germ.*, 26: 1247–1249.

Plapinger, L. (1973) *Ontogeny of Presumptive Estrogen Receptors in the Rat Brain*, Ph.D. Thesis, New York University.

Plapinger, L. and McEwen, B.S. (1973) Ontogeny of estradiol-binding sites in rat brain. I. Appearance of presumptive adult receptors in cytosol and nuclei. *Endocrinology*, 93: 1119–1128.

Plapinger, L. and McEwen, B.S. (1976) Gonadal steroid–brain interactions in sexual differentiation. In *Biological Determinants of Sexual Behavior*, J. Hutchinson (Ed.), Wiley, New York, in press.

Portanova, R. and Sayers, G. (1974) Corticosterone suppression of ACTH secretion: actinomycin D sensitive and insensitive components of the response. *Biochem. biophys. Res. Commun.*, 56: 928–933.

Quadagno, D.M. and Ho, G.K.W. (1975) The reversible inhibition of steroid-induced sexual behavior by intracranial cycloheximide. *Horm. Behav.*, 6: 19–26.

Raynaud, J.P. (1973) Influence of rat estradiol binding plasma protein (EBP) on uterotrophic activity. *Steroids*, 21: 249–258.

Reddy, V.V.R., Naftolin, F. and Ryan, K.J. (1974) Conversion of androstenedione to estrone by neural tissues from fetal and neonatal rats. *Endocrinology*, 94: 117–121.

Resko, J.A., Goy, R.W. and Phoenix, C.H. (1967) Uptake and distribution of exogenous testosterone-1,2-^{3}H in neural and genital tissues of the castrate guinea pig. *Endocrinology*, 80: 490–498.

Rhees, R.W., Abel, J.H., Jr. and Haack, D.W. (1972) Uptake of tritiated steroids in the brain of the duck (*Anas platyrhynchos*). An autoradiographic study. *Gen. comp. Endocrinol.*, 18: 292–300.

Rhees, R.W., Grosser, B.I. and Stevens, W. (1975) Effect of steroid competition and time on the uptake of [^{3}H]corticosterone in the rat brain: an autoradiographic study. *Brain Res.*, 83: 293–300.

Robinson, J.A. and Leavitt, W.W. (1971) Estrogen related changes in anterior pituitary RNA levels. *Proc. Soc. exp. Biol. (N.Y.)*, 139: 471–475.

Rotsztejn, W.H., Normand, M., LaLonde, J. and Fortier, C. (1975) Relationship between ACTH release and corticosterone binding by the receptor sites of the adenohypophysis and dorsal hippocampus following infusion of corticosterone at a constant rate in the adrenalectomized rat. *Endocrinology*, 97: 223–230.

Roy, S.K., Jr. and Laumas, K.R. (1969) 1,2-³H-testosterone: distribution and uptake in neural and genital tissues of intact male, castrate male and female rats. *Acta endocr. (Kbh.)*, 61: 629–640.

Samperez, S., Thieulant, M.L. et Jouan, P. (1969a) Mise en évidence d'une association macromoléculaire de la testostérone 1-2-³H dans l'hypophyse antérieure et l'hypothalamus du rat normal et castré. *C.R. Acad. Sci. (Paris)*, 268: 2965–2968.

Samperez, S., Thieulant, M.L., Poupon, R., Duval, J. et Jouan, P. (1969b) Étude de la pénétration de la testostérone-1,2-³H dans l'hypophyse antérieure, l'hypothalamus, et le cortex cérébral du rat castré et normal. *Bull. Soc. Chim. Biol.*, 51: 117–131.

Sar, M. and Stumpf, W.E. (1972) Cellular localization of androgen in the brain and pituitary after the injection of tritiated testosterone. *Experientia (Basel)*, 28: 1364–1366.

Sar, M. and Stumpf, W.E. (1973a) Autoradiographic localization of radioactivity in the rat brain after the injection of 1,2-³H-testosterone. *Endocrinology*, 92: 251–256.

Sar, M. and Stumpf, W.E. (1973b) Pituitary gonadotrophs: nuclear concentration of radioactivity after injection of ³H-testosterone. *Science*, 179: 389–391.

Sar, M. and Stumpf, W.E. (1973c) Cellular and subcellular localization of radioactivity in the rat pituitary after injection of 1,2-³H-testosterone using dry-autoradiography. *Endocrinology*, 92: 631–635.

Sar, M. and Stumpf, W.E. (1973d) Effects of progesterone or cyproterone acetate on androgen uptake in the brain, pituitary and peripheral tissues. *Proc. Soc. exp. Biol. (N.Y.)*, 144: 26–29.

Sar, M. and Stumpf, W.E. (1973e) Neurons of the hypothalamus concentrate ³H-progesterone or its metabolites. *Science*, 182: 1266–1268.

Sayers, G. and Portanova, R. (1974) Secretion of ACTH by isolated anterior pituitary cells: kinetics of stimulation by corticotropin-releasing factor and of inhibition by corticosterone. *Endocrinology*, 94: 1723–1730.

Schwarzel, W.C., Kruggel, W.G. and Brodie, H.J. (1973) Studies on the mechanism of estrogen biosynthesis. VII. The development of inhibitors of the enzyme system in human placenta. *Endocrinology*, 92: 866–880.

Seiki, K. and Hattori, M. (1973) In vivo uptake of progesterone by the hypothalamus and pituitary of the female ovariectomized rat and its relationship to cytoplasmic progesterone-binding protein. *Endocr. jap.*, 20: 111–119.

Shirley, B., Wolinsky, J. and Schwartz, N.B. (1968) Effects of a single injection of an estrogen antagonist on the estrous cycle of the rat. *Endocrinology*, 82: 959–968.

Södersten, P. (1974) Effects of an estrogen antagonist, MER-25, on mounting behavior and lordosis behavior in the female rat. *Horm. Behav.*, 5: 111–121.

Spona, J. (1974) LH-RH interaction with the pituitary plasma membrane is affected by sex steroids. *FEBS Lett.*, 39: 221–225.

Stern, J. and Eisenfeld, A. (1971) Distribution and metabolism of ³H-testosterone in castrated male rats: effects of cyproterone, progesterone and unlabelled testosterone. *Endocrinology*, 88: 1117–1125.

Stevens, W., Reed, D.J., Erickson, S. and Grosser, B.I. (1973) The binding of corticosterone to brain proteins: diurnal variation. *Endocrinology*, 93: 1152–1156.

Stumpf, W.E. (1968a) Estradiol-concentrating neurons: topography in the hypothalamus by dry mount autoradiography. *Science*, 162: 1001–1003.

Stumpf, W.E. (1968b) Cellular and subcellular ³H-estradiol localization in the pituitary by autoradiography. *Z. Zellforsch.*, 92: 23–33.

Stumpf, W.E. (1970) Estrogen-neurons and estrogen-neuron systems in the periventricular brain. *Amer. J. Anat.*, 129: 207–217.

Stumpf, W.E. (1972) Estrogen, androgen and glucocorticosteroid concentrating neurons in the amygdala, studied by autoradiography. In *Advances in Behavioral Biology, Vol. 2, The Neurobiology of the Amygdala*, B.E. Eleftheriou (Ed.), Plenum Press, New York, pp. 763–774.

Stumpf, W.E. and Roth, L.J. (1966) High resolution autoradiography with dry-mounted, freeze-dried, frozen sections. Comparative study of six methods using two diffusible compounds, ^{3}H-estradiol and ^{3}H-mesobilirubinogen. *J. Histochem. Cytochem.*, 14: 274–287.

Swaneck, G.E., Highland, E. and Edelman, I.S. (1969) Stereospecific nuclear and cytosol aldosterone-binding proteins of various tissues. *Nephron*, 6: 297–316.

Terkel, A.S., Shryne, J. and Gorski, R.A. (1973) Inhibition of estrogen facilitation of sexual behavior by the intracerebral infusion of actinomycin-D. *Horm. Behav.*, 4: 377–386.

Thieulant, M.L., Mercier, M.L., Samperez, S. et Jouan, P. (1974) Mise en évidence d'un récepteur spécifique de la testostérone dans le cytoplasme de l'hypophyse antérieure du rat mâle prépubère. *C.R. Acad. Sci. (Paris)*, 278: 2569–2572.

Thoenen, H., Mueller, R.A. and Axelrod, J. (1969) Trans-synaptic induction of adrenal tyrosine hydroxylase. *J. Pharmacol. exp. Ther.*, 169: 249–254.

Tuohimaa, P. (1971) The radioautographic localization of exogenous tritiated dihydrotestosterone, testosterone, and oestradiol in the target organs of female and male rats. In *Basic Actions of Sex Steroids on Target Organs*, P.O. Hubinot, F. Leroy and P. Galand (Eds.), Karger, Basel, pp. 208–214.

Vertes, M. and King, R.J.B. (1971) The mechanism of oestradiol binding in rat hypothalamus: effect of androgenization. *J. Endocr.*, 51: 271–282.

Vilchez-Martinez, J.A., Arimura, A., Debeljuk, L. and Schally, A.V. (1974) Biphasic effect of estradiol benzoate on the pituitary responsiveness to LH-RH. *Endocrinology*, 94: 1300–1303.

Warembourg, M. (1970) Fixation de l'oestradiol-^{3}H au niveau des noyaux amygdaliens, septaux et du système hypothalamo-hypophysaire chez la souris femelle. *C.R. Acad. Sci. (Paris)*, 270: 152–154.

Warembourg, M. (1974) Etude radioautographique des rétroactions centrales des corticostéroides-^{3}H chez le rat et le cobaye. In *Neuroendocrinologie de l'Axe Corticotrope, Brain-Adrenal Interactions*, P. Dell (Ed.), INSERM, Paris, pp. 41–66.

Warembourg, M. (1975) Radioautographic study of the rat brain after injection of [1,2-^{3}H]corticosterone. *Brain Res.*, 89: 61–70.

Watanabe, H. (1975) Dexamethasone-binding receptor in bovine pituitary cytosol. *J. Steroid Biochem.*, 6: 27–33.

Watanabe, H., Orth, D.N. and Toft, D.O. (1973) Glucocorticoid receptors in pituitary tumor cells. I. Cytosol receptors. *J. biol. Chem.*, 248: 7625–7630.

Watanabe, H., Orth, D.N. and Toft, D.O. (1974) Glucocorticoid receptors in mouse pituitary tumor cells. II. Nuclear binding. *Biochemistry*, 13: 332–337.

Weisz, J. and Gibbs, C. (1974a) Conversion of testosterone and androstenedione to estrogen in vitro by the brain of female rats. *Endocrinology*, 94: 616–620.

Weisz, J. and Gibbs, C. (1974b) Metabolites of testosterone in the brain of the newborn female rat after an injection of tritiated testosterone. *Neuroendocrinology*, 14: 72–86.

Whalen, R.E. and Gorzalka, B.B. (1973) Effects of an estrogen antagonist on behavior and on estrogen retention in neural and peripheral target tissues. *Physiol. Behav.*, 10: 35–40.

Whalen, R.E., Gorzalka, B.B., DeBold, J.F., Quadagno, D.M., Ho, K.-W.G. and Hough, J.C., Jr. (1974) Studies on the effects of intracerebral actinomycin-D implants on estrogen-induced receptivity in rats. *Horm. Behav.*, 5: 337–343.

Wiggert, B. and Chader, G. (1974) Studies on the specific glucocorticoid-hormone receptor in the developing chick retina. *Exp. Eye Res.*, 18: 477–484.

Wiggert, B. and Chader, G. (1975) A glucocorticoid and progesterone receptor in the chick optic tectum. *J. Neurochem.*, 24: 585–586.

Yamamoto, K.R. (1974) Characterization of the 4S and 5S forms of the estradiol receptor protein and their interaction with deoxyribonucleic acid. *J. biol. Chem.*, 249: 7068–7075.

Yamamoto, K.R., Stampfer, M.R. and Tomkins, G.M. (1974) Receptors from glucocorticoid-sensitive lymphoma cells and two classes of insensitive clones: physical and DNA-binding properties. *Proc. nat. Acad. Sci. (Wash.)*, 71: 3901–3905.

Zigmond, R.E. (1975) Binding and metabolism of steroid hormones in the central nervous system. In *Handbook of Psychopharmacology*, L.L. Iversen, S.D. Iversen and S.H. Snyder (Eds.), Plenum Press, New York, pp. 239–328.

Zigmond, R.E. and McEwen, B.S. (1970) Selective retention of oestradiol by cell nuclei in specific brain regions of the ovariectomized rat. *J. Neurochem.*, 17: 889–899.

Zigmond, R., Stern, J. and McEwen, B.S. (1972) Retention of radioactivity in cell nuclei in the hypothalamus of the ring dove after injections of ^{3}H-testosterone. *Gen. comp. Endocr.*, 18: 450–453.

Zigmond, R.E., Nottebohm, F. and Pfaff, D.W. (1973) Androgen Concentrating cells in the midbrain of a songbird. *Science*, 179: 1005–1007.

Chapter 15

Progesterone metabolism by neuroendocrine tissues

H.J. KARAVOLAS and K.M. NUTI

Department of Physiological Chemistry, The Endocrinology-Reproductive Physiology Program, and the Waisman Center on Mental Development, University of Wisconsin, Madison, Wisc. 53706 (U.S.A.)

I. INTRODUCTION

During the past few years our interest in the molecular mechanisms by which progesterone exerts its effects on neuroendocrine function has prompted us to investigate the uptake and metabolism of progesterone (and related progestins) by neuroendocrine tissues. It seemed reasonable to postulate inter alia that the conversion of progesterone by these tissues into one or more metabolites might be an important component of the mechanism(s) by which progesterone exerts its effects on gonadotropin regulation, sexual behavior, and other progesterone-sensitive neural functions.

The concept that metabolic conversion of a steroid hormone in target tissues may be requisite for its intracellular action is not new and is usually proffered as one possible mode of action (Robel, 1971; Baulieu, 1973). The best example of this is the conversion of testosterone to 17β-hydroxy-5α-androstan-3-one (5α-dihydrotestosterone) (Wilson and Gloyna, 1970). On the other hand, it is estradiol itself, and not a metabolite, that modulates estrogenic end-points in the uterus (Baulieu, 1973). Progesterone, of course, has a number of biological effects. Either alone, or in combination with other hormones, it affects uterine function, gonadotropin production and/or release, ovulation, sexual behavior, lactation, sleep patterns — anesthetic effects, blood formation — erythropoiesis, thermogenicity, protein metabolism, etc. (Goldman and Zarrow, 1973; Goy and Goldfoot, 1973). These multiple and varied effects could theoretically result from the action(s) of progesterone itself and/or specific metabolites.

In this article, we will summarize and discuss recent findings by ourselves and others concerning the metabolism of progesterone by neuroendocrine tissues, with particular emphasis on the hypothalamus and pituitary. We shall consider the qualitative and quantitative aspects of in vitro and in vivo metabolite production, tissue specificity, and some characteristics of the enzyme systems involved in these metabolic transformations including their steroid specificity. The biological relevance of the metabolites and their possible roles in mediating some of the neuroendocrine effects of progesterone will also be discussed. In the sections that follow, experimental details will not be presented other than those needed for a general understanding since these details can be found in the original publications.

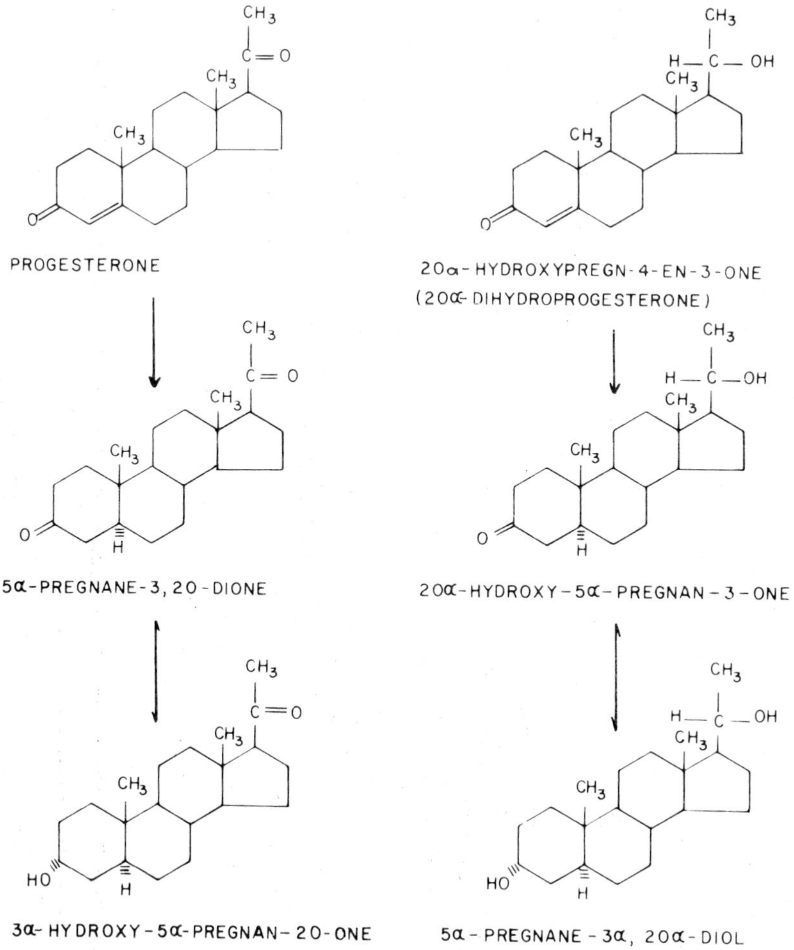

Fig. 1. Proposed pathways for the metabolism of progesterone and 20α-dihydroprogesterone by rat hypothalamus and anterior pituitary.

II. IN VITRO METABOLISM OF PROGESTERONE (AND 20α-DIHYDROPROGESTERONE) BY MEDIAL BASAL HYPOTHALAMUS AND ANTERIOR PITUITARY

The in vitro metabolism of progesterone has been studied with tissue slices of rat medial basal hypothalamus and anterior pituitary which were removed at various times during the estrous cycle and then incubated with radiolabeled progesterone (Karavolas and Herf, 1971; Cheng and Karavolas, 1973; Robinson and Karavolas, 1973). After reverse isotopic dilution analyses and purification to constant specific activity, three radioactive compounds accounted for almost all the radioactivity. These were identified as progesterone, 5α-pregnane-3,20-dione (5α-dihydroprogesterone; 5α-DHP), and 3α-hydroxy-5α-pregnan-20-one. Structures of these steroids and a proposed metabolic pathway are presented in Fig. 1. Recent reports from several other laboratories have also described the presence of Δ^4-steroid (progesterone) 5α-reductase and 3α-hydroxy-steroid dehydrogenase in the rat anterior pituitary and hypothalamus (Rommerts and Van der Molen, 1971; Massa et al., 1972b; Snipes and Shore, 1972).

In our studies, other metabolites of progesterone, if any, were present in very small amounts, and there was no appreciable conversion to 20α-dihydroprogesterone (20α-DHP). This contrasts with the report by Tabei et al. (1974) who found that pituitary tissues from both male and female rats and male hypothalamic tissues convert progesterone principally to 5α-dihydroprogesterone and 20α-DHP.

Similar patterns of metabolism by pituitary and hypothalamus have been observed when [^{3}H]20α-DHP is used as substrate (Nowak and Karavolas, 1974a, b). In these studies, the principal in vitro metabolic products were 20α-hydroxy-5α-pregnan-3-one and 5α-pregnane-3α, 20α-diol (Fig. 1). These two metabolites plus the original substrate ([^{3}H]20α-DHP) accounted for nearly all the radioactivity. There was no appreciable conversion to progesterone. Thus, with either progesterone or 20α-DHP as substrate, the principal enzymatic activities in the rat hypothalamus and anterior pituitary appear to be Δ^4-steroid 5α-reduction and 3α-hydroxy-steroid dehydrogenation. It is likely that the same enzymes catalyze the metabolism of either progesterone or 20α-DHP.

If these enzymatic conversions play a role in the regulation of reproductive events, there might be qualitative or quantitative changes during the estrous cycle or in response to exogenous hormones. When qualitative aspects of these conversions were examined, it was found that rat hypothalamic and adenohypophyseal tissue removed at different stages of the estrous cycle metabolize progesterone to the same two major metabolites (Massa et al., 1972b; Cheng and Karavolas, 1973; Robinson and Karavolas, 1973). Limited quantitative studies suggest cyclic variation in hypothalamic 5α-reductase activity (sum of both 5α-reduced products) with highest levels on the mornings of diestrus 2 and estrus (Fig. 2) (Cheng and Karavolas, 1973). The conversion of progesterone to 5α-DHP by hypothalamus from ovariectomized rats was enhanced slightly following estrogen priming. These results do not agree with those of Massa et al. (1972b) who did not detect any changes in hypothalamic enzymic activity during the estrous cycle or in response to exogenous estrogen. However, Massa et al. (1972b) did find that pituitary 5α-reductase activity was cyclic, exhibiting a peak at estrus, and could be inhibited by exogenous estrogen.

Some further quantitative studies on 5α-reductase activity were performed using [^{3}H]20α-DHP as substrate. The in vitro conversion of 20α-DHP to 20α-hydroxy-5α-pregnan-3-one and 5α-pregnane-3α, 20α-diol was investigated throughout the day of proestrus in the hypothalamus and pituitary of 4-day cyclic rats (Fig. 3) (Nowak et al., 1975). Conversion of 20α-DHP to its 5α-reduced metabolites by the pituitary was constant throughout proestrus except for a significant decrease at 16.00 hr, near the end of the critical period. Although 5α-reduction of 20α-DHP by the hypothalamus fluctuated, it was relatively high at 16.00 hr and was lowest at 14.00 hr. Small amounts of progesterone (<2%) were formed and there was no variation with time. The decrease in pituitary

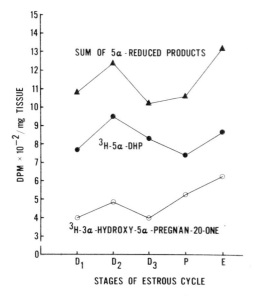

Fig. 2. Hypothalamic conversion of [^{3}H]progesterone to 3α-hydroxy-5α-pregnan-20-one and 5α-pregnane-3,20-dione at different stages of the estrous cycle (Cheng and Karavolas, 1973).

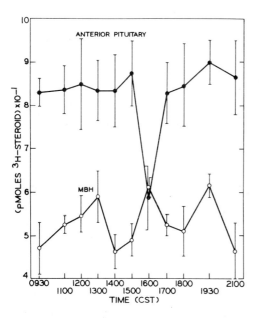

Fig. 3. Total 5α-reductase activity (sum of 20α-hydroxy-5α-pregnan-3-one and 5α-pregnane-3α,20α-diol) in extracts of anterior pituitary or hypothalamus removed from rats at the indicated time on the day of proestrus and then incubated with [^{3}H]20α-dihydroprogesterone. Each point represents the mean ± S.E.M. (N=5).

enzymic activity coincided closely with the time when serum levels of LH, FSH, and progesterone were increasing (Fig. 4) but not with later times when these elevated serum levels were maintained.

Thus, it appears from these initial studies by us and others that there may be endogenous regulation of 5α-reductase and possibly 3α-hydroxysteroid dehydrogenase activity in rat pituitary and hypothalamus.

III. TISSUE LOCALIZATION OF PROGESTERONE 5α-REDUCTASE

In addition to the hypothalamus and pituitary, other neural tissues (i.e., midbrain, cerebellum, pineal, preoptic area) have been implicated in the neuroendocrine control of gonadotropin releasing hormones and/or have been postulated to contain progesterone receptors (Clemens et al., 1971; Kaasjager et al., 1971; Wade and Feder, 1972; Piva et al., 1973; Velasco and Rothchild, 1973; Whalen and Gorzalka, 1974; White et al., 1974). Preliminary results from studies in progress indicate that female rat thalamus, midbrain, cerebellum, medulla oblongata, and pineal are also capable of the in vitro conversion of progesterone to 5α-DHP, and 3α-hydroxy-5α-pregnan-20-one (Hanukoglu and Karavolas, unpublished observations). The observation of 5α-reduction and 3α-hydroxysteroid dehydrogenase activity in the cerebellum agrees with earlier findings of Rommerts and Van der Molen (1971) with male rats. The cerebral cortex of female (Karavolas and Herf, 1971) and male (Rommerts and Van der Molen, 1971) rats also converts progesterone to

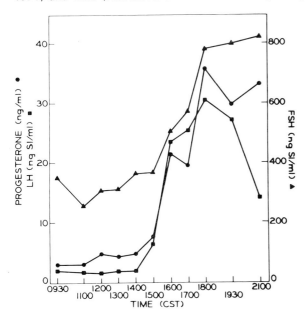

Fig. 4. Proestrous serum levels of progesterone, LH and FSH. Each point represents the mean for 5 individuals.

TABLE I

Reverse isotopic dilution analysis of progesterone and metabolites after incubation of dissociated anterior pituitary cells with [³H]progesterone followed by separation into enriched fractions by velocity sedimentation *

³H-steroid	Fraction** (dpm/cell × 10⁷)			Ratio of fraction 1 to other fractions	
	Gonadotropes 1	Somatotropes 3	Chromophobes 5	Fr 1/Fr 3	Fr 1/Fr 5
Progesterone	1900 ± 236*** (59.4 ± 2.9)§	625 ± 127 (67.6 ± 2.9)	635 ± 105 (69.2 ± 3.9)	3.0	3.0
3α-Hydroxy-5α-pregnan-20-one	150 ± 32 (4.8 ± 1.1)	45 ± 8.2 (5.3 ± 1.6)	35 ± 5.9 (3.7 ± 0.6)	3.3	4.2
5α-Pregnane-3,20-dione	366 ± 40.1 (11.5 ± 0.8)	99.6 ± 19.1 (10.9 ± 2.0)	68.6 ± 6.8 (7.6 ± 0.6)§§	3.7	5.2
20α-Hydroxy-4-pregnen-3-one	62 ± 16 (2.1 ± 0.7)	19 ± 7.7 (2.4 ± 1.4)	40 ± 21 (4.3 ± 1.8)	3.3	1.6

* Taken from Lloyd and Karavolas (1975).
** Fractions 1, 3, and 5 are fractions of enriched gonadotropes, somatotropes and chromophobes, respectively.
*** Mean ± S.E.M. (N = 4).
§ Percentage of radioactivity associated with cells after reverse isotopic dilution.
§§ $P < 0.05$ compared to fraction 1.

5α-DHP. More recently Tabei et al. (1974) have reported 5α-reductase activity in cell-free homogenates of the cerebral cortex of adult male and female rats. Thus, the 5α-reduction of progesterone appears to be a common property of several neural tissues of the rat.

The in vitro metabolism of progesterone has been studied with whole brain homogenates of human fetus (Mickan, 1972) and with subcellular fractions of dog brain (105,000 × *g* supernatant) (Kawahara et al., 1975). The major metabolite with human fetal brain was 5α-DHP, while 5β-DHP was identified as a major metabolic product of progesterone in dog brain.

If the 5α-reduced progestin metabolites play a role in regulating gonadotropin production and/or secretion, one might anticipate that the pituitary 5α-reductase activity would be localized in specific cell types of the anterior pituitary. When pronase-dissociated pituitary cells from male rats were incubated for 30 min with [³H]progesterone and then separated by velocity sedimentation into enriched fractions containing gonadotropic, somatotropic, or chromophobic cells, uptake of [³H]progesterone by the enriched gonadotropic cell fraction was approximately three-fold greater than by the other two fractions (Table I) (Lloyd and Karavolas, 1975). In addition, the gonadotropic cell fraction, which possesses almost 60% of the total LH activity (Lloyd and McShan, 1973), metabolized progesterone to proportionately more 5α-reduced products. The principal metabolites identified after purification to constant specific activity were 5α-DHP and 3α-hydroxy-5α-pregnan-20-one with lesser amounts of 20α-dihydroprogesterone, except with the enriched chromophobic cell fraction where comparable and lower amounts of all three metabolites were found. Analogous results were obtained when testosterone was similarly incubated. 5α-Dihydrotestosterone and 5α-androstane-3α, 17β-diol were the principal metabolites (Lloyd and Karavolas, 1975).

Progesterone 5α-reductase activity is not, of course, unique to neural and pituitary tissue since a variety of target tissues in a number of species also appear to possess 5α-reductase activity. These include chick oviduct (Morgan and Wilson, 1970; Schrader et al., 1972), rat uterine (Wiest, 1963; Armstrong and King, 1971) and decidual tissue (Reel et al., 1971), human endometrium (Bryson and Sweat, 1967), and mammary glands of rabbits (Chatterton et al., 1969). Additionally, progesterone 5α-reductase activity is not unique to progesterone-sensitive target tissues, since it has also been found inter alia in human skin (Frost et al., 1969), rat testes (Inano and Tamaoki, 1966) and epididymis (Inano et al., 1969; Tabei et al., 1974), human adrenal gland (Charreau et al., 1968) and the placenta of the mare (Ainsworth and Ryan, 1969).

The almost ubiquitous presence of 5α-reductase activity in body tissues would seem to militate against the notion that the formation of 5α-reduced metabolites may be a specific target tissue mechanism for mediating some of progesterone's actions unless, of course, the biological role of progesterone is greater than presently envisaged. Nonetheless, mechanistic specificity could still be conferred via these metabolic conversions in target tissues if inter alia there were: (a) differences in the catalytic and regulatory properties of the various tissue 5α-reductases and/or 3α-hydroxysteroid dehydrogenases, (b) relative differences in enzymatic levels and/or activity, (c) differences in the flux of metabolites, and/or (d) differences in the tissue levels of these progesterone metabolites which would ultimately affect subsequent intracellular events in a concentration-dependent manner. Furthermore, specificity could be endowed by the differentiated function of a particular tissue whereby only a few of all the progesterone-sensitive processes are being expressed by the genetic machinery of that differentiated cell.

We will consider several of these aforementioned possibilities in subsequent sections of this article.

IV. PROPERTIES AND SUBCELLULAR DISTRIBUTION OF RAT HYPOTHALAMIC AND ANTERIOR PITUITARY Δ^4-STEROID 5α-REDUCTASE(S)

The subcellular distribution and properties of the hypothalamic and pituitary Δ^4-steroid (progesterone) 5α-reductases have been investigated using ^{3}H-labeled substrate and a reverse isotopic dilution assay system (Cheng and Karavolas, 1975a, b). With both tissue sources, 5α-reductase activity was stimulated by NADPH but not NADH (Table II). The hypothalamic and anterior pituitary enzymes exhibited similar K_ms for progesterone of $4.8 \pm 1.4 \times 10^{-7} M$ and $2.7 \pm 0.9 \times 10^{-7} M$, respectively (Fig. 5). These values are in the range of plasma progesterone concentrations in the adult female rat ($10^{-9}-10^{-7} M$) (Hashimoto et al., 1968; Goldman et al., 1969; Piacsek et al., 1971).

Another common feature of these two enzymic activities was a broad substrate specificity for Δ^4-3-keto-steroids as well as an inhibitory effect of 17β-estradiol (in vitro) on the 5α-reduction of progesterone. 20α-Dihydroprogesterone was more reactive than progesterone, and testosterone was less reactive (Table III). Our findings that 20α-DHP is more reactive than progesterone is consistent with results found with the Δ^4-steroid 5α-reductase from rat prostate (Frederiksen and Wilson, 1971). Likewise, the observation that progesterone is a better substrate for 5α-reductase than testosterone is in agreement with Massa et al. (1972a). Kinetic studies also indicate that 20α-DHP is a potent competitive inhibitor (substrate) of progesterone 5α-reduction (Fig. 6.). Conversely, since both progesterone and 20α-DHP are substrates for this enzymic activity, progesterone might also be a competitive inhibitor (substrate) of the 5α-reduction of 20α-DHP. The inhibitory

TABLE II

Pyridine nucleotide requirement of progesterone 5α-reductase activity in rat hypothalamus and anterior pituitary*

Cofactor added	Sum of 5α-reduced products formed	
	Hypothalamus (fmole/25 mg tissue/30 min)	Anterior pituitary (fmole/24 mg tissue/30 min)
None	4	17
NADPH	496	455
NADPH-generating system	604	541
NADH	4	16

* Taken from Cheng and Karavolas (1975a,b).

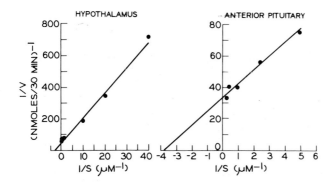

Fig. 5. Lineweaver–Burk plots of the effect of increasing concentrations of progesterone on the rate of 5α-reduced steroid formation by hypothalamic and anterior pituitary Δ^4-steroid 5α-reductases (Cheng and Karavolas, 1975a, b).

TABLE III

Substrate specificity of 5α-reductase activity in rat hypothalamus and anterior pituitary *

Δ^4-3-ketosteroid substrate	Sum of 5α-reduced products formed	
	Hypothalamus (fmole/30 mg tissue/30 min)	Anterior pituitary (fmole/15 mg tissue/30 min)
Testosterone	58	57
Progesterone	1557	833
20α-Dihydroprogesterone	5270	2640

* Taken from Cheng and Karavolas (1975a,b).

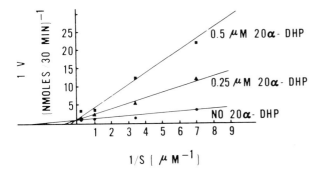

Fig. 6. Lineweaver–Burk plots of the inhibitory effect of 20α-dihydroprogesterone on the rate of 5α-reduction of progesterone by hypothalamic Δ^4-steroid 5α-reductase.

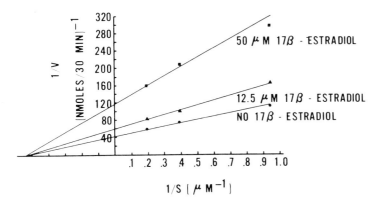

Fig. 7. Lineweaver–Burk plots of the inhibitory effect of 17β-estradiol on the rate of 5α-reduction of progesterone by hypothalamic Δ^4-steroid 5α-reductase.

effect of estradiol (in vitro) on the hypothalamic or pituitary 5α-reduction of progesterone was explored further with the hypothalamic enzyme and shown to be non-competitive (Fig. 7). This inhibitory effect may be related to the ability of estradiol to modulate pituitary sensitivity to LRF in response to progesterone (Schally et al., 1973; Martin et al., 1974).

At present, the only difference we observed between the hypothalamic and pituitary enzymes is in their subcellular distribution. Hypothalamic enzymic activity was associated with a cell debris-membranes fraction derived from the 1000 x g pellet (Table IV). Little or no activity was associated with purified nuclei. However, studies on the subcellular distribution of 5α-reductase activity in the pituitary (Table IV) indicated that the activity was widely distributed among particulates sedimenting at 1000, 15,000 and 100,000 x g. The 15,000 x g pellet contained the most enzymic activity, and after further fractionation of the 1000 x g pellet, the activity was distributed equally between the purified nuclear and cell debris-membranes fractions. Tabei et al. (1974) have also found progesterone 5α-reductase activity in rat pituitary associated with particulates sedimenting at 10,000 and 105,000 x g but no data are reported for their 700 x g particulates.

TABLE IV

Subcellular distribution of progesterone 5α-reductase activity in rat hypothalamus and anterior pituitary*

Fraction	Sum of 5α-reduced products formed	
	Hypothalamus (fmole/50 mg tissue/hr)	Anterior pituitary (fmole/50 mg tissue/30 min)
Whole homogenate	1521	815
1000 x g pellet	1415	176
Subfractionation of the 1000 x g pellet		
1. Cell debris-membranes fraction (top layer)	429	60
2. Middle layer		
a. Top third	0	5
b. Middle third	3	9
c. Bottom third	5	11
3. Purified nuclei (bottom layer)	17	63
1000 x g supernatant	105**	682
Subfractionation of the 1000 x g supernatant		
1. 15,000 x g pellet	----	272
2. 100,000 x g pellet	----	113
3. 100,000 x g supernatant	----	18

* Taken from Cheng and Karavolas (1975a,b).
** Not fractionated further since activity was low.

V. QUALITATIVE AND QUANTITATIVE DETERMINATIONS OF THE IN VIVO UPTAKE AND METABOLISM OF PROGESTERONE AND 5α-DIHYDROPROGESTERONE BY PITUITARY, HYPOTHALAMUS AND OTHER TISSUES

Concomitant with studies on the properties of progesterone 5α-reductase, we also investigated the in vivo uptake and metabolism of [^{3}H]progesterone and [^{3}H]5α-DHP 10 min after an i.v. injection in ovariectomized rats (Karavolas et al., 1975, 1976) with a view towards an analysis of the identity and concentrations of ^{3}H-steroids present after injection. These studies were focused upon the two neuroendocrine feedback tissues (hypothalamus and pituitary), the uterus (as a well established progestin target tissue), the muscle (as a non-target tissue), and cerebral cortex (as a putative non-target neural tissue), as well as plasma. Comparable results were obtained with adrenalectomized-ovariectomized rats (Karavolas et al., 1975, 1976) but will not be presented here.

Tissue and plasma ^{3}H concentrations are given in Table V. With progesterone, none of the tissues had higher ^{3}H contents than plasma. This is consistent with previous reports by others who could not demonstrate selective uptake of [^{3}H]progesterone in rat anterior pituitary or hypothalamus (Wade and Feder, 1972; Seiki and Hattori, 1973; Wade et al., 1973; Whalen and Gorzalka, 1974). However, ^{3}H concentrations in the pituitary and hypothalamus were significantly greater than in muscle.

The above results expressed as ^{3}H concentrations merely represent, of course, unidentified ^{3}H. These comparisons of tissue and plasma ^{3}H levels are often misleading since, as shown in Fig. 8, most of the radioactivity in the tissue and plasma samples is not the original injected ^{3}H-steroid, i.e., [^{3}H]progesterone or [^{3}H]5α-DHP. We have found that the ^{3}H is associated with several steroidal compounds which have been shown previously

TABLE V

Total tritium content of selected tissues from ovariectomized rats 10 min after injection of [^{3}H]progesterone or [^{3}H]5α-DHP*

Tissue	Progesterone (dpm/mg tissue or μl plasma)	5α-DHP (mean ± S.E.M.)
Cerebral cortex	150 ± 18	205 ± 18***
Hypothalamus (MBH)	189 ± 18***	336 ± 30§§
Anterior pituitary	209 ± 24**	274 ± 15***
Uterus	144 ± 15	93 ± 4
Muscle	102 ± 9	104 ± 12
Plasma	289 ± 13	234 ± 22

* Taken from Karavolas et al. (1976).
** Significantly ($P < 0.05$) higher than muscle samples.
*** $P < 0.01$.
§ Significantly ($P < 0.05$) higher than cerebral cortical samples.
§§ $P < 0.01$.

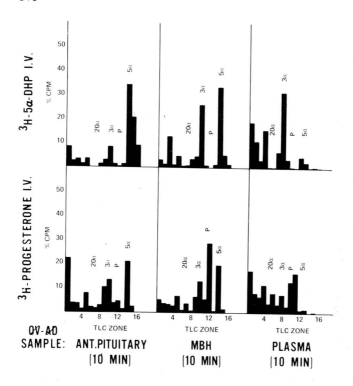

Fig. 8. Distribution of radioactivity after thin-layer chromatography (TLC) of extracts of pituitary, hypothalamic and plasma samples removed from ovariectomized-adrenalectomized rats 10 min after an injection of [^{3}H]progesterone or [^{3}H]5α-DHP. Approximately 1 cm lengths of the silica gel tracts (TLC zones) were removed for scintillation counting; origin is designated zone 1. Bar heights represent the per cent of total recovered radioactivity. The abbreviations used are: 5α, 5α-DHP; P, progesterone; 3α, 3α-hydroxy-5α-pregnan-20-one; 20α, 20α-DHP. Similar ^{3}H distribution profiles were seen after TLC of extracts from ovariectomized rats (Karavolas et al., 1976).

to be the principal metabolites of progesterone in these tissues.

Thus, in order to make meaningful tissue and plasma comparisons, the concentrations of specific ^{3}H-steroids (Fig. 1) were determined by reverse isotopic dilution analyses. After injection of progesterone (Table VI), most of the tissue ^{3}H was associated with progesterone, 5α-DHP, and 3α-hydroxy-5α-pregnan-20-one. Progesterone was the predominant ^{3}H-steroid in hypothalamus, cerebral cortex, uterus, and muscle. [^{3}H]5α-DHP was the other major ^{3}H-steroid in hypothalamus and the predominant one in the pituitary. Although [^{3}H]progesterone levels in hypothalamus, cerebral cortex, and uterus were higher than those in plasma, the differences were not significant. [^{3}H]Progesterone concentrations in these three tissues were also greater than those in muscle, but only the hypothalamic values were significantly higher. In contrast, [^{3}H]5α-DHP levels in pituitary, hypothalamus, and cerebral cortex were many times greater than muscle, uterus, or plasma. Pituitary and hypothalamic levels of 5α-DHP were also significantly

TABLE VI

Distribution of ^{3}H-steroids in tissues and plasma of ovariectomized rats after an i.v. injection of [^{3}H]progesterone[a]

^{3}H-steroid[a]	Anterior pituitary		Hypothalamus (MBH)		Cerebral cortex		Muscle		Uterus		Plasma
	dpm/mg[b]	t/p[c]	dpm/mg	t/p	dpm/mg	t/p	dpm/mg	t/p	dpm/mg	t/p	dpm/μl
5α-DHP	70.2 ± 3.3[g,k] (34%)[d]	21.9	38.8 ± 2.9[g,k,l] (21%)	12.1	19.3 ± 3.7[f,j] (12%)	6.0	2.8 ± 0.3 (3%)	0.9	5.6 ± 1.0 (4%)	1.8	3.2 ± 0.4 (1.1%)
Progesterone	40.0 ± 0.4 (19%)	0.7	102.8 ± 16.4[h] (54%)	1.9	79.0 ± 12.0 (53%)	1.5	41.8 ± 5.8 (41%)	0.8	68.0 ± 11.4 (48%)	1.3	53.9 ± 7.7 (19%)
3α-OH	38.4 ± 4.4[f,j] (18%)	4.1	14.1 ± 1.2[e,h] (7%)	1.5	14.4 ± 3.1 (9%)	1.5	8.4 ± 1.1 (8%)	0.9	32.5 ± 4.3[f,j] (23%)	3.5	9.4 ± 1.1 (3.2%)
20α-5α	26.3 ± 5.5[j] (12%)	5.8	7.9 ± 1.8[h] (4%)	1.8	5.0 ± 1.4 (3%)	1.1	2.3 ± 0.3 (2%)	0.5	(<1%)[d]		4.5 ± 1.0 (1.6%)
20α-DHP	12.0 ± 0.5 (6%)	0.5	9.0 ± 2.6 (5%)	0.4	6.8 ± 0.8 (4%)	0.3	29.1 ± 7.1 (29%)	1.3	3.0 ± 0.4 (2%)	0.1	22.4 ± 4.3 (7.8%)
5α-Diol	26.2 ± 4.8[e,j] (11%)	2.9	4.1 ± 0.6 (2%)	0.5	(<1%)[d]		3.3 ± 0.6 (3%)	0.4	2.9 ± 0.3 (2%)	0.3	9.0 ± 0.7 (3.4%)

[a] Steroid abbreviations: 3α-OH, 3α-hydroxy-5α-pregnan-20-one; 20α-5α, 20α-hydroxy-5α-pregnan-3-one; 5α-diol, 5α-pregnane-3α, 20α-diol.
[b] Concentrations of each ^{3}H-steroid on a dpm/mg of tissue or per μl of plasma basis. Values given are means of measurements from 4 individual tissue or plasma samples ± S.E.M.
[c] Ratios of $\frac{dpm/mg\ tissue}{dpm/\mu l\ plasma}$.
[d] Percentage of total tissue tritium associated with a particular carrier steroid. Values given as (<1%) indicate specific activity analyses where dpm were too low to permit meaningful determinations.
[e] Significantly > plasma ($P < .05$); [f] $P < .01$; [g] $P < .001$.
[h] Significantly > muscle ($P < .05$); [j] $P < .01$; [k] $P < .001$.
[l] Significantly > cortex ($P < .01$).

TABLE VII

Distribution of ³H-steroids in tissues and plasma of ovariectomized rats after an i.v. injection of [³H]5α-DHP

³H-steroid[a]	Anterior pituitary		Hypothalamus (MBH)		Cerebral cortex		Muscle		Uterus		Plasma
	dpm/mg[b]	t/p[c]	dpm/mg	t/p	dpm/mg	t/p	dpm/mg	t/p	dpm/mg	t/p	dpm/μl
5α-DHP	148.2 ± 14.6[f,g] (57%)[d]	18.3	200.7 ± 21.1[f,g,h] (61%)	24.8	68.7 ± 5.7[f,g] (34%)	8.5	2.8 ± 0.2 (3%)	0.3	3.6 ± 0.3 (4%)	0.4	8.1 ± 1.5 (3.5%)
Progesterone	N.D.[e]		N.D.		N.D.		N.D.		N.D.		N.D.
3α-OH	52.9 ± 7.3 (20%)	0.9	76.3 ± 7.8 (23%)	1.3	75.6 ± 5.8 (37%)	1.3	51.3 ± 10.6 (50%)	0.9	60.7 ± 4.2 (65%)	1.0	59.3 ± 7.7 (25%)
20α-5α	8.0 ± 0.7 (3%)	10.0	4.8 ± 0.7 (<1%)	6.0	8.1 ± 0.8 (4%)	10.1	3.6 ± 1.8 (4%)	4.5	2.1 ± 1.8 (2%)	2.6	0.8 ± 0.3 (<0.3%)
20α-DHP	N.D.		N.D.		N.D.		N.D.		N.D.		N.D.
5α-Diol	16.5 ± 0.7 (6%)	0.6	13.2 ± 0.9 (4%)	0.5	10.0 ± 1.2 (5%)	0.4	14.0 ± 1.9 (14%)	0.5	5.8 ± 0.8 (6%)	0.2	26.7 ± 3.6 (11%)

[a] Steroid abbreviations: 3α-OH, 3α-hydroxy-5α-pregnan-20-one; 20α-5α, 20α-hydroxy-5α-pregnan-3-one; 5α-diol, 5α-pregnane-3α, 20α-diol.
[b] Concentrations of each ³H-steroid on a dpm/mg of tissue or per μl of plasma basis. Values given are means of measurements from 4 individual tissue or plasma samples ± S.E.M.
[c] Ratios of $\frac{dpm/mg\ tissue}{dpm/\mu l\ plasma}$
[d] Percentage of total tissue tritium associated with a particular carrier steroid.
[e] Values given as N.D. indicate specific activity analyses where dpm values were equivalent to background levels or too few to permit meaningful determinations.
[f] Significantly > plasma ($P < .001$).
[g] Significantly > muscle ($P < .001$).
[h] Significantly > cortex ($P < .01$).

greater than cerebral cortex. Muscle and uterine contents of this steroid were low and comparable to plasma.

The lack of a definitive accumulation of [^{3}H]progesterone by uterus and pituitary above plasma and muscle levels may be due to a lack of estrogen or other hormonal priming which would have increased uterine uptake (Milgrom et al., 1970; Faber et al., 1972; Wiest, 1972; Cooper et al., 1974; Freifeld et al., 1974) and, presumably, pituitary responsiveness (Aiyer et al., 1974). Nonetheless, hypothalamic levels of [^{3}H]progesterone were greater than plasma or muscle levels. In contrast to the slight differences between tissue and plasma [^{3}H]progesterone concentrations, the [^{3}H]5α-DHP tissue/plasma comparisons were quite striking, particularly since significant concentrations were found in neuroendocrine target tissues such as pituitary and hypothalamus but not in uterus.

Following injections of [^{3}H]5α-DHP, the concentrations of ^{3}H in neural and pituitary tissues were generally higher than the progesterone group (Table V) while uterine samples contained little ^{3}H. When tissue ^{3}H contents were compared with muscle, significant increases were found with pituitary, hypothalamus, and cerebral cortex. However, as with the progesterone experiments, much of the radioactivity in these samples was not associated with the original steroidal compound injected (Fig. 8).

After isotopic dilution analyses, most tissue ^{3}H was shown to be associated with 5α-DHP and/or 3α-hydroxy-5α-pregnan-20-one (Table VII). No [^{3}H]progesterone was detected. [^{3}H]5α-DHP predominated in pituitary and hypothalamus, while [^{3}H]3α-hydroxy-5α-pregnan-20-one predominated in the other tissues. As with the progesterone experiments, there were no significant accumulations of [^{3}H]5α-DHP in uterus or muscle. In terms of tissue and plasma comparisons, hypothalamic and pituitary concentrations of [^{3}H]5α-DHP were 2–3-fold greater than those in cerebral cortex, and 15–65-fold greater than plasma, muscle or uterine levels. Although the levels of [^{3}H]5α-DHP in cerebral cortex were lower than in hypothalamus, they were substantially higher than plasma or muscle, which may simply be due to the high lipid content of neural tissues or may possibly be related to some other CNS effects of progesterone.

The only other ^{3}H-steroid present in appreciable amounts after the injection of either [^{3}H]progesterone or [^{3}H]5α-DHP, was [^{3}H]3α-hydroxy-5α-pregnan-20-one (Tables VI and VII). There was, however, no consistent pattern of selective tissue accumulation. In the [^{3}H]progesterone experiment, the concentrations of 3α-hydroxy-5α-pregnan-20-one in pituitary, hypothalamus and uterus were significantly higher than muscle or plasma levels, but not in the [^{3}H]5α-DHP study.

Thus, after an injection of either [^{3}H]progesterone or [^{3}H]5α-DHP, large quantities of [^{3}H]5α-DHP were accumulated in progestin feedback tissues such as pituitary and hypothalamus but not in uterus, a progestin-responsive target tissue. These data are consistent with a hypothesis that 5α-dihydroprogesterone may be functionally important in governing progesterone-sensitive processes in neuroendocrine tissues such as pituitary and hypothalamus but not in uterus.

VI. THE BIOLOGICAL RELEVANCE OF THE CONVERSION OF PROGESTERONE TO 5α-REDUCED METABOLITES

5α-Dihydroprogesterone does not have the same effect(s) as progesterone on uterine parameters such as endometrial proliferation, pregnancy maintenance, parturition inhibition and the decidual reaction (Stucki and Glenn, 1961; Wiest, 1972). Unfortunately, the absence of a uterine progestational effect with 5α-DHP has led to the unwarranted conclusion that 5α-DHP has no effect on biological end-points and has diverted attention from its possible role in other progesterone-mediated events. Recently, however, several groups of investigators have reported on some progesterone-like effects of 5α-dihydroprogesterone which are summarized in Table VIII.

Both 5α-DHP and progesterone facilitate ovulation in immature rats treated with PMS (Sanyal and Todd, 1972; Sridharan et al., 1974). Bosley and Leavitt (1972) have also reported that 5α-DHP as well as progesterone facilitate lordosis, ovulation and the loss of uterine luminal fluid in cyclic hamsters after the ovulatory surge of gonadotropins has been suppressed by phenobarbital. 5β-DHP and 20α-DHP were ineffective at the doses tested. 5α-DHP, like progesterone, has been shown to have a biphasic effect on sexual behavior in ovariectomized estrogen-primed guinea pigs (Czaja et al., 1974). Although 3α-hydroxy-5α-pregnan-20-one and 5β-DHP have moderate facilitative effects, they are not as effective as 5α-DHP. 5α-DHP also exerts a progesterone-like effect on avidin production in the chick oviduct (Strott, 1974). Schally et al. (1973) have reported that 5α-DHP is more potent than progesterone in inhibiting the release of LH and FSH in response to LRF in vitro.

More recently, however, Nuti and Karavolas (unpublished observations) have found that a single s.c. injection of 4 mg 5α-DHP or progesterone into ovariectomized rats, three

TABLE VIII

Reported progesterone-like effects of 5α-DHP*

Non-uterine progesterone-like effects	
Facilitation of ovulation	yes**
Facilitation and inhibition of lordosis	yes**
Inhibition of LRF-induced LH release in vitro	yes***
Stimulation of chick oviduct avidin production	yes***
Uterine progesterone-like effects	
Endometrial proliferation	no
Pregnancy maintenance	no
Decidual reaction	no
Parturition inhibition	no

* See text for references.
** Not as potent as progesterone.
*** More potent than progesterone.

TABLE IX

Effects of progesterone and 5α-DHP on serum LH and FSH levels in ovariectomized estrogen-primed rats

Day of treatment*	Treatment	(μg NIAMD std/ml serum)**	
		LH	FSH
3	corn oil	1.66 ± .22	2.43 ± .21
	4 mg progesterone	7.59 ± .36 §	6.81 ± .42 §,§§
	4 mg 5α-DHP	7.48 ± .93 §	4.66 ± .38 §
3 and 4	corn oil	1.47 ± .19	2.38 ± .29
	4 mg progesterone	0.50 ± .02 §,§§§	2.94 ± .39
	4 mg 5α-DHP	1.24 ± .30	2.74 ± .50

* Primed with 20 μg estradiol benzoate on day 0.
** Mean ± S.E.M. (N = 8).
§ $P < .005$ vs. corn oil.
§§ $P < .005$ vs. 5α-DHP.
§§§ $P < .025$ vs. 5α-DHP.

days after a priming dose of 20 μg estradiol benzoate, stimulated a significant increase in serum LH and FSH within 6 hr (Table IX). Unlike progesterone (Caligaris et al., 1971; Nuti and Karavolas, unpublished observations), two injections of 5α-DHP 24 hr apart did *not* inhibit serum LH in ovariectomized estrogen-primed rats.

Thus, these progesterone-like effects of 5α-dihydroprogesterone on end-points that are ostensibly functions of the hypothalamus and pituitary strongly support the hypothesis made earlier that conversion of progesterone to 5α-DHP may be important in governing progesterone-sensitive processes in these neuroendocrine feedback tissues.

The relationship of 3α-hydroxy-5α-pregnan-20-one to the action(s) of progesterone is difficult to assess, for unlike 5α-DHP, there was no consistent pattern of selective tissue accumulation and it has no appreciable progesterone-like effects on ovulation (Sridharan et al., 1974), lordosis (Czaja et al., 1974), or uterine progestational end-points (Robel, 1971; Wiest, 1972). These observations, as well as the fact that large amounts of this steroid are formed by the liver (Crane et al., 1970), suggest that the formation of this particular metabolite represents a destruction of potency. However, since the 3α-hydroxysteroid dehydrogenase can catalyze either the oxidation or the reduction of the C-3 oxygen, the possibility exists that 3α-hydroxy-5α-pregnan-20-one may function in some unrecognized role such as in modulating the effects of 5α-DHP.

VII. SUMMARY

The metabolism of progesterone by rat neuroendocrine tissues is principally via a reductive pathway to 5α-dihydroprogesterone (5α-DHP) and 3α-hydroxy-5α-pregnan-20-one. Generally, the formation of 5α-DHP predominates and 5α-reduction appears to occur before 3α-hydroxy-steroid dehydrogenation. The two enzymic activities involved in these transformations (Δ^4-steroid (progesterone) 5α-reductase and 3α-hydroxysteroid dehydrogenase) are not unique to neuroendocrine tissues but are rather widely distributed. The Δ^4-steroid (progesterone) 5α-reductase activities of the hypothalamus and pituitary exhibit several common properties which are also common to Δ^4-steroid 5α-reductases of some other tissues; i.e., a requirement for NADPH, a broad substrate specificity for several Δ^4-3-ketosteroids, and a similar order of reactivity with 20α-dihydroprogesterone, progesterone, and testosterone. However, some differences have emerged with respect to subcellular localization, reactivity with testosterone, and inhibition with estradiol. Limited studies by us and others suggest that these enzymic activities may be subject to endogenous regulation by hormones or other factors since the level of enzymic activity appears to fluctuate with various physiological and hormonal states.

The physiological significance of these metabolic transformations of progesterone by neuroendocrine tissues, and their relationship, if any, to progesterone's mode of action is difficult to assess since the two enzymic activities are seen in many tissues. However, significant differences between tissues emerged when we investigated the tissue and plasma accumulations of 5α-reduced metabolites after an injection of radiolabeled progesterone. As reported above, large amounts of [^{3}H]5α-DHP were accumulated in progestin feedback tissues such as pituitary and hypothalamus. These concentrations were several-fold higher than the levels found in plasma, non-target tissues (muscle and cerebral cortex) or in a non-neural target tissue such as the uterus. Comparable results were obtained when [^{3}H]5α-DHP was similarly injected and studied. The biological relevance of these findings draws support from the afore mentioned progesterone-like effects of 5α-DHP on gonadotropin regulation, ovulation and lordosis, which are ostensibly functions of neuroendocrine tissues such as the hypothalamus and pituitary. The absence of a uterine progestational effect with 5α-DHP is also consistent with the above findings.

The progesterone-like effects of 5α-DHP cannot be attributed to the back conversion of 5α-DHP to progesterone since no back-conversion has been observed in mammalian tissues (Wiest, 1972; Strott, 1974; Karavolas et al., 1976). These results suggest that this metabolite of progesterone may be involved in governing progesterone-sensitive processes in these two neuroendocrine feedback tissues. The role of 5α-DHP may be biologically analogous to the role of 5α-dihydrotestosterone in mediating some of testosterone's effects.

The present findings, together with those of many others on progesterone's mode of action, support a working hypothesis that the multiple and varied effects of progesterone on different end-organs or within the same organ may result from progesterone per se (e.g., as in the uterus) and/or certain of its metabolites (e.g., as with 5α-DHP in gonadotropin regulation).

ACKNOWLEDGEMENTS

Our experimental work was supported by USPHS Grants No. HD-00104, HD-05414, and HD-03352 from the NICHD and a Ford Foundation Grant No. 630-0505A. The senior author (H.J. Karavolas) is the recipient of a Research Career Development Award

from the National Institutes of Child Health and Human Development (No. HD-70,006).
The authors should like to acknowledge the valuable contributions to this project made by the following colleagues: Dr. Y. Cheng, Dr. J.A. Czaja, P.A. Gilles, Dr. D. Goldfoot, Dr. R.W. Goy, I. Hanukoglu, Dr. S. Herf, D. Hodges, J. Krause, Dr. R. Lloyd, C.J. Mapletoft, Dr. R.K. Meyer, Dr.F. Nowak, D. O'Brien, Dr. J. Robinson, D. Smith, and Dr. B. Sridharan.

REFERENCES

Ainsworth, L. and Ryan, K.J. (1969) Steroid hormone transformations by endocrine organs from pregnant mammals. III. Biosynthesis and metabolism of progesterone by mare placenta in vitro. *Endocrinology*, 84: 91–97.

Aiyer, M.S., Fink, G. and Greig, F. (1974) Changes in the sensitivity of the pituitary gland to luteinizing hormone releasing factor during the oestrous cycle of the rat. *J. Endocr.*, 60: 47–64.

Armstrong, D.T. and King, E.R. (1971) Uterine progesterone metabolism and progestational response: effects of estrogens and prolactin. *Endocrinology*, 89: 191–197.

Baulieu, E. (1973) Mode of action of steroid hormones. In *Proc. 4th Int. Congr. Endocrinology, Int. Congr. Ser. 273,* R.D. Scow (Ed.), Excerpta Medica, Amsterdam, pp. 30–62.

Bosley, C.G. and Leavitt, W.W. (1972) Specificity of progesterone action during the preovulatory period in the cyclic hamster. *Fed. Proc.*, 31: 257.

Bryson, M.J. and Sweat, M.L. (1967) Metabolism of progesterone in human proliferative endometrium. *Endocrinology*, 81: 729–734.

Caligaris, L., Astrada, J.J. and Taleisnik, S. (1971) Biphasic effect of progesterone on the release of gonadotropins in rats. *Endocrinology*, 89: 331–337.

Charreau, E.H., Dufour, M.L., Villee, D.B. and Villee, C.A. (1968) Synthesis of 5α-pregnane-3,20-dione by fetal and adult human adrenals. *J. clin. Endocr.*, 28: 629–632.

Chatterton, R.T., Chatterton, A.J. and Hellman, L. (1969) Metabolism of progesterone by the rabbit mammary gland. *Endocrinology*, 85: 16–24.

Cheng, Y.J. and Karavolas, H.J. (1973) Conversion of progesterone to 5α-pregnane-3,20-dione and 3α-hydroxy-5α-pregnan-20-one by rat medial basal hypothalami and the effects of estradiol and stage of estrous cycle on the conversion. *Endocrinology*, 93: 1157–1162.

Cheng, Y.J. and Karavolas, H.J. (1975a) Properties and subcellular distribution of Δ^4-steroid (progesterone) 5α-reductase in rat anterior pituitary. *Steroids*, 26: 57–72.

Cheng, Y.J. and Karavolas, H.J. (1975b) Subcellular distribution and properties of progesterone (Δ^4-steroid) 5α-reductase in rat medial basal hypothalamus. *J. biol. Chem.*, 250: 7997–8003.

Clemens, J.A., Sharr, C.J., Kleber, J.W. and Tandy, W.A. (1971) Areas of the brain stimulatory to LH and FSH secretion. *Endocrinology*, 88: 180–184.

Cooper, K.J., Fawcett, C.P. and McCann, S.M. (1974) Inhibitory and facilitatory effects of estradiol-17β on pituitary responsiveness to a luteinizing hormone-follicle stimulating hormone releasing factor (LH-RF/FSH-RF) preparation in the ovariectomized rat. *Proc. Soc. exp. Biol. (N.Y.)*, 145: 1422–1426.

Crane, M., Loring, J. and Villee, C.A. (1970) Progesterone metabolism in regenerating rat liver. *Endocrinology*, 87: 80–83.

Czaja, J.A., Goldfoot, D.A. and Karavolas, H.J. (1974) Comparative facilitation and inhibition of lordosis in the guinea pig with progesterone, 5α-pregnane-3,20-dione, or 3α-hydroxy-5α-pregnan-20-one. *Hormone Behav.*, 5: 261–274.

Faber, L.E., Sandmann, M.L. and Stavely, H.E. (1972) Progesterone binding in uterine cytosols of the guinea pig. *J. biol. Chem.*, 247: 8000–8004.

Frederiksen, D.W. and Wilson, J.D. (1971) Partial characterization of the nuclear reduced nicotinamide adenine dinucleotide phosphate: Δ^4-3-ketosteroid 5α-oxidoreductase of rat prostate. *J. biol. Chem.*, 246: 2584–2593.

Freifeld, M.L., Feil, P.D. and Bardin, C.W. (1974) The in vivo regulation of the progesterone "receptor" in guinea pig uterus: dependence on estrogen and progesterone. Steroids, 23: 93—103.

Frost, P., Gomez, E.C., Weinstein, G.D., Lamas, J. and Hsia, S.L. (1969) Metabolism of progesterone-4-^{14}C in vitro in human skin and vaginal mucosa. Biochemistry, 8: 948—952.

Goldman, B.D. and Zarrow, M.X. (1973) The physiology of progestins. In Handbook of Physiology-Endocrinology, Vol. 2, Part 1, R.O. Greep (Ed.), American Physiological Society, Washington, D.C., pp. 547—572.

Goldman, B.D., Kamberi, I.A., Siiteri, P.K. and Porter, J.C. (1969) Temporal relationship of progestin secretion, LH release and ovulation in rats. Endocrinology, 85: 1137—1143.

Goy, R.W. and Goldfoot, D.A. (1973) Hormonal influences on sexually dimorphic behavior. In Handbook of Physiology-Endocrinology, Vol. 2, Part 1, R.O. Greep (Ed.), American Physiological Society, Washington, D.C., pp. 169—186.

Hashimoto, I., Henricks, D.M., Anderson, L.L. and Melampy, R.M. (1968) Progesterone and pregn-4-en-20α-ol-3-one in ovarian venous blood during various reproductive states in the rat. Endocrinology, 82: 333—341.

Inano, H. and Tamaoki, B. (1966) Bioconversion of steroids in immature rat testes in vitro. Endocrinology, 79: 579—590.

Inano, H., Machino, A. and Tamaoki, B. (1969) In vitro metabolism of steroid hormones by cell-free homogenates of epididymides of adult rats. Endocrinology, 84: 997—1003.

Kaasjager, W.A., Woodbury, D.M., Van Dieten, J.A.M.J. and Van Rees, G.P. (1971) The role played by the preoptic region and the hypothalamus in spontaneous ovulation and ovulation induced by progesterone. Neuroendocrinology, 7: 54—64.

Karavolas, H.J. and Herf, S.M. (1971) Conversion of progesterone by rat medial basal hypothalamic tissue to 5α-pregnane-3,20-dione. Endocrinology, 89: 940—942.

Karavolas, H.J., Hodges, D. and O'Brien, D. (1975) Uptake of ^{3}H-progesterone and ^{3}H-5α-dihydroprogesterone by rat tissues in vivo and analysis of accumulated radioactivity: selective accumulation of 5α-dihydroprogesterone by hypothalamic and pituitary tissues. Endocrine Soc. Abstr., June: No. 26.

Karavolas, H.J., Hodges, D. and O'Brien, D. (1976) Uptake of ^{3}H-progesterone and ^{3}H-5α-dihydroprogesterone by rat tissues in vivo and analysis of accumulated radioactivity: accumulation of 5α-dihydroprogesterone by pituitary and hypothalamic tissues. Endocrinology, in press.

Kawahara, F.S., Berman, M.L. and Green, O.C. (1975) Conversion of progesterone-1,2-^{3}H to 5β-pregnane-3,20-dione by brain tissue. Steroids, 25: 459—463.

Lloyd, R.V. and Karavolas, H.J. (1975) Uptake and conversion of progesterone and testosterone to 5α-reduced products by enriched gonadotropic and chromophobic rat anterior pituitary cell fractions. Endocrinology, 97: 517—526.

Lloyd, R.V. and McShan, W.H. (1973) Study of the anterior pituitary cells separated by velocity sedimentation at unit gravity. Endocrinology, 92: 1639—1651.

Martin, J.E., Tyrey, L., Everett, J.W. and Fellows, R.E. (1974) Estrogen and progesterone modulation of the pituitary response to LRF in the cyclic rat. Endocrinology, 95: 1664—1673.

Massa, R., Stupnicka, E., Kniewald, Z. and Martini, L. (1972a) The transformation of testosterone into dihydrotestosterone by the brain and the anterior pituitary. J. Steroid Biochem., 3: 385—399.

Massa, R., Stupnicka, E. and Martini, L. (1972b) Metabolism of progesterone in the anterior pituitary, the hypothalamus and the uterus of female rats. In 4th Int. Congr. of Endocrinology, Washington, D.C., p. 118 (abstract)

Mickan, H. (1972) Metabolism of 4-^{14}C-progesterone and 4-^{14}C-testosterone in brain of the previable human fetus. Steroids, 19: 659—665.

Milgrom, E., Atger, M. and Baulieu, E.-E. (1970) Progesterone in uterus and plasma. IV. Progesterone receptor(s) in guinea pig uterus cytosol. Steroids, 16: 741—754.

Morgan, M.D. and Wilson, J.D. (1970) Intranuclear metabolism of progesterone-1,2-^{3}H in the hen oviduct. J. biol. Chem., 245: 3781—3789.

Nowak, F.V. and Karavolas, H.J. (1974a) Conversion of 20α-hydroxypregn-4-en-3-one to 20α-hydroxy-5α-pregnen-3-one and 5α-pregnane-3α, 20α-diol by rat medial basal hypothalamus. *Endocrinology*, 94: 994–997.

Nowak, F.V. and Karavolas, H.J. (1974b) Conversion of 20α-hydroxy-4-pregnan-3-one to 20α-hydroxy-5α-pregnen-3-one and 5α-pregnane-3α, 20α-diol by rat anterior pituitary. *Steroids*, 24: 351–357.

Nowak, F.V., Nuti, K.M. and Karavolas, H.J. (1975) Quantitative changes in the conversion of 20α-hydroxy-4-pregnen-3-one by rat medial basal hypothalamus and anterior pituitary during the day of proestrus. (Ph. D. Thesis, University of Wisconsin, Madison, Wis.) In preparation.

Piacsek, B.E., Schneider, T.C. and Gay, V.L. (1971) Sequential study of luteinizing hormone (LH) and "progestin" secretion on the afternoon of proestrus in the rat. *Endocrinology*, 89: 39–45.

Piva, F., Kalra, P.S. and Martini, L. (1973) Participation of the amygdala and of the cerebellum in the feedback effects of progesterone. *Neuroendocrinology*, 11: 229–239.

Reel, J.R., Van Dewark, S.D., Shih, Y. and Callantine, M.R. (1971) Macromolecular binding and metabolism of progesterone in the decidual and pseudopregnant rat and rabbit uterus. *Steroids*, 18: 441–461.

Robel, P. (1971) Steroid hormone metabolism in responsive tissues in vitro. In *In Vitro Methods in Reproductive Cell Biology, Karolinska Symposia on Research Methods in Reproductive Endocrinology*, E. Diczfalusy (Ed.), Bogtrykkeriet Forum, Copenhagen, pp. 279–292.

Robinson, J.A. and Karavolas, H.J. (1973) Conversion of progesterone by rat anterior pituitary tissue to 5α-pregnane-3,20-dione and 3α-hydroxy-5α-pregnan-20-one. *Endocrinology*, 93: 430–435.

Rommerts, F.F.G. and Van der Molen, H.J. (1971) Occurrence and localization of 5α-steroid reductase, 3α- and 17β-hydroxysteroid dehydrogenases in hypothalamus and other brain tissues of the male rat. *Biochim. biophys. Acta (Amst.)*, 248: 489–502.

Sanyal, M.K. and Todd, R.B. (1972) 5α-Dihydroprogesterone influence on ovulation of prepuberal rats. *Proc. Soc. exp. Biol. (N.Y.)*, 141: 622–624.

Schally, A.V., Redding, T.W. and Arimura, A. (1973) Effect of sex steroids on pituitary responses to LH- and FSH-releasing hormone in vitro. *Endocrinology*, 93: 893–902.

Schrader, W.T., Toft, D.O. and O'Malley, B.W. (1972) Progesterone-binding protein of chick oviduct. IV. Interaction of purified progesterone-receptor components with nuclear constituents. *J. biol. Chem.*, 247: 2401–2407.

Seiki, K. and Hattori, M. (1973) In vivo uptake of progesterone by the hypothalamus and pituitary of the female ovariectomized rat and its relationship to cytoplasmic progesterone-binding protein. *Endocr. jap.*, 20: 111–119.

Snipes, C.A. and Shore, L.S. (1972) Metabolism of progesterone in vitro by neural and uterine tissues. *Fed. Proc.*, 31: 236.

Sridharan, B.N., Meyer, R.K. and Karavolas, H.J. (1974) Effect of 5α-dihydroprogesterone, pregn-5-ene-3,20-dione, pregnenolone and related progestins on ovulation in PMSG-treated immature rats. *J. Reprod. Fertil.*, 36: 83–90.

Strott, C.A. (1974) Metabolism of progesterone in the chick oviduct: relation to the progesterone receptor and biological activity. *Endocrinology*, 95: 826–837.

Stucki, J.C. and Glenn, E.M. (1961) Endometrial proliferation, pregnancy maintenance, parturition inhibition and myometrial block production with various steroids. In *Progesterone, Brook Lodge Symposium, Augusta, Mich.,* A.C. Barnes (Ed.), Brook Lodge Press, Augusta, Mich., pp. 25–36.

Tabei, T., Haga, H., Heinrichs, W.L. and Herrmann, W. (1974) Metabolism of progesterone by rat brain, pituitary gland and other tissues. *Steroids*, 23; 651–666.

Velasco, M.E. and Rothchild, I. (1973) Factors influencing the secretion of luteinizing hormone and ovulation in response to electrochemical stimulation of the preoptic area in rats. *J. Endocr.*, 58: 163–176.

Wade, G.N. and Feder, H.H. (1972) [1,2-^{3}H]Progesterone uptake by guinea pig brain and uterus: differential localization, time-course of uptake and metabolism, and effects of age, sex, estrogen-priming and competing steroids. *Brain Res.*, 45: 525–543.

Wade, G.N., Harding, C.F. and Feder, H.H. (1973) Neural uptake of [1,2-^{3}H]progesterone in ovariectomized rats, guinea pigs and hamsters: correlation with species differences in behavioral responsiveness. *Brain Res.*, 61: 357–367.

Whalen, R.E. and Gorzalka, B.B. (1974) Estrogen-progesterone interactions in uterus and brain of intact and adrenalectomized immature and adult rats. *Endocrinology*, 94: 214–223.

White, W.F., Hedlund, M.T., Weber, G.F., Rippel, R.H., Johnson, E.S. and Wilbur, J.F. (1974) The pineal gland: a supplemental source of hypothalamic-releasing hormones. *Endocrinology*, 94: 1422–1426.

Wiest, W.G. (1963) In vitro metabolism of progesterone and 20α-hydroxypregn-4-en-3-one by tissues of the female rat. *Endocrinology*, 73: 310–316.

Wiest, W.G. (1972) The distribution and metabolism of progesterone in the uterus. In *The Sex Steroids*, K.W. McKerns (Ed.), Appleton-Century-Crofts, New York, pp. 295–313.

Wilson, J.D. and Gloyna, R.E. (1970) The intranuclear metabolism of testosterone in the accessory organs of reproduction. *Recent Progr. Hormone Res.*, 26: 309–336.

Chapter 16

Androgen reduction by neuroendocrine tissues: physiological significance

LUCIANO MARTINI

Department of Endocrinology, Institute of Endocrinology, University of Milan, 20129 Milan (Italy)

I. TRANSFORMATION OF TESTOSTERONE INTO ITS 5α-REDUCED METABOLITES IN THE ANTERIOR PITUITARY AND IN THE CENTRAL NERVOUS SYSTEM

There is a lot of evidence to suggest that, in the majority of the peripheral androgen-sensitive structures (prostate, seminal vesicles, etc.), testosterone (T) is transformed into 5α-androstan-17β-ol-3-one (androstanolone, dihydrotestosterone, DHT), 5α-androstan-3β, 17β-diol (3β-diol) and 5α-androstan-3α, 17β-diol (3α-diol) (Baulieu et al., 1968; Bruchowsky and Wilson, 1968; Wilson and Gloyna, 1970; Robel, 1971; Massa and Martini, 1974). The intracellular formation of these metabolites seems to be an important step for the appearance of the biological activities of T.

On the basis of these findings, the hypothesis that T might be transformed into DHT and the corresponding diols in the structures, on which the hormone exerts its feedback effect on gonadotropin secretion and influences sexual behavior, has been submitted for experimental verification. Kniewald et al. (1971) and Massa et al. (1972a) have incubated ^{14}C-labeled T "in vitro" in the presence of fragments of the pituitary gland, of the hypothalamus, of the amygdala, of the cerebral cortex and of the prostate taken from adult male rats. After 3 hr of incubation, the metabolites formed have been extracted from the incubation medium, purified and identified. In order to have, for each tissue examined, a quantitative and separate evaluation of the activity of the enzymes involved in the transformation of T into its metabolites (5α-reductase, 3-hydroxysteroid-dehydrogenases), two parameters have been selected: (1) the total amounts of 5α-reduced metabolites formed, i.e., the sum of the amounts of DHT plus the amounts of the diols; this figure provides a clear indication of the 5α-reductase activity present in the different tissues; and (2) the per cent quantities of the diols contributing to the total amounts of 5α-reduced metabolites; this index provides a satisfactory estimation of the efficiency of the 3-hydroxysteroid-dehydrogenases which convert DHT into the diols. It must be noted that, at the prostatic level, both the 3α- and the 3β-diols are formed. On the contrary, the anterior pituitary and all central nervous structures studied apparently possess only a 3α-hydroxysteroid-dehydrogenase and consequently form only the 3α-diol (Kniewald et al., 1971; Massa et al., 1972a).

The results obtained using tissues from normal untreated rats are summarized in Table I. It is clear that T is converted into 5α-reduced metabolites by all structures

TABLE I

Conversion of testosterone into its 5α-reduced metabolites (dihydrotestosterone, 5α-androstan-3α, 17β-diol) by different tissues*

Tissue**		5α-reduced metabolites (pg/mg)***	% of diols
Prostate	(5)	5535.10 ± 1083.60	7.64 ± 1.06
Anterior pituitary	(18)	1211.00 ± 96.50 §	25.50 ± 1.84
Hypothalamus	(14)	582.90 ± 107.50 §, §§	42.40 ± 2.64
Amygdala	(5)	341.70 ± 93.10	42.40 ± 5.70
Cerebral cortex	(5)	281.70 ± 11.50	39.20 ± 2.30

* Values are means ± S.E.
** Number of experiments performed in parentheses.
*** Picograms of 5α-reduced metabolites formed per mg of wet tissue following a 3-hr incubation with 160 ng of [^{14}C]testosterone (specific activity: 56.6 mCi/mmole).
§ $p < 0.001$ vs. prostate, amygdala and cerebral cortex.
§§ $p < 0.005$ vs. anterior pituitary.

examined. The rate of conversion, however, varies considerably from tissue to tissue. In confirmation of previous findings (Baulieu et al., 1968; Bruchowsky and Wilson, 1968; Wilson and Gloyna, 1970; Robel, 1971), the formation of 5α-reduced metabolites is very elevated in the prostate. Low amounts of 5α-reduced metabolites are formed by the cerebral cortex and by the amygdala; these cerebral zones have been included in this study as "control" tissues, since they are supposed not to be androgen-sensitive. T is converted into 5α-reduced metabolites also by the anterior pituitary: the amounts of metabolites formed by the gland are significantly higher than those formed by the two "control" tissues; however, they do not reach the very elevated levels made by the prostate. The hypothalamus is also able to reduce T in the 5α-position to an extent significantly higher than that of the "control" tissues; however, the 5α-reducing capacity of the hypothalamus appears to be significantly lower than that of the anterior pituitary. The hydroxysteroid-dehydrogenase activity of the prostate appears to be rather low, the diols representing only 7% of the total amounts of 5α-reduced metabolites formed by this organ. The hydroxysteroid-dehydrogenase activity of the anterior pituitary and of the structures of the central nervous system (CNS) is significantly higher than that of the prostate. It has already been mentioned that the end product, in the case of the anterior pituitary and of the central structures, is almost exclusively the 3α-diol. The question may be asked whether the high 3α-hydroxysteroid-dehydrogenase activity of these tissues reflects the presence of an "inactivation" process through which the effectiveness of DHT is diminished, or whether it represents a mechanism through which compounds more active on the CNS-pituitary complex are obtained. An answer to this question will be provided in the next sections of this paper.

These results agree with the findings of several other authors, even if they have used tissues from other species of animals and different incubation procedures (Jaffe, 1969; Perez-Palacios et al., 1970; Rommerts and Van der Molen, 1971; Kniewald and Milkovic,

1973; Thieulant et al., 1973; Denef et al., 1974; Weisz and Gibbs, 1974). The conversion of T into its 5α-reduced metabolites in the CNS and in the anterior pituitary has been shown to occur also "in vivo" (Stern and Eisenfeld, 1971; Sholiton et al., 1974; Sholl et al., 1975).

After having shown that the anterior pituitary and the hypothalamus are able to convert T into DHT and 3α-diol, it was deemed of interest to investigate whether the conversion process might be modified by experimental manipulations which activate or inhibit the hypothalamic-pituitary axis (Kniewald et al., 1971; Massa et al., 1972a). First of all, the conversion occurring in the anterior pituitary and in the hypothalamus of orchidectomized adult male rats was studied. Animals were killed 2, 7, 14, 21 and 90 days following castration. Gonadectomy considerably activated the transformation of T into 5α-reduced metabolites at the pituitary level. The activation was evident 2 days after the operation, and reached its maximum 2–3 weeks after gonadectomy. The rate of conversion of T was still significantly increased in the pituitaries of animals which had been castrated 90 days before. The conversion process seems to be activated by castration also at the hypothalamic level; however, the increase in the activity of the 5α-reductase in this tissue was not as great as that found at the pituitary level. Castration did not modify the conversion capacity of the cerebral cortex and of the amygdala.

The next step was that of studying whether the "in vivo" administration of T to castrated animals might bring back to normal the ability of the anterior pituitary and of the hypothalamus to convert the hormone into its 5α-reduced metabolites (Kniewald et al., 1971; Massa et al., 1972a). Testosterone propionate (TP) was injected subcutaneously, in a daily dose of 2 mg/rat; treatment was initiated immediately after castration. Tissues to be examined were taken 2, 7, 14 and 21 days after the operation. The administration of TP was followed by a significant decrease of the 5α-reductase activity of the anterior pituitary of castrated animals at all intervals considered. The depressing effect of the substitution therapy with TP on the formation of the 5α-reduced metabolites appeared also at the hypothalamic level. The effects, however, were less dramatic than those in the anterior pituitary. Treatment of castrated animals with TP did not influence the conversion of labeled T into its "active" metabolites at the level of the cerebral cortex and of the amygdala. Similar results have been obtained by Kniewald and Milkovic (1973), by Denef et al. (1973, 1974) and by Thieulant et al. (1974).

It is not clear at the moment why castration and androgen administration should exert opposite effects on the 5α-reductase activity of the anterior pituitary. The most likely explanation is that the post-castration increase of the 5α-reductase activity reflects the changes in the composition of pituitary cell populations which follows the operation. It is known that gonadotrophs increase in size and proliferate after gonadectomy. Should this hypothesis be correct, the data might be taken as providing evidence for a specific localization of the 5α-reductase in the gonadotrophs. The possibility that anterior pituitary 5α-reductase might be exclusively located in the gonadotrophs is presently under investigation in the Milan laboratory. This hypothesis is supported by the observation that the 5α-reductase activity of the anterior pituitary of castrated animals reverts to normal after the administration of T, a steroid known to be able to restore normal pituitary histology.

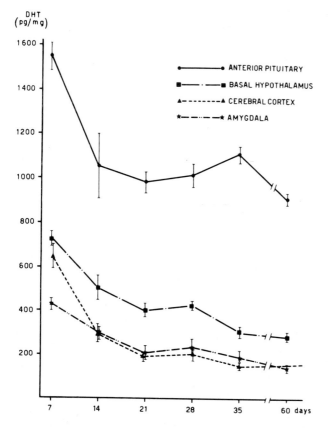

Fig. 1. Conversion of testosterone into dihydrotestosterone (DHT) by the basal hypothalamus, the amygdala, the cerebral cortex, and the anterior pituitary of maturing male rats. The data are expressed in terms of picograms of dihydrotestosterone formed per mg of wet tissue, following a 3-hr incubation with 160 ng of [^{14}C]testosterone (specific activity, 56.6 mCi/mmole).

The data discussed so far suggest that, in male animals, the transformation of T into DHT and 3α-diol is probably a necessary step for the appearance of androgen-induced feedback and behavioral responses. The results also indicate that T probably exerts its feedback effects both on the hypothalamus and on the anterior pituitary. From the data one would probably be inclined to assign a prominent role in this process to the anterior pituitary. The observation that the anterior pituitary and the hypothalamus form considerable amounts of 3α-diol may indicate that this steroid plays a major role in feedback processes.

It has been repeatedly shown that the centers regulating gonadotropin secretion are markedly more sensitive to the negative feedback influences of androgens in prepuberal than in adult animals (Ramirez and McCann, 1965; Critchlow and Bar-Sela, 1967; Davidson and Smith, 1967, Negro-Vilar et al., 1973). The hypersensitivity to androgens of the immature animal might be due to the fact that, before puberty, the central structures convert testosterone into DHT more efficiently than after adulthood has been reached. This hypothesis was verified by incubating "in vitro" the hypothalamus, the amygdala, the cerebral cortex, the anterior pituitary and the prostate of immature male rats in the presence of labeled testosterone; in this experiment animals were killed at 7, 14, 21, 28, 35 and 60 days of age (Massa et al., 1971, 1972a). It appears from Fig. 1 that the ability to convert testosterone into DHT of all brain structures considered and of the pituitary is inversely related to the age of the animal. On the contrary, the activity of the 5α-reductase of the prostate is not linked to the age of the animal. These data certainly support the idea that the hypersensitivity to androgen of the prepuberal male rat is due to its ability to "utilize" T better than the adult. They also provide a biochemical basis for explaining the change in the sensitivity of the cerebral "gonadostat" which occurs at the time of sexual maturation, and which seems to be crucial for initiating puberty in male animals (Ramirez and McCann, 1965; Negro-Vilar et al., 1973).

Recent studies have indicated that the 5α-reductase and the 3α-hydroxysteroid-dehydrogenase are present also in several structures of the female endocrine system. Using labeled progesterone (P) as the substrate, it has been shown that, in females, these enzymes exist in the uterus (Armstrong and King, 1971), in the hypothalamus and in the anterior pituitary (Karavolas and Herf, 1971; Massa et al., 1972b; Cheng and Karavolas, 1973; Robinson and Karavolas, 1973; Tabei and Henricks, 1974; Tabei et al., 1974). It has been postulated that these enzymes might convert P into more effective metabolites.

A series of "in vitro" experiments have consequently been designed: (a) to analyze whether the 5α-reductase present in the central structures (anterior pituitary, hypothala-

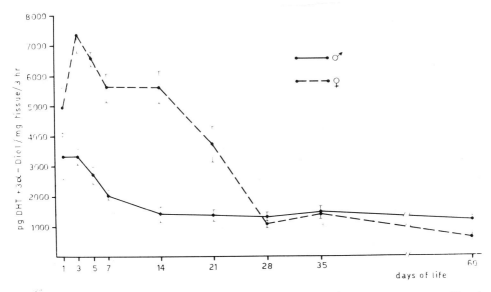

Fig. 2. Conversion of testosterone into its 5α-reduced metabolites (dihydrotestosterone, DHT and 5α-androstan-3α, 17β-diol, 3α-diol) by the anterior pituitary of maturing male and female rats. The data are expressed in terms of picograms of steroids formed (DHT+3α-diol) per mg of wet tissue, following a 3-hr incubation with 160 ng of [^{14}C]testosterone (specific activity, 59.4 mCi/mmole).

mus, etc.) of female animals is able to convert T as well as P; (b) to study whether the activity of this enzyme shows modifications with increasing age similar to those previously found in males (Massa et al., 1971, 1972a); and (c) to clarify some of the factors which control these modifications (Massa et al., 1975).

Special attention has been paid to a possible role played by the presence of androgens during the neonatal period, since it is known that, at this time, androgens exert crucial effects on the sexual differentiation of the hypothalamic-pituitary axis of the rat (Barraclough, 1967; Gorski, 1971).

Fig. 2 summarizes the data on the 5α-reductase activity of the anterior pituitary of female and male animals studied from day 1 up to day 60 of life. The data are expressed in terms of "total" 5α-reduced metabolites (DHT+3α-diol) formed per mg of fresh tissue in the 3 hr of the incubation period. It is evident, first of all, that the 5α-reducing activity of the female anterior pituitary is higher than that of the male up to 21 days of age. It also appeared that the pattern exhibited by the 5α-reducing activity of the anterior pituitary at different ages is different in the two sexes.

In males, the ability of the anterior pituitary to transform T into DHT and 3α-diol progressively decreases from birth to day 60. On the contrary, the 5α-reductase activity of the anterior pituitary of normal females shows a rapid and significant increase ($p < 0.05$) between days 1 and 3 of life. Subsequently, a drop of such an activity is observed between days 3 and 7. This decrease is followed by a plateau up to day 14, and by a further progressive decline from day 14 to day 28. This final decline brings the 5α-reductase activity of the female anterior pituitary to the levels found in the anterior pituitary of male animals of comparable age. These data are similar, in several respects, to those of Denef et al. (1974).

Fig. 3 shows the data obtained studying the 5α-reducing activity of the hypothalamus of female and male rats at various intervals after birth. It must be noted first of all, that at any age that is considered, the 5α-reducing ability of the hypothalamus of male and female animals is, in quantitative terms, much lower than that of the anterior pituitary. This is confirmatory of the data previously obtained in adult animals. Contrary to what happens in the anterior pituitary, there is an increase of the 5α-reductase activity of the hypothalamus between days 1 and 3 of life in both males and females. Beginning from day 3, the ability to transform T into DHT and 3α-diol progressively declines in male animals. On the contrary, in females, a further increase is observed up to day 7. This increase is followed by a decline between days 7 and 14, by a further elevation around day 21, and by a subsequent drop. This drop brings the 5α-reductase level of the female hypothalamus to become similar to that of the male, beginning from day 28.

The observations here reported agree, in general, with the data of Denef et al. (1974). However, their samples were collected beginning from day 5 of life; this has prevented them from registering the increase of 5α-reductase activity which occurs in the first few days after birth. The data obtained using the cerebral cortex of female and male animals are shown in Fig. 4. In both sexes, the 5α-reductase activity of the cerebral cortex is higher in the first period of life, with a parallel increment between days 1 and 7. Subsequently, there is a progressive decline of such an activity. There are no differences between males and females at any age considered.

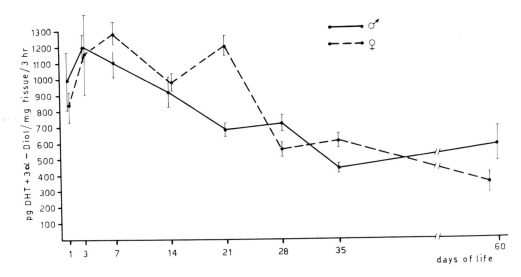

Fig. 3. Conversion of testosterone into its 5α-reduced metabolites (dihydrotestosterone, DHT and 5α-androstan-3α, 17β-diol, 3α-diol) by the basal hypothalamus of maturing male and female rats. The data are expressed in terms of picograms of steroids formed (DHT+3α-diol) per mg of wet tissue, following a 3-hr incubation with 160 ng of [^{14}C]testosterone (specific activity, 59.4 mCi/mmole).

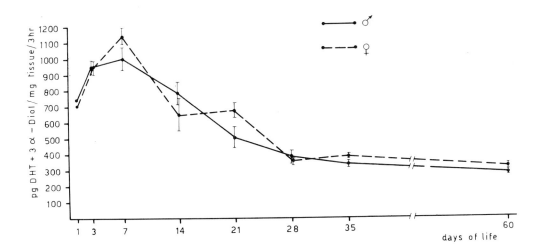

Fig. 4. Conversion of testosterone into its 5α-reduced metabolites (dihydrotestosterone, DHT and 5α-androstan-3α, 17β-diol, 3α-diol) by the cerebral cortex of maturing male and female rats. The data are expressed in terms of picograms of steroids formed (DHT+3α-diol) per mg of wet tissue, following a 3-hr incubation with 160 ng of [^{14}C]testosterone (specific activity, 59.4 mCi/mmole).

The reason and the biological significance of the sex-linked differences observed in the developmental pattern of the 5α-reductase activity in the anterior pituitary and in the hypothalamus of male and female animals are not understood at the present moment. The working hypothesis has been formulated that the sexual dimorphism here reported might be linked to the neonatal exposure to androgens, which occurs in male animals and which does not occur in females. It is known that the administration of exogenous T to the neonatal female rat permanently transforms the activity of its hypothalamic-pituitary axis to a pattern which resembles that normally found in the male (Barraclough, 1967; Gorski, 1971). It has also been shown that castration performed in neonatal male rat brings about a permanent transformation of its hypothalamic-pituitary axis, which becomes more similar to that of the female (Barraclough, 1967; Gorski, 1971).

In order to study the possible role played by androgenic steroids circulating in the neonatal period in inducing the sexual dimorphism here reported, it has been decided to study the 5α-reductase activity of the anterior pituitary, of the hypothalamus and of the cerebral cortex in the following 6 groups of animals: (a) normal males; (b) normal females; (c) females androgenized on the third day of life with a low dose of TP (100 μg/rat); (d) females androgenized on the third day of life with a high dose of TP (1 mg/rat); (e) females androgenized on the ninth day of life with a high dose of TP (1 mg/rat); and (f) males castrated on the first day of life. The 5α-reductase activity of the three structures considered has been evaluated at the age of 14 days. This particular period was selected for this preliminary study mainly because, at this age, there are clear-cut differences in the 5α-reductase activity of the anterior pituitaries of the two sexes.

Fig. 5 shows once more that, at the age of 14 days, the anterior pituitary of normal female rats is able to convert T into DHT and 3α-diol at an extent much greater than the anterior pituitary of normal male animals. It is also apparent that the administration of 1 mg of TP on day 3 of life brings the 5α-reductase activity of the anterior pituitary of female animals to levels as low as those found in the anterior pituitary of normal males of comparable age. The lower dose of TP (100 μg) administered on day 3 of life is also effective in decreasing the 5α-reductase activity of the anterior pituitary of female animals. However, the decrease is lower than that observed following administration of the larger dose of TP. This dose-related effect is in accordance with previous evidence indicating that, when TP is given on day 3 of life, higher doses of the steroid provoke a stronger masculinizing effect on the hypothalamic-pituitary axis than lower doses (Barraclough, 1967; Gorski, 1971). The data shown in Fig. 5 indicate that female animals treated with 1 mg of TP on day 9 of life still exhibit a decrease of anterior pituitary 5α-reductase activity. However, the decrease is not as big as that obtained with the same dose of TP given on day 3. This is perfectly in line with the observations of Barraclough (1967) and Gorski (1971) which indicate that the same dose of TP is less efficient in masculinizing the hypothalamic-pituitary complex if given on day 10 than on previous days of life. Finally, Fig. 5 indicates that castration of male animals, performed on day 1 of life, brings about a significant increase in the 5α-reductase of the anterior pituitary measured at day 14. The 5α-reductase activity of the anterior pituitary of male animals castrated on day 1 of life becomes similar to that of the anterior pituitary of normal females of the same age.

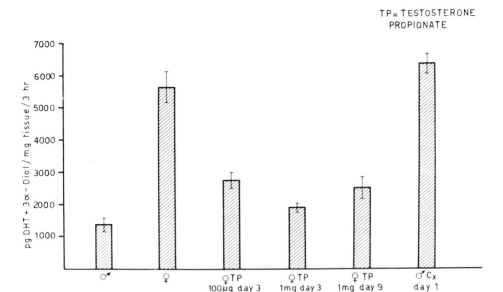

Fig. 5. Conversion of testosterone into its 5α-reduced metabolites (dihydrotestosterone, DHT and 5α-androstan-3α, 17β-diol, 3α-diol) by the anterior pituitary of male and female rats of 14 days of age. The data are expressed in terms of picograms of steroids formed (DHT+3α-diol) per mg of wet tissue, following a 3-hr incubation with 160 ng of [^{14}C]testosterone (specific activity, 59.4 mCi/mmole).
Females, TP, 100 μg, day 3, vs. normal females, $p < 0.0005$.
Females, TP, 1 mg, day 3, vs. normal females, $p < 0.0005$.
Females, TP, 1 mg, day 9, vs. normal females, $p < 0.0025$.
Castrated males, day 1, vs. normal males, $p < 0.0005$.

The data obtained evaluating the 5α-reductase activity of the hypothalami of the animals submitted to the different types of neonatal treatments are shown in Fig. 6. As previously indicated, at this particular age there are not significant sex-linked differences in the 5α-reductase activity of this structure. However, neonatal androgenization performed with 1 mg of TP (given on day 3 or day 9 of life) brings about a significant decrease in the 5α-converting ability of the hypothalamus of female animals. Castration performed in male animals on day 1 of life does not seem to significantly modify the 5α-reductase measured at 14 days. The different treatments had no influence on the 5α-reductase activity of the cerebral cortex.

These results seem to indicate that neonatal exposure to androgens may be one of the factors which create and entertain the sexual dimorphism on the 5α-reductase activity of the anterior pituitary and of the hypothalamus. Unfortunately, it has not been possible so far to perform a longitudinal study (at 21, 28 days, etc.) to confirm whether the effect of neonatal administration of androgens is permanent or transitory. Such a study (which is presently in progress) is made desirable by the observation, already mentioned in this chapter, that, in adult animals, castration enhances and the administration of T diminishes the levels of the 5α-reductase in the anterior pituitary (Kniewald et al., 1971; Massa

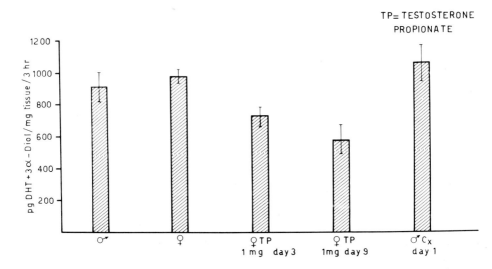

Fig. 6. Conversion of testosterone into its 5α-reduced metabolites (dihydrotestosterone, DHT and 5α-androstan-3α, 17β-diol, 3α-diol) by the basal hypothalamus of male and female rats of 14 days of age. The data are expressed in terms of picograms of steroids formed (DHT+3α-diol) per mg of wet tissue following a 3-hr incubation with 160 ng of [^{14}C]testosterone (specific activity, 59.4 mCi/mmole).
Females, TP, 1 mg, day 3, vs. normal females, $p < 0.005$.
Females, TP, 1 mg, day 9, vs. normal females, $p < 0.0025$.

et al., 1972a; Kniewald and Milkovic, 1973; Denef et al., 1974). There are, however, a few reasons for believing that the effect here reported is different from the one which is observed in adults. First of all, when castration is performed in mature animals, the increase of the 5α-reductase of the anterior pituitary levels observed 14 days following the operation is much smaller than that found following neonatal castration (this fact has been confirmed also by Denef et al., 1974). Secondly, it is expected that T administered on day 3 of life might still be circulating and exerting biological effects on day 14.

II. EFFECTS OF SYSTEMIC INJECTIONS OF TESTOSTERONE AND OF ITS 5α-REDUCED METABOLITES ON LH AND FSH RELEASE

The results presented in the preceding section of this paper prompted an investigation on the effects the 5α-reduced metabolites of T might exert on the release of pituitary gonadotropins. Serum levels of LH and of FSH have been measured in castrated adult male rats using specific radioimmunoassays (Niswender et al., 1968; Daane and Parlow, 1971) given one single subcutaneous injection of T, DHT and 3α-diol 24 hr before sacrifice. Two doses of each steroid (free alcohol form) were used: 2 mg/rat and 4 mg/rat (Zanisi et al., 1973a and b).

Table II indicates that, under the conditions of the present experiment, T, in a dose of

2 mg, was unable to modify serum levels of LH. On the contrary, DHT, when given in the same dose, induced a significant reduction of this gonadotropin. This observation is in line with the findings of Swerdloff et al. (1972), of Naftolin and Feder (1973), and of Verjans et al. (1974). A much bigger decrease of serum LH was obtained after treatment with the 2 mg dose of 3α-diol. According with these results, the respective activities of the three androgens in suppressing LH secretion appears to be: 3α-diol>DHT>T. Little help in quantitating the effects of the three steroids on LH secretion is provided by the data obtained with the 4 mg dose, since at this dose T, DHT and 3α-diol are all maximally effective.

At the dose of 2 mg, none of the three androgens was able to induce significant variations of serum levels of FSH (Table II). When the higher dose of 4 mg was used, T, DHT and 3α-diol induced a moderate but significant reduction of this gonadotropin. T appeared to be more effective than DHT and DHT more effective than 3α-diol. It is interesting to underline that, at variance with what had been observed in the case of LH, none of the steroids was able to totally suppress FSH secretion. It is also interesting to note that the two steroids which were more effective in inhibiting FSH secretion were those which exhibited the poorest activity on LH release. The observation that DHT and 3α-diol are more inhibitory than T itself on LH release may indicate that T influences the release of this gonadotropin following conversion into 5α-reduced metabolites. The very high potency of 3α-diol as a suppressor of LH release suggests that this steroid might be the physiological intracellular mediator of the feedback activity of T on LH secretion. This hypothesis is certainly supported by the results summarized in Section I of this paper. The observation that, in the case of FSH secretion, T is more effective than its androstane metabolites probably indicates that the process of 5α-reduction is less im-

TABLE II

Effect of acute treatment with testosterone and with its 5α-reduced metabolites (dihydrotestosterone, DHT, and 5α-androstan-3α,17β-diol, 3α-diol) on serum LH and FSH levels of adult castrated male rats*

Groups**			LH (ng/ml of serum) (NIH-LH S-16)	FSH (ng/ml of serum) (NIAMD-Rat FSH-RP-1)
Castrated controls		(8)	31.57 ± 1.84	1325.00 ± 1.03
Testosterone	(2 mg)	(7)	30.57 ± 1.22	1357.14 ± 64.71
DHT	(2 mg)	(7)	26.79 ± 1.08***	1283.33 ± 15.28
3α-Diol	(2 mg)	(7)	2.71 ± 0.75 §	1278.57 ± 17.62
Testosterone	(4 mg)	(7)	< 1.0	612.50 ± 50.52 §
DHT	(4 mg)	(7)	< 1.0	950.00 ± 96.50 §
3α-Diol	(4 mg)	(7)	2.84 ± 0.74 §	1207.50 ± 33.96***

* Values are means ± S.E.
** Number of animals in parentheses.
*** $P < 0.025$ vs. castrated controls.
§ $P < 0.0005$ vs. castrated controls.

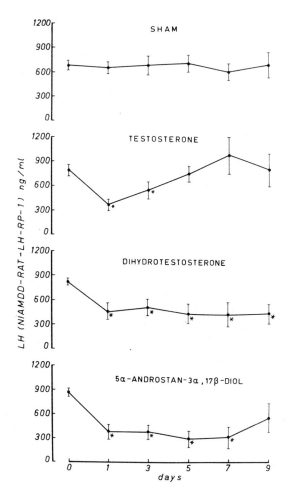

Fig. 7. Effect of median eminence implants of testosterone and of its 5α-reduced metabolites (dihydrotestosterone, 3α-androstan-3α,17β-diol) on the secretion of LH in adult castrated male rats. *Significantly different from time 0.

portant for the feedback control of this gonadotropin. It is possible that, with regard to the inhibition of FSH release, T acts either as such or, after having been converted, locally or systemically into estrogenic molecules. The aromatization of androgenic molecules to estrogen has been reported to occur in the hypothalamus and in other CNS structures (Naftolin et al., 1971, 1972; Flores et al., 1973; Reddy et al., 1973; Weisz and Gibbs, 1974).

III. EFFECTS OF INTRAHYPOTHALAMIC AND INTRAPITUITARY IMPLANTS OF TESTOSTERONE AND OF ITS 5α-REDUCED METABOLITES ON LH AND FSH RELEASE

A subsequent portion of the study on the effects of the 5α-reduced metabolites of T on gonadotropin release was aimed at clarifying whether T, DHT and 3α-diol might influence LH and FSH secretion after having been implanted in the median eminence region of the hypothalamus and in the anterior pituitary. The experiments have been performed in adult castrated male rats. The stereotaxic implantation of the different steroids was made three weeks following castration using a technique previously described

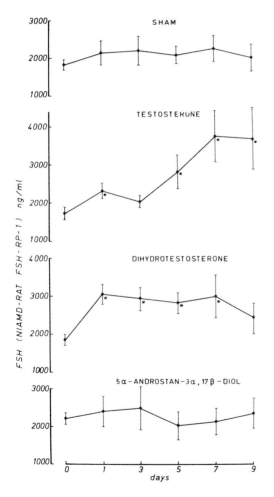

Fig. 8. Effect of median eminence implants of testosterone and of its 5α-reduced metabolites (dihydrotestosterone, 5α-androstan-3α,17β-diol) on the secretion of FSH in adult castrated male rats.
* Significantly different from time 0.

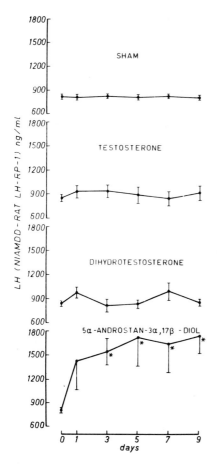

Fig. 9. Effect of intrapituitary implants of testosterone and of its 5α-reduced metabolites (dihydrotestosterone, 5α-androstan-3α,17β-diol) on the secretion of LH in adult castrated male rats. *Significantly different from time 0.

(Corbin et al., 1965). Sham-operated animals received a median eminence or an intrapituitary implant of the vehicle alone (saturated sucrose solution). Blood was collected immediately before submitting the animals to the implantation procedure (in order to obtain basal levels of LH and FSH) and 1, 3, 5, 7 and 9 days after implantation. Serum levels of gonadotropins were quantitated with specific radioimmunoassays as recommended in the directions provided with the NIAMDD kits with minor modifications (Daane and Parlow, 1971).

Fig. 7 shows that serum levels of LH are not modified throughout the experiment by sham-implantations performed at hypothalamic level. Rats bearing a T implantation in the median eminence showed a reduction of serum levels of LH immediately after implantation. However, the effect of T was transient, so that serum levels of LH returned to control values beginning 5 days after implantation. Median eminence implants of DHT induced a more prolonged decrease of serum LH levels. These were significantly lower than departure values up to 9 days after the implantation of this steroid. These results are in perfect agreement with the observation of Smith and Davidson (1974). Median eminence implants of 3α-diol also exerted a prolonged suppressory effect on LH release, serum levels of the hormone remaining low up to 7 days following implantation. Fig. 8 summarizes the results obtained in the same animals while evaluating serum levels of FSH. It is apparent that sham-implantation did not alter serum FSH titers throughout the experiment. On the contrary, the intrahypothalamic placement of either T or DHT significantly increased the serum levels of this gonadotropin above control values at all post-implantation times considered. 3α-Diol was totally ineffective.

The results presented in Fig. 9 indicate that the direct intrapituitary implantation of cannulae containing only sucrose and of cannulae containing either T or DHT did not significantly influence serum levels of LH during the 9 days of the experiment. The intrapituitary implantation of 3α-diol induced a pronounced and long-lasting increase in serum LH levels. Sham-implants and intrapituitary implants of DHT did not change serum FSH

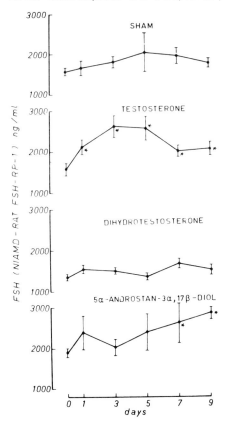

Fig. 10. Effect of intrapituitary implants of testosterone and of its 5α-reduced metabolites (dihydrotestosterone, 5α-androstan-3α,17β-diol) on the secretion of FSH in adult castrated male rats. *Significantly different from time 0.

levels throughout the experiment (Fig. 10). Intrapituitary implants of T resulted in a significant increase of serum FSH, which lasted for 9 days and which was maximal 3 and 5 days after implantation. An increase of serum FSH was also observed 7 and 9 days following intrapituitary placement of 3α-diol.

The effects exerted on LH release by the intrahypothalamic implants of the three steroids tested are comparable to those obtained after systemic administration of the same hormones (see Section II of this paper). These results are compatible with the hypothesis that, when placed in the hypothalamus, T operates on the feedback mechanism inhibiting LH secretion after having been reduced in the 5α-position. The fact that DHT suppresses LH release after having been placed in the hypothalamus and does not inhibit LH secretion when implanted in the anterior pituitary might be taken as an argument in favor of an intrahypothalamic site of action of its inhibitory effect on this gonadotropin. With regard to 3α-diol, this argument is reinforced by the observation that this steroid exerts opposite effects when placed on the two sites (inhibition of LH release following intrahypothalamic placement, stimulation of LH release following intrapituitary implantation). However, it is not felt that the data presented definitely prove this point. Additional experiments involving the administration of LH-RH to animals bearing implants of the two steroids in the median eminence and in the anterior pituitary are needed. It is hoped that such experiments, which are presently in progress, might also provide an explanation for the stimulatory effect exerted on LH release by intrapituitary implants of 3α-diol.

The stimulatory effect exerted by intrahypothalamic implants of T and DHT and by intrapituitary implants of T and 3α-diol on FSH release are surprising, since they seem to underline the fact that, under particular circumstances, androgenic molecules might exert a "positive" feedback effect on FSH release. Additional studies are obviously needed in order to clarify why T, DHT and 3α-diol, depending on the experimental conditions, might exert either a "negative" (see Section II of this paper) or a "positive" feedback effect on FSH release. These studies are presently underway. However, it seems pertinent to recall that there are a few literature data which seem to support the possibility that androgens might exert a "positive" feedback effect on FSH release. Johnson and Naqvi (1969) and Naqvi and Johnson (1970), using an indirect method for evaluating plasma levels of FSH, have reported that T is able to induce cyclic increases of FSH release in castrated female rats. Swerdloff et al. (1972), Kalra et al. (1973) and Gay (1973) have found that the systemic administration of small doses of T and of DHT is followed by a rise of serum FSH levels in castrated male and female rats. In "in vitro" studies, Mittler (1972, 1974) and Schally and his coworkers (1973) have demonstrated that the addition of either T or DHT to the culture media containing anterior pituitary tissue results in a stimulation of FSH output. Zanisi et al. (1973c) and Debeljuk et al. (1974) have reported that pretreatment with either T or DHT enhances the release of FSH in response to the administration of synthetic LH-RH both in normal and in castrated male rats. Finally, there is plenty of evidence indicating that systemic, intrahypothalamic or intrapituitary administrations of androgens result in an increase of pituitary FSH stores in normal as well as in castrated animals (Gay and Bogdanove, 1969; Kamberi and McCann, 1972; Kingsley and Bogdanove, 1973).

ACKNOWLEDGEMENTS

The experimental work described in this paper has been supported by grants of the Ford Foundation, New York, of the Population Council, New York and of the Consiglio Nazionale delle Ricerche, Rome.

The assistance of Drs. A.R. Midgley Jr., L.E. Reichert Jr., G.D. Niswender and of the National Institutes of Arthritis, Metabolism and Digestive Diseases NIAMDD), Rat Pituitary Hormone Distribution Program, in providing LH and FSH radioimmunoassay materials is gratefully acknowledged.

Thanks are also due to Mr. M. Vigliani and Mr. L. Guadagni for their skilful technical assistance.

REFERENCES

Armstrong, D.T. and King, E.R. (1971) Uterine progesterone metabolism and progestational response: effects of estrogens and prolactin. *Endocrinology*, 89: 191–197.

Barraclough, C.A. (1967) Modifications in reproductive function after exposure to hormone during the prenatal and early postnatal period. In *Neuroendocrinology, Vol. 2,* L. Martini and W.F. Ganong (Eds.), Academic Press, New York, pp. 61–99.

Baulieu, E.E., Lasnitzki, I. and Robel, P. (1968) Metabolism of testosterone and action of metabolites on prostate glands grown in organ culture. *Nature (Lond.)*, 219: 1155–1156.

Bruchowsky, N. and Wilson, J.D. (1968) The conversion of testosterone to 5α-androstan-17β-ol-3-one by rat prostate in vivo and in vitro. *J. biol. Chem.*, 243: 2012–2021.

Cheng, Y.J. and Karavolas, H.J. (1973) Conversion of progesterone to 5α-pregnane-3,20-dione and 3α-hydroxy-5α-pregnan-20-one by rat medial basal hypothalami and the effects of estradiol and stage of estrous cycle on the conversion. *Endocrinology*, 93: 1157–1162.

Corbin, A., Mangili, G., Motta, M. and Martini, L. (1965) Effect of hypothalamic and mesencephalic steroid implantation on ACTH feedback mechanisms. *Endocrinology*, 76: 811–818.

Critchlow, V. and Bar-Sela, M.E. (1967) Control of the onset of puberty. In *Neuroendocrinology, Vol. 2,* L. Martini and W.F. Ganong (Eds.), Academic Press, New York, pp. 101–162.

Daane, T.A. and Parlow, A.F. (1971) Periovulatory patterns of rat serum follicle stimulating hormone and luteinizing hormone during the normal estrous cycle: effects of pentobarbital. *Endocrinology*, 88: 653–663.

Davidson, J.M. and Smith, E.R. (1967) Testosterone feedback in the control of somatic and behavioral aspects of male reproduction. In *Hormonal Steroids,* L. Martini, F. Fraschini and M. Motta (Eds.), Excerpta Medica, Amsterdam, pp. 805–813.

Debeljuk, L., Vilchez-Martinez, J.A., Arimura, A. and Schally, A.V. (1974) Effect of gonadal steroids on the response to LH-RH in intact and castrated male rats. *Endocrinology*, 94: 1519–1524.

Denef, C., Magnus, C. and McEwen, B.S. (1973) Sex differences and hormonal control of testosterone metabolism in rat pituitary and brain. *J. Endocr.*, 59: 605–621.

Denef, C., Magnus, C. and McEwen, B.S. (1974) Sex dependent changes in pituitary 5α-dihydrotestosterone and 3α-androstanediol formation during post-natal development and puberty in the rat. *Endocrinology*, 94: 1265–1274.

Flores, F., Naftolin, F. and Ryan, K.J. (1973) Aromatization of androstenedione and testosterone by rhesus monkey hypothalamus and limbic system. *Neuroendocrinology*, 11: 177–182.

Gay, V.L. (1973) Inhibition of LH secretion concomitant with hypersecretion of FSH in castrated rats bearing silastic implants of dihydrotestosterone. *Program 55th Meet. Endocrine Soc.*, p. 116.

Gay, V.L. and Bogdanove, E.M. (1969) Plasma and pituitary LH and FSH in the castrated rat following short-term steroid treatment. *Endocrinology*, 84: 1132–1142.

Gorski, R.A. (1971) Gonadal hormones and the perinatal development of neuroendocrine function. In *Frontiers in Neuroendocrinology*, L. Martini and W.F. Ganong (Eds.), Oxford University Press, New York, pp. 237–290.

Jaffe, R.B. (1969) Testosterone metabolism in target tissues: hypothalamic and pituitary tissues of the adult rat and human fetus, and the immature rat epiphysis (1). *Steroids*, 14: 483–499.

Johnson, D.C. and Naqvi, R.H. (1969) A positive feedback action of androgen on pituitary follicle stimulating hormone: induction of a cyclic phenomenon. *Endocrinology*, 85: 881–885.

Kalra, P.S., Fawcett, C.P., Krulich, L. and McCann, S.M. (1973) The effect of gonadal steroids on plasma gonadotropins and prolactin in the rat. *Endocrinology*, 92: 1256–1265.

Kamberi, I.A. and McCann, S.M. (1972) Effects of implants of testosterone in the median eminence and pituitary on FSH secretion. *Neuroendocrinology*, 9: 20–29.

Karavolas, H.J. and Herf, S.M. (1971) Conversion of progesterone by rat medial basal hypothalamic tissue to 5α-pregnane 3,20-dione. *Endocrinology*, 89: 940–942.

Kingsley, T.R. and Bogdanove, E.M. (1973) Direct feedback of androgens: localized effects of intrapituitary implants of androgens on gonadotropic cells and hormone stores. *Endocrinology*, 93: 1398–1409.

Kniewald, Z. and Milkovic, S. (1973) Testosterone: a regulator of 5α-reductase activity in the pituitary of male and female rats. *Endocrinology*, 92: 1772–1775.

Kniewald, Z., Massa, R. and Martini, L. (1971) Conversion of testosterone into 5α-androstan-17β-ol-3-one at the anterior pituitary and hypothalamic levels. In *Hormonal Steroids*, V.H.T. James and L. Martini (Eds.), Excerpta Medica, Amsterdam, pp. 784–791.

Massa, R. and Martini, L. (1974) Testosterone metabolism: a necessary step for activity?. *J. Steroid Biochem.*, 5: 941–947.

Massa, R., Stupnicka, E., Villa, A. and Martini, L. (1971) The conversion of testosterone (T) into dihydrotestosterone (DHT) in the brain and in the anterior pituitary (AP) of immature rats. *Program 53rd Meet. Endocrine Soc.*, p. 229.

Massa, R., Stupnicka, E., Kniewald, Z. and Martini, L. (1972a) The transformation of testosterone into dihydrotestosterone by the brain and the anterior pituitary. *J. Steroid Biochem.*, 3: 385–399.

Massa, R., Stupnicka, E. and Martini, L. (1972b) Metabolism of progesterone in the anterior pituitary, the hypothalamus and the uterus of female rats. In *Abstract Book 4th International Congress of Endocrinology*, Excerpta Medica, Amsterdam, p. 118.

Massa, R., Justo, S. and Martini, L. (1975) Conversion of testosterone into 5α-reduced metabolites in the anterior pituitary and in the brain of maturing rats. *J. Steroid Biochem.*, 6: 567–571.

Mittler, J.C. (1972) Androgen effect on gonadotropin secretion by organ-cultured anterior pituitary. *Proc. Soc. exp. Biol. (N.Y.)*, 140: 1140–1142.

Mittler, J.C. (1974) Androgen effects on follicle-stimulating hormone (FSH) secretion in organ culture. *Neuroendocrinology*, 16: 265–272.

Naftolin, F. and Feder, H.H. (1973) Suppression of luteinizing hormone secretion in male rats by 5α-androstan-17β-ol-3-one (dihydrotestosterone) propionate. *J. Endocr.*, 56: 155–156.

Naftolin, F., Ryan, K.J. and Petro, Z. (1971) Aromatization of androstenedione by the diencephalon. *J. clin. Endocr.*, 33: 368–370.

Naftolin, F., Ryan, K.J. and Petro, Z. (1972) Aromatization of androstenedione by the anterior hypothalamus of adult male and female rats. *Endocrinology*, 90: 295–298.

Naqvi, R.H. and Johnson, D.C. (1970) Effect of progesterone on androgen or estrogen-induced increases in endogenous FSH in immature female rats. *Endocrinology*, 87: 418–421.

Negro-Vilar, A., Ojeda, S.R. and McCann, S.M. (1973) Evidence for changes in sensitivity to testosterone negative feedback on gonadotropin release during sexual development in the male rat. *Endocrinology*, 93: 729–735.

Niswender, G.D., Midgley, A.R., Jr., Monroe, S.E. and Reichert, L.E., Jr. (1968) Radioimmunoassay for rat luteinizing hormone with antiovine LH serum and ovine LH-[131]I. *Proc. Soc. exp. Biol. (N.Y.)*, 128: 807–811.

Perez-Palacios, G., Castaneda, E., Gomez-Perez, F., Perez, A.E. and Gual, C. (1970) In vitro metabolism of androgens in dog hypothalamus, pituitary and limbic system. *Biol. Reprod.,* 3: 205–214.

Ramirez, V.D. and McCann, S.M. (1965) Inhibitory effect of testosterone on luteinizing hormone secretion in immature and adult rats. *Endocrinology*, 76: 412–417.

Reddy, V.V.R., Naftolin, F. and Ryan, K.J. (1973) Aromatization in the central nervous system of rabbits: effects of castration and hormone treatment. *Endocrinology*, 92: 589–594.

Robel, P. (1971) Steroid hormone metabolism in responsive tissues in vitro. *Acta endocr. (Kbh.)*, Suppl. 153: 279–294.

Robinson, J.A. and Karavolas, H.J. (1973) Conversion of progesterone by rat anterior pituitary tissue to 5α-pregnane-3,20-dione and 3α-hydroxy-5α-pregnan-20-one. *Endocrinology*, 93: 430–435.

Rommerts, F.F.G. and Van der Molen, H.J. (1971) Occurrence and localization of 5α-steroid reductase, 3α- and 17β-hydroxysteroid dehydrogenase in hypothalamus and other brain tissues of the male rat. *Biochim. biophys. Acta (Amst.)*, 248: 489–497.

Schally, A.V., Redding, T.W. and Arimura, A. (1973) Effect of sex steroids on pituitary responses to LH- and FSH-releasing hormone in vitro. *Endocrinology,* 93: 893–902.

Sholiton, L.S., Taylor, B.B. and Lewis, H.P. (1974) The uptake and metabolism of labelled testosterone by the brain and pituitary of the male rhesus monkey (*Macaca mulatta*). *Steroids*, 24: 537–547.

Sholl, S.A., Robinson, J.A. and Goy, R.W. (1975) Neural uptake and metabolism of testosterone and dihydrostestosterone in the guinea pig. *Steroids*, 25: 203–215.

Smith, E.R. and Davidson, J.M. (1974) Location of feedback receptors: effects of intracranially implanted steroids on plasma LH and LRF response. *Endocrinology*, 95: 1566–1573.

Stern, J.M. and Eisenfeld, A.J. (1971) Distribution and metabolism of ^{3}H-testosterone in castrated male rats; effects of cyproterone, progesterone and unlabelled testosterone. *Endocrinology,* 88: 1117–1125.

Swerdloff, R.S., Walsh, P.C. and Odell, W.D. (1972) Control of LH and FSH secretion in the male: evidence that aromatization of androgens to estradiol is not required for inhibition of gonadotropin secretion. *Steroids*, 20: 13–18.

Tabei, T. and Heinrichs, W.L. (1974) Metabolism of progesterone by the brain and pituitary gland of subhuman primates. *Neuroendocrinology*, 15: 281–289.

Tabei, T., Haga, H., Heinrichs, W.L. and Herrmann, W.L. (1974) Metabolism of progesterone by rat brain, pituitary gland and other tissues. *Steroids*, 23: 651–666.

Thieulant, M.L., Samperez, S. and Jouan, P. (1973) Binding and metabolism of [^{3}H]testosterone in the nuclei of rat pituitary in vivo. *J. Steroid Biochem.*, 4: 677–685.

Thieulant, M.L., Pelle, G., Samperez, S. et Jouan, M.P. (1974) Augmentation de l'activité de la 5α-stéroide réductase des noyaux cellulaires purifiés d'hypophyses de rats mâles castrés. *C.R. Acad. Sci. (Paris)*, 278: 1281–1284.

Verjans, H.L., Eik-Nes, K.B., Aafjes, J.H., Vels, F.J.M. and Van der Molen, H.J. (1974) Effects of testosterone propionate, 5α-dihydrotestosterone propionate and oestradiol benzoate on serum levels of LH and FSH in the castrated adult male rat. *Acta endocr. (Kbh.)*, 77: 643–654.

Weisz, J. and Gibbs, C. (1974) Conversion of testosterone and androstenedione to estrogens *in vitro* by the brain of female rats. *Endocrinology*, 94: 616–620.

Wilson, J.D. and Gloyna, R.E. (1970) The intranuclear metabolism of testosterone in the accessory organs of reproduction. *Recent Progr. Hormone Res.*, 26: 309–336.

Zanisi, M., Motta, M. and Martini, L. (1973a) Feedback activity of testosterone and of its 5α-reduced metabolites. In *The Endocrine Function of the Human Testis, Vol. 1,* V.H.T. James, M. Serio and L. Martini (Eds.), Academic Press, New York, pp. 431–438.

Zanisi, M., Motta, M. and Martini, L. (1973b) Inhibitory effect of 5α-reduced metabolites of testosterone on gonadotropin secretion. *J. Endocr.*, 56: 315–316.

Zanisi, M., Motta, M. and Martini, L. (1973c) Recent findings on the physiology of the gonadotropin-releasing factors. In *Hypothalamic Hypophysiotropic Hormones*, C. Gual and E. Rosemberg (Eds.), Excerpta Medica, Amsterdam, pp. 24–32.

Subcellular Mechanisms in Reproductive Neuroendocrinology, edited by
F. Naftolin, K.J. Ryan and J. Davies
© 1976 Elsevier Scientific Publishing Company—Amsterdam, The Netherlands

Chapter 17

Androgen aromatization by neuroendocrine tissues

F. NAFTOLIN, K.J. RYAN and I.J. DAVIES

Department of Obstetrics and Gynecology, McGill University Faculty of Medicine and The Royal Victoria Hospital, Montreal, Que. H3A 1A1 (Canada) and Department of Obstetrics and Gynecology and Laboratory for Human Reproduction and Reproductive Biology, Harvard Medical School and The Boston Hospital for Women, Boston, Mass. 02115 (U.S.A.)

INTRODUCTION

It is not surprising that biologists should be interested in steroid hormones and neuroendocrine tissues (NET). In fact approaches to the broad issues are exhaustive. While studies have been limited by methodologic considerations, there have also been a number of key conceptual deficits, which until recently have constricted our overview and severely retarded progress. With increasing understanding of the mechanism of action of steroid hormones, the idea of pre- or prohormones and clarification of intermediary steroid hormone metabolism, the entire field of endocrinology has been revolutionized. It is the animation of these concepts into a more vital picture of neuroendocrine function which has recently focused attention on androgens as prehormones in neuroendocrine tissues, and especially on the question of aromatization as a mechanism or prerequisite for this activity (Naftolin and Ryan, 1975; Naftolin et al., 1975a).

II. AROMATIZATION

For many years the source of estrogens (C_{18} phenolic steroids) in animals and humans has been known to be androgens (C_{19} Δ^4 or Δ^5 neutral steroids) (Fig. 1). Androgens are secreted mainly by the adrenal gland and gonad. In addition to the gonad, which efficiently aromatizes androgens, there are increasing numbers of other tissues proven capable of this metabolism, including the placenta and fetal tissues, as well as adult organs (Table I). Aromatization is the result of a multiple reaction process, which explains the lack of product inhibition of aromatization by estrogens (Engel, 1974). It entails the hydroxylation of carbon 19, and ultimately its loss to form an 18 carbon steroid. The attendant loss of two hydrogens which results in the unsaturated A ring may require enzymatic steps or may be a concomitant of the instability that attends the loss of C_{19}. Although the estrogens formed are quite stable, they are regularly subjected to considerable interaction and metabolism, such as sulfation and hydroxylation, in vivo. More about this will be discussed in the next paper and, to be sure, in our discussions.

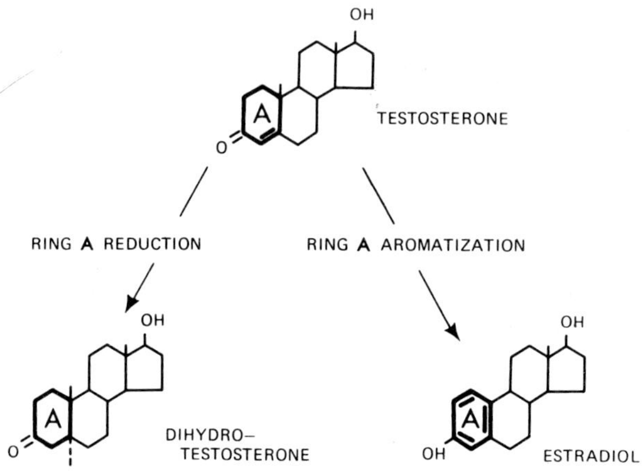

Fig. 1. Metabolic pathways of androgens. (From Naftolin et al., 1974.)

TABLE I

Results of in vitro aromatization studies in various tissues

HF = human fetus; Rh = rhesus; Rb = rabbit; Rt = rat; M = mouse. (From Naftolin et al., 1975a.)

Tissues studied	Estrogen formation proven	
	Yes	No
Hypothalamus	HF, Rh, Rb, Rt, M	—
Limbic system	HF, Rh, Rb, Rt	—
Pituitary gland	HF*	Rb, Rt
Cortex	HF*	Rb, Rt**
Septum	—	Rb
Olfactory bulbs	—	Rb
Skin	—	HF
Seminal vesicles	—	Rt
Liver	HF	—
Placenta	HF	—

* Very low activities, seen only through the use of the cold estrogen trap.
** Rat cortex activity found only during experiments on newborn rats where dissection may have included parts of limbic system.

III. CENTRAL ESTROGEN ACTIONS

Whatever their origins, estrogens are very potent at the *central* (brain and pituitary as opposed to *peripheral*, e.g., mullerian tissues) level. The central effects of estrogens may extend to neuronal and myelin growth, brain differentiation, sexual maturation, sexual and emotional behavior and gonadotropin control (Table II). Although we have previously reviewed the data regarding a number of these actions, a brief and pointed capsule of significant findings should be of illustrative use.

TABLE II

Central estrogen actions

(From Naftolin and Ryan, 1975.)

Rat brain differentiation
  $E > 19\ OH\text{-}T > T$
  Anti-E blocks T effect

Gonadotropin control
  $E > T/DHT$ (DHT not effective in all species)
  Anti-E blocks T effect

Sex behavior
  $E > T/DHT$ (DHT not effective in all species)
  $E \rightarrow \male$ or $\female$ sex behavior
  Anti-E blocks T effect

Timing of puberty
  $E > T$/ring A reduced androgens
  DHT no effect on puberty

IV. BRAIN DIFFERENTIATION

Brain programming (i.e., characteristic gonadotropin control mechanisms and sexual behavior) in lower animals is a testis-dependent phenomenon which occurs during or shortly after gestation (Harris and Naftolin, 1970). There is plentiful evidence that the active factors in brain differentiation are actually estrogens which are formed from (testicular origin) androgens (Reddy et al., 1974).

V. SEXUAL MATURATION

While androgens can effect the timing of sexual maturation in female animals (and humans), in each case tested experimentally estrogens are more potent in causing these effects (Reddy et al., 1974).

VI. SEXUAL BEHAVIOR

Under a number of different experimental situations, both androgens and estrogens are capable of causing gender appropriate or inappropriate sexual behavior in animals. In each case tested estrogens are more potent than androgens (Ryan et al., 1972).

VII. ADULT GONADOTROPIN CONTROL

Estrogens and androgens can both control LH and FSH under experimental conditions. In general, estrogens are far more potent than androgens in this action (McCann and Ramirez, 1964; Sherins and Loriaux, 1973).

In many cases, the above experimental effects of androgens have been blocked by pre-treatment with anti-estrogens such as MER-25 (Beyer and Vidal, 1971; McDonald and Daughty, 1972; Naftolin et al., 1973). Because of the apparent discrepancy in mechanism of action of androgens at the peripheral level (testosterone forms ring A reduced C_{19} compounds which are the active factors in Wolffian duct growth and secondary sex characteristics (Wilson and Gloyna, 1970)) and central level (viz. supra), it has been suggested that the term "androgen action" should be reconsidered and defined (Naftolin and Ryan, 1975).

VIII. AROMATIZATION IN NET

Estrogens are centrally potent hormones. They are also formed in considerable quantities in many non-neuroendocrine tissues, including the gonads and fat of adults (McDonald et al., 1967; Schindler et al., 1972). However, the drawbacks of a system which relies upon production of active hormones outside of NET for neuroendocrine actions include dilution of the active factor in the circulation, untoward effects of the circulating active factor, inappropriate metabolism and binding of the active factor, poor transfer into the NET and the requirement for long-loop control systems to govern production and secretion of active factor (Naftolin et al., 1974). Accordingly, efforts have been made to prove and quantitate the presence of aromatizing systems in NET. Presently, in vitro aromatization of various androgen substrates by hypothalami from 5 species has been reported (Table III). This has included the aromatization of 17-ethynyl 19-nor testosterone, a prototype progestin used in the oral contraceptives. Using androstenedione as a substrate, a species hierarchy of aromatizing activity has been shown as well as a very consistent sex difference, males having greater in vitro aromatizing activity than females (Table IV). While hypothalamus is the most potent aromatizing area of the NET, there has been measurable activity found in the limbic tissues (and probably also some activity in pituitary and cortex of human fetuses) (Table I). The fetal brain has been of special interest in these studies. Along these lines human fetal (Naftolin et al., 1974) and rat fetal neonatal (Reddy et al., 1974; Weisz and Gibbs, 1974a) tissues have been found active in vitro aromatizers. The activity appears to peak during the critical postnatal period in the rat (Fig. 2).

Proof of in vivo NET aromatization has been obtained in rhesus monkeys (Flores et al., 1973). Through the use of the living isolated rhesus monkey brain the formation of both free and conjugated estrogens from androgens has been quantitated, nanograms of estrogen having been produced from micrograms of prehormone androgen during 1–2 hr of perfusion.

The localization of central nervous system aromatization has, thus far, corresponded with prior studies localizing the uptake of radiolabel after the administration of radiolabeled sex steroids (Pfaff, 1968; Stumpf et al., 1971). The one marked difference between the labeling and in vitro studies is that although the pituitary avidly takes up label, it does not form notable quantities of estrogen. The pattern of aromatizing activity fits

TABLE III

Substrates and products in central aromatization

(From Naftolin et al., 1975a.)

	Substrates			
	Δ^4	T	17α-ethinyl-19-nor-T	E_1
Human fetus	E_1, E_2	—	EE_2	—
Rhesus	E_1	E_2	—	—
	E_2*	E_1*		
Rabbit	E_1, T	E_2, Δ^4	—	E_2
Rat	E_1	—	—	—
Mouse	E_1	—	—	—

* During in vivo experiments both E_1 and E_2 were formed from Δ^4 or T.

TABLE IV

In vitro conversion of Δ^4 to E_1 by hypothalamus (nmoles × 10^2/g tissue)

(From Naftolin et al., 1975a.)

	Human	Rhesus	Rat	Mouse	Rabbit
Male	2.00		0.43	0.41	0.055
Female	0.75		0.16	0.14	0.020
Pooled (♂ + ♀)		0.53*			

* Tissues from 3 animals pooled after estrogen separation because of low activity present.

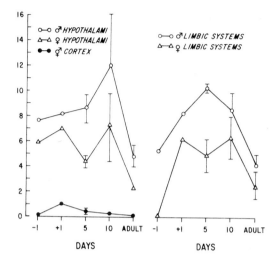

Fig. 2. In vitro aromatization by rat brain tissues. Single data points represent single experiments, bars represent the range of two experiments with the average per cent conversion (x 10^3) shown as the data point for that age. (From Reddy et al., 1974.)

well with both the known experimental data for activity of locally administered steroids (Nadler, 1972; Marcus et al., 1975) and for studies of estrogen binding by brain cytosol (Eisenfeld, 1972). Moreover, such manipulations of animals as castration or steroid hormone pretreatment change the in vitro aromatizing activity of both male and female rabbits (Reddy et al., 1973). This is especially of interest vis-à-vis the known effect of FSH on in vitro NET ring A steroid reductase (Massa et al., 1972).

IX. INDIRECT CORRELATIVE STUDIES

In addition to the previously mentioned implantation and pharmacologic studies, two other related lines of evidence support the importance of NET estrogen formation.

(a) Although radiolabel is taken up by the many cells after administration of [^{3}H]testosterone to the rat, only a comparatively few cell nuclei show labeling. Furthermore, pretreatment of the animals with aromatizable (testosterone) or an aromatized (estradiol) steroid blocked the nuclear labeling, while pretreatment with ring A reduced steroid (dihydrotestosterone) did not affect nuclear uptake (Sheridan et al., 1974).

(b) Injection of radiolabeled testosterone has resulted in the finding of [^{3}H]estradiol in the nuclear pellet from the hypothalamus-limbic system dissections in these experimental animals (Lieberburg and McEwen, 1975). Similar results were also reported using less specific localization techniques (Knapstein et al., 1968; Flores et al., 1973; Weisz and Gibbs, 1974b).

X. FORMATION OF ESTROGENS IN NET

The implications of local in situ formation of active primary estrogens in NET are remarkable. From the standpoint of economy, this system has the benefits and lacks the liabilities as described above. It does not exclude a direct contribution of non-NET derived estrogen in neuroendocrine control. However, there are times when even this must be the case; during the rat's fetal and newborn life when there are present proteins in blood (Reddy et al., 1974) and NET (Plapinger and McEwen, 1973; Plapinger et al., 1973) which bind estrogens and not androgens. Under these circumstances the prehormone levels and aromatizing activity in NET become critical (Reddy et al., 1974). The possible effects of local available estrogen must be considered in light of the present understanding of steroid hormone action. The recent demonstration of steroid sterilization by 0.125 pg estradiol microelectrophoresed into the neonatal female rat's brain (Marcus et al., 1975) and the 10,000-fold higher potency of estradiol to cause precocious vaginal opening compared to testosterone (Eckstein et al., 1973) attest to this. The effect of estrogens on dendritic sprouting as described by observations on hypothalamic blocks dissected from neonatal rats (Toran-Allerand, 1975) is further evidence along these lines. Estrogen is well known to effect the response of the pituitary gland to releasing factor (Rosemberg and Gual, 1972). Since the adenohypophysis receives its major blood supply from the portal vessels, central nervous system estrogen must reach adenohypophyseal cells. The proof of limbic as well as hypothalamic aromatization has implications for rhythmic neuroendocrine functions and the control of the onset of puberty (Reddy et al., 1974). Recently, the possibility of central estrogen actions via hydroxylated metabolites has been raised (Naftolin et al., 1975b; Parvizi and Ellendorf, 1975).

In short, it appears that estrogens may be fundamental to many central neuroendocrine and reproductive processes. In situ central aromatization, therefore, assumes an importance disproportionate to the small proportion of the total estrogen production it would appear to represent.

REFERENCES

Beyer, C. and Vidal, N. (1971) Inhibitory action of MER-25 on androgen-induced oestrous behavior in the ovariectomized rabbit. *J. Endocr.*, 51: 401.

Eckstein, B., Golan, R. and Shani, J. (1973) Onset of puberty in the immature female rat induced by 5α-androstane-3β, 17β-diol. *Endocrinology*, 92: 941.

Eisenfeld, A.J. (1972) Interaction of estrogens, progestational agents, and androgens with brain and pituitary and their role in the control of ovulation. In *Perspectives in Neuropharmacology*, S.H. Snyder (Ed.), Oxford Univ. Press, London, p. 113.

Engel, L.L. (1974) The biosynthesis of estrogens. In *Handbook of Physiology, Endocrinology II, Part I*, R.O. Greep (Ed.), Amer. Physiol. Soc., Washington, D.C., Chapter 20, pp. 467.

Flores, F., Naftolin, F., Ryan, K.J. and White, R.J. (1973) Estrogen formation by the isolated perfused rhesus monkey brain. *Science*, 180: 1074.

Harris, G.W. and Naftolin, F. (1970) The hypothalamus and control of ovulation. *Brit. med. Bull.*, 26: 3.

Knapstein, P., David, A., Wu, C.-H., Archer, D.F., Flickinger, G.L. and Touchstone, J.C. (1968) Metabolism of free and sulfo-conjugated DHEA in brain tissue *in vivo* and *in vitro*. *Steroids*, 11: 885.

Lieberburg, I. and McEwen, B.S. (1975) Estradiol-17β: a metabolite of testosterone recovered in cell nuclei from limbic areas of neonatal rat brains. *Brain Res.*, 85: 165–170.

McCann, S.M. and Ramirez, V.D. (1964) The neuroendocrine regulation of hypophyseal luteinizing hormone secretion. *Recent Progr. Hormone Res.*, 20: 131.

McDonald, P.G. and Daughty, C. (1972) Inhibition of androgen-sterilization in the female rat by administration of an antioestrogen. *J. Endocr.*, 55: 455.

McDonald, P.C., Rombaut, R.P. and Siiteri, P.K. (1967) Plasma precursors of estrogen. I. Extent of conversion of plasma Δ^4-androstenedione to estrone in normal males and non-pregnant normal, castrate and adrenalectomized females. *J. clin. Endocr.*, 27: 1103.

Marcus, D., Schuler, H. and Josimovich, J. (1975) Improved discrete injection of estradiol into anterior hypothalamus of newborn rats for induction of oligo-ovulation. Program Endocrine Soc. 57th Annual meeting, Abstract No. 257.

Massa, R., Stupnicka, E., Kniewald, Z. and Martini, L. (1972) The transformation of testosterone into dihydrotestosterone by the brain and the anterior pituitary. *Steroid Biochem.*, 3: 385.

Nadler, R.D. (1972) Intrahypothalamic exploration of androgen-sensitive brain loci in neonatal female rats. *Trans. N.Y. Acad. Sci.*, 34: 572.

Naftolin, F. and Ryan, K.J. (1975) The metabolism of androgens in central neuroendocrine tissues. *J. Steroid Biochem.*, 6: 993.

Naftolin, F., Judd, H.L. and Yen, S.S.C. (1973) Pulsatile patterns of gonadotrophins and testosterone in man: the effects of clomiphene with and without testosterone. *J. clin. Endocr.*, 36: 285.

Naftolin, F., Ryan, K.J., Davies, I.J., Petro, Z. and Kuhn, M. (1974) The formation and metabolism of estrogens in brain tissues. *Advanc. Biosci.*, 15: 105.

Naftolin, F., Ryan, K.J., Davies, I.J., Reddy, V.V., Flores, F., Petro, Z., Kuhn, M., White, R.J., Wolin, L. and Takaoka, Y. (1975a) The formation of estrogens by central neuroendocrine tissues. *Recent Progr. Hormone Res.*, 31: 295.

Naftolin, F., Morishita, H., Davies, I.J., Todd, R. and Ryan, K.J. (1975b) 2-Hydroxyestrone induced rise in serum luteinizing hormone in the immature male rat. *Biochem. biophys. Res. Commun.*, 64: 905–910.

Parvizi, N. and Ellendorf, F. (1975) 2-Hydroxy-oestradiol-17β as a possible link in steroid brain interaction. *Nature (Lond.)*, 256: 59.

Pfaff, D.W. (1968) Autoradiographic localization of radioactivity in rat brain after injection of tritiated sex hormones. *Science*, 161: 1355.

Plapinger, L. and McEwen, B.S. (1973) Ontogeny of estradiol-binding sites in rat brain. I. Appearance of presumptive adult receptors in cytosol and nuclei. *Endocrinology*, 93: 1119–1128.

Plapinger, L., McEwen, B.S. and Clemens, L.E. (1973) Ontogeny of estradiol-binding sites in rat brain. II. Characteristics of a neonatal binding macromolecule. *Endocrinology*, 93: 1129–1139.

Reddy, V.V.R., Naftolin, F. and Ryan, K.J. (1973) Aromatization in the central nervous system of rabbits: effects of castration and hormone treatment. *Endocrinology*, 92: 589–594.

Reddy, V.V.R., Naftolin, F. and Ryan, K.J. (1974) Conversion of androstenedione to estrone by neural tissues from fetal and neonatal rats. *Endocrinology*, 94: 117.

Ryan, K.J., Naftolin, F., Reddy, V., Flores, F. and Petro, Z. (1972) Estrogen formation in the brain. *Amer. J. Obstet. Gynec.*, 114: 454.

Rosemberg, E. and Gual, C. (Eds.) (1972) *Hypothalamic Hypophysiotropic Hormones. Int. Congr. Ser. No. 263*, Excerpta Medica, Amsterdam.

Schindler, A.E., Ebert, A. and Friedrich, E. (1972) Conversion of androstenedione to estrone by human fat tissue. *J. clin. Endocr.*, 35: 627.

Sheridan, P.J., Sar, M. and Stumpf, W.E. (1974) Interaction of exogenous steroids in the developing rat brain. *Endocrinology*, 95: 1749–1753.

Sherins, R.J. and Loriaux, D.L. (1973) Studies on the role of sex steroids in the feedback control of FSH concentrations in men. *J. clin. Endocr.*, 36: 886.

Stumpf, W.E., Baerwaldt, C. and Sar, M. (1971) Autoradiographic cellular and subcellular localization of sexual steroids. In *Basic Actions of Sex Steroids on Target Organs*, P.O. Hubinont and F. Leroy (Eds.), Karger, Basel, pp. 3—20.

Toran-Allerand, C.D. (1975) Sex hormones and the development of long-term cultures of the newborn mouse hypothalamic/preoptic region. *Proc. Endocrine Soc. 59th Ann. Meeting*, Abstr. 9, p. 55.

Weisz, J. and Gibbs, C. (1974a) Conversion of testosterone and androstenedione to estrogens *in vitro* by the brain of female rats. *Endocrinology*, 94: 2.

Weisz, J. and Gibbs, C. (1974b) Metabolism of testosterone in the brain of the newborn female rat after an injection of tritiated testosterone. *Neuroendocrinology*, 14: 72—86.

Wilson, J.D. and Gloyna, R.E. (1970) The intranuclear metabolism of testosterone in the accessory organs of reproduction. *Recent Progr. Hormone Res.*, 26: 309.

Chapter 18

Estrogen metabolism by neuroendocrine tissues

JACK FISHMAN

Institute for Steroid Research and Department of Biochemistry, Montefiore Hospital and Medical Center, Albert Einstein College of Medicine, Bronx, N.Y. 10467 (U.S.A.)

Overwhelming evidence now exists that the mammalian central nervous system including that of man is a target tissue for the female sex hormone. The evidence consists not only of clear-cut behavioral and biochemical changes induced by estrogens administered directly into the central system (Ferin et al., 1974; Luttge et al., 1975) but includes also the documented presence of specific estrogen binding components or receptors in neuroendocrine tissues (Eisenfeld, 1969; Leavitt et al., 1969; Davies et al., 1975a). Recent examples of the expression or modulation of the activity of a hormone at the target sites via its in situ transformations (Bruchowsky and Wilson, 1968; Blunt and Deluca, 1969; Ryan et al., 1973) direct attention to the metabolism of estrogens in central nervous tissues as well as the pituitary. Of particular interest is the possible existence of differential biotransformations in separate central sites which may help to rationalize the variety of effects elicited by the estrogens.

The study of the metabolism of estrogens in neuroendocrine tissues under in vivo conditions is complicated by the difficult if not impossible task of identifying whether the metabolites found in the tissue were formed there or were produced elsewhere before crossing the blood—brain barrier. This consideration appears to have been neglected in evaluating the results of several such in vivo studies. Much of the dependable information on estrogen metabolism in neuroendocrine tissues is therefore derived from in vitro incubations.

The general pathways of estradiol metabolism includes an initial and rapid oxidation to estrone and a considerably slower reverse reduction (Fishman et al., 1960a). Estrone, which is the central compound in estrogen metabolism, is then hydroxylated either at C-16 to yield estriol or at C-2 to give the catecholestrogen 2-hydroxyestrone (Fishman et al., 1960b; Ball et al., 1975). These two hydroxylations are in a large sense competitive and mutually exclusive (Fishman et al., 1963). In addition to the above oxidative transformations extensive conjugation and deconjugation with sulfuric and glucuronic acids is also a dominant feature of estrogen metabolism (Diczfalusy and Levitz, 1970). In searching for biotransformations of estradiol in neuroendocrine tissues attention was first directed to those alterations which characterize its general metabolism and several of these have now been demonstrated to take place within central sites.

Incubation of minced tissue obtained from the anterior pituitary, the hypothalamus and the cerebral cortex of female sheep with [^{3}H]estrone sulfate led to the formation of free [^{3}H]estrone in every case (Payne et al., 1973). Although arylsulfatases are known to be widely present in central nervous tissues (Perumal and Robins, 1973), this experiment demonstrated that one or several of these enzymes are capable of hydrolyzing estrogen aryl sulfates. The deconjugation of estrone sulfate occurred in all three tissues studied to nearly the same extent and was about 70% complete after 1 hr. When the same studies were repeated with tissues obtained from castrated sheep there was a sharp decrease in the deconjugation of estrone sulfate by the hypothalamic and cerebral cortex tissues but there was no change in the hydrolysis by the anterior pituitary.

Of even greater significance was the finding in the same study (Payne et al., 1973) that the estrone sulfate substrate gave not only free estrone but also free estradiol. This transformation which occurred in all three tissues demonstrated the presence of the 17β-oxidoreductase in these neuroendocrine sites. The relative estrone to estradiol concentrations were about 10 to 1 in the hypothalamus and cerebral cortex but only 2 to 1 in the pituitary indicating the particularly effective reduction of estrone in this organ. Castration did not affect these ratios, but the addition of the cofactor NADPH increased the estradiol yield in the hypothalamus disproportionately to the other tissues.

These results of Payne et al. (1973) were confirmed and extended by Jenkin et al. (1975) who studied the metabolism of [^{3}H]estrone sulfate in minces of hypothalami, cerebral cortex and pituitaries obtained from fetal and adult pregnant sheep. They found that the estrone sulfate was transformed to both estrone and estradiol, with the fetal pituitary being the most active tissue. Similarly when [^{3}H]estrone was used as the substrate in the incubations, the fetal pituitary was 5 times more effective than in the corresponding adult tissue in converting estrone to estradiol while the hypothalamus or the cerebral cortex whether from fetal or adult origin were equal but less effective in making this conversion.

The enzymatic hydrolysis of estrone sulfate has recently been also explored by in vivo procedures by Kishimoto (1973). He administered doubly labeled [^{3}H]estrone-[^{35}S]sulfate to rats via intracardiac injections and measured the free and conjugated isotope content on both sides of the blood—brain barrier. His results showed that the blood contained virtually only conjugated and no free estrogens with the reverse situation occurring in the brain except that a small amount of ^{35}S-containing estrone sulfate was also present. When the same experiment was repeated except that the dose was injected intracerebrally, the brain contained both conjugated and free estrogens in a ratio of about 2 to 1, while the blood contained equal but very small quantities of both free and conjugated estrogen. Kishimoto interprets his results as indicating that hydrolysis of the estrone sulfate participates in the transfer of the hormone through the blood—brain barrier and that the hydrolysis most likely occurs at the barrier inside the capillary membrane. It should be noted that this investigator (Kishimoto, 1973) also observed the conversion of estrone sulfate to unidentified polar compounds in the brain.

The existence of an estradiol 17β-oxidoreductase in the brain and pituitary was recently also confirmed in the elegant in vivo studies of Kazuma and Longcope (1974). These authors injected either tritium labeled estradiol or estrone into the jugular vein of ewes and analyzed blood samples obtained from the ipsilateral jugular vein and the common carotid artery. They were thus able to measure the transformation of these substrates by the cranial tissue. Their findings demonstrated the conversion of estradiol to estrone and vice versa and in addition several polar metabolites postulated as estradiol-17α and estrone sulfate were also found. Calculations of the interconversion of estradiol and estrone in the brain and pituitary revealed that they are approximately equal with a factor of 0.1 in either direction (Kazama and Longcope, 1974), which is different from the general situation where the oxidation predominates over the reduction (Fishman et al., 1960a).

Indirect evidence for the in vivo interconversion of estradiol and estrone in the monkey can also be derived from the studies of Flores et al. (1973) in which rhesus monkey brains were perfused with androstenedione and estrone and estradiol were among the products obtained. While the estradiol might have been derived from preformed testosterone, it is more probable that its source was primarily from estrone produced by the aromatization of androstenedione.

In the general metabolic scheme of estradiol its oxidation to estrone is followed by hydroxylations either at C-16 or C-2. At this time there is no convincing evidence available of the presence of C-16 hydroxylation of estrogens in neuroendocrine tissues. Presl et al. (1973) report the isolation of labeled estradiol from the rat brain after the peripheral administration of labeled estradiol. This metabolite, however, need not have been formed within the brain but could have equally well been formed elsewhere and entered the brain subsequently. In contrast, sufficient data are now available to conclude that C-2 hydroxylation of the estrogens does occur in the neuroendocrine tissues, that it is substantial and that it is highly localized. Because the products of C-2 hydroxylation are catecholestrogens which are extremely labile their isolation in the limited quantities expected would have been very difficult and therefore the first evidence of the existence of the reaction was obtained from the displacement of tritium from [2-^{3}H]estradiol. It has already been previously shown (Fishman et al., 1970) that C-2 hydroxylation of [2-^{3}H]estrogens proceeds with transfer of the isotope into water, that the reactions are not accompanied by an isotope effect or an NIH shift and that the isotope content of the water can be used as an index of the presence and extent of the reaction. Incubations of homogenized central tissues and pituitaries obtained fresh from female rats with [2-^{3}H]estradiol in the presence of cofactors were followed by lyophilization of incubation water and determination of its specific activity (Fishman and Norton, 1975a, b). Blank control incubations were also run to determine background non-enzymatic tritium transfer and liver incubations served as positive controls. Repeat experiments using [2-^{3}H]estrone and [2-^{3}H]estrone sulfate gave results quite similar to those obtained with [2-^{3}H]estradiol. In view of the previously described transformations in central tissues such as desulfation and estrone-estradiol interconversion no judgement as to which is the preferred substrate could be made. Incubations with tissues from ovariectomized animals on the other hand produced evidence of significantly greater 2-hydroxylation. The results with [2-^{3}H]estradiol summarized in Table I show that the hypothalamus was the most effective 2-hydroxylating tissue examined. Indeed on a weight basis it was more effective than the highly active liver tissue.

The nature of the products formed as a result of the tritium transfer from C-2 was conclusively proven in other experiments where the tissues were incubated with [^{14}C]estrone. The incubations were terminated by the addition of carrier 2-hydroxyestrone to

TABLE I

Formation of ^3H_2O from [2-^{3}H]estradiol during 25 min incubations

Tissue	Intact rats			Ovariectomized rats		
	Total (counts/min)	% Conversion	Tissue (counts/min/mg)	Total (counts/min)	% Conversion	Tissue (counts/min/mg)
Control	1,540*	0.24	--	1,620*	0.25	--
Liver	31,100	4.78	53	--	--	--
C. cortex	46	0.01	0	989	0.15	4
Pituitary	0	0	0	1,690	0.26	37
Hypothalamus	3,960	0.6	140	8,520	1.30	289

*These counts have been subtracted from all of the other results in this table.

serve as a trap. The material was then converted to the phenazine derivative (Gelbke and Knuppen, 1973) and recrystallized to constant specific activity. The radioactivity remaining with the 2-hydroxyestrone confirmed the production of this material by hypothalamic tissue, and its virtual absence from the cerebral cortex incubations. Experiments, the results of which will be published shortly (Fishman et al., 1976), have now also confirmed 2-hydroxylation of estrogens by human fetal neuroendocrine tissues.

The securely established transformations of estrogens in neuroendocrine tissues described above include hydrolysis of estrone sulfate, interconversion of estradiol and estrone and 2-hydroxylation to catechol estrogens, and it is necessary to consider their significance. It is possible to speculate that the reactions catalyzed by the sulfatase and 17β-oxidoreductase are necessary prerequisites to subsequent biotransformations since they lead to free estrone which, as has already been noted, is a central compound in estrogen metabolism. The localized biosynthesis of catecholestrogens in specific neuroendocrine sites may have critical implications in the estrogen control of gonadotrophin release and in other estrogen central effects. Experiments already concluded show that catecholestrogens in immature male rats induce a large increase in plasma gonadotrophins (Naftolin et al., 1975). When administered intracerebrally 2-hydroxyestradiol has been reported (Parvizi and Ellendorf, 1975) to lower gonadotrophins in immature pigs, and experiments are under way in other species. Several plausible mechanisms can be put forward for this action of the catecholestrogens. One explanation hinges on the fact that the catecholestrogens are excellent competitive inhibitors of catechol-O-methyl transferase mediated O-methylation of catecholamines (Ball et al., 1972), which are themselves implicated in gonadotrophin control mechanisms (Rubinstein and Sawyer, 1970). Another explanation resides in the effective competition of the catecholestrogens for the estrogen receptors (Davies et al., 1975b). Since these compounds are not estrogenic themselves (Gordon et al., 1964), their binding confers on them antiestrogenic properties which may be involved in their central nervous system functions.

The results already obtained from the study of estrogen biotransformations in neuroendocrine tissues, and some of the consequences which have already ensued from these studies lend impetus to further investigations to elucidate the further metabolic pathways of the estrogens in all of the sites composing the neuroendocrine system. It is hoped that such knowledge will aid immeasurably in the study of the intricate and varied mechanism of neuroendocrine controls wich still elude our understanding.

REFERENCES

Ball, P., Knuppen, R., Haupt, M. and Breuer, H. (1972) Interactions between estrogens and catecholamines. *J. clin. Endocr.*, 34: 736–746.

Ball, P., Gelbke, H.P. and Knuppen, R. (1975) The excretion of 2-hydroxyestrone during the menstrual cycle. *J. clin. Endocr.*, 40: 406–408.

Blunt, J.W. and DeLuca, H.F. (1969) The synthesis of 25-hydroxycholerol-ciferol a biologically active metabolite of vitamin D_3. *Biochemistry*, 8: 671–674.

Bruchowsky, N. and Wilson, J.D. (1968) The conversion of testosterone to 5α-androstan-17β-ol-3-one by rat prostate in vivo and in vitro. *J. biol. Chem.*, 243: 2012–2021.

Davies, I.J., Siu, J., Naftolin, F. and Ryan, K.J. (1975a) Cytoplasmic binding of steroids in brain tissues and pituitary. *Advanc. Biosci.*, 15: 89–103.

Davies, I.J., Naftolin, F., Ryan, K.J., Fishman, J. and Siu, J. (1975b) The affinity of catecholstrogens for estrogen receptors in the pituitary and anterior hypothalamus of the rat. *Endocrinology*, 97: 554–557.

Diczfalusy, E. and Levitz, M. (1970) Formation, metabolism and transport of estrogen conjugates. In *Chemical and Biological Aspects of Steroid Conjugation*, S. Bernstein and S. Solomon (Eds.), Springer, Berlin.

Eisenfeld, A.J. (1969) Hypothalamic estradiol binding macromolecules. *Nature (Lond.)*, 224: 1202–1203.

Ferin, M., Carmel, P.W., Zimmerman, E.A., Warren, M. and Van de Wiele, R.L. (1974) Location of intrahypothalamic estrogen responsive sites influencing LH secretion in the female rhesus monkey. *Endocrinology*, 95: 1059–1068.

Fishman, J. and Norton, B. (1975a) Brain catecholestrogens. Formation and possible function. *Advanc. Biosci.*, 15: 123–131.

Fishman, J. and Norton, B. (1975b) Catecholestrogen formation in the central nervous system of the rat. *Endocrinology*, 96: 1054–1059.

Fishman, J., Bradlow, H.L. and Gallagher, T.F. (1960a) Oxidative metabolism of estradiol. *J. biol. Chem.*, 235: 3104–3107.

Fishman, J., Cox, R.I. and Gallagher, T.F. (1960b) 2-Hydroxyestrone: a new metabolite of estradiol in man. *Arch. Biochem. Biophys.*, 90: 318–319.

Fishman, J., Hellman, L., Zumoff, B. and Gallagher, T.F. (1963) Effect of thyroid on hydroxylation of estrogen in man. *J. clin. Endocr.*, 25: 365–368.

Fishman, J., Guzik, H. and Hellman, L. (1970) Aromatic ring hydroxylation of estradiol in man. *Biochemistry*, 9: 1593–1598.

Fishman, J., Naftolin, F., Davies, I.J., Ryan, K.J. and Petro, Z. (1976). *J. clin. Endocr.*, in press.

Flores, F., Naftolin, F., Ryan, K.J. and White, R.J. (1973) Estrogen formation by the isolated perfused rhesus monkey brain. *Science*, 180: 1074.

Gelbke, H.P. and Knuppen, R.P. (1973) Synthesis of specific phenazine derivatives of 2-hydroxyestrogens. *Steroids*, 21: 689–702.

Gordon, S., Cantrall, E.W., Cekleniak, W.P., Albers, H.J., Maner S., Stolar, S.M. and Bernstein, S. (1964) The hypocholesteremic effect of estrogen metabolites. *Steroids*, 4: 267–271.

Jenkin, G., Henville, A. and Heap, R.B. (1975) Metabolism of estrone sulphate and binding of estrogens by the brain and pituitary of fetal and adult sheep. *J. Endocr.*, 62: 22.

Kazama, N. and Longcope, C. (1974) In vivo studies on the metabolism of estrone and estradiol-17β by the brain. *Steroids*, 23: 469–481.

Kishimoto, Y. (1973) Estrone sulphate in rat brain: uptake from blood and metabolism in vivo. *J. Neurochem.*, 20: 1489–1492.

Leavitt, W.W., Friend, J.P. and Robinson, J.A. (1969) Estradiol: specific binding by pituitary nuclear fraction in vitro. *Science*, 165: 496–498.

Luttge, W.G., Hall, N.R. and Wellis, C.J. (1975) Physiologic and pharmacologic actions of hormonal steroids in sexual behavior. In *Sexual Behavior: Pharmacology and Biochemistry*, M. Sandler and G.L. Gessa (Eds.), Raven Press, New York, pp. 209–219.

Naftolin, F., Morishita, H., Davies, I.J., Todd, R., Ryan, K.J. and Fishman, J. (1975) 2-Hydroxyestrone induced rise in serum LH in the immature male rat. *Biochem. biophys. Res. Commun.*, 64: 905–910.

Parvizi, N. and Ellendorf, F. (1975) 2-Hydroxyestradiol as a possible link in steroid brain interaction. *Nature (Lond.)*, 256: 59–60.

Payne, A.H., Lawrence, C.C., Foster, D.L. and Jaffe, R.B. (1973) Intranuclear binding of estradiol and estrone in female ovine pituitaries following incubation with estrone sulfate. *J. biol. Chem.*, 248: 1598–1602.

Perumal, A.S. and Robins, E. (1973) Arylsulphatases in human brain: purification and characterization of an insoluble arylsulphatase. *J. Neurochem.*, 21: 459–471.

Presl, J., Herzmann, J., Rohling, S. and Horsky, J. (1973) Regional distribution of estrogenic metabolites in the female rat hypothalamus. *Endocr. Exp.*, 7: 119.

Rubinstein, L. and Sawyer, L.H. (1970) Role of catecholamines in stimulating the release of pituitary ovulating hormones in rats. *Endocrinology*, 86: 988–995.

Ryan, K.J., Naftolin, F., Reddy, V., Flores, F. and Petro, Z. (1973) Estrogen formation in the brain. *Amer. J. Obstet. Gynec.*, 114: 454–460.

Chapter 19

Oxidative metabolism in neuroendocrine tissue

OMAR SCHIAFFINI

Collegio Universitario, Division de Medicina, Departamento de Fisiologia, Las Palmas de Gran Canaria (Spain)

The metabolic activity of cerebral tissue is mainly based on the biotransformation of glucose. It has been shown that the amounts of oxygen used and lactic acid produced are directly proportional to the quantity of glucose made available; other energy sources are also used but to a much lower extent (Libertun, 1969).

It has been also observed that 60—70% of glucose reaching the brain is converted to lactic acid, about 10% is used for the production of amino acids and their derivatives and that 20% enters the respiratory cycle (Lowry, 1966; Libertun, 1969).

It is also known that the brain is one of the body structures avid for oxygen; the human brain alone consumes about 20% of the oxygen made available, and among the cerebral structures the hypothalamus is one of the more active in utilizing oxygen (Greville, 1962).

On the basis of this kind of observation, and since neuroendocrine processes also require energy, it was deemed of interest to measure the variations in oxygen consumption (QO_2) taking place in the brain in different endocrine situations (Moguilevsky and Malinow, 1964). In fact from what has been said above, it is acceptable to conclude that evaluation of oxygen uptake is a good index of the metabolic transformations realized in the brain.

Other parameters might be taken into consideration, such as lactic acid production, CO_2 release, etc., but our previous experiments have shown that lactic acid production varies only when profound endocrine changes occur or are induced in the experimental animal, for example, when it is rendered diabetic (Schiaffini et al., 1968). CO_2 production could be a more useful parameter, but only CO_2 coming from the direct glycolytic pathway undergoes changes according to the different neuroendocrine mechanisms (Schiaffini et al., 1970).

Such kind of studies requests particular techniques which are not easily reproducible. On the contrary, oxygen uptake measurement is based on an easy and very reliable methodology.

We have therefore carried out a series of studies on the metabolic activity of the hypothalamus, and of some extrahypothalamic areas, of female and male rats in various endocrine situations, based on the measurement of oxidative activity (QO_2) of these brain structures according to Warburg manometric technique (Umbreit et al., 1959), modified for the evaluation of QO_2 of small fragments of tissue (Libertun, 1969; Schiaffini, 1974).

We have often found it possible to explain in terms of bioenergetic processes the neuroendocrine phenomena taking place in the areas to which our interest was directed. Consequently a series of comparisons with data more properly pertaining to neuroendocrinology have been done. For sake of brevity only key papers, mainly reviews, will be quoted in this survey.

Our first series of experiments was directed to check whether hypothalamic QO_2 was varying according to the different phases of the estrous cycle of adult female rats.

For this purpose, adult female rats with a history of regular 4-day estrous cycles were sacrificed in the afternoon during the different phases of their cycle. They were sacrificed by decapitation, their hypothalami were carefully removed and divided into 3 portions following a rostrocaudal progression. The anterior part was represented by the portion of hypothalamic tissue anterior to a cut performed along the posterior border of the optic chiasm. The middle region was that going from the posterior aspect of the optic chiasm to the posterior border of the infundibulum, this portion was consequently represented essentially by the median eminence; finally the posterior hypothalamus was the tissue included between the infundibulum and the mammillary bodies (Moguilevsky et al., 1966; Libertun, 1969; Schiaffini, 1974).

Fragments of anterior cerebral cortex were used as control. A further comparison was done with hypothalami similarly dissected from normal adult male rats.

As can be seen in Table I, oxygen consumption of the anterior hypothalamus of cycling females varied significantly during the different phases of the estrous cycle, being at the lowest level during diestrus and reaching a maximum at estrus. QO_2 at proestrus appeared to be lower than at estrus but such a difference was not statistically significant. A similar pattern was also evident in the posterior hypothalamus. Nothing significant occurred at the level of the medial hypothalamus and of the cerebral cortex. The QO_2 of the different areas of the male hypothalami was always comparable to that of female rats in diestrus.

These findings can be considered to be a direct proof that the pituitary depends on the hypothalamus for displaying its activity.

It is an old observation that the anterior pituitary secretes high amounts of gonadotropins during proestrus and estrus while its secretory activity is reduced during diestrus (Everett, 1956, 1964; Goldman and Mahesh, 1968; Goldman et al., 1969; Gay et al., 1970).

Our data showed that the hypothalamus was also undergoing similar cyclic fluctuations in terms of oxidative activity, in accordance with pituitary secretion during the different phases of the female rat estrous cycle.

It could be possible to conclude that these fluctuations of hypothalamic metabolic activity reflect energy requirements according to the higher or lower necessities of synthesizing gonadotropin releasing hormone(s).

Such a point of view is corroborated by preliminary results from our laboratory: incorporation of labeled amino acids by the anterior and posterior hypothalamus and anterior pituitary increases at proestrus and estrus (Moguilevsky et al., 1972). This could indicate a higher biosynthetic activity correlated with a higher secretion of pituitary gonadotropins.

Also the metabolic specialization of the different areas of the hypothalamus is in agreement with the literature suggesting that gonadotropin releasing hormone(s) are mainly produced in the anterior and posterior hypothalamus (Barraclough, 1966; Mess et al., 1967; Martini et al., 1968; Gual and Rosemberg, 1973; Motta et al., 1975). The hypothesis that the anterior and posterior hypothalamus are the loci of the regulation of anterior pituitary gonadotropin secretion, while the middle hypothalamus (the median eminence) should be considered only the storage site of such compound(s) (Martini et al., 1968), could account for the lack of metabolic activity of the middle portion of the hypothalamus.

Our observations also confirm a broad literature showing that profound neuroendocrine differences exist between female and male rats. While in females, changes in QO_2 were seen, according to the various stages of the estrous cycle, in males the parameter under study in the different regions of the hypothalamus was always not significantly different from the levels of oxygen consumption of females in diestrus.

If our assumption that metabolic activity is an index of hypothalamic production of releasing hormone(s) is valid, we could also conclude from this point of view that in males only a basal, low hypothalamic activity is necessary and sufficient to regulate anterior pituitary function in terms of gonadotropin secretion. In this context our hypothesis is corroborated by the data reported in Table II. Female rats, in which a persistent estrus was induced by neonatal treatment with testosterone, were killed in adulthood and their hypothalamic oxygen uptake was evaluated.

TABLE I

Oxidative activity of the hypothalamus and of the frontal cerebral cortex of adult male rats and of adult female rats in different phases of the estrous cycle *

Groups	QO_2 ($\mu l\ O_2$/mg wet tissue/hr)			
	Hypothalamus			Cerebral cortex
	Anterior	Medial	Posterior	
Males (M)	1.63 ± 0.04 (16)	1.67 ± 0.08 (16)	1.53 ± 0.08 (16)	1.62 ± 0.02 (14)
Diestrus (D)	1.57 ± 0.04 (17)	1.73 ± 0.04 (15)	1.61 ± 0.04 (14)	1.71 ± 0.07 (13)
Proestrus (P)	1.71 ± 0.04 (10)	1.67 ± 0.08 (12)	1.81 ± 0.08 (11)	1.59 ± 0.06 (14)
Estrus (E)	1.81 ± 0.07 (14)	1.76 ± 0.06 (15)	1.85 ± 0.06 (16)	1.58 ± 0.07 (13)
Statistical significance** $p < 0.01$	M vs. E D vs. E	NS	M vs. P M vs. E D vs. E	NS

* Values are mean ± S.E.M. Figures in parentheses are number of determinations.
** According to the multiple comparisons test of Tukey (1953). NS, not significant.

TABLE II

Oxidative activity of the hypothalamus and of the frontal cerebral cortex of female rats in persistent estrus by neonatal testosterone treatment*

Groups	QO_2 ($\mu l\ O_2$/mg wet tissue/hr)			
	Hypothalamus			Cerebral cortex
	Anterior	Medial	Posterior	
(E) Estrus	1.80 ± 0.06 (20)	1.70 ± 0.06 (9)	1.85 ± 0.06 (10)	1.59 ± 0.07 (13)
(D) Diestrus	1.55 ± 0.05 (21)	1.73 ± 0.04 (15)	1.60 ± 0.04 (14)	1.62 ± 0.06 (15)
(A) Androgenized females	1.57 ± 0.06 (23)	1.72 ± 0.05 (13)	1.84 ± 0.06 (11)	1.60 ± 0.06 (9)
(M) Males	1.59 ± 0.04 (24)	1.74 ± 0.06 (10)	1.55 ± 0.04 (12)	1.58 ± 0.06 (14)
Statistical significance** $P < 0.01$	E vs. D E vs. A E vs. M	NS	E vs. D E vs. M D vs. A A vs. M	

* Values are mean ± S.E.M. Figures in parentheses are number of determinations.
** According to the multiple comparisons test of Tukey (1953). NS, not significant.

It was evident that QO_2 of the anterior hypothalamus of androgenized females did not differ from that of normal males and both the values again were significantly lower than those found in normal females during estrus; in both the cases the data obtained were similar to those found during the diestrus phase of normally cycling females. Again these findings are in good agreement with the literature showing that females androgenized early after birth do not cycle and do not ovulate (Barraclough, 1966). In particular we could infer that the effects of male steroids are also reflected by metabolic activity.

In particular we have seen that the anterior hypothalamus is affected by androgen treatment while the posterior portion is not (Table II). This finding could explain the results of the studies by Barraclough (1966, 1967) and by Halász (1969) which tend to show that the "trigger" for ovulation is localized in the anterior hypothalamus. In the light of our results it could be inferred once more that the anterior hypothalamus is responsible for the cyclic secretion of gonadotropins; and furthermore it could be proposed that the mechanisms which cause tonic secretion of LH and FSH are located in the rat posterior hypothalamus. Recent reports in fact show that lesions of the anterior hypothalamus or hypothalamic cuts reduce but do not abolish gonadotropin secretion (Blake et al., 1972).

It has also been shown that LH and FSH of rat anterior pituitary are not particularly affected by such manipulations (Mess et al., 1967; Tima et al., 1969; Blake et al., 1972). These findings might be interpreted as indicating that the posterior hypothalamus has still the capacity for inducing the production of gonadotropins but that the signal for their release from the anterior pituitary is lacking.

Alternatively it could be taken into consideration that, in the absence of the stimulatory signal coming from the preoptic area, other inhibitory parts of the brain take over.

For instance Gallo et al. (1971) have published results tending to attribute to hippocampal efferents impinging on the arcuate nucleus an inhibitory role on the activity of this hypothalamic structure. Mess et al. (1971) have reported that female rats with lesions of the anterior hypothalamus show a sort of ovulatory cycle after pinealectomy.

These observations directed our interest to study the metabolic activity of some extrahypothalamic areas which are known to be correlated with the hypothalamus: more precisely we have chosen the limbic system.

From an anatomical point of view, it has been ascertained that at least two components of the limbic system, the amygdala and the hippocampus, send efferents to the hypothalamus (Raisman and Field, 1971; Moguilevsky and Schiaffini, 1972).

Furthermore, studies exist showing that chemical, surgical and electrical manipulations of these brain structures bring about a series of modifications of the reproductive system, even if the results obtained by the various authors are conflicting and controversial (for a review see Piva et al., 1973; Schiaffini and Martini, 1973).

In an attempt to get more information on the role of these brain areas on neuroendocrine processes in the female rat, QO_2 of the amygdala and hippocampus from female rate in diestrus and estrus was evaluated. The whole amygdala was collected, while only the ventral portion of the hippocampus was dissected out and used for this study.

The ablation of the amygdala and of the basal aspect of the hippocampus was done according to the method of Foglia et al. (1973). Basically, the amygdala was represented by the parts of cerebral tissue extending among the following limits: anteriorly up to the ideal lateral prolongation of the optic chiasm, posteriorly until the entorhinal fossa, medially to the hypothalamic lateral fissure and dorsally until the entorhinal fissure. The medial end was another ideal line connecting the entorhinal and hypothalamic lateral fissures.

The basal hippocampus was represented by all the cerebral tissue extending from the posterior border of the amygdala to the "subiculum". After dissection of the two structures, the cortical tissue connected with these two limbic structures was carefully removed.

Table III shows that the amygdala and the hippocampus were also undergoing cyclic variations according to the phases of female rat estrous cycle but in an inverse fashion. The amygdala was behaving like the hypothalamus: it showed the lowest QO_2 at diestrus and the highest at estrus. In contrast the hippocampus had its maximum oxygen uptake during diestrus, while the minimal activity took place in estrus. As already shown above the metabolic activity of the frontal cerebral cortex did not change throughout the estrous cycle.

These results are in line with previous observations which attribute to the limbic system a role in the modulation of the neuroendocrine machinery leading to gonadotropin release.

What is interesting is that the hippocampus and the amygdala work in opposition. We have no data to try to explain why this is happening.

These results confirm other findings. Teresawa and Timiras (1968) have reported that

TABLE III

Oxidative activity of limbic structures and of frontal cortex in different phases of the estrous cycle of adult female rats*

Groups	QO_2 ($\mu l\ O_2$/mg wet tissue/hr)		
	Amygdala	Hippocampus	Cerebral cortex
Diestrus	1.08 ± 0.11 (15)	1.43 ± 0.10 (26)	1.31 ± 0.12 (13)
Estrus	1.51 ± 0.14 (17)	1.18 ± 0.08 (24)	1.30 ± 0.11 (13)
P value	< 0.02	< 0.05	NS

* Values are mean ± S.E. Figures in parentheses are number of determinations. NS, not significant.

the electrical activity of the amygdala and of the hippocampus is opposite in the various phases of the estrous cycle. There is also evidence indicating that the amygdala has a stimulatory role and the hippocampus an inhibitory one on ACTH secretion (Mangili et al., 1966). It has also been proposed that the hippocampus may have an inhibitory influence on gonadotropin secretion (Gallo et al., 1971).

Data are now available suggesting that the limbic system may be a station for the realization of sex steroid feedback in female rats (Piva et al., 1973).

Also, other structures of the brain are known to exert an influence on the reproductive system (for instance the pineal) and it might be that we are observing a series of events not yet finalized but still to be integrated before giving their real effect on the hypothalamic–pituitary unit, as has been proposed by Motta et al. (1970).

Alternatively it could be proposed that one or other cerebral structure takes over and develops its particular actions on the hypothalamic–pituitary unit according to the changes taking place in the endocrine milieu.

Finally we have not studied with our approach what is happening in the dorsal hippocampus during the different phases of the female rat estrous cycle.

The possibility could also exist that the differences we observed between the hypothalamus and the limbic system occurred by chance. But such a hypothesis had to be rejected when we studied the effect of neonatal androgenization by testosterone.

The effect of neonatal androgenization on the amygdala and the hippocampus is shown in Table IV.

It is possible to see that treatment with the androgen decreased amygdalar QO_2 to the levels already seen in the anterior hypothalamus during diestrus (see Table II), while there was a significant enhancement of hippocampal metabolism. This could make attractive the hypothesis that the amygdala and the ventral hippocampus direct their effects on different parts of the hypothalamus, but when castration was performed in female rats the data obtained were difficult to be interpreted in such a way.

In fact Table V shows that in adult female rats, ovariectomized 30 days previously, there was a dramatic decrease of hypothalamic oxidative activity. On the contrary, amygdala and hippocampus showed an increase comparable to that found when QO_2 reaches its maximum in the two structures (Table III).

To try to find an explanation of these results, more experiments are needed but they have not been done yet, even if they are in project. For instance it would be of interest to check whether this "blockade" of the activity of these two structures can be manipulated by different sex steroids present in the female reproductive system; how LH, FSH and

TABLE IV

Oxidative activity of the limbic system of androgenized female rats*

Groups	QO_2 ($\mu l\ O_2$/mg wet tissue/hr)	
	Amygdala	Hippocampus
D. Diestrus	1.07 ± 0.10 (15)	1.41 ± 0.09 (24)
E. Estrus	1.57 ± 0.12 (11)	1.21 ± 0.08 (21)
A. Androgenized females	1.16 ± 0.12 (19)	1.46 ± 0.08 (20)
Statistical significance** $P < 0.05$	D vs. E E vs. A	D vs. E E vs. A

* Values are mean ± S.E. Figures in parentheses are number of determinations.
** According to the multiple comparisons test of Tukey (1953). NS, not significant.

TABLE V

Oxidative activity of hypothalamus and of the limbic system of adult female rats. Effect of ovariectomy*

Groups	QO_2 (O_2/mg wet tissue/hr)		
	Hypothalamus	Amygdala	Hippocampus
D. Diestrus	1.30 ± 0.07 (14)	1.09 ± 0.09 (19)	1.44 ± 0.08 (31)
E. Estrus	1.59 ± 0.08 (17)	1.58 ± 0.11 (28)	1.21 ± 0.06 (35)
O. Ovariectomized	0.99 ± 0.11 (11)	1.43 ± 0.07 (15)	1.61 ± 0.08 (15)
Statistical significance** $P < 0.05$	D vs. E D vs. O E vs. O	D vs. E D vs. O	D vs. E E vs. O

* Values are mean ± S.E. Figures in parentheses are number of determinations.
** According to the multiple comparisons test of Tukey (1953).

TABLE VI

Effect of castration and substitutive therapy on the oxidative activity of the hypothalamus and of the pituitary of adult male rats*

Groups	QO_2 ($\mu l\ O_2$/mg wet tissue/hr)			
	Hypothalamus			Pituitary
	Anterior	Medial	Posterior	
C. Control	1.10 ± 0.05 (12)	1.00 ± 0.06 (11)	1.20 ± 0.06 (12)	0.40 ± 0.05 (10)
Cx. Castrated	0.70 ± 0.08 (17)	1.10 ± 0.08 (13)	0.90 ± 0.07 (17)	0.81 ± 0.10 (13)
TP. Castrated + TP (150 µg twice a week)	1.08 ± 0.10 (12)	1.10 ± 0.07 (13)	1.15 ± 0.07 (12)	0.40 ± 0.06 (12)
Statistical significance** $P < 0.01$	C vs. Cx Cx vs. TP	NS	C vs. Cx Cx vs. TP	C vs. Cx Cx vs. TP

* Values are mean ± S.E. Figures in parentheses are number of determinations.
** According to the multiple comparisons test of Tukey (1953). NS, not significant. TP, testosterone propionate.

prolactin could affect the limbic system and conversely how the limbic system might affect gonadotropin secretion.

For this reason we directed our studies to the metabolic activity of different brain structures of the male rat in different endocrine situations, with the hope that without the complication of the female estrous cycle it would be easier to have an idea of how the limbic system, the hypothalamus and the pituitary can be correlated.

Table VI shows how the hypothalamic and pituitary oxidative metabolism of adult male rats is affected by the lack or presence of male sex steroids. The experimental animals were represented by adult males castrated 30 days previously and treated until the moment of sacrifice either with testosterone propionate suspended in oil or with the vehicle alone.

It can be seen that the anterior and posterior hypothalamus of castrated animals have a significant decrease of QO_2 when compared to that of normal rats, and that testosterone propionate given to castrated animals restores the normal situation. In contrast pituitary metabolic activity is enhanced by castration and becomes normal again after substitution therapy.

If we assume that metabolic activity parallels neuroendocrine phenomena these data would suggest that sex steroids inversely affect the hypothalamus and pituitary: in fact castration decreases hypothalamic, but increases pituitary, activity. Testosterone propionate restores the normal situation, increasing hypothalamic QO_2 and decreasing pituitary oxygen consumption.

In terms of feedback one would be inclined to interpret these data in the sense that two kinds of long feedback sustained by testosterone exist in the male: a positive one regarding the hypothalamus and a negative one affecting the pituitary.

From these data it is not possible to conclude which of the two effects is more important in terms of gonadotropin release, since the plasma titers of these hormones were not studied. Considering anyway that pituitary QO_2 was doubled after castration we could assume that the inhibitory action of testosterone is more important; consequently very likely we are observing the metabolic aspect of the negative feedback effect of testosterone on gonadotropin secretion, already reported by others (Martini et al., 1968; Davidson, 1969; Gay and Bogdanove, 1969).

The experiments to check whether two sites sensitive to testosterone could exist in the hypothalamic—pituitary unit did not confirm the hypothesis outlined above.

The initial series of results are reported in Table VII. In this experiment the QO_2 of either hypothalamic or pituitary tissue from normal and castrated animals was measured; the oxygen uptake of hypothalami and pituitaries of castrated animals with testosterone propionate present in the medium ("in vitro") was also evaluated. One can see that testosterone was active in reducing the enhanced pituitary QO_2 induced by castration, but did not significantly affect hypothalamic activity. With all the reservations necessary to compare "in vivo" with "in vitro" results, one could explain these findings in view of a differing sensitivity of the hypothalamus and of the pituitary to androgens.

It has been recently demonstrated that testosterone is not only acting "per se" on its target organs but also through its 5-alpha-reduced metabolites (Baulieu et al., 1968; Wilson and Gloyna, 1970); that the hypothalamus and pituitary also have the enzymatic equipment for such transformations (Kniewald et al., 1971) and that the pituitary, in particular, converts testosterone at a higher rate than the hypothalamus (Massa et al., 1972).

It could then be proposed that the concentration of testosterone able to stimulate pituitary metabolic activity "in vitro" would not instead affect the hypothalamus. More precisely one could think that the pituitary "in vitro" could be able to react directly to testosterone, while the hypothalamus is much less able to convert testosterone (Massa et al., 1972), once isolated from its afferents.

On the other hand the results reported in Table VII can be interpreted also in another way, i.e. they could indicate the existence of a negative short feedback between the pituitary and the hypothalamus: when the metabolic activity of the pituitary is high, that of the hypothalamus is decreased. Again we require a metabolic confirmation of the general theory on short feedback (Motta et al., 1969).

This second interpretation was tested by the following experiments.

The oxygen consumption of the hypothalamus of animals hypophysectomized 15 days previously was measured. The effect of replacement with either LH or FSH was also tested. The results obtained are shown in Tables VIII and IX.

It is evident that hypophysectomy significantly enhances the metabolic activity of the anterior and posterior hypothalamus and that the administration of either LH or FSH brought the oxygen uptake of the hypothalamus of hypophysectomized animals back to control values. These findings would indicate that, of the two theories about the mechanism of action of testosterone put forward above, the one emphasizing the importance of the pituitary is more attractive: the hypothalamus would not be sensitive to testosterone

TABLE VII

Effect of castration and testosterone "in vitro" on the oxygen uptake of hypothalamic and pituitary tissue of adult male rats*

Groups	QO_2 ($\mu l\ O_2$/mg wet tissue/hr)			
	Hypothalamus			Pituitary
	Anterior	Medial	Posterior	
C. Control	1.12 ± 0.04 (11)	1.00 ± 0.03 (12)	1.20 ± 0.08 (10)	0.40 ± 0.05 (12)
Cx. Castrated	0.75 ± 0.08 (9)	1.00 ± 0.05 (9)	0.90 ± 0.06 (9)	0.80 ± 0.09 (12)
TP. Castrated + TP (30 µg/ml "in vitro")	0.80 ± 0.07 (11)	1.10 ± 0.07 (10)	0.90 ± 0.07 (11)	0.50 ± 0.07 (10)
Statistical significance** $P < 0.01$	C vs. Cx C vs. TP	NS	C vs. Cx C vs. TP	C vs. Cx

* Values are mean ± S.E. Figures in parentheses are number of determinations.
** According to the multiple comparisons test of Tukey (1953). NS, not significant. TP, testosterone propionate.

TABLE VIII

Effect of hypophysectomy on the oxidative metabolism of the hypothalamus of adult male rats*

⁰ values are mean ± S.E. Figures in parentheses are number of determinations. NS, not significant.

Groups	QO_2 ($\mu l\ O_2$/mg wet tissue/hr)		
	Hypothalamus		
	Anterior	Medial	Posterior
Control	1.47 ± 0.04 (15)	1.44 ± 0.07 (13)	1.41 ± 0.07 (17)
Hypophysectomized	1.80 ± 0.05 (15)	1.31 ± 0.05 (13)	1.70 ± 0.09 (17)
P value	< 0.01	NS	< 0.05

* Values are mean ± S.E. Figures in parentheses are number of determinations. NS, not significant.

"per se", but to the modifications that the androgen is inducing in the pituitary.

The hypothalamus did not react "in vitro" to testosterone because it is sensitive, in terms of feedback, only to gonadotropins which were not present in the system chosen (Table VIII).

Other interesting information is presented in Table IX: FSH affected only the posterior hypothalamus, and LH only the anterior hypothalamus. Neither of the two hormones altered the activity of the medial portion of the hypothalamus. These findings could speak again in favor of a specialization of the different hypothalamic areas in controlling anterior pituitary function.

Furthermore these observations could be the spectacular representation of data which at present are accumulating to suggest that, at least to a certain extent, different mechanisms exist to control the secretion of LH and the secretion of FSH (Zanisi et al., 1973; Borrel, Piva and Martini, personal communication).

In view of the results on females, the possible role of the limbic system in male rats in the control of gonadotropin secretion was also tested. For this purpose QO_2 of the amygdala and of the hippocampus was studied in normal and castrated males. Also the effect of testosterone "in vitro" and "in vivo" was observed in an initial series of experiments. Table X shows that the QO_2 of the amygdala is augmented by castration similar to that seen for the female amygdala (Table V). This is contrary to what has been observed for the male hypothalamus (Table VII).

Treatment with testosterone "in vitro" did not affect the high values of amygdalar QO_2 induced by castration. Finally testosterone given "in vivo" restored the metabolic activity of the amygdala to normal. The QO_2 of the hippocampus was not affected in any of the experiments.

Comparison of Tables VII and X gives two conclusions.

(1) The metabolic activity of the male hippocampus is not affected by manipulations of the endocrine system, contrary to what has been observed in the hypothalamus. Furthermore this situation also differs from that which occurs in the female limbic system (see Tables III–V). This result could be considered another proof of the difference

TABLE IX

Effect of hypophysectomy and of gonadotropin replacement on the oxidative metabolism of the hypothalamus of adult male rats[*]

Groups	QO_2 ($\mu l\, O_2$/mg wet tissue/hr)		
	Hypothalamus		
	Anterior	Medial	Posterior
C. Control	1.46 ± 0.04 (15)	1.44 ± 0.07 (13)	1.41 ± 0.07 (17)
Hx. Hypophysectomized	1.82 ± 0.06 (15)	1.49 ± 0.08 (12)	1.75 ± 0.10 (15)
FSH. Hypophysectomized + 250 µg of FSH	1.79 ± 0.10 (20)	1.40 ± 0.09 (15)	1.39 ± 0.08 (13)
LH. Hypophysectomized + 250 µg of LH	1.38 ± 0.06 (16)	1.39 ± 0.09 (18)	1.80 ± 0.10 (17)
Statistical significance[**] $P < 0.01$	C vs. Hx C vs. FSH Hx vs. LH FSH vs. LH	NS	C vs. Hx C vs. LH Hx vs. FSH FSH vs. LH

[*] Values are mean ± S.E. Figures in parentheses are number of determinations.

[**] According to the multiple comparisons test of Tukey (1953). NS, not significant.

TABLE X

Effect of castration and of testosterone (TP) replacement "in vitro" and "in vivo" on the oxidative metabolism of the amygdala and of the hippocampus of adult male rats*

Groups	QO_2 ($\mu l\ O_2$/mg wet tissue/hr)	
	Amygdala	Hippocampus
- C. Control	1.07 ± 0.08 (20)	1.44 ± 0.03 (15)
Cx. Castrated	1.44 ± 0.04 (12)	1.47 ± 0.07 (11)
Cxa. Castrated + TP "in vitro" (30 µg/ml)	1.44 ± 0.07 (10)	1.44 ± 0.05 (10)
Cxb. Castrated + TP "in vivo" (200 µg/rat) every second day for 4 weeks	1.09 ± 0.09 (10)	1.41 ± 0.09 (10)
Statistical significance** $P < 0.01$	C vs. Cx C vs. Cxa Cx vs. Cxb Cxa vs. Cxb	NS

* Values are mean ± S.E. Figures in parentheses are number of determinations.
** According to the multiple comparisons test of Tukey (1953). NS, not significant.

existing between male and female rats from an endocrinological point of view. It may be that in the female the complex series of events which leads to the estrous cycle needs a series of intermediate steps. One of them can be represented by events taking place in the hippocampus.

Being apparently simpler, it could be hypothesized that the hippocampus is not necessary in the neuroendocrine mechanisms controlling the reproductive system in the male, while the amygdala must still be involved.

(2) The oxygen uptake of the amygdala, as well as of the hypothalamus, in the male rat, was instead reacting to the effect of castration; contrary to the hypothalamus amygdalar metabolic activity was enhanced as a consequence of orchidectomy.

In the light of the findings regarding hypothalamic metabolic activity, suggesting that the effect of testosterone is mediated through gonadotropins, it remained to be investigated how the amygdala was affected by these changes of the endocrine milieu. In other words whether the amygdala was directly sensitive to the changes of circulating levels of testosterone or to the consequent variations of gonadotropin plasma titers.

The results of Table X already indicate the second hypothesis, according to information obtained from experiments with the hypothalamus. There was no reaction to testosterone in an in vitro system, but the metabolic activity of the amygdala was affected when testosterone was given in vivo.

The results presented in Tables XI and XII reinforce such a point of view. In fact it is

TABLE XI

Effects of hypophysectomy and of gonadotropins "in vivo" on the oxidative metabolism of the limbic system of adult male rats*

Groups	QO_2 ($\mu l\ O_2$/mg wet tissue/hr)	
	Amygdala	Hippocampus
C. Control	1.07 ± 0.08 (20)	1.44 ± 0.03 (15)
Hx. Hypophysectomized	0.74 ± 0.09 (12)	1.46 ± 0.06 (11)
LH. Hypophysectomized + LH "in vivo" (50 μg) daily doses for 5 days	1.00 ± 0.07 (11)	1.32 ± 0.08 (11)
FSH. Hypophysectomized + FSH "in vivo" (100 μg) daily doses for 5 days	1.21 ± 0.07 (11)	1.34 ± 0.06 (10)
G. Hypophysectomized + LH (50 μg) + FSH (100 μg) daily doses for 5 days	1.19 ± 0.09 (11)	1.36 ± 0.05 (12)
Statistical significance** $P < 0.01$	C vs. Hx Hx vs. LH Hx vs. FSH Hx vs. G	NS

* Values are mean ± S.E. Figures in parentheses are number of determinations.
** According to the multiple comparisons test of Tukey (1953). NS, not significant.

TABLE XII

"In vitro" effect of LH and FSH on the oxidative metabolism of the limbic system of adult male rats*

Groups	QO_2 ($\mu l\ O_2$/mg wet tissue/hr)	
	Amygdala	Hippocampus
N. Normal controls	1.07 ± 0.08 (20)	1.44 ± 0.03 (15)
LH. Normal + LH "in vitro" (50 μg/ml)	1.35 ± 0.03 (11)	1.38 ± 0.04 (11)
FSH. Normal + FSH "in vitro" (200 μg/ml)	1.58 ± 0.07 (10)	1.46 ± 0.11 (10)
Statistical significance** $P < 0.01$	N vs. LH N vs. FSH	NS

* Values are mean ± S.E. Figures in parentheses are number of determinations.
** According to the multiple comparisons test of Tukey (1953). NS, not significant.

TABLE XIII

Effect of electrolytic lesions of the cortical and of the basolateral nucleus of the amygdala on the oxidative metabolism of the hypothalamus and of the frontal cerebral cortex of adult male rats*

Groups	QO_2 ($\mu l\ O_2$/mg wet tissue/hr)	
	Hypothalamus	Cerebral cortex
C. Control	1.55 ± 0.02 (15)	1.42 ± 0.02 (15)
S. Sham	1.56 ± 0.07 (18)	1.44 ± 0.06 (17)
CN. Cortical nucleus lesioned	1.27 ± 0.06 (10)	1.45 ± 0.07 (10)
BN. Basolateral nucleus lesioned	1.19 ± 0.01 (15)	1.44 ± 0.01 (15)
Statistical significance** $P < 0.01$	C vs. CN C vs. BN S vs. CN S vs. BN	NS

* Values are mean ± S.E. Figures in parentheses are number of determinations.
** According to the multiple comparisons test of Tukey (1953). NS, not significant.

shown that hypophysectomy induces a significant decrease of amygdalar QO_2 which is brought back to normal by exogenous administration of gonadotropins; and this is also true for the castrated—hypophysectomized animals.

A further proof is given in Table XII: LH and FSH when added "in vitro" to amygdalar tissue of normal male rats significantly enhanced the amygdalar metabolic activity, while testosterone "in vitro" (see Table X) did not display any effect.

Again in no case was the hippocampus affected.

All the experiments reported here seem to indicate that changes of circulating levels of gonadotropins inversely affect the hypothalamus and the limbic system.

To have some information on the possible interrelations between the amygdala, the hypothalamus and the pituitary, we studied the effects of amygdalar electrolytic lesions on the oxidative metabolism of the hypothalamus and of the cerebral cortex; the latter tissue was used as a control.

Adult male rats were bilaterally lesioned either in the basolateral or in the cortical nucleus of their amygdalae, by the insertion of a platinum electrode through which a direct current of 3 mA was applied for 10 sec. Thirty-five days later the animals were killed, and the QO_2 of their hypothalamus and frontal cerebral cortex was measured.

As seen in Table XIII, it was shown that, as usual, the frontal cerebral cortex was not affected by the lesions performed in the two nuclei selected. On the contrary, lesion of either of the two nuclei induced a significant decrease of hypothalamic QO_2 (Schiaffini and Fernandez, 1974; Schiaffini and Marin, 1974).

These results, even if preliminary in nature, are in line with other reports attributing a stimulatory role in hypothalamic activity to the amygdala.

We do not at all exclude the possibility that the destruction of other areas of the amygdala, which we have not examined, would bring about other, and possibly opposite, effects on hypothalamic oxygen consumption. As already pointed out above, reports exist attributing to the amygdala an inhibitory influence on gonadotropin secretion (for a review see Schiaffini and Martini, 1973).

Our lesion results suggest that the amygdalar regions selected have a stimulatory role on the hypothalamus. The hypothalamus, from the point of view of the control of pituitary gonadotropin secretion, has a stimulatory action on the pituitary. The pituitary then becomes the controlling device of the system amygdala–hypothalamus–pituitary. According to the signals (changes of plasma titers of testosterone?) it stimulates the "circuit" acting on the amygdala or blocks it from inhibiting the hypothalamus.

More simply it can be imagined that the positive "circuit", amygdala–hypothalamus–pituitary, is working automatically and only when necessary is its activity interrupted by the short negative feedback exerted by the pituitary on the hypothalamus.

Considering all together the data presented in this review, we are of the opinion that they confirm that the changes of metabolic variations reflect what is happening in the neuroendocrine system from a secretory point of view.

The results reported show that, in the female rat, there are hypothalamic areas highly specialized for the control of gonadotropin secretion: probably the anterior hypothalamus is devoted to the modulation of cyclic secretion of LH and FSH, while very likely the posterior hypothalamus is in charge of the regulation of the tonic secretion of the two hormones. The medial portion, the median eminence, probably represents only the storage site of the hypothalamic hormones.

Our findings suggest furthermore that the mechanisms responsible for the regulation of gonadotropin secretion are not limited to the hypothalamus, but that the limbic system is also involved.

Finally both the hypothalamus and the limbic system are sensitive to steroid feedbacks.

The experiments performed in the male rat can give some more information on the differences existing between females and males, and on how the limbic system and the hypothalamus are interrelated.

First of all the lack of variation in the activity of the hypothalamus of the male rat seems to suggest that only a tonic secretion of gonadotropins is present in the male.

Secondly it seems probable that the mechanisms controlling gonadotropin secretion are simpler in the male than in the female: the hippocampus in fact in the male does not seem to be involved in the mechanisms modulating anterior pituitary gonadotropin function.

Testosterone feedback seems to be limited to the pituitary while the effects observed on the hypothalamus and on the amygdala can be explained in terms of short feedback of gonadotropins directed toward both the hypothalamus and the amygdala, as has been evidenced by the experiments based on hypophysectomy and replacement with exogenous LH and FSH.

In particular these experiments suggest the existence of a negative short feedback of gonadotropins on the hypothalamus and of a short positive feedback directed towards the amygdala.

Furthermore these results show that FSH exerts its action selectively on the posterior hypothalamus and LH acts preferentially on the anterior portion of this structure.

Lesion experiments suggest that the amygdala exerts a stimulatory role on the hypothalamus.

On the basis of the evidence reported here and of the literature available on the role of the hypothalamus in stimulating the secretion of LH and FSH from the anterior pituitary, it could be proposed that a stimulatory "circuit", in terms of gonadotropin secretion, exists among the amygdala, the hypothalamus, the pituitary and the amygdala again. Probably the switch to open such a circuit is represented by the short negative feedback existing between the pituitary and the hypothalamus.

ACKNOWLEDGEMENTS

Thanks are due to the Pituitary Hormone Distribution Program of the National Institute of Arthritis and Metabolic Diseases, Bethesda, Md., for the generous gift of LH and FSH preparations.

REFERENCES

Barraclough, C.A. (1966). *Recent Progr. Hormone Res.*, 22: 503.
Barraclough, C.A. (1967). In *Neuroendocrinology, Vol. 2*, L. Martini and W.F. Ganong (Eds.), Academic Press, New York, p. 61.
Baulieu, E.E., Lasntzki, I. and Robel, P. (1968). *Nature (Lond.)*, 219: 1155.
Blake, C.A., Weiner, R.I., Gorsky, R.A. and Sawyer, C.H. (1972) *Endocrinology*, 90: 855.
Davidson, J.M. (1969). In *Frontiers in Neuroendocrinology*, W.F. Ganong and L. Martini (Eds.), Oxford University Press, New York, p. 343.
Everett, J.W. (1956). *Endocrinology*, 59: 580.
Everett, J.W. (1964). *Physiol. Rev.*, 44: 373.
Foglia, V.G., Schiaffini, O. and Marin, B. (1973). *Pren. méd. argent.*, 60: 842.
Gallo, R.V., Johnson, J.H., Goldman, B.D., Whitmoyer, D.I. and Sawyer, C.H. (1971). *Endocrinology*, 89: 704.
Gay, V.L. and Bogdanove, E.M. (1969). *Endocrinology*, 84: 112.
Gay, V.L., Midgley, A.R. and Niswender, G.D. (1970). *Fed. Proc.*, 29: 1880.
Goldman, B.D. and Mahesh, V.B. (1968). *Endocrinology*, 83: 97.
Goldman, B.D., Kamberi, I.A., Siiteri, P.K. and Porter, J. (1969). *Endocrinology*, 85: 1137.
Greville, G.D. (1962). In *Neurochemistry*, K.A.C. Elliot, I.H. Page and J.H. Quastel (Eds.), Thomas, Springfield, Ill., p. 218.
Gual, C. and Rosemberg, E. (Eds.) (1973) *Hypothalamic Hypophysiotropic Hormones,* Excerpta Medica, Amsterdam.
Kniewald, Z., Massa, R. and Martini, L. (1971). In *Hormonal Steroids,* V.H.T. James and L. Martini (Eds.), Excerpta Medica, Amsterdam, p. 784.
Halász, B. (1969). In *Frontiers in Neuroendocrinology*, W.F. Ganong and L. Martini (Eds.), Oxford University Press, New York, p. 307.

Libertun, C., Moguilevsky, J.A., Schiaffini, O. and Christot, J. (1969). *J. Endocr.*, 43: 317.
Libertun, C. (1969) M.D. Thesis, School of Medicine, University of Buenos Aires.
Lowry, O.H., (1966). In *Nerve as a Tissue*, K. Rodahl and B. Issekutz (Eds.), Haipa, New York p. 261.
Mangili, G., Motta, M. and Martini, L. (1966). In *Neuroendocrinology*, L. Martini and W.F. Ganong (Eds.), Academic Press, New York, p. 297.
Martini, L., Fraschini, F. and Motta, M. (1968). *Recent Progr. Hormone Res.*, 24: 439.
Massa, R., Stupnicka, E., Kniewald, Z. and Martini, L. (1972). *J. Steroid Biochem.*, 3: 385.
Mess, B., Fraschini, F., Motta, M. and Martini, L. (1967). In *Hormonal Steroids*, L. Martini, F. Fraschini and M. Motta (Eds.), Excerpta Medica, Amsterdam, p. 1004.
Mess, B., Heizer, A., Toth, A. and Tima, L. (1971). In *The Pineal Gland*, G.E.W. Wolstenholme and J. Knight (Eds.), Churchill, London, p. 229.
Moguilevsky, J.A. and Malinow, M.R. (1964). *Amer. J. Physiol.*, 206: 855.
Moguilevsky, J.A. and Schiaffini, O. (1972). In *Hipofisis*, J.A. Moguilevsky and O. Schiaffini (Eds.), Lopez Libreros Editores, Buenos Aires.
Moguilevsky, J.A., Schiaffini, O. and Foglia, V.G. (1966). *Life Sci.*, 5: 447.
Moguilevsky, J.A., Libertun, C., Schiaffini, O. and Szwarcfarb, B. (1968). *Neuroendocrinology*, 3: 193.
Moguilevsky, J.A., Christot, J. and Schiaffini, O. (1972). Excerpta Med. int. Congr. Ser., 256: 205.
Motta, M., Fraschini, F. and Martini, L. (1969). In *Frontiers in Neuroendocrinology*, W.F. Ganong and L. Martini (Eds.), Oxford University Press, New York, p. 211.
Motta, M., Piva, F. and Martini, L. (1970). In *The Hypothalamus* L. Martini, M. Motta and F. Fraschini (Eds.), Academic Press, New York, p. 463.
Motta, M., Crosignani, P.G. and Martini, L. (Eds.), (1975) *Hypothalamic Hypophysiotropic Hormones: Chemistry, Physiology, Pharmacology and Clinical Studies,* Academic Press, London.
Piva, F., Kalra, P.S. and Martini, L. (1973). *Neuroendocrinology*, 11: 229.
Raisman, G. and Field, P.M. (1971). In *Frontiers in Neuroendocrinology*, L. Martini and W.F. Ganong (Eds.), Oxford University Press, New York, p. 3.
Schiaffini, O. (1974). Ph.D. Thesis, School of Biochemistry, University of Buenos Aires.
Schiaffini, O. and Fernandez, L. (1974). *Rev. esp. Fisiol.*, 30: 49.
Schiaffini, O. and Marin, B. (1974). *Reproduccion*, 1: 299.
Schiaffini, O. and Martini, L. (1973). *Acta endocr. (Kbh.)*, 70: 209.
Schiaffini, O., Moguilevsky, J.A., Libertun, C. and Foglia, V.G. (1968). *Acta physiol. latinoamer.*, 18: 257.
Schiaffini, O., Herrera, E. and Gallego, A. (1970). *Rev. esp. Fisiol.*, 26: 27.
Teresawa, E. and Timiras, P.S. (1968). *Endocrinology*, 83: 207.
Tima, L., Motta, M. and Martini, L. (1969). Program. 51st Mt. Endocrine Soc., p. 196.
Tukey, J.W. (1953). *Trans. N.Y. Acad. Sci.*, 2: 16.
Umbreit, W.W., Burris, R.H. and Stauffer, J. (1959). In *Manometric Techniques and Tissue Metabolism*, Burgess Publ., Mineapolis, Minn.
Wilson, J.D. and Gloyna, R.E. (1970). *Recent Progr. Hormone Res.*, 26: 309.
Zanisi, M., Motta, M. and Martini, L. (1973). *Hypophysiotropic Hormones*, C. Gual and E. Rosemberg (Eds.), Excerpta Medica, Amsterdam, p. 24.

Subcellular Mechanisms in Reproductive Neuroendocrinology, edited by
F. Naftolin, K.J. Ryan and J. Davies
© 1976 Elsevier Scientific Publishing Company—Amsterdam, The Netherlands

Chapter 20

Cyclic nucleotides in the limbic system

G. ALAN ROBISON

Department of Pharmacology, University of Texas, Health Science Center, Houston, Texas 77025 (U.S.A.)

MacLean suggested the term "limbic system" as a designation for the limbic cortex and related structures that seem to be involved in guiding behavior required for self-preservation and the preservation of the species (see MacLean, 1975, for a recent review). Even if we accept this proposition as defining the role of the limbic system, we should confess at the outset that we know very little about the role of cyclic nucleotides in it. Perhaps the most definitive thing we can say at the moment is that cyclic nucleotides do in fact exist there (Schmidt et al., 1972). Somewhat more is known about the role of cyclic AMP in the general area of self-preservation, and perhaps a brief discussion of that will put us in a better position to discuss its possible role in the limbic system per se.

It may be worth noting that self-preservation and the preservation of the species do not always go hand in hand, and physiological mechanisms favoring the one do not necessarily favor the other. It is obvious that in order to reproduce, a pair of animals must live to at least the age of puberty, so that in that sense any mechanism favoring the survival of one or both parents will be conducive to the survival of the species. This is the area in which the role of cyclic AMP is best understood. We also know, however, from population biology, that other mechanisms exist which favor the death of individuals but which are good for the species as a whole. An example of this can be seen in certain populations of wild rabbits. When the population density reaches a point beyond which (it can be seen in retrospect) the food supply would be endangered, the livers of the lowest ranking members of the population become depleted of glycogen, gluconeogenesis fails, and the animals die in a state of hypoglycemic shock. Thus, instead of the whole population dying of starvation or disease, a few are sacrificed so that the rest can survive. Numerous other examples of the same basic phenomenon could be cited. Many of them will involve reproductive failure either in addition to or instead of the death of adult animals.

I was stimulated to contemplate this phenomenon by Burns' and Langley's discovery that human fat cells possess both *alpha* and *beta* adrenergic receptors, which we now know mediate divergent effects on the level of cyclic AMP in these cells (Robinson et al., 1972a). Pancreatic beta cells also contain both types of receptors, also mediating divergent effects on the level of cyclic AMP (Montague and Howell, 1975). *Beta* receptors ordinarily predominate in fat cells, so that the result of sympathoadrenal discharge in these cells is an increase in cyclic AMP, leading to the mobilization of free fatty acids. Conversely, *alpha* receptors predominate in pancreatic beta cells, so that the result in these cells is a reduction in the level of cyclic AMP, leading to inhibition of insulin release.

It is interesting to realize that this state of affairs is highly conducive to the survival of the individual, under conditions leading to sympathoadrenal discharge, whereas the opposite receptor predominance would be tantamount to having a built-in suicide button ready to be pushed. I hope it will be clear from what I have already said, however, that such a situation might not necessarily be detrimental to the species as a whole. I have speculated elsewhere (Robinson, 1975) that certain psychosomatic disease processes in humans may reflect this sort of change, but, since the role of cyclic AMP in psychosomatic disease states has yet to be seriously investigated, I will say nothing further about it in this lecture.

In discussing the role of cyclic AMP in survival, I have occasionally found it useful to refer to Hess' distinction between the ergotropic system and the trophotropic system. Almost everything we know about cyclic AMP, from studies not only of mammalian systems but of lower organisms as well (Strada and Robison, 1974; Tomkins, 1975), leads us to believe that cyclic AMP functions primarily if not exclusively as a mediator of ergotropic responses. To illustrate this, let us consider the changes in cyclic AMP that occur in response to the presence or absence of food. Since the trappings of civilization or the laboratory may partially obscure their survival value, it may be especially useful to consider these changes as they occur in wild animals, or as they must have occurred in our prehistoric ancestors. The situation will be one of relatively long periods of famine interrupted by occasional periods of plenty. During starvation, survival will depend on the appropriate utilization of stored sources of energy (mainly glycogen, triglycerides, and proteins), and during this time the plasma ratio of glucagon to insulin will be relatively high. As a result of this and other mechanisms, including catecholamine release (Robison et al., 1972b), the level of cyclic AMP rises in almost all tissues except the pancreas. As a result, glycogen stored in the liver will be converted to glucose, which can be utilized by other peripheral tissues as well as the brain, and triglycerides stored in adipose tissue will be converted to free fatty acids, which can be utilized by most tissues except the brain. Protein throughout the body will gradually be converted to amino acids, some of which will in turn be converted to glucose by the liver. Also as a result of the high hepatic level of cyclic AMP, free fatty acids will, to an increasing degree, be converted to ketone bodies, as the brain gradually learns to utilize them. The appearance of another animal, viewed either as a potential predator or as a potential source of food, will signal a massive release of catecholamines from the adrenal gland and sympathetic nerve endings, and this will lead, among other things, to the rapid mobilization of whatever stored sources of energy may remain. Other things being equal, it is clear that the victor in the ensuing struggle will be whichever animal had the best mechanisms for storing energy when food was available and for utilizing it when the food was scarce. An interesting feature of these mechanisms is that they tend to insure a steady supply of glucose or other utilizable substrates to the brain, such that the brain is invariably the last organ to deteriorate during starvation.

The response to feeding will differ somewhat depending on the chemical nature of the food being eaten. A meal rich in protein but deficient in carbohydrates will lead to increased levels of cyclic AMP in both the alpha and beta cells of the pancreas, leading to the release of glucagon as well as insulin. This insures that a substantial fraction of the ingested amino acids will be converted to glucose by the liver. Gradually, as blood glucose levels rise (or more suddenly, if the meal was rich in carbohydrate as well as protein), the release of glucagon is suppressed and the plasma ratio of insulin to glucagon becomes very high. As a result, cyclic AMP levels fall in almost all peripheral tissues, with the pancreas again being an exception. The liver is thus converted from an organ of glucose production to an organ of glucose storage, free fatty acids are converted to triglycerides in adipose tissue and elsewhere, and almost all of the ingested amino acids are converted to protein instead of to glucose.

It is possible that as we learn more about the role of cyclic AMP in the immune response, reproduction, and other vital processes, the foregoing analysis will come to seem grossly oversimplified and totally inadequate. In the meantime, however, the generalization that cyclic AMP functions primarily as an ergotropic mediator seems to be a reasonable one. It is important to realize that the nature of the hormones involved in response to environmental stimuli seem much less important than the resulting change in cyclic AMP. In birds, for example, glucagon is a potent lipolytic hormone while the catecholamines are impotent, whereas in humans it is just the other way around. This is because avian fat cells contain receptors for glucagon but not for catecholamines, whereas human fat cells contain receptors for catecholamines but not for glucagon (Burns et al., 1975)*. The hormone–receptor interaction leads in either case to the stimulation of adenylyl cyclase, and the physiological significance of it seems roughly equivalent in both birds and men. The impression one has is that nature is relatively indifferent as to which hormone does the job, just as long as the resulting change in cyclic AMP is in the right direction. This may be important when considering a role for cyclic AMP in the limbic system. Some of the neurotransmitter agents thought to play a role in this system are known to have an effect on cyclic AMP in certain other cells and tissues, but this fact alone does not tell us that they have a similar effect within the limbic system. The ability of the catecholamines and the prostaglandins to alter cyclic AMP levels in either direction in some cells adds a further note of complexity to the situation.

Less is known about the biological role of cyclic GMP than about that of cyclic AMP, but evidence is accumulating to suggest that in many cases cyclic GMP acts antagonistically to cyclic AMP (Goldberg et al., 1975). This might suggest a role for cyclic GMP within the trophotropic system, and indeed there is some evidence to support this view. The parasympathetic nervous system, along with insulin and some of the gastrointestinal hormones, represents an important component of the trophotropic system, and all of the muscarinic responses to cholinergic stimulation that have been studied to date have been found to be associated with increased levels of cyclic GMP. Similarly, some of the responses to prostaglandins which are the opposite of those mediated by cyclic AMP, such as contraction of smooth muscle, are also associated with increased levels of cyclic GMP. Whether any of these responses are actually *mediated* by cyclic GMP is far from clear, however. Since the presence of calcium is required in all cases (for the rise in

---

* The ability of glucagon to increase free fatty acid levels in human patients under certain conditions (Liljenquist et al., 1974) is most likely a result of glucagon's ability to stimulate catecholamine release (Lewis, 1975).

cyclic GMP as well as the functional response), and since calcium ions but not acetylcholine or prostaglandins are capable of stimulating guanylyl cyclase activity in broken cell preparations, another suggestion has been that calcium rather than cyclic GMP should be regarded as the second messenger for these responses (Berridge, 1975; Schultz and Hardman, 1975). According to this view, cyclic GMP levels rise as a result of the increased availability of calcium, and may then function as part of a mechanism for reducing intracellular calcium to pre-stimulation levels.

The material summarized to this point might suggest that cyclic AMP could function within the limbic system to promote behavior requiring the expenditure of energy. Stimuli leading to the accumulation of cyclic AMP in the appropriate neurons would stimulate behavior appropriate to the killing of prey rather than the eating of it, to aggressive sexual activity rather than the nursing of offspring, and to situations requiring wakefulness rather than sleep, for example. Reducing the level of cyclic AMP (or perhaps raising the level of cyclic GMP) in the same neurons might be expected to lead to the opposite patterns of behavior. But these ideas represent nothing more than extrapolations from relatively simple peripheral systems, and may be so grossly oversimplified as to be heuristically useless. It is clear, for example, that situations requiring the expenditure of similar amounts of energy may require entirely different patterns of behavior, and also that the same stimuli may provoke different reactions according to the animal's perception of what the stimuli mean. It is certain that the limbic system controls many of these reactions, and it is highly probable that the system will function differently if the intracellular level of cyclic AMP changes, even if the change occurs in only a few neurons. But the complexity of the system is such that it is much easier to imagine these changes than to prove that they occur.

Studies with limbic and other systems have provided clues to what the neuronal role of cyclic AMP may be. Many substances known or thought to play important roles within the limbic system, including norepinephrine, dopamine, serotonin, and histamine, have been shown to stimulate the formation of cyclic AMP in one brain area or another (Rall, 1972; Daly et al., 1972; Drummond and Ma, 1973; Palmer et al., 1973; Clement-Cormier et al., 1974; Iversen, 1975). Although it is seldom possible in these experiments to distinguish changes occurring in neurons from those occurring in glial cells, other experiments have provided evidence that cyclic AMP probably does play a role in regulating neuronal function. Bloom et al. (1975) have shown that norepinephrine released from axons arising in the locus ceruleus affects cerebellar Purkinje cells and also hippocampal pyramidal cells by causing hyperpolarization associated with generally increased membrane resistance. Dopamine-containing axons arising in the substantia nigra seem to affect neurons in the caudate nucleus in a similar way, and the result in all three cases is to inhibit the rate of firing of the affected cells. This combination of biophysical changes differs from that produced by other inhibitory agents, such as GABA, which hyperpolarize their target cells by *decreasing* membrane resistance. These effects of norepinephrine (on cerebellar and hippocampal cells) and dopamine (on caudate nucleus neurons) can be mimicked to a remarkable degree by the iontophoretic application of cyclic AMP itself and are enhanced by the administration of phosphodiesterase inhibitors. In the case of cerebellar Purkinje cells, evidence for increased cyclic AMP formation in response to locus ceruleus stimulation or iontophoretically applied norepinephrine was obtained by the use of an immunocytochemical technique. Greengard and his colleagues (see Greengard, 1975, for a recent review) were able to show that the application of cyclic AMP to isolated superior cervical ganglia led to hyperpolarization of postganglionic neurons, thus mimicking the effect of dopamine in this system. Interestingly enough, they also showed that exogenous cyclic GMP could produce a prolonged depolarization, thus mimicking the muscarinic component of the effect of acetylcholine. On the basis of these and other studies, it seems reasonable to conclude that when the level of cyclic AMP rises within a neuron of one of the structures of the limbic system, as it probably does in response to dopamine, the result will be to inhibit transmission within the affected pathway (Drummond and Ma, 1973). The net result of this will depend on whether the affected pathway was itself excitatory or inhibitory, among other possible factors. It is possible to at least imagine how this could ultimately lead to certain observable mechanical changes, such as penile erection or yawning (MacLean, 1975). It seems more difficult to imagine how it could lead to feelings of anger or despair. I mention this to emphasize the depth of our ignorance, although it may reflect nothing more than my own lack of imagination.

How might cyclic AMP act to produce membrane hyperpolarization? Many of the effects of cyclic AMP are known to result from the phosphorylation of one or more proteins, secondary to the activation of a protein kinase. In the case of the glycogenolytic effect (Soderling and Park, 1974), the affected proteins are phosphorylase kinase, the activity of which is increased as a result of phosphorylation, and glycogen synthetase, the activity of which is decreased by phosphorylation. Now there is evidence that neuronal membranes contain two proteins which are rapidly phosphorylated in response to cyclic AMP (Greengard, 1975), and it seems entirely possible that one or both of these proteins are involved in regulating membrane permeability.

Chlorpromazine and other antipsychotic drugs are capable of inhibiting dopamine-stimulated adenylyl cyclase in broken cell preparations of several limbic structures (Clement-Cormier et al., 1974), and the correlation between their ability to do this and their activities as antipsychotic agents is remarkably good. Iversen (1975) has discussed the potential importance of these studies as an approach towards understanding the biochemical basis of mental illness. It is nevertheless clear that even if we knew that the excessive formation of cyclic AMP in certain limbic neurons was responsible for this or that psychotic state, we would still have a long way to go. It is sobering to realize that after some 20 years of relatively intensive investigation, even such a relatively simple response as the positive inotropic response of the heart to epinephrine is still not understood in any kind of detail. Some progress is being made, however, and we can at least hope that some of the resulting insights can be applied to the studies of the limbic system, and that a better understanding of the role of cyclic AMP in this system will lead to a better understanding of the system itself.

Some recent developments in morphine research may also contribute to a better understanding of the role of cyclic nucleotides in the limbic system. Morphine and related narcotic analgesics interact reversibly and stereospecifically with receptors which are present in certain neurons, including those in the limbic system and related structures such as the corpus striatum (Pert et al., 1975; Simon et al., 1975). Now there is evidence that one of the results of the interaction of morphine with these receptors, known as opiate receptors, is to inhibit the stimulation of adenylyl cyclase by dopamine, prostaglandins, and possibly other agents (Collier et al., 1975; Klee et al., 1975; Traber et al., 1975). We have ourselves had trouble seeing this effect in rat brain homogenates, but have observed it using cultured brain cells (Van Inwegen et al., 1975). The puzzling question to many pharmacologists has been why receptors for a plant alkaloid should develop in the mammalian central nervous system. Now an endogenous agonist for these receptors, an endogenous morphine-like factor (MLF), has been discovered (Hughes et al., 1975; Pasternak et al., 1975; Terenius and Wahlström, 1975). It appears to be a polypeptide with a molecular weight of about 1000, and it represents a discovery of potentially great significance, even though at the moment its existence has raised many more questions than answers. In the normal role of this factor to suppress the accumulation of cyclic AMP in neurons that are sensitive to it? Traber et al. (1975) have reported that morphine also raises the level of cyclic GMP in these neurons, and have suggested that this may be more important than the suppression of cyclic AMP. Why would the ingestion of a plant alkaloid with affinity for these receptors lead to such striking effects as euphoria and freedom from pain, and what is the relationship between these neurons and those which are apparently responsible for mediating the rather different effects of the antipsychotic drugs, such as chlorpromazine? What happens to MLF to cause animals to become dependent on morphine for normal function? When answers to these and other questions are obtained, it is almost certain that we will know more about the role of cyclic nucleotides in the limbic system than we know at present.

The main problem in understanding the role of cyclic nucleotides in the limbic system or in any other brain area has been the incredible complexity and cellular heterogeneity of the brain. This is a problem because both cyclic nucleotides probably exist in all cells, and it is very difficult to distinguish changes occurring in other cells from those occurring only in cells whose function is being studied. It seems to me that this problem is much worse in brain than in other complex organs, because in most other organs it can be assumed that any significant functional change will reflect the concerted action of many cells acting together, whereas this may not be true of the brain. Instead, it seems probable that many significant behavioral changes may reflect biochemical changes occurring in only one or a few neurons, and these changes may be very difficult to measure.

Many approaches will continue to be useful as we gradually arrive at a better understanding of the biochemistry of brain function, and it is always possible that some new discovery, undreamt of at present, will lead to a quantum leap in the degree of our understanding. In the meantime, studies of cultured cells in vitro and of primitive organisms with relatively simple nervous systems should continue to provide clues that will lead to hypotheses that can be tested in more complex systems. The development of better histochemical assay procedures may be especially useful in this regard. I personally think the study of species differences has been a vastly underutilized approach, especially since the aim of most of our studies in this area has been to throw light on human brain function, and this will forever be a difficult thing to study biochemically. Different neurotransmitters affect cyclic AMP in brain tissue in strikingly different ways, depending on the species (e.g., Krishna et al., 1971; Schmidt and Robison, 1971), and it is probably a serious mistake to continue to rely on only one or a few species as providing an adequate model for the study of brain function in general. For example, the distribution of serotonin in the limbic system of the rat has now been studied in considerable detail (Saavedra et al., 1974), but it is hard to know what these data mean in the absence of other information. We see that there are some striking differences in the distribution of serotonin as we move from one part of the limbic system to another, but would these differences also be seen in the brains of rabbits, guinea pigs, and monkeys? If so, this might tell us something potentially important. If not, might it then be possible to correlate the behavioral effects of drugs that affect the formation or action of serotonin with the known distribution of serotonin in the brains of the various species? I rather suspect that if we knew the distribution within the limbic system of all known or suspected neurotransmitters in as few as four species, say, and if we knew how each of these agents affected cyclic nucleotide formation in each area of all four species, then we would be in a much better position than we are now to generate intelligent testable hypotheses about the roles these various substances may play in the regulation of brain function.

In summary, I have tried to speculate, on the basis of known information, about what the role of cyclic AMP and cyclic GMP in the limbic system might be. I have been forced to the conclusion that valid generalizations are not possible at this time.

REFERENCES

Berridge, M.J. (1975) The interaction of cyclic nucleotides and calcium in the control of cellular activity. *Advanc. Cyclic Nucleotide Res.*, 6: 1–98.

Bloom, F.E., Siggins, G.R., Hoffer, B.J., Segal, M. and Oliver, A.P. (1975) Cyclic nucleotides in the central synaptic actions of catecholamines. *Advanc. Cyclic Nucleotide Res.*, 5: 603–618.

Burns, T.W., Langley, P.E. and Robison, G.A. (1975) Site of free fatty acid inhibition of lypolysis by human adipocytes. *Metabolism,* 24: 265–276.

Clement-Cormier, Y.C., Kebabian, J.W., Petzold, G.L. and Greengard, P. (1974) Dopamine-sensitive adenylate cyclase in mammalian brain: a possible site of action of antipsychotic drugs. *Proc. nat. Acad. Sci. (Wash.),* 71: 1113–1117.

Collier, H.O.J., Francis, D.L., McDonald-Gibson, W.J., Roy, A.C. and Saeed, S.A.(1975) Prostaglandins, cyclic AMP, and the mechanisms of opiate dependence. *Life Sci.,* 17: 85–90.

Daly, J.W., Huang, M. and Shimizu, H. (1972) Regulation of cyclic AMP levels in brain tissue. *Advanc. Cyclic Nucleotide Res.,* 1: 375–387.

Drummond, G.I. and Ma, Y. (1973) Metabolism and functions of cyclic AMP in nerve. In *Progress in Neurobiology, Vol. 2, Part 2,* G.A. Kerkut and J.W. Phillis (Eds.), Pergamon Press, New York, pp. 119–176.

Goldberg, N.D., Haddox, M.K., Nicol, S.E., Glass, D.B., Sanford, C.H., Kuehl, F.A. and Estensen, R. (1975) Biological regulation through opposing influences of cyclic GMP and cyclic AMP: the yin yang hypothesis. *Advanc. Cyclic Nucleotide Res.,* 5: 307–338.

Greengard, P. (1975) Cyclic nucleotides, protein phosphorylation, and neuronal function. *Advanc. Cyclic Nucleotide Res.*, 5: 585–601.

Hughes, J., Smith, T., Morgan, B. and Fothergill, L. (1975) Purification and properties of enkephalin — the possible endogenous ligand for the morphine receptor. *Life Sci.*, 16: 1753–1758.

Iversen, L.L. (1975) Dopamine receptors in the brain. *Science*, 188: 1869–1874.

Klee, W.A., Sharma, S.K. and Nirenberg, M. (1975) Opiate receptors as regulators of adenylate cyclase. *Life Sci.*, 16: 1869–1874.

Krishna, G., Forn, J., Voigt, K., Paul, M. and Gessa, G.L. (1971) Dynamic aspects of neurohormonal control of cyclic AMP synthesis in brain. *Advanc. Biochem. Psychopharmacol.*, 3: 155–172.

Lewis, G.P. (1975) Physiological mechanisms controlling secretory activity of adrenal medulla. In *Handbook of Physiology, Section 7, Endocrinology, Vol. 6, Adrenal Gland*, H. Blaschko, G. Sayers and A.D. Smith (Eds.), American Physiological Society, Washington, D.C., pp. 309–319.

Liljenquist, J.E., Bomboy, J.D., Lewis, S.B., Sinclair-Smith, B.C., Felts, P.W., Lacy, W.W., Crofford, O.B. and Liddle, G.W. (1974) Effects of glucagon on lipolysis and ketogenesis in normal and diabetic men. *J. clin. Invest.*, 53: 190–197.

MacLean, P.D. (1975) Brain mechanisms of primal sexual function and related behavior. In *Sexual Behavior: Pharmacology and Biochemistry*, M. Sandler and G.L. Gessa (Eds.), Raven Press, New York, pp. 1–11.

Montague, W. and Howell, S.L. (1975) Cyclic AMP and the physiology of the islets of Langerhans. *Advanc. Cyclic Nucleotide Res.*, 6: 201–243.

Palmer, G.C., Sulser, F. and Robison, G.A. (1973) Effects of neurohumoral and adrenergic agents on cyclic AMP levels in various areas of the rat brain in vitro. *Neuropharmacology*, 12: 327–337.

Pasternak, G.W., Goodman, R. and Snyder, S.H. (1975) An endogenous morphine-like factor in mammalian brain. *Life Sci.*, 16: 1766–1769.

Pert, C.B., Kuhar, M.J. and Snyder, S.H. (1975) Autoradiographic localization of the opiate receptor in rat brain. *Life Sci.*, 16: 1849–1854.

Rall, T.W. (1972) Role of cyclic AMP in actions of catecholamines. *Pharmacol. Rev.*, 24: 399–409.

Robison, G.A. (1975) Cyclic AMP and disease: an overview. In *Molecular Pathology*, R.A. Good, S.B. Day and J.J. Yunis (Eds.), Thomas, Springfield, Ill., pp. 394–404.

Robison, G.A., Langley, P.E. and Burns, T.W. (1972a) Adrenergic receptors in human adipocytes: divergent effects on cyclic AMP and lipolysis. *Biochem. Pharmacol.*, 21: 589–592.

Robison, G.A., Butcher, R.W. and Sutherland, E.W. (1972b) The catecholamines. In *Biochemical Actions of Hormones, Vol. 2*, G. Litwack (Ed.), Academic Press, New York, pp. 81–111.

Saavedra, J.M., Brownstein, M. and Palkovits, M. (1974) Serotonin distribution in the limbic system of the rat. *Brain Res.*, 79: 437–441.

Schmidt, M.J. and Robison, G.A. (1971) The effect of norepinephrine on cyclic AMP levels in discrete regions of the developing rabbit brain. *Life Sci.*, Part I, 10: 459–463.

Schmidt, M.J., Schmidt, D.E. and Robison, G.A. (1972) Cyclic AMP in the rat brain: microwave irradiation as a means of tissue fixation. *Advanc. Cyclic Nucleotide Res.*, 1: 425–434.

Schultz, G., and Hardman, J.G. (1975) Regulation of cyclic GMP levels in the ductus deferens of the rat. *Advanc. Cyclic Nucleotide Res.*, 5: 339–351.

Simon, E.J., Hiller, J.M., Edelman, J., Groth, J. and Stahl, K.D. (1975) Opiate receptors and their interactions with agonists and antagonists. *Life Sci.*, 16: 1795–1800.

Soderling, T.R. and Park, C.R. (1974) Recent advances in glycogen metabolism. *Advanc. Cyclic Nucleotide Res.*, 4: 283–333.

Strada, S.J. and Robison, G.A. (1974) Cyclic AMP as a mediator of hormonal effects. In *Endocrine Physiology*, S.M. McCann (Ed.), University Park Press, Baltimore, Md., pp. 309–336.

Terenius, L. and Wahlström, A. (1975) Morphine-like ligand for opiate receptors in human CSF. *Life Sci.*, 16: 1759–1764.

Tomkins, G.M. (1975) The metabolic code. *Science*, 189: 760–763.

Traber, J., Gullis, R. and Hamprecht, B. (1975) Influence of opiates on the levels of cyclic AMP in neuroblastoma x glioma hybrid cells. *Life Sci.*, 16: 1863–1868.

Van Inwegen, R.G., Strada, S.J. and Robison, G.A. (1975) Effects of prostaglandins and morphine on brain adenylyl cyclase. *Life Sci.*, 16: 1875–1876.

ADDENDUM

The foregoing manuscript was meant to be read by the symposium participants prior to the symposium, and was written not so much for posterity as for the purpose of stimulating discussion. The slide I used to illustrate the point of the second paragraph was from an old paper by C.M. Breder and C.W. Coates (*Copeia,* 1932, pp. 147—155). Since it did stimulate a certain amount of discussion (over cocktails if not during the regular discussion period) and since the journal in which it appeared is not readily available in all libraries, I am including it here as Fig. 1. In brief summary, Breder and Coates took two

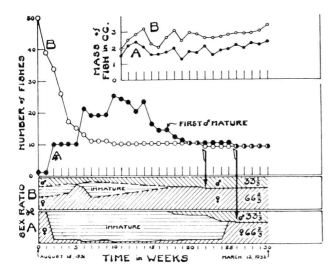

Fig. 1.

tanks of equal size, arranged in such a way that lighting conditions were identical. Into one tank (A) they placed a single pregnant guppy, and into the other (B) they placed 50 guppies, including approximately equal numbers of males, females, and immature guppies. Then under conditions in which neither food nor oxygen availability was a limiting factor, they measured the population at weekly intervals. At the end of the experiment, as shown in the figure, there were 9 guppies in either tank, 6 females and 3 males, which is the proportion in which guppies happen to be born. This particular example probably involves cannibalism more than reproductive failure, but does illustrate the point that there are some interesting regulatory mechanisms which are not conducive to the survival of individuals but which may be conducive to the survival of the species. This experiment was previously discussed by Robert Ardrey (*The Social Contract*, Atheneum, New York, 1970, pp. 174—176).

Chapter 21

Role of cyclic AMP in neuroendocrine control

F. LABRIE, P. BORGEAT, N. BARDEN, M. BEAULIEU, L. FERLAND, J. DROUIN, A. DE LEAN and O. MORIN

Medical Research Council Group in Molecular Endocrinology, Centre Hospitalier de l'Université Laval, Quebec G1V 4G2 (Canada)

I. INTRODUCTION

The role of the hypothalamus as main site of control of anterior pituitary function has been recognized for many years on the basis of detailed physiological and anatomical evidence (Green and Harris, 1947; Schally et al., 1968; McCann and Porter, 1969; Burgus and Guillemin, 1970). It is only recently, however, that the concept of neurohormonal control of adenohypophyseal function could be translated into biochemical and chemical terms.

This new era of neuroendocrinology really started with the elucidation of the structure of porcine and ovine TRH[*], the neurohormone controlling the activity of the TSH-secreting cells, as being (pyro)Glu-His-Pro-NH$_2$ (Bøler et al., 1969; Burgus et al., 1969). This achievement was soon followed by the isolation of LH-RH, the neurohormone which stimulates the release of both LH and FSH and its characterization as a decapeptide having the following structure: (pyro)Glu-His-Trp-Ser-Tyr-Gly-Leu-Arg-Pro-Gly-NH$_2$ (Burgus et al., 1971; Matsuo et al., 1971). More recently, the tetradecapeptide H-Ala-Gly-Cys-Lys-Asn-Phe-Phe-Trp-Lys-Thr-Phe-Thr-Ser-Cys-OH has been isolated from ovine and porcine hypothalami (Brazeau et al., 1973; Schally et al., 1975) on the basis of its ability to inhibit GH release and called somatostatin or GH-RIH.

The relative ease of synthesis of these peptides and their analogs has opened new possibilities for studies of their mechanism of action and has already permitted many interesting structure-function studies. It has in fact led to a rapid expansion of our knowledge of the physiology of the hypothalamo-pituitary complex.

This presentation will attempt to summarize the evidence obtained so far on the effect of three synthetic hypothalamic hormones, namely TRH, LH-RH and somatostatin and also of prostaglandins on cyclic AMP accumulation in anterior pituitary gland. Mention will also be made of the stimulatory effect of purified GH-RH on cyclic AMP accumulation and on the specificity of the effect of prostaglandins on pituitary cyclic AMP accumulation and hormone release. Since systemically administered prostaglandins appear

---
[*] The abbreviations used are: TSH, thyrotropin; LH, luteinizing hormone; FSH, follicle-stimulating hormone; GH, growth hormone; PRL, prolactin; TRH, TSH-releasing hormone; LH-RH, LH-releasing hormone; GH-RH, GH-releasing hormone; GH-RIH or somatostatin, GH-release inhibiting hormone; cyclic AMP, adenosine 3',5'-cyclic phosphate; PG, prostaglandin.

to stimulate the release of at least some hypothalamic hormones, data will also be presented on the changes of cyclic AMP levels in the median eminence and pituitary after in vivo administration of PGs.

II. STIMULATORY EFFECT OF LH-RH ON CYCLIC AMP ACCUMULATION IN ANTERIOR PITUITARY GLAND

The first suggestive evidence of a role of cyclic AMP as mediator of the action of the hypothalamic regulatory hormones in the anterior pituitary gland originated from the observations that cyclic AMP derivatives or theophylline, an inhibitor of cyclic nucleotide phosphodiesterase, stimulate the release of all 6 main anterior pituitary hormones (Schofield 1967; Wilber et al., 1968; Lemay and Labrie, 1972; Labrie et al., 1973, 1975a). Stimulation or inhibition of adenylate cyclase activity in specific pituitary cell types by the corresponding stimulatory and inhibitory hypothalamic regulatory hormones could thus provide a mechanism of control of adenohypophyseal hormone secretion. Proof of the role of cyclic AMP had, however, to be obtained from measurements of changes of adenylate cyclase activity or cyclic AMP concentrations under the influence of pure or synthetic neurohormones.

Much recent evidence indicates that cyclic AMP is involved as mediator of the action of LH-RH in the anterior pituitary gland. It is in fact well recognized that addition of LH-RH leads to a stimulation of cyclic AMP accumulation in rat anterior pituitary gland in vitro (Borgeat et al., 1972, 1974a; Jutisz et al., 1972; Makino, 1973; Kaneko et al., 1973; Labrie et al., 1973). The concentration of LH-RH required for half-maximal stimulation of cyclic AMP accumulation is between 0.1 and 1.0 ng/ml or between 1×10^{-10} and 1×10^{-9} M LH-RH (Borgeat et al., 1972). A close correlation is always observed between rates of LH and FSH release and changes of intracellular cyclic AMP concentrations, both as a function of time of incubation and concentration of the neurohormone.

Moreover, when LH-RH analogs having a spectrum of biological activity ranging from 0.001% to 500—1000% the activity of LH-RH itself were used, the same close parallelism between stimulation of cyclic AMP accumulation and both LH and FSH release was found under all experimental conditions (Borgeat et al., 1974a). That LH-RH exerts its action by activation of adenylate cyclase and not by inhibition of cyclic nucleotide phosphodiesterase is indicated by the observation that a similar effect of the neurohormone is observed in the presence or absence of theophylline (Borgeat et al., 1972).

The possibility of developing a contraceptive method based on inhibitory analogs of LH-RH has led to the synthesis of many LH-RH analogs, some of which being potent inhibitors of LH-RH action both in vivo (Ferland et al., 1976) and in vitro (Labrie et al., 1976). The availability of such good LH-RH antagonists offered the possibility of investigating the correlation between their inhibitory effect on LH-RH-induced cyclic AMP accumulation and LH and FSH release.

As an example, Fig. 1 shows the inhibitory effect of increasing concentrations of [D-Phe^2, D-Leu^6] LH-RH on cyclic AMP accumulation and LH and FSH release in rat anterior pituitary gland in vitro. The present observation of such a close correlation

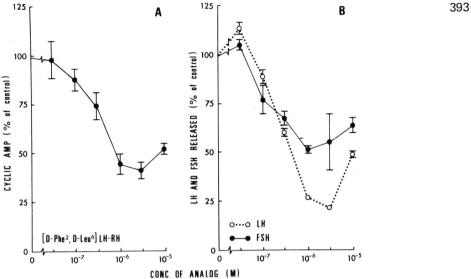

Fig. 1. Effect of increasing concentrations of [D-Phe^2, D-Leu^6]LH-RH on 3×10^{-9} M LH-RH-induced cyclic AMP accumulation (A) and LH and FSH release (B) in rat hemipituitaries in vitro.

between inhibition of LH-RH-induced cyclic AMP accumulation and LH and FSH release adds strong support to the already obtained evidence for an obligatory role of the adenylate cyclase system as mediator of the action of LH-RH in the anterior pituitary gland.

III. STIMULATORY EFFECT OF TRH ON CYCLIC AMP ACCUMULATION IN ANTERIOR PITUITARY GLAND

Although there was no doubt that theophylline and cyclic AMP derivatives stimulate TSH release from anterior pituitary gland in vitro (Cehovic, 1969; Wilber et al., 1969), more convincing proof of the role of the adenylate cyclase system in the action of TRH awaited the demonstration of a stimulation of cyclic AMP accumulation in anterior pituitary gland by synthetic TRH. Although the changes of cyclic AMP levels are of relatively small magnitude than those observed under stimulation by LH-RH, a significant increase (30% over control) is measured after 15 min of incubation with 10^{-6} M TRH, while a maximal effect at 50% over control is found after 2 hr of incubation (Labrie et al., 1975b).

IV. STIMULATORY EFFECT OF PURIFIED GH-RH ON CYCLIC AMP ACCUMULATION IN ANTERIOR PITUITARY GLAND

Much evidence shows that the hypothalamus contains GH-releasing activity (Deuben and Meites, 1964; Schally et al., 1968; McCann and Porter, 1969). Although synthetic

GH-RH was not yet available, it was felt important to obtain some information about the mechanism of action of GH-RH using a purified fraction of the neurohormone prepared from porcine hypothalami up to the gel filtration step on Sephadex G-25 (Schally et al., 1969). Using a submaximal dose of purified GH-RH, a 2.5-fold stimulation of cyclic AMP accumulation was measured in the adenohypophyseal tissue 2 min after addition of the purified neurohormone, while a 4-fold increase of cyclic AMP concentration was measured after 4 min of incubation (Borgeat et al., 1973). As observed previously for LH-RH and TRH, the release of GH closely paralleled the changes of cyclic AMP concentration, a 7–10-fold stimulation of GH release being measured after 2 min of incubation in the presence of purified GH-RH. Although these data obtained with a purified preparation of GH-RH remain to be verified using the pure or synthetic hormone, they suggest clearly a role of cyclic AMP in the control of GH secretion.

V. INHIBITORY EFFECT OF SOMATOSTATIN ON CYCLIC AMP ACCUMULATION IN ANTERIOR PITUITARY GLAND

Somatostatin is a potent inhibitor of GH release in man (Siler et al., 1973) and rat (Brazeau et al., 1973; Bélanger et al., 1974) and has also been found to decrease the TSH response to TRH both in vivo (Drouin et al., 1976) and in vitro (Vale et al., 1974; Drouin et al., 1976). Moreover, under certain experimental conditions, the tetradecapeptide has been reported to inhibit the basal and TRH-induced release of PRL (Vale et al., 1974; Drouin et al., 1976).

Following our observations of a stimulatory effect of purified GH-RH on cyclic AMP accumulation, we felt it interesting to study a possible inhibitory effect of somatostatin on pituitary cyclic AMP accumulation. The finding of an inhibitory effect of pituitary cyclic AMP levels would give further support for a role of the cyclic nucleotide in the control of GH secretion. As illustrated in Fig. 2, a maximal inhibitory effect of somatostatin on cyclic AMP accumulation in rat hemipituitaries is observed 10 min after addition of the peptide. This inhibitory effect of somatostatin is accompanied by a marked inhibition of both GH and TSH release (Borgeat et al., 1974b).

Since GH- and TSH-secreting cells account for 50–70% of the total adenohypophyseal cell population in adult male rats, the 50% inhibition of cyclic AMP accumulation in total pituitary tissue suggests an almost complete inhibition of cyclic AMP accumulation in the GH- and TSH-secreting cells. The inhibitory effect of somatostatin is observed under both basal and prostaglandin E_2- or theophylline-induced conditions, thus suggesting that somatostatin exerts its action by inhibiting adenylate cyclase activity.

The data summarized so far show clearly that two synthetic stimulatory hypothalamic hormones, TRH and LH-RH, as well as purified GH-RH, lead to parallel stimulation of cyclic AMP accumulation and specific hormone release while one inhibitory peptide, somatostatin, leads to parallel inhibition of cyclic AMP accumulation and GH and TSH release. Such findings suggest strongly that changes of adenylate cyclase activity are involved in the mechanism of action of these peptides in the anterior pituitary gland.

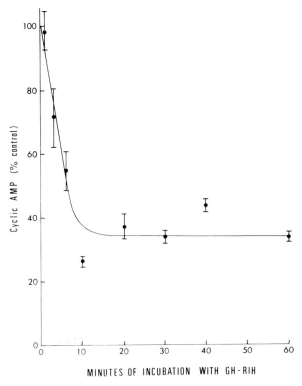

Fig. 2. Time course of the effect of somatostatin (GH-RIH) on cyclic AMP accumulation in anterior pituitary gland. Incubation and assay were performed as described (Borgeat et al., 1974b).

VI. CALCIUM REQUIREMENT FOR LH-RH ACTION

Since calcium (Ca^{2+}) was known to be required for the hypothalamic extract-induced release of LH (Samli and Geschwind, 1968; Wakabayashi et al., 1969), and FSH (Wakabayashi et al., 1969; Jutisz and Paloma de la Llosa, 1970), it was felt important to study a possible requirement of Ca^{2+} at a step preceding activation of adenylate cyclase by LH-RH (Borgeat et al., 1975a). Such an early site of action in the LH and FSH secretory cells would then be added to the already suspected late site of Ca^{2+} requirement observed during high K^+-induced release of LH and FSH (Samli and Geschwind, 1968; Wakabayashi et al., 1969; Jutisz and Paloma de la Llosa, 1970).

The interest of such a study was strengthened by the observations that Ca^{2+} is required for hormonal activation of adenylate cyclase in many systems. These studies pertain to the stimulatory action of ACTH in fat cells (Bar and Hoechter, 1969), adrenal cells (Sayers et al., 1972) and adrenal cell membrane particles (Lefkowitz et al., 1970). The effect of increasing concentrations of free Ca^{2+} on LH-RH-induced cyclic AMP accumulation is illustrated in Fig. 3. These data show that Ca^{2+} is required at a step preceding stimulation of cyclic AMP accumulation by LH-RH. The site of Ca^{2+} requirement could be on the LH-RH receptor, on the adenylate cyclase, on cyclic nucleotide phosphodieste-

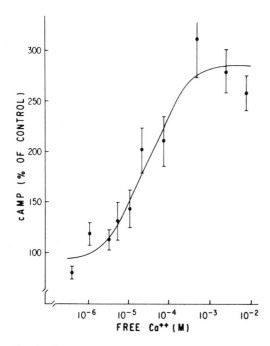

Fig. 3. Effect of increasing concentrations of free Ca^{2+} on LH-RH-induced cyclic AMP accumulation in male rat anterior pituitary gland. Adenohypophyseal tissue was first incubated for 60 min before 4 successive incubations of 1, 1, 0.75 and 0.75 hr in the presence or absence of the indicated concentrations of free Ca^{2+}. 10^{-7} M LH-RH, 10^{-2} M theophylline and 2×10^{-3} M EGTA were present in all groups.

rase, at some intermediate step between binding of LH-RH and activation of adenylate cyclase or at a combination of these sites. Since the binding of another releasing hormone, TRH (Labrie et al., 1972), is not increased by Ca^{2+} and 1 mM EDTA does not affect fluoride-stimulated adenohypophyseal adenylate cyclase activity (Poirier and Labrie, unpublished), it appears more likely that Ca^{2+} is required at some step(s) between binding of LH-RH and activation of adenylate cyclase, although an action at other sites remains possible.

VII. STIMULATORY EFFECT OF PROSTAGLANDINS ON PITUITARY CYCLIC AMP ACCUMULATION

The stimulatory effect of PGE_1 and PGE_2 on cyclic AMP accumulation in anterior pituitary tissue is well substantiated (MacLeod and Lehmeyer, 1970; Zor et al., 1970; Borgeat et al., 1975b). However, as illustrated in Fig. 4, the various PGs exhibit markedly different potencies to induce cyclic AMP accumulation measured in rat anterior pituitary tissue after 120 min of incubation (Fig. 4).

At 1×10^{-7} M, the stimulatory effect of PGE_2 is already significant at 2 min and

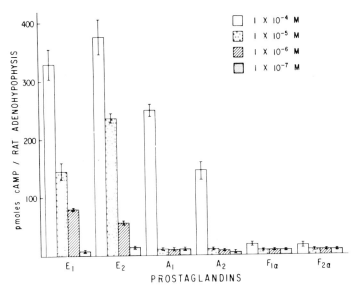

Fig. 4. Effect of increasing concentrations of PGE_1, E_2, A_1, A_2, $F_{1\alpha}$ and $F_{2\alpha}$ on cyclic AMP accumulation in rat anterior pituitary gland in vitro. Cyclic AMP was measured after 120 min of exposure to the indicated concentrations of prostaglandins.

becomes maximal at 30 min (Borgeat et al., 1975b). This maximal effect is followed by a progressive decrease to basal levels at 120 min. Upon addition of higher concentrations of PGE_2, cyclic AMP levels remain higher than controls at least up to 180 min. When studied at short time intervals (15 and 30 min), PGA_1 stimulates cyclic AMP accumulation only at high concentrations (10^{-4} M) while $PGF_{1\alpha}$ has only a minimal effect (Borgeat et al., 1975b). PGs of the E type are thus the most potent activators of cyclic AMP accumulation in anterior pituitary tissue. PGE_1 and E_2 are in fact more potent than A_1 and A_2, while a slight stimulation by $PGF_{1\alpha}$ and $PGF_{2\alpha}$ is only observed at high concentrations (10^{-4} M).

VIII. SPECIFICITY OF THE STIMULATORY EFFECT OF PROSTAGLANDINS ON THE RELEASE OF ANTERIOR PITUITARY HORMONES

Since PGE-induced increases of pituitary cyclic AMP were measured in total anterior pituitaries, it was felt important to study the specificity of PG action by measuring changes of specific hormone release. In fact, cyclic AMP being a potent stimulator of the release of all 6 pituitary hormones, changes of specific hormone release should be a good indication of the cell types responsible for the increased cyclic AMP levels observed after addition of PGs.

As illustrated in studies performed with anterior pituitary cells in culture (Fig. 5), PGE_1 has an important stimulatory effect on GH release only, a half-maximal (ED_{50}) stimulation of GH release being measured at 1×10^{-7} M PGE_1. Up to 3×10^{-4} M, PGE_1

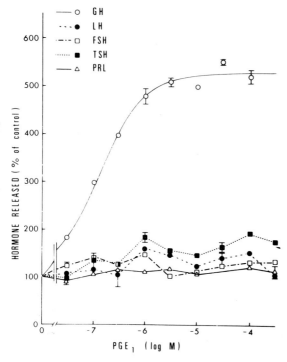

Fig. 5. Effect of increasing concentrations of PGE_1 on hormone release by anterior pituitary cells in primary culture. The incubation was performed for 5 hr.

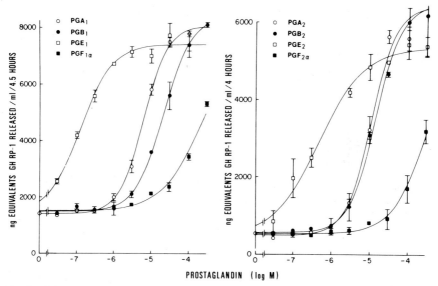

Fig. 6. Effect of increasing concentrations of the 8 primary prostaglandins on GH release by adenohypophyseal cells. The experiment was performed as described in Fig. 5.

has no effect on LH, FSH and PRL release while TSH release is slightly but reproducibly stimulated (170% of control value) at concentrations higher than 10^{-6} M.

Stimulation of GH release does not show an absolute specificity for PGEs since, as shown in Fig. 6, all 8 primary prostaglandins are able to elicit GH release at high concentrations. PGEs are the most potent stimulators of GH release, a half-maximal stimulation by PGE_1 and E_2 being measured at $1-4 \times 10^{-7}$ M while values of $3-10 \times 10^{-6}$, $1-3 \times 10^{-5}$ and approximately 3×10^{-4} M are obtained for PGAs, PGBs and PGFs, respectively. The order of potency of the various PGs to stimulate GH release from anterior pituitary cells in culture closely parallels the potency previously observed on cyclic AMP accumulation (Fig. 5).

IX. IN VIVO EFFECTS OF PROSTAGLANDINS AT THE HYPOTHALAMIC LEVEL ON LH-RH RELEASE

It is well known that PGs of the E type stimulate LH release after in vivo administration. PGEs have in fact been found to increase plasma LH levels in pentobarbital-blocked proestrous rats (Tsafiri et al., 1973) as well as in steroid-primed ovariectomized (Harms et

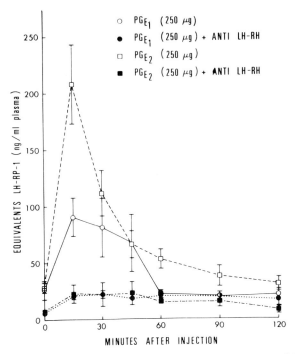

Fig. 7. Effect of 1 ml anti-LH-RH serum on PGE_1- or PGE_2-induced LH release in anesthetized female rats on the afternoon of proestrus. The anti-serum was injected 1 hr before PGEs through a cannula inserted into the right superior vena cava. Plasma LH concentrations are plotted as mean ± S.E.M. of 6–8 rats.

al., 1974; Sato et al., 1974) and in intact male (Ratner et al., 1974) animals. Evidence for a stimulatory effect of $PGF_{2\alpha}$ on in vivo LH release has also been obtained in sheep (Carlson et al., 1973; McCracken et al., 1973). These data do not, however, differentiate between an effect of PGs at the hypothalamic level on LH-RH release and a direct pituitary site of action on LH secretion.

Since the reports concerning the in vitro effects of PGs on LH release are conflicting (Zor et al., 1970; Ratner et al., 1974), it was felt important to examine the in vivo site of action of PGs using a LH-RH antiserum. In the event of a hypothalamic site of action of PGs, administration of the antiserum should neutralize the LH-RH released by PGs and thus prevent the PG-induced increase of plasma LH levels.

As clearly illustrated in Fig. 7, not only the basal plasma LH concentration was reduced by approximately 75% one hour after injection of the antiserum, but the treatment almost completely obliterated the plasma LH rise observed after injection of PGE_1 and PGE_2. This observation of an almost complete inhibition of the PGE_1- or PGE_2-induced rise of plasma LH in animals treated one hour previously with sheep anti-LH-RH serum leaves little doubt that the increased plasma LH levels observed in vivo after PGE administration are secondary to an effect of PGs on LH-RH release at the hypothalamic level.

Since PGE_1 and PGE_2 raise plasma LH levels in pentobarbital or thyamilal-treated animals, it is possible that PGs act at a site close to the nerve endings containing LH-RH in the external layer of the median eminence. In support of this, the centers controlling LH release are inhibited by the anesthetic (McCann and Porter, 1969) and are thus likely to be unresponsive to stimuli, while entry into the brain of parenterally administered PGs has been found to be very limited (Holmes and Horton, 1968). Moreover, the median eminence does not show such a barrier to small molecules (Pelletier, unpublished data). The present data make it thus likely that parenterally administered PGs act directly on the nerve endings containing LH-RH (Pelletier et al., 1974) in the median eminence to induce release of the neurohormone.

X. IN VIVO EFFECTS OF PROSTAGLANDINS ON ANTERIOR PITUITARY AND HYPOTHALAMIC CYCLIC AMP LEVELS

The previous data clearly show that, at the pituitary level, PGs of the E type lead to parallel stimulation of cyclic AMP accumulation and GH release accompanied by a small but consistent stimulation of TSH release while the release of LH, FSH and PRL remains unchanged. As far as the hypothalamic site of action of PGs is concerned, data illustrated in Fig. 7 suggest that they might well be involved in the control of LH-RH secretion. We have also recently obtained evidence for a similar stimulatory effect of PGE_1 on TRH secretion (Ferland and Labrie, unpublished data).

Since PGs have been found to stimulate cyclic AMP accumulation in brain (Ramwell and Shaw, 1970; Wellmann and Schwabe, 1973) and there is much evidence for a role of cyclic AMP in neuronal function (Bloom et al., 1975; Greengard, 1975), it was felt of interest to study a possible stimulatory effect of systemically administered PGs on cyclic AMP accumulation in the hypothalamus and anterior pituitary. In order to prevent the

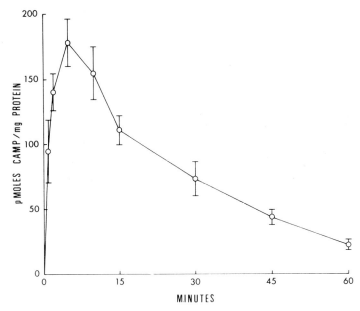

Fig. 8. Time course of the effect of intravenous administration of 100 μg of PGE_1 on anterior pituitary cyclic AMP levels in rats under Surital anesthesia. Animals were killed by microwave irradiation.

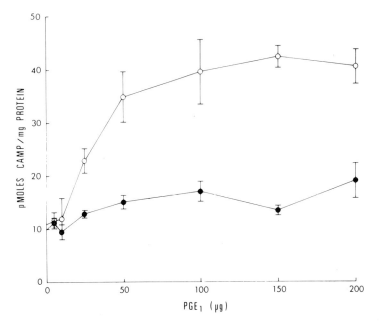

Fig. 9. Effect of increasing doses of PGE_1 administered intravenously on the cyclic AMP levels in total hypothalamus and median eminence in rats under Surital anesthesia and killed by microwave irradiation 5 min after injection of PGE_1.

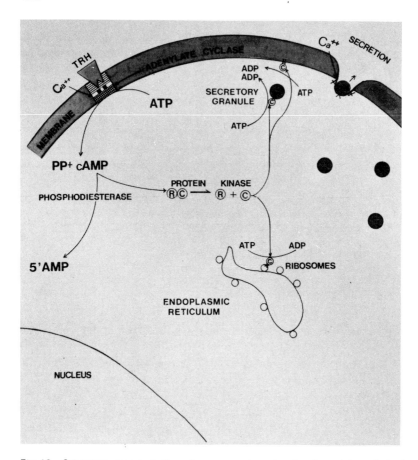

Fig. 10. Schematic representation of a proposed mode of action of hypothalamic regulatory hormones in the adenohypophyseal cell. Binding of the neurohormone to a receptor located in the plasma membrane would modulate adenylate cyclase activity. After stimulation of its synthesis, cyclic AMP would then bind to its receptor, thus releasing the activated catalytic subunit of protein kinase. Altered levels of phosphorylation of different intracellular protein substrates under the influence of cyclic AMP could then lead to changes of the activity of the various specialized processes of the adenohypophyseal cell.

important rise of cyclic AMP levels occurring at the pituitary and hypothalamic levels after decapitation, rats were killed by microwave irradiation. Both anterior pituitary and hypothalamic cyclic AMP levels were increased after injection of PGE_1. A significant stimulation of cyclic AMP accumulation was apparent at a dose of 1 μg of PGE_1 while a maximal effect was seen at the dose of 100 μg.

As illustrated in Fig. 8, the effect is rapid, maximal levels of pituitary cyclic AMP being reached 5 min after the injection. Whilst the injection of 100 μg of PGE_1 was found to stimulate anterior cyclic AMP levels 100-fold, only a 2-fold stimulation was seen in total hypothalamic tissue (Fig. 9). Measurements of cyclic AMP in the median eminence

(Fig. 9) that show the stimulation of cyclic AMP are restricted to this area and are about 4-fold at the highest dose of PGE_1 used.

These data show that after in vivo injections, PGs can stimulate cyclic AMP accumulation in the median eminence and anterior pituitary gland. While the effect at the level of the median eminence is probably direct and might be involved with the PG-induced release of LH-RH and TRH, the effect on pituitary cyclic AMP may either be direct or secondary to PG-induced changes of hypothalamic hormone secretion (Fig. 10).

XI. SUMMARY

First suggestive evidence for a role of cyclic AMP as mediator of the action of the hypothalamic regulatory hormones in the anterior pituitary gland originated from the observations that cyclic AMP derivatives and theophylline stimulate the release of all 6 main anterior pituitary hormones. Convincing proof of the role of cyclic AMP had, however, to be obtained in studies of the effect of LH-RH, TRH and somatostatin on cyclic AMP accumulation in adenohypophyseal tissue. Using LH-RH, its agonistic and antagonistic analogs and TRH, a parallel stimulation of cyclic AMP accumulation and specific hormone release was observed under all experimental conditions. Somatostatin, on the other hand, leads to a marked decrease of pituitary cyclic AMP levels accompanied by inhibition of GH and TSH release.

Prostaglandins (PGs) lead in vitro to parallel stimulation of pituitary cyclic AMP accumulation and GH release accompanied by a small but consistent increase of TSH release while the release of LH, FSH and PRL remains unchanged. The order of potency of PGs is $E_1 = E_2 > A_1 = A_2 > F_{1\alpha} = F_{2\alpha}$. Using an anti-LH-RH serum, good evidence has been obtained for a stimulation of LH-RH secretion after systemic administration of PGE_1 or PGE_2. Such in vivo treatment with PGE_1 is accompanied by a 100-fold increase of pituitary cyclic AMP levels while the median eminence cyclic AMP content is increased 4-fold.

REFERENCES

Bar, H.P. and Hoechter, O. (1969) Adenyl cyclase and hormone action. 3. Calcium requirement for ACTH stimulation of adenyl cyclase. *Biochem. biophys. Res. Commun.,* 34: 681–686.

Bélanger, A., Labrie, F., Borgeat, P., Savary, M., Côté, J., Drouin, J., Schally, A.V., Coy, D.H., Coy, E.J., Immer, H., Sestanj, K., Nelson, V. and Gotz, M. (1974) Inhibition of growth hormone and thyrotropin release by growth hormone release-inhibiting hormone. *J. molec. cell. Endocr.,* 1: 329–339.

Bloom, F.E., Siggins, G.R., Hoffer, B.J., Segal, M. and Oliver, A.P. (1975) Cyclic nucleotides in the central synaptic actions of catecholamines. In *Advances in Cyclic Nucleotide Research,* G.I. Drummond, P. Greengard and G.A. Robison (Eds.), Raven Press, New York, pp. 603–618.

Bøler, J., Enzman, F., Folkers, K., Bowers, C.Y. and Schally, A.V. (1969) The identity of chemical and hormonal properties of the thyrotropin releasing hormone and pyroglutamyl-histydyl-proline amide. *Biochem. biophys. Res. Commun.,* 37: 705–710.

Borgeat, P., Chavancy, G., Dupont, A., Labrie, F., Arimura, A. and Schally, A.V. (1972) Stimulation of adenosine 3′,5′-cyclic monophosphate accumulation in anterior pituitary gland *in vitro* by synthetic luteinizing hormone-releasing hormone/follicle-stimulating hormone-releasing hormone (LH-RH/FSH-RH). *Proc. nat. Acad. Sci. (Wash.),* 69: 2677–2681.

Borgeat, P., Labrie, F., Poirier, G., Chavancy, G. and Schally, A.V. (1973) Stimulation of adenosine 3′,5′-monophosphate accumulation in anterior pituitary gland by purified growth hormone-releasing hormone. *Trans. Ass. Amer. Phys.,* 86: 284–299.

Borgeat, P., Labrie, F., Côté, J., Ruel, F., Schally, A.V., Coy, D.H., Coy, E.J. and Yanaihara, N. (1974a) Parallel stimulation of cyclic AMP accumulation and LH and FSH release by analogs of LH-RH *in vitro. J. molec. cell. Endocr.,* 1: 7–20.

Borgeat, P., Labrie, F., Drouin, J., Bélanger, A., Immer, H., Sestanj, K., Nelson, V., Gotz, M., Schally, A.V., Coy, D.H. and Coy, E.J. (1974b) Inhibition of adenosine 3′,5′-monophosphate accumulation in anterior pituitary gland *in vitro* by growth hormone release-inhibiting hormone. *Biochem. biophys. Res. Commun.,* 56: 1052–1059.

Borgeat, P., Garneau, P. and Labrie, F. (1975a) Calcium requirement for stimulation of cyclic AMP accumulation in anterior pituitary by LH-RH. *Cell. Endocr.,* 2: 117–124.

Borgeat, P., Garneau, P. and Labrie, F. (1975b) Characteristics of action of prostaglandins on cyclic AMP accumulation in rat anterior pituitary gland. *Canad. J. Biochem.,* 53: 455–460.

Brazeau, P., Vale, W., Burgus, R., Ling, N., Butcher, M., Rivier, J. and Guillemin, R. (1973) Hypothalamic polypeptide that inhibits the secretion of immunoreactive pituitary growth hormone. *Science,* 179: 77–79.

Burgus, R. and Guillemin, R. (1970) Hypothalamic releasing factors. *Ann. Rev. Biochem.,* 39: 499–526.

Burgus, R., Dunn, T.F., Desiderio, D. et Guillemin, R. (1969) Structure moléculaire du facteur hypothalamique TRF d'origine ovine: mise en évidence par spectrométrie de masse de la séquence PCA-His-Pro-NH$_2$. *C.R. Acad. Sci. (Paris),* 269: 1870–1873.

Burgus, R., Butcher, M., Ling, N., Monahan, M., Rivier, J., Fellows, R., Amoss, M., Blackwell, R., Vale, W. et Guillemin, R. (1971) Structure moléculaire du facteur hypothalamique (LRF) d'origine ovine contrôlant la sécrétion de l'hormone de gonadotrope hypophysaire de lutéinisation. *C.R. Acad. Sci. (Paris),* 273: 1611–1613.

Carlson, J.C., Barcikowski, B. and McCracken, J.A. (1973) Prostaglandin $F_{2\alpha}$ and the release of LH in sheep. *J. Reprod. Fertil.,* 34: 357–361.

Cehovic, G. (1969) Rôle de l'adénosine de 3′-5′-monophosphate cyclique dans la libération de TSH hypophysaire. *C.R. Acad. Sci. (Paris),* 268: 2929–2931.

Deuben, R.R. and Meites, J. (1964) Stimulation of pituitary growth hormone release by a hypothalamic extract *in vitro. Endocrinology,* 74: 408–414.

Drouin, J., De Léan, A., Rainville, D., Lachance, R. and Labrie, F. (1976) Characteristics of the interaction between TRH and somatostatin for thyrotropin and prolactin release. *Endocrinology,* in press.

Ferland, L., Labrie, F., Savary, M., Beaulieu, M., Coy, D.H., Coy, E.J. and Schally, A.V. (1976) Inhibitory activity of analogs of luteinizing hormone-releasing hormone in vitro and in vivo. *Clin. Endocr.,* in press.

Green, J.D. and Harris, G.W. (1947) The neurovascular link between the neurohypophysis and adenohypophysis. *J. Endocr.,* 5: 136–146.

Greengard, P. (1975) Cyclic nucleotide protein phosphorylation and neuronal function. In *Advances in Cyclic Nucleotide Research,* G.I. Drummond, P. Greengard and G.A. Robison (Eds.), Raven Press, New York, pp. 585–602.

Harms, P.G., Ojeda, S.R. and McCann, S.M. (1974) Prostaglandin-induced release of pituitary gonadotropin: central nervous system and pituitary sites of action. *Endocrinology,* 94: 1459–1464.

Holmes, S.W. and Horton, E.W. (1968) The identification of four prostaglandins in dog brain and their regional distribution in the central nervous system. *J. Physiol. (Lond.),* 195: 731–741.

Jutisz, M. and Paloma de la Llosa, M. (1970) Requirement of Ca^{++} and Mg^{++} ions for the *in vitro* release of follicle-stimulating hormone from rat pituitary gland and its subsequent biosynthesis. *Endocrinology,* 86: 761–768.

Jutisz, M., Kerdelhue, G., Berauld, A. and Paloma de la Llosa, M. (1972) On the mechanism of action of the hypothalamic gonadotropin releasing factors. In *Gonadotropins,* B.B. Saxena, C.G. Beling and H.M. Gandy (Eds.), Wiley Interscience, New York, pp. 64—71.

Kaneko, T., Saito, S., Oka, H., Oda, T. and Yanaihara, N. (1973) Effects of synthetic LH-RH and its analogs on rat anterior pituitary cyclic AMP and LH and FSH release. *Metabolism,* 22: 77—78.

Labrie, F., Barden, N., Poirier, G. and De Léan, A. (1972) Characteristics of binding of [^{3}H]thyrotropin-releasing hormone to plasma membranes of bovine anterior pituitary gland. *Proc. nat. Acad. Sci. (Wash.),* 69: 283—287.

Labrie, F., Pelletier, G., Lemay, A., Borgeat, P., Barden, N., Dupont, A., Savary, M., Côté, J. and Boucher, R. (1973) Control of protein synthesis in anterior pituitary gland. In *Karolinska Symp. Res. Meth. Reprod. Endocr.,* E. Diczfalusy (Ed.), pp. 301—340.

Labrie, F., Pelletier, G. and Barden, N. (1975a) Aspects of the mechanism of action of hypothalamic releasing hormones in the anterior pituitary gland. In *Advances in Sex Hormone Research, Vol. 1,* J.A. Thomas and R.L. Singhal (Eds.), University Park Press, Baltimore, Md., pp. 77—127.

Labrie, F., Borgeat, P., Lemay, A., Lemaire, S., Barden, N., Drouin, J. and Bélanger, A. (1975b) Role of cyclic AMP in the action of hypothalamic regulatory hormones in the anterior pituitary gland. In *Advances in Cyclic Nucleotide Research, Vol. V,* G.I. Drummond and G.A. Robison (Eds.), Raven Press, New York, pp. 787—801.

Labrie, F., Savary, M., Coy, D.H., Coy, E.J. and Schally, A.V. (1976) Inhibition of LH release by analogs of LH releasing hormone (LH-RH) *in vitro. Endocrinology,* 98: 287—292.

Lefkowitz, R.J., Roth, J. and Pastan, I. (1970) Effects of calcium on ACTH stimulation of the adrenal-separation of hormone binding from adenyl cyclase activation. *Nature (Lond.),* 228: 864—866.

Lemay, A. and Labrie, F. (1972) Calcium-dependent stimulation of prolactin release in rat anterior pituitary in vitro by N^6-monobutyryl adenosine 3',5'-monophosphate. *FEBS Lett.,* 20: 7—10.

MacLeod, R.M. and Lehmeyer, J.E. (1970) Release of pituitary growth hormone by prostaglandins and dibutyryl adenosine cyclic 3',5'-monophosphate in the absence of protein synthesis. *Proc. nat. Acad. Sci. (Wash.),* 67: 1172—1179.

Makino, T. (1973) Study of the intracellular mechanism of LH release in the anterior pituitary. *Amer. J. Obstet. Gynec.,* 115: 606—614.

Matsuo, H., Baba, Y., Nair, R.M.G., Arimura, A. and Schally, A.V. (1971) Structure of the porcine LH- and FSH-releasing hormone. I. The proposed amino acid sequence. *Biochem. biophys. Res. Commun.,* 43: 1334—1339.

McCann, S.M. and Porter, J.C. (1969) Hypothalamic pituitary stimulating and inhibiting hormones. *Physiol. Rev.,* 49: 240—284.

McCracken, J.A., Barcikowski, B., Carlson, J.C., Green, K. and Samuelsson, B. (1973) The physiological role of prostaglandin $F_{2\alpha}$ in corpus luteum regression. *Advanc. Biosci.,* 9: 599—605.

Pelletier, G., Labrie, F., Puviani, R., Arimura, A. and Schally, A.V. (1974) Immunohistochemical localization of luteinizing hormone-releasing hormone in the rat median eminence. *Endocrinology,* 95: 314—317.

Ramwell, P.W. and Shaw, J.E. (1970) Biological significance of the prostaglandins. *Recent Progr. Hormone Res.,* 26: 139—173.

Ratner, A., Wilson, M.C., Srivastava, L. and Peake, G.T. (1974) Stimulatory effects of prostaglandin E_1 on rat anterior pituitary cyclic AMP and luteinizing hormone release. *Prostaglandins,* 5: 165—171.

Samli, M.H. and Geschwind, I.I. (1968) Some effects of energy-transfer inhibition and of Ca^{++}-free or K^+-enhanced media on the release of luteinizing hormone (LH) from the rat pituitary gland *in vitro. Endocrinology,* 82: 225—231.

Sato, T., Taya, K., Jyuno, T., Hirono, M. and Igarashi, M. (1974) The stimulatory effect of prostaglandins on luteinizing hormone release. *Amer. J. Obstet. Gynec.,* 118: 875—876.

Sayers, G., Beall, R.J. and Seeling, S. (1972) Isolated adrenal cells: adrenocorticotropin hormone, calcium steroidogenesis and cyclic adenosine monophosphate. *Science,* 175: 1131—1133.

Schally, A.V., Arimura, A., Bowers, C.Y., Kastin, A.J., Sawano, S. and Redding, T.W. (1968) Hypothalamic neurohormones regulating anterior pituitary function. *Recent Progr. Hormone Res.,* 24: 497–588.

Schally, A.V., Sawano, S., Arimura, A., Barrett, J.F., Wakabayashi, I. and Bowers, D.Y. (1969) Isolation of growth hormone-releasing hormone (GRH) from porcine hypothalami. *Endocrinology,* 84: 1493–1506.

Schally, A.V., Dupont, A., Arimura, A., Redding, T.W. and Linthicom, G.L. (1975) Isolation of porcine GH-release inhibiting hormone (GH-RIH): the existence of 3 hours of GH-RIH. *Fed. Proc.,* 34: 584–586.

Schofield, J.G. (1967) Measurement of growth hormone released by ox anterior pituitary slices *in vitro. Biochem. J.,* 103: 331–341.

Siler, T.M., VendenBerg, G., Yen, S.S.C., Brazeau, P., Vale, W. and Guillemin, R. (1973) Inhibition of growth hormone release in humans by somatostatin. *J. clin. Endocr.,* 37: 632–634.

Tsafiri, A., Koch, Y. and Lindner, H.R. (1973) Ovulation rate and serum LH levels in rats treated with indomethacin or prostaglandin E_2. *Prostaglandins,* 3: 461–467.

Vale, W., Rivier, C., Brazeau, P. and Guillemin, R. (1974) Effects of somatostatin on the secretion of thyrotropin and prolactin. *Endocrinology,* 95: 968–977.

Wakabayashi, K., Kamberi, I.A. and McCann, S.M. (1969) *In vitro* response of the rat pituitary to gonadotropin-releasing factors and to ions. *Endocrinology,* 85: 1046–1056.

Wellmann, W. and Schwabe, U. (1973) Effects of prostaglandins E, E_2 and $F_{2\alpha}$ on cyclic AMP levels in brain *in vivo. Brain Res.,* 59: 371–378.

Wilber, J.G., Peake, G.T., Mariz, I., Utiger, R.D. and Daughaday, W.H. (1968) Theophylline and epinephrine effect upon the secretion of growth hormone (GH) and thyrotropin (TSH) *in vitro. Clin. Res.,* 16: 277–280.

Wilber, J., Peake, G.T. and Utiger, R. (1969) Thyrotropin release *in vitro:* stimulation by cyclic 3′,5′-adenosine monophosphate. *Endocrinology,* 84: 758–760.

Zor, U., Kaneko, T., Schneider, H.P.G., McCann, S.M. and Field, J.B. (1970) Further studies of stimulation of anterior pituitary cyclic adenosine 3′,5′-monophosphate formation by hypothalamic extract and prostaglandins. *J. biol. Chem.,* 245: 2883–2888.

Chapter 22

Control of adenohypophyseal hormone secretion by prostaglandins*

S.M. McCANN, S.R. OJEDA, P.G. HARMS, J.E. WHEATON, D.K. SUNDBERG and C.P. FAWCETT

Department of Physiology, The University of Texas Health Science Center, Southwestern Medical School, Dallas, Texas 72535 (U.S.A.)

I. INTRODUCTION

Because of the ubiquitous distribution and the many faceted biological actions of the prostaglandins, it was logical to study the possible effects of these agents on the hypothalamic-pituitary unit. In this paper we will examine the data which indicate that prostaglandins have actions both on the pituitary and on the hypothalamus to alter secretion of adenohypophyseal hormones. We will summarize the present status of the evidence for a role of prostaglandins in the control of ACTH, TSH and growth hormone, but will evaluate in more detail the role of the prostaglandins in the secretion of the gonadotropins and prolactin since we have been particularly concerned with the regulation of these hormones in our laboratory.

The first evidence that prostaglandins could alter the function of the adenohypophysis was obtained in the experiments of Zor et al. (1969, 1970) in which it was demonstrated that prostaglandins can increase adenylate cyclase and cyclic AMP levels in adenohypophyses incubated in vitro. In these early studies, none of the prostaglandins increased the release of LH. Many later studies have demonstrated that the systemic administration of prostaglandins can alter adenohypophyseal hormone release and that these alterations are brought about by actions not only directly on the pituitary gland itself but also on the hypothalamus which in turn alters release of pituitary hormones. In this review we will examine in sequence the evidence for effects of prostaglandins on the release of ACTH, TSH, growth hormone, prolactin, FSH and LH. In each instance we will divide the consideration into the effects of systemic administration of prostaglandins on the release of the hormone in question and then turn to experiments in which prostaglandins have been incubated with pituitaries in vitro or injected into the gland in vivo in order to determine if these agents can affect the gland directly. Then, studies in which the prostaglandins have been injected into hypothalamic tissue or into the ventricular system to examine their possible effects on the CNS will be considered. Lastly, the possible physiological role of the prostaglandins in pituitary hormone release will be evaluated by examining the

---

* This work was supported in part by grants from the NIH (AM 10073 and HD 05151) and the Ford Foundation.

results of experiments in which inhibitors of prostaglandin synthesis or action have been used.

II. THE EFFECTS OF PROSTAGLANDINS ON ACTH RELEASE

De Wied et al. (1969) were the first to evaluate the effects of prostaglandins on ACTH secretion. Prostaglandin E_1 and E_2 but not $F_{1\alpha}$ or $F_{2\alpha}$ stimulated ACTH release as indicated by a rise in corticosterone secretion and the response was not completely blocked by CNS-depressant drugs, such as pentobarbital-chlorpromazine. The response was largely inhibited by pentobarbital-morphine but was not blocked by dexamethasone-pentobarbital. The ACTH release induced by prostaglandins persisted in neurohypophysectomized animals but there was little effect in animals with median eminence lesions. Since PGE_1 was also inactive in increasing ACTH release from pituitaries incubated in vitro, it was concluded that the action was exerted on the CNS to release corticotropin-releasing factor. Similarly, Peng et al. (1970) observed that intravenous injection of PGE_1 could increase ACTH release in vivo. Since the response was blocked by treatment with Nembutal plus morphine, the action was thought to be via an effect on the CNS. Hedge (1972) observed that injection of prostaglandins directly into the anterior pituitary had no effect on ACTH secretion in the rat as measured by changes in plasma corticosterone titers. He was the first to inject prostaglandins into hypothalamic tissue and found that PGE_1, $PGF_{1\alpha}$ or $PGF_{2\alpha}$ stimulated ACTH secretion when micro-injected into the median eminence but not into neighboring portions of the hypothalamus in the rat. Therefore, the results of these studies clearly indicate that prostaglandins can act on the hypothalamus to stimulate a release of corticotropin-releasing factor which then activates ACTH release.

III. EFFECTS OF PROSTAGLANDINS ON THE RELEASE OF TSH

Although TSH release was not altered by systemic administration of prostaglandins in the experiments of Brown and Hedge (1974), it was observed that they could enhance the response to thyrotropin-releasing hormone (TRH). Vale et al. (1972) has earlier observed increased TSH release from cultured anterior pituitary cells after addition of PGE_2 to the incubation media. Similarly, Dupont and Chavancy (1972) observed increased TSH release from hemipituitaries incubated in vitro in the presence of PGE_1; however, Tal et al. (1974) and Sundberg et al. (1975) found no effects of prostaglandins on release of TSH by pituitaries incubated in vitro. Sundberg et al. did observe in confirmation of the in vivo results of Brown and Hedge that prostaglandins could potentiate the stimulatory effects of TRH on TSH release in vitro.

IV. EFFECTS OF PROSTAGLANDINS ON THE RELEASE OF GROWTH HORMONE

Either PGE_1 or PGE_2 have been found to be potent prostaglandins in releasing growth hormone after systemic administration to animals and man (Ito et al., 1971; Hertelendy et al., 1972). In contrast to the relative lack of effect of prostaglandins on the in vitro release of ACTH and TSH is their dramatic stimulation of the release and synthesis of growth hormone in vitro (MacLeod and Lehmeyer, 1970; Schofield, 1970; Hertelendy, 1971; Sundberg et al., 1975) (Table I). It is probable that the effects of prostaglandins, particularly of the E series, in releasing growth hormone are mediated by adenylate cyclase since the prostaglandins increased not only adenylate cyclase but also cyclic AMP levels in the pituitary (Zor et al., 1970) and this increase was correlated with their effectiveness in increasing growth hormone release (MacLeod and Lehmeyer, 1970; Sundberg et al., 1975).

TABLE I

Effect of prostaglandins E_1, E_2, $F_{1\alpha}$ and $F_{2\alpha}$ on basal hormone release in vitro *

(From Sundberg et al. (1975) Proc. Soc. exp. Biol. (N.Y.), 148: 55.)

	μg hormone/ml/4 hr				
	LH	FSH	PRL	TSH	GH
Experiment 1					
Control	0.21 ± 0.03	5.5 ± 0.2	16.4 ± 1.1	30 ± 5	50 ± 4
PGE_1 0.03 mM	0.18 ± 0.01	5.8 ± 0.2	16.4 ± 0.9	30 ± 7	173 ± 15***
Experiment 2					
Control	0.17 ± 0.02	6.1 ± 0.3	14.4 ± 0.5	27 ± 1	34 ± 3
PGE_2 0.015 mM	0.18 ± 0.01	6.3 ± 0.5	16.3 ± 0.7	26 ± 1	166 ± 12***
$PGF_{1\alpha}$ 0.015 mM	0.20 ± 0.02	6.9 ± 0.3	15.2 ± 1.5	24 ± 1	88 ± 14**
$PGF_{2\alpha}$ 0.015 mM	0.19 ± 0.02	6.2 ± 0.4	15.6 ± 1.3	24 ± 3	84 ± 3***

*   Each value represents the mean ± S.E. of 4 flasks per experimental group.
** $P < 0.05$.
*** $P < 0.01$ vs. the respective control.

Unfortunately, there have been no studies in vivo to determine the site of action of prostaglandins in stimulating growth hormone release, but one could speculate on the basis of the in vitro studies that injections of prostaglandins into the pituitary or into cannulated portal vessels would stimulate growth hormone release. Whether they would have any action on the hypothalamus to alter growth hormone release analogous to that described for ACTH secretion remains to be determined.

V. EFFECTS OF PROSTAGLANDINS ON THE RELEASE OF PROLACTIN

Although neither PGE_1 nor E_2 increased plasma prolactin titers in ovariectomized rats following their intravenous injection (Ojeda et al., 1974a), higher doses of PGEs or PGFs have induced prolactin release in ovarietomized, estrogen-primed rats (Sato et al., 1974).

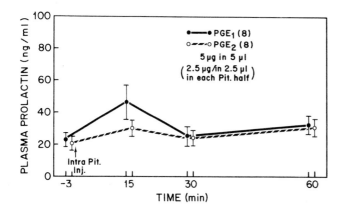

Fig. 1. Effect on plasma prolactin of intrapituitary injection of PGE$_1$ or PGE$_2$ in ovariectomized rats. 2.5 μg in 2.5 μl of each prostaglandin were injected into each lobe. (From Ojeda et al., *Endocrinology*, 96 (1974) p. 616.)

PGF$_{2\alpha}$ has recently been found to be a powerful stimulus for prolactin release in the bovine (Louis et al., 1974; Tucker et al., 1975) and in the human female (Yue et al., 1974). As in the case of experiments with systemic administration of prostaglandins and ACTH secretion, the widespread physiological and pharmacological effects of the compounds may produce non-specific stress which could then produce an alteration in prolactin release. Stress has been found to be a potent stimulus for prolactin release in all species examined. Therefore, to localize the action of prostaglandins on prolactin release, it was essential to perform experiments in which the drugs were applied directly to tissue in either in vivo or in vitro studies. Sundberg et al. (1975) observed no effect of prostaglandins on the release of prolactin from pituitaries incubated in vitro (Table I), but intrapituitary injection of PGE$_2$ produced a slight increase in prolactin titers in ovariectomized, but not in ovariectomized, estrogen-primed rats (Ojeda et al., 1974a) (Fig. 1). There was no effect on prolactin release when PGE$_2$ was injected into a cannulated portal vessel in male rats (Eskay et al., 1975).

In contrast to the relative lack of effect of prostaglandins on the pituitary to release prolactin was the effect obtained by intraventricular injection of these agents. Injection of PGE$_1$, but not PGE$_2$, PGF$_{1\alpha}$ or PGF$_{2\alpha}$ into the third ventricle evoked prolactin release in ovariectomized rats whether or not they were primed with estrogen (Ojeda et al., 1974a) (Fig. 2). The response was more readily observed in the ovariectomized animals and in this situation was directly related to the dose injected. Injection of PGE$_2$ into the lateral ventricle released prolactin in intact male rats as well (Eskay et al., 1975).

The action of PGE$_1$ to release prolactin has been related to the dopaminergic inhibitory control over prolactin release. For example, intraventricular injection of dopamine can lower plasma prolactin. This effect of intraventricular dopamine was completely blocked by the injection of PGE$_1$ at a dose which by itself had no effect on prolactin release (Fig. 3). These observations suggest that PGE$_1$ may elevate prolactin by blocking the stimulation of prolactin-inhibiting factor (PIF) release by dopamine; however, a stim-

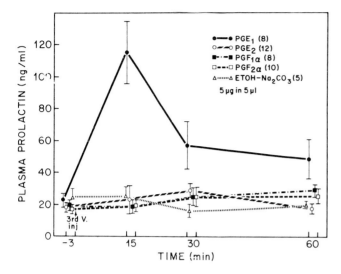

Fig. 2. Effect of third ventricular injections of prostaglandins on plasma prolactin in ovariectomized rats. Controls (injected with 5 µl, 95% ethanol–0.02% Na_2CO_3, 1:9) are also represented. Vertical lines represent the standard error of the mean and figures in parentheses indicate the number of animals used. (From Ojeda et al., *Endocrinology* 96 (1974) p. 615.)

ulatory effect of PGE_1 on prolactin-releasing factor discharge cannot be ruled out (Ojeda et al., 1974b).

Since intraventricular cyclic AMP can drastically lower plasma prolactin titers, it has been postulated that the PIF-releasing action of dopamine is mediated by an increase in intracellular cyclic AMP in the PIF-secreting elements. Intraventricular PGE_1 failed to alter the lowering of prolactin which followed the intraventricular injection of dibutyryl cyclic AMP, which suggests that prostaglandins may act prior to the cyclic AMP step in this sequence of events (Fig. 3) (Ojeda et al., 1974b).

VI. THE EFFECT OF PROSTAGLANDINS ON THE RELEASE OF GONADOTROPINS

Both PGE_2 and E_1 can stimulate gonadotropin release in either male or female rats following their intravenous injection (Tsafrini et al., 1973; Harms et al., 1974; Sato et al., 1974). $PGF_{2\alpha}$ may also stimulate LH release in the sheep (Carlson et al., 1973) although this effect seems to be mediated at least in part by the ovaries (Chamley and Christie, 1973). In order to localize the site of action of prostaglandins in releasing gonadotropins, these agents have been incubated with pituitaries in vitro, but the results are controversial. Although Makino (1973), Ratner et al. (1974) and Sato et al. (1975a) reported increased LH release following incubation of pituitaries with prostaglandins, Sundberg et al. (1975) and Chobsieng et al. (1975) found no effect of prostaglandins on either basal or LHRH-stimulated release of FSH or LH (Table I).

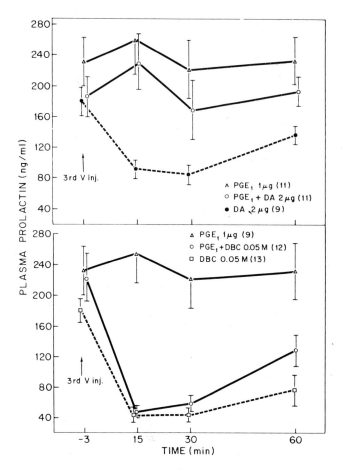

Fig. 3. Effect of intraventricular injection of PGE_1 on the decrease in plasma prolactin induced by dopamine (DA) (upper panel) or dibutyryl cyclic AMP (DBC) (lower panel) in ovariectomized, estrogen-treated rats. PGE_1 was administered 5 min before in a 2 μl vol. (From Ojeda et al., *Endocrinology*, 95 (1974) p. 1698.)

When either PGE_2 or PGE_1 were injected directly into the pituitary of ovariectomized animals, there was similarly no effect on gonadotropin release, but PGE_2 produced a small increase in plasma LH in ovariectomized, estrogen-primed rats (Harms et al., 1974) (Fig. 4). When PGE_2 was infused into cannulated portal vessels in male rats, no alteration in FSH and LH titers was observed (Eskay et al., 1975). In summary, it would appear that prostaglandins have relatively little action on the anterior pituitary itself to stimulate gonadotropin release.

The results were completely different when the prostaglandins were micro-injected into the third ventricle. In ovariectomized animals, a dramatic increase in plasma LH titers and a slight increase in plasma FSH followed injections of PGE_2 into the third ventricle (Harms et al., 1973) (Fig. 5). Since the release of LH was much greater than the release of FSH, the effect could be accounted for by the presumed release of the decapeptide LHRH by the prostaglandins as LHRH has a much greater effect on LH than on FSH release in the ovariectomized rat. In ovariectomized, estrogen-primed animals, the effect was even more pronounced. The response to intraventricular PGE_2 was dose-related in these animals (Fig. 6), and in this situation PGE_1 was active although less so (Harms et al., 1974). This observation is in agreement with the findings of Spies and Norman (1973) who found that PGE_1 was active to release LH in the Nembutalized, proestrous rat.

Since all of these results were obtained in anesthetized animals, it was important to determine if prostaglandins could act in conscious rats. Prostaglandin E_2 was the only prostaglandin effective to release LH in castrate males, but PGE_1 also exerted a small effect in intact animals (Ojeda et al., 1974c). $PGF_{1\alpha}$ and $PGF_{2\alpha}$ were ineffective to alter gonadotropin release in any of these studies.

In subsequent studies, we evaluated the effect of the prostaglandin derivative, 15-

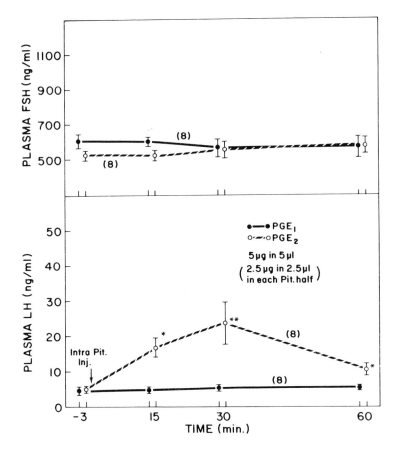

Fig. 4. Concentration of plasma LH and FSH before and after intrapituitary injections of PGE_1 or PGE_2 in ovariectomized rats treated with estrogen. (From Harms et al., *Endocrinology*, 94 (1974) p. 1463.)

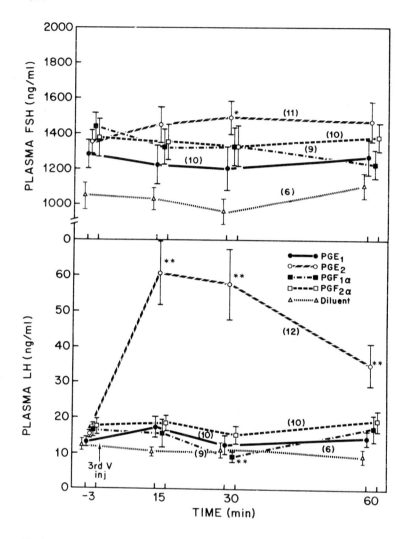

Fig. 5. Concentrations of plasma LH and FSH before and after 3rd V injection of PGs in ovariectomized rats. Values are means ± S.E., with number of animals in parentheses. Arrow indicates time of injection. Statistical significance: * = $P < 0.05$; ** = $P < 0.01$; *** = $P < 0.001$. (From Harms et al., Endocrinology, 94 (1974) p. 1461.)

methyl PGE_2, and prostaglandins of the A series (Ojeda et al., 1975b). The 15-methyl PGE_2 which is more slowly degraded than the natural product had a slightly greater effect in stimulating FSH release following its intravenous injection than the natural product, but following intraventricular injection, it was less active than PGE_2. The greater effect on FSH release when the analog was administered intravenously may be related to its slow destruction which caused a more prolonged, if less potent, action. This coupled with a slower disappearance of FSH from the plasma could account for its greater effect on FSH release. By contrast, the analog was less potent than PGE_2 in stimulating LH release. PGA_1 was completely ineffective to alter gonadotropin release following its intraventricular injection but PGA_2 had a slight effect.

From these studies it is clear that the presumed prostaglandin receptor which activates gonadotropin release in the rat has very precise structural requirements. The omission of the double bond in the 5,6 position in PGE_2 to produce PGE_1 is associated with considerable loss of activity. The loss of a hydroxyl group from position 11 is associated with an even further reduction in activity in the case of PGA_2, and the conversion of the carbonyl group to a hydroxyl group in position 9 as in $PGF_{1\alpha}$ and $F_{2\alpha}$ associated with complete loss of activity.

To determine if intraventricular injection of PGE_2 was indeed acting by releasing LHRH, peripheral plasma LHRH titers were measured in conscious, ovariectomized rats by radioimmunoassay (Ojeda et al., 1975c). Intraventricular injection of PGE_2 rapidly increased peripheral plasma LHRH titers in these animals (Fig. 7). Similarly, PGE_2 increased the concentration of LHRH in hypophyseal portal blood of anesthetized male rats (Eskay et al., 1975). Furthermore, Sato et al. (1975b) observed a decline in hypothalamic LHRH within 5 min after massive intravenous injection of PGE_2 in ovariectomized, estrogen-progesterone-treated rats. Additional evidence for an action of prostaglandins via a release of LHRH is provided by the observations of Chobsieng et al. (1975)

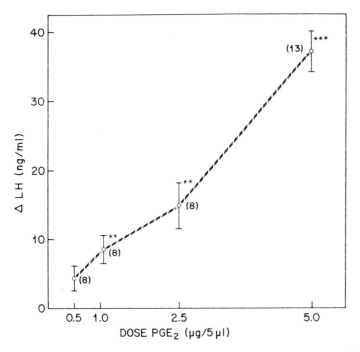

Fig. 6. Increases in plasma LH 30 min following 3rd V injections of varying doses of PGE_2 in ovariectomized estrogen-primed rats. (From Harms et al., *Endocrinology*, 94 (1974) p. 1462.)

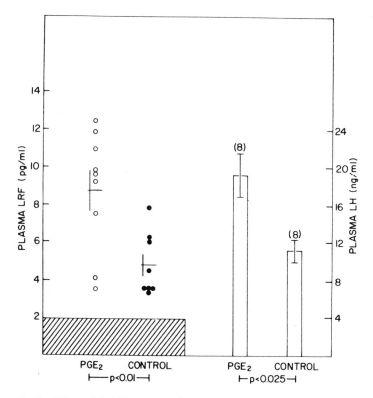

Fig. 7. Effect of 3rd V injection of PGE_2 on plasma LRF and LH concentrations of ovariectomized rats decapitated 5 min after injection. The shaded area represents the limit of sensitivity of the LRF assay. The numbers is parentheses indicate the numbers of animals used. Vertical lines are the S.E.M. (From Ojeda et al., Neuroendocrinology, 17 (1975) p. 285.)

who reported that an antiserum to LHRH completely blocked the effect of systemic PGE_2 on the release of LH.

It is thought that PGE_2 acts directly on the LHRH-secreting neurons to stimulate release of the neurohormone since a variety of receptor-blocking agents failed to alter the LH release produced by intraventricular PGE_2 (Harms et al., 1975). Thus, phentolamine, an alpha receptor blocker, Pimozide, a dopamine receptor blocker, methysergide and Cinanserin, serotonin receptor blockers, and atropine, a muscarinic cholinergic receptor blocker, were all unable to alter PGE_2-induced LH release.

That hypothalamic tissue has the intrinsic ability to synthesize prostaglandins is suggested by the presence of detectable amounts of these compounds in the tissue (Holmes and Horton, 1968). Moreover, it has been found recently that arachidonic acid, the precursor of PGE_2, elicited LH release in a dose-related manner when injected intraventricularly to ovariectomized, estrogen-treated rats. The interpretation of these latter results, however, is complicated by the fact that other fatty acids not bioconvertable into prostaglandins, such as 11,14,17-eicosatetraenoic acid and linoleic acid, were also able to increase LH release, though less efficiently. Docosahexaenoic acid was ineffective.

VII. EFFECTS OF INHIBITORS OF PROSTAGLANDIN SYNTHESIS ON PITUITARY HORMONE RELEASE

Inhibitors of prostaglandin synthesis, such as aspirin and indomethacin, have either been injected systemically, implanted into the hypothalamus, or added to media for incubation of pituitary glands in order to determine the importance of prostaglandins in

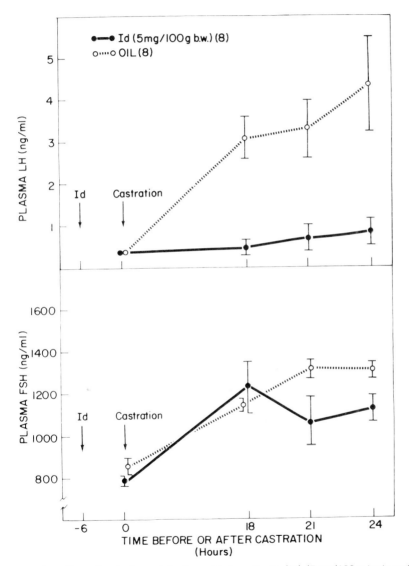

Fig. 8. Effect of a single s.c. injection of indomethacin (Id) (5 mg/100 g body weight) on the postcastration rise in plasma gonadotropins of male rats. (From Ojeda, S.R., Harms, P.G. and McCann, S.M., *Endocrinology*, in press.)

the natural release of pituitary hormones. The prostaglandin antagonist, 7-oxa-13 prostanoic acid, and the inhibitor of prostaglandin synthetase, indomethacin, have been effective to suppress release of growth hormone from pituitaries incubated in vitro (Ratner et al., 1974; Sundberg et al., 1975). In contrast to the effect of these inhibitors in blocking growth hormone release in vitro was the absence of action of indomethacin on gonadotropin release from incubated pituitaries. Incubation of glands with this inhibitor of prostaglandin synthetase either had no effect on basal or LHRH-stimulated release of gonadotropins at low doses, or actually augmented responses to LHRH at high doses of the drug (Sundberg et al., 1975).

In vivo studies with inhibitors of prostaglandin synthesis have resulted in controversial findings. Some groups (Carlson et al., 1973; Tsafrini et al., 1973) found that indomethacin could block ovulation in the absence of a suppression of LH release, but Behrman et al. (1972) observed that aspirin-induced blockade of ovulation could be reversed by LH or LHRH which suggested that this inhibitor of prostaglandin synthetase blocked LHRH release. In more recent studies, it has been found that there is a latency of more than 8 hr after systemic injection of indomethacin prior to suppression of LH release (Ojeda et al., 1975a). After this latent period, the drug was found to depress LH titers in ovariectomized rats, to inhibit the circhoral rhythm of LH release in these animals and to prevent the post-castration elevation of plasma LH concentrations in males (Fig. 8). The progesterone-induced LH release in estrogen-primed, ovariectomized rats was also blocked by indomethacin (Ojeda et al., 1975a) and estrogen-induced LH release failed to occur in sheep (Carlson et al., 1974). In contrast to the suppression of release of LH by indomethacin in these studies was the lack of effect of this agent on FSH release except for the partial suppression of the release of FSH in ovariectomized, estrogen-primed rats.

A central nervous system site of action of indomethacin is indicated since the drug did not inhibit the LH release in response to synthetic LHRH in ovariectomized rats and only partially blocked the response in estrogen-progesterone-treated animals (Fig. 9) (Ojeda et al., 1975a; Sato et al., 1975a). In this latter situation, in agreement with previously cited in vitro results of Sundberg et al. (1975), indomethacin actually potentiated the FSH release in response to LHRH (Ojeda et al., 1975a).

Further evidence that indomethacin was acting centrally was obtained in experiments in which the drug was injected into the third ventricle or implanted in the medial basal hypothalamus. Following introduction of indomethacin into either the ventricle or the medial basal hypothalamus, plasma LH was suppressed in ovariectomized animals at 1–6 hr. Similarly, 5,8,11,14-eicosatetraenoic acid, another inhibitor of prostaglandin synthesis, lowered plasma LH in ovariectomized rats within 1–4 hr following its injection into the third ventricle.

Since the doses of indomethacin injected systemically in these studies were higher than the minimal effective dose to lower prostaglandin levels in tissue, it is still conceivable that these effects were exerted by another action of the drug; however, these experiments provide evidence that prostaglandins may play an essential role in control of LH release (Ojeda et al., 1975c).

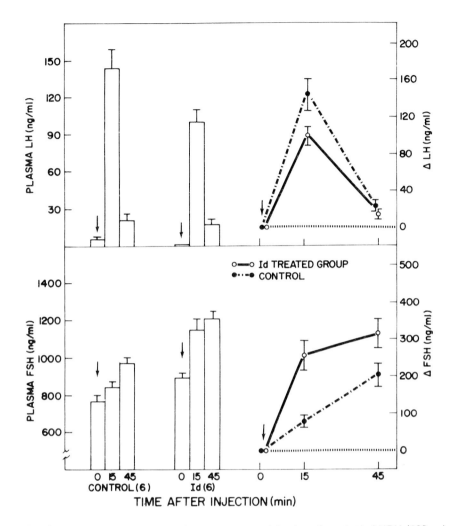

Fig. 9. Release of gonadotropins in response to an injection of synthetic LHRH (100 ng/rat, i.v.) in ovariectomized, estrogen-progesterone-treated rats injected with a single s.c. dose of indomethacin (Id) (5 mg/100 g body weight) 28 hr before LHRH. Arrows indicate time of LHRH injection. (From Ojeda, S.R., Harms, P.G. and McCann, S.M., *Endocrinology*, in press.)

VIII. CONCLUSIONS

From the results presented above, it is clear that prostaglandins have widespread effects on the release of pituitary hormones. They can cause a release of corticotropin-releasing factor which in turn stimulates secretion of adrenocorticotropin from the adenohypophysis. They appear to act directly on the pituitary to increase the responsiveness of the thyrotrophs to TRH. On the basis of in vitro studies only, it appears that prostaglandins increase growth hormone release mainly by an action directly on the pituitary.

This effect may be causally related to an increase in cyclic AMP within the gland. PGE_1 appears to have an action to stimulate prolactin release, possibly via an inhibition of dopamine-induced PIF discharge. The primary action on gonadotropin secretion appears to be mediated via the hypothalamus, probably via a direct stimulation of the LHRH-secreting neurons to increase the discharge of LHRH. Both in the case of growth hormone and LH, inhibitors of prostaglandin synthesis can decrease the release of the hormone in question, at least under certain conditions, which provides evidence that the prostaglandins may play an essential role in the releasing process; however, additional experimental work is required to substantiate this hypothesis.

REFERENCES

Behrman, H.R., Orczyk, G.P. and Greep, R.O. (1972) Effect of synthetic gonadotrophin-releasing hormone (Gn-RH) on ovulation blockade by aspirin and indomethacin. *Prostaglandins*, 1: 245–257.

Brown, M. and Hedge, G.A. (1974) In vivo effects of prostaglandins on TRH-induced TSH secretion. *Endocrinology*, 95: 1392–1397.

Carlson, J.C., Barcikowski, B. and McCracken, J.A. (1973) Prostaglandin $F_{2\alpha}$ and the release of LH in sheep. *J. Reprod. Fertil.*, 34: 357–361.

Carlson, J.C., Barcikowski, B., Cargill, V. and McCracken, J.A. (1974) The blockade of LH release by indomethacin. *J. clin. Endocr.*, 39: 399–402.

Chamley, W.A. and Christie, M. (1973) Failure of prostaglandin $F_{2\alpha}$ to affect LH secretion in the ovariectomized ewe. *Prostaglandins*, 3: 405–412.

Chobsieng, P., Naor, Z., Zor, U. and Lindner, H.R. (1975) Stimulatory effect of prostaglandin E_2 on LH release in the rat: evidence for hypothalamic site of action. *Neuroendocrinology*, 17: 12–17.

De Wied, D., Witter, A., Versteeg, D.H.G. and Mulder, A.H. (1969) Release of ACTH by substances of central nervous system origin. *Endocrinology*, 85: 651–669.

Dupont, A. and Chavancy, G. (1972) Prostaglandins and cyclic AMP as mediators of thyrotropin-releasing hormone action. *Progr. IV int. Congr. Endocrinol.*, p. 84.

Eskay, R.L., Warberg, J., Mical, R.S. and Porter, J.C. (1975) Prostaglandin E_2-induced release of LHRH into hypophysial portal blood. *Endocrinology*, 97: 816–824.

Harms, P.G., Ojeda, S.R. and McCann, S.M. (1973) Prostaglandin involvement in hypothalamic control of gonadotropin and prolactin release. *Science*, 181: 760–761.

Harms, P.G., Ojeda, S.R. and McCann, S.M. (1974) Prostaglandin-induced release of pituitary gonadotropins: central nervous system and pituitary sites of action. *Endocrinology*, 94: 1459–1464.

Harms, P.G., Ojeda, S.R. and McCann, S.M. (1975) Failure of monoaminergic and cholinergic receptor blockers to prevent prostaglandin E_2-induced LH release. *Endocrinology*, in press.

Hedge, G.A. (1972) The effects of prostaglandins on ACTH secretion. *Endocrinology*, 91: 925–933.

Hertelendy, F. (1971) Studies on growth hormone secretion. II. Stimulation by prostaglandins in vitro. *Acta endocr. (Kbh.)*, 68: 355–362.

Hertelendy, F., Todd, H., Ehrhart, K. and Blute, R. (1972) Studies on growth hormone secretion. IV. In vivo effects of prostaglandin E_1. *Prostaglandins*, 2: 79.

Holmes, S.W. and Horton, E.W. (1968) The identification of four prostaglandins in dog brain and their regional distribution in the central nervous system. *J. Physiol. (Lond.)*, 195: 731–741.

Ito, H., Momse, G., Katayama, T., Takagishi, H., Ito, L., Nakajima, H. and Takei, Y. (1971) Effect of prostaglandin on the secretion of human growth hormone. *J. clin. Endocr.*, 32: 857–859.

Louis, T.M., Stellflug, J.N., Tucker, H.A. and Hafs, H.D. (1974) Plasma prolactin, growth hormone, luteinizing hormone and glucocorticoids after prostaglandin $F_{2\alpha}$ in heifers. *Proc. Soc. exp. Biol. (N.Y.)*, 147: 128–133.

MacLeod, R.M. and Lehmeyer, J.E. (1970) Release of pituitary growth hormone by prostaglandins and dibutyryl adenosine cyclic $3':5'$-monophosphate in the absence of protein synthesis. *Proc. nat. Acad. Sci. (Wash.)*, 67: 1172–1179.

Makino, T. (1973) Study on the intracellular mechanism of LH release in the anterior pituitary. *Amer. J. Obstet. Gynec.*, 115: 606–614.

Ojeda, S.R., Harms, P.G. and McCann, S.M. (1974a) Central effect of prostaglandin E_1 (PGE_1) on prolactin release. *Endocrinology*, 96: 613–618.

Ojeda, S.R., Harms, P.G. and McCann, S.M. (1974b) Possible role of cyclic AMP and prostaglandin E_1 in the dopaminergic control of prolactin release. *Endocrinology*, 95: 1694–1703.

Ojeda, S.R., Harms, P.G. and McCann, S.M. (1974c) Effect of third ventricular injections of prostaglandins (PG's) on gonadotropin release in conscious free moving male rats. *Prostaglandins*, 8: 545–552.

Ojeda, S.R., Harms, P.G. and McCann, S.M. (1975a) Effect of inhibitors of prostaglandin synthesis on gonadotropin release in the rat. *Endocrinology*, 97: 843–854.

Ojeda, S.R., Jameson, H.E. and McCann, S.M. (1975b) Plasma prolactin levels in maturing intact and cryptorchid male rats: development of stress response. *Proc. Soc. exp. Biol. (N.Y.)*, in press.

Ojeda, S.R., Wheaton, J.E. and McCann, S.M. (1975c) Prostaglandin E_2-induced release of luteinizing hormone-releasing factor (LRF). *Neuroendocrinology*, 17: 283–287.

Peng, T.C., Six, K.M. and Munson, P.L. (1970) Effects of prostaglandin E_1 on the hypothalamo-hypophyseal-adrenocortical axis in rats. *Endocrinology*, 86: 202–206.

Ratner, A., Wilson, M.C., Srivastava, L. and Peake, G.T. (1974) Stimulatory effects of prostaglandin E_1 on rat anterior pituitary cyclic AMP and luteinizing hormone release. *Prostaglandins*, 5: 165–171.

Sato, T., Jyujo, T., Iesaka, T., Ishikawa, J. and Igarashi, M. (1974) Follicle-stimulating hormone and prolactin release induced by prostaglandins in rat. *Prostaglandins*, 5: 483–490.

Sato, T., Hirono, M., Jyujo, T., Iesaka, T., Taya, K. and Igarashi, M. (1975a) Direct action of prostaglandins on the rat pituitary. *Endocrinology*, 96: 45–49.

Sato, T., Jyujo, T., Kawarai, Y. and Asai, T. (1975b) Changes in LH-releasing hormone content of the hypothalamus and electron microscopy of the anterior pituitary after prostaglandin E_2 injection in rats. *Amer. J. Obstet. Gynec.*, 122: 637–641.

Schofield, J.C. (1970) Prostaglandin E_1 and the release of growth hormone *in vitro*. *Nature (Lond.)*, 228: 179–180.

Spies, H.G. and Norman, R.L. (1973) Luteinizing hormone release and ovulation induced by the intraventricular infusion of PGE_1 into pentobarbital blocked rats. *Prostaglandins*, 4: 131–141.

Sundberg, D.K., Fawcett, C.P., Illner, P. and McCann, S.M. (1975) The effect of various prostaglandins and a prostaglandin synthetase inhibitor on rat anterior pituitary cyclic AMP levels on hormone release *in vitro*. *Proc. Soc. exp. Biol. (N.Y.)*, 148: 54–63.

Tal, E., Szabo, M. and Burke, G. (1974) TRH and prostaglandin action on rat anterior pituitary: dissociation between cyclic AMP levels and TSH release. *Prostaglandins*, 5: 175–182.

Tsafrini, A., Koch, Y. and Lindner, H.R. (1973) Ovulation rate and serum LH levels in rats treated with indomethacin or prostaglandin E_2. *Prostaglandins*, 3: 461–467.

Tucker, H.A., Vines, D.T., Stellflug, J.N. and Convey, E.M. (1975) Milking, thyrotropin-releasing hormone and prostaglandin induced release of prolactin and growth hormone in cows. *Proc. Soc. exp. Biol. (N.Y.)*, 149: 462–469.

Vale, W., Grant, G., Amoss, M., Blackwell, R. and Guillemin, R. (1972) Culture of enzymatically dispersed anterior pituitary cells: functional validation of a method. *Endocrinology*, 91: 562–572.

Yue, D.K., Smith, I.D., Turtle, J.R. and Shearman, R.P. (1974) Effect of $PGF_{2\alpha}$ on the secretion of human prolactin. *Prostaglandins*, 8: 387–395.

Zor, U., Kaneko, T., Schneider, H.P.G., McCann, S.M., Lowe, I.P., Bloom, G., Borland, B. and Field, J.B. (1969) Stimulation of anterior pituitary adenyl cyclase activity and adenosine $3':5'$-cyclic phosphate by hypothalamic extract and prostaglandin E_1. *Proc. nat. Acad. Sci. (Wash.)*, 63: 918–925.

Zor, U., Kaneko, T., Schneider, H.P.G., McCann, S.M. and Field, J.B. (1970) Further studies of stimulation of anterior pituitary cyclic adenosine $3':5'$-monophosphate formation by hypothalamic extract and prostaglandins. *J. biol. Chem.*, 245: 2883–2888.

Chapter 23

Limbic-preoptic responses to estrogens and catecholamines in relation to cyclic LH secretion

MASAZUMI KAWAKAMI and FUKUKO KIMURA

2nd Department of Physiology, Yokohama City University School of Medicine, Yokohama (Japan)

I. INTRODUCTION

Since Marshal and Verney (1936) first succeeded in inducing ovulation in the estrous rabbit by passing electrical current through the brain, intense experimental work followed to clarify the CNS mechanism controlling reproductive activity. Among these works, the epoch-making findings were: the stimulatory implication of ovarian hormones on the gonadotropin release (Hohlweg, 1934), and the hypothesis over those substances in the brain that are correlated with the release of gonadotropin; the releasing factors (Harris, 1961; McCann, 1962; Szentágothai et al., 1962) and the neurotransmitters (Sawyer et al., 1949). The recent major problems on the CNS mechanism are, thus, at what nerve cells ovarian hormones act, and by which neural transmitters the information from the hormone responsive cells is transmitted to the cells producing releasing factors. These also include a further problem as to the mode of functional interaction between ovarian hormones and neural transmitters in exhibiting their actions.

In the following presentation, an emphasis has been placed, with special reference to these problems, on the roles of limbic forebrain and preoptic system, including the organum vasculosum laminae terminalis in the mechanisms for cyclic gonadotropin release. Most recent experimental results as well as reports from our laboratory are cited.

II. MECHANISMS FOR CYCLIC GONADOTROPIN RELEASE WITH REFERENCE TO STIMULATORY ACTION OF ESTRADIOL AND ESTRONE

Cyclical reproductive activity is one among the most dramatic biological rhythms representing in vertebrate organisms. In the rat, ovulation occurs spontaneously every 4 or 5 days. The CNS stimulates, thus, the pituitary gland to release ovulatory gonadotropin every 4 or 5 days as well. Furthermore, the CNS mechanisms responsible for this gonadotropin release have circadian periodicity, activating only during a limited period of a day, "the critical period". Thus the CNS mechanisms which exert control over reproductive activity include both 4- or 5-day, and circadian clock factors.

It is accepted now that estradiol is the most important for the 4- or 5-day cyclic activation of the CNS mechanisms concerning control of ovulatory gonadotropin (Everett, 1948; Ferin et al., 1969). During the estrous cycle, secretion of estradiol-17β starts to increase in the afternoon before the day of proestrus and reaches the maximum preceding the ovulatory surge of luteinizing hormone (LH), and administration of antibodies to estradiol inhibits the LH surge. Furthermore, administration of estradiol at an appropriate time during the estrous cycle induces early ovulation (Ying and Greep, 1972; Krey and Everett, 1973). Studies of uptake and subcellular distribution of tritiated estradiol in the brain demonstrated the estradiol-concentrating or estradiol-sensitive neurons in the preoptic-hypothalamic structures as well as in part of the limbic periventricular structure (Stumpf, 1971; Pfaff, 1973).

On the other hand, either estrone or estriol was also effective to induce ovulation (Kobayashi et al., 1969). Estrone, in blood, fluctuates with a periodicity on each day of the estrous cycle and shows its highest level on the day of proestrus in the rat (Horikoshi and Suzuki, 1974; Kawakami et al., 1975c). Furthermore, recent evidences support that estradiol taken up in the brain tissue is transformed into estrone (Luttge and Whalen, 1970; Kazama and Longcope, 1972). In the present chapter, we report some findings obtained from experiments analyzing the stimulatory feedback sites of estradiol and estrone in the brain of the Wistar female rat, with special reference to their roles in the manifestation of 4- or 5-day and circadian rhythmicity.

II.1. Medial amygdala and bed nucleus of stria terminalis as the key structures for 4- or 5-day rhythmic release of gonadotropin

II.1.1. Dependence on estradiol. The crucial roles of the medial amygdala (m-AMYG) and the bed nucleus of stria terminalis (BST) in the mechanisms responsible for stimulating the release of ovulatory LH are suggested by the following observations: (1) electrical stimulation of the m-AMYG was effective to induce ovulation in the rats under various experimental conditions (Bunn and Everett, 1957; Velasco and Taleisnik, 1969; Kawakami and Terasawa, 1972a). The stimulation increased serum concentration of LH and follicle-stimulating hormone (FSH) in the pentobarbital-blocked proestrous rat, but not in the rat on the other days (Kawakami et al., 1973f). Similarly electrical stimulation of the BST induced ovulation and LH release when it was applied in proestrus (Kawakami and Kimura, 1974); (2) acute horizontal cut at the level of the anterior commissure or transection of the stria terminalis or lesion of the amygdaloid nucleus, which eliminated the AMYG or BST influences on the morning of proestrus, prevented ovulation (Velasco and Taleisnik, 1971; Kawakami and Terasawa, 1972c; Van Rees, 1972); (3) acute transection of the m-AMYG afferents and destruction of the BST on the day of diestrus 1 also prevented the progress of gonadal activity and ovulation (Kawakami and Terasawa, 1972c), and in the rat with the chronic transection, serum concentration of FSH was significantly low on the days of diestrus 2 and proestrus (Kawakami and Kimura, 1975). In addition, electrical stimulation of the m-AMYG or BST on the day of diestrus 2 induced FSH release (Kawakami and Terasawa, 1974; Kawakami and Kimura, 1974); (4) the electrical activity in the m-AMYG started to increase in the afternoon of diestrus 1 or early morning of diestrus 2 and maintained the high level until the morning of proestrus preceding the maximum elevation in the arcuate nucleus (ARC) (Kawakami et al., 1970, 1973c); (5) electrical stimulation of the m-AMYG or the BST in the pentobarbital-blocked proestrous rat, which stimulation was expected to induce ovulation on the following day, increased the multiunit activity of the medial preoptic area (MPO) and the ARC by more than 30 min (Kawakami et al., 1973d, f). It is therefore apparent that the neural stimuli, which are responsible for gonadotropin release leading to ovulation, originate at least in the m-AMYG and BST and activate the preoptic-hypothalamic system.

Evidence, that supports that the factor potentiating the m-AMYG and BST to stimulate anterior pituitary gland with a periodicity of 4 or 5 days is estradiol, was obtained by implantation experiment. Terasawa and Kawakami (1974) showed that a minute amount of estradiol benzoate, implanted in the m-AMYG or the BST on the night of diestrus 2 in the 4-day cyclic rat treated with progesterone to prevent expected ovulation, overcame the inhibitory effect of progesterone on LH release during the critical period of proestrus. In the 5-day cyclic rat, similar implantation in the m-AMYG or the BST at 12:00 on the day of diestrus 2 advanced the ovulation by 1 day (Kawakami et al., 1975a). In the same experiment, implantation in the ARC was also effective, but was not in the MPO. Multiunit activity in the ARC showed afternoon elevation on the day of implantation in the m-AMYG or the BST and as well as on the day following, while it did not show such elevation on the day of implantation in the MPO. Those observations clearly indicate that the m-AMYG and BST are highly sensitive to estradiol, and are subsequently activated to stimulate the preoptic-hypothalamus. In fact, the stimulatory effect of estradiol injection (s.c.) on ovulation in the progesterone treated 4-day cyclic rat in diestrus 2 or in the 5-day cyclic rat in diestrus 2 was blocked by acute destruction of the amygdaloid fibers or of the BST (Terasawa and Kawakami, 1974; Kawakami and Kimura, 1974). Furthermore, estradiol implantation in the m-AMYG or in the BST in the ovariectomized-estrogen primed rat induced LH release (Kawakami et al., 1975a), and electrical stimulation of the m-AMYG in the ovariectomized rat increased serum LH and FSH under the condition with estradiol dominance in blood but not without it (Kawakami and Terasawa, 1974), indicating that the facilitation of activity in the m-AMYG and BST is dependent on the estradiol. Taking into consideration a monophasic increase of estradiol in blood during the estrous cycle, the m-AMYG and BST are highly active on the day of proestrus, and thus may trigger the preoptic-hypothalamic system to induce LH surge. In this sense, the estradiol-concentrating neurons in those areas become functional every 4 or 5 days by stimulatory feedback action of estradiol.

Furthermore, we have obtained an evidence supporting that the stimulation of folliculotropin release before the day of proestrus by the m-AMYG and BST, as indicated in the brain transection experiment, is also dependent on stimulatory estradiol action on those areas (Kawakami et al., 1975a, b). It was observed that subcutaneous injection of estradiol at 12:00 on the day of diestrus 2 in the 5-day cyclic rat, which was effective in advancing ovulation, induced acute release of FSH and prolactin concomitantly with an inhibition of LH release in the afternoon of the same day. This estradiol effect was mimicked by the implantation of estradiol into the m-AMYG or the BST, although an effect on LH release was not observed. As a result, implantation advanced cyclic changes in vaginal smears, indicating an acute estrogen release, and only those rats which showed advancement of vaginal cycle ovulated early. More interestingly, when the implantation in the BST were placed for shorter period of time on the day of diestrus 2 (from 12:00 to 20:00), advancement of those parameters were similarly observed. It may be said that the m-AMYG and the BST in mid-diestrus in the 5-day cycle, responding to estradiol, stimulated FSH and prolactin release without affecting LH, which release triggers a chain of reproductive events including release of ovarian steroid responsible for early ovulatory gonadotropin release. The activity of the m-AMYG and the BST elevates in accordance with estradiol rise in blood, which is a single spread during the estrous cycle, and thus stimulate the preoptic-hypothalamic system with 4- or 5-day periodicity.

II.2. Medial preoptic area and arcuate median eminence regions as the key structures for circadian rhythmic release of gonadotropin

II.2.1. Dependence on estrone. We speculated that the role of the MPO in the neural apparatus for ovulatory LH surge is to integrate and serve the neural stimuli from the m-AMYG and BST with the limitation of timing. Therefore, the limbic neuronal impulses reach the hypothalamus only during a period of its circadian excitation. The background for this speculation is the following: (1) as mentioned before, the MPO is not the site of positive feedback action of estradiol on gonadotropin release. Acute transection of the limbic afferents to the MPO—ARC axis prevented the stimulating action of the estradiol, and implantation of estradiol into the MPO was almost ineffective to cause the release of LH in the cycling and estrogen-primed ovariectomized rats within 6 hr after the implantation (Terasawa and Kawakami, 1974; Kawakami and Kimura, 1975). Also the implantation did not facilitate the ARC multiunit activity (Kawakami and Kimura, 1974); (2) nevertheless, electrical stimulation of the MPO could induce activation of the ARC multiunit and unit activities, LH release and subsequent ovulation on each day of the cycle, and the effectiveness was higher when stimulation was applied in the afternoon (Kawakami et al., 1971). These data may suggest that the MPO may undergo less response to circulating estradiol, which fluctuates with 4- or 5-day periodicity, but rather manifests its feature of circadian excitation. It has also been found that circadian fluctuation of electrical activity in the limbic structures as well as in the preoptic and hypothalamic areas related to gonadotropin release (Kawakami et al., 1970; Kawakami et al., 1973b).

Circadian or diurnal rhythmicity has been observed over the activities of almost every organ of the body (Richter, 1971). The circadian clock involving CNS mechanism regulating release of ovulatory surge of LH, which was suggested by Everett and Sawyer (1950, 1953) and Everett et al. (1949), was more recently demonstrated in the LH and FSH release on other days during the estrous cycle. During the days following proestrus, serum LH and FSH showed a low amplitude circadian rhythm (Gay et al., 1970), LH-RF potency in stalk median eminence (SME) and LH-RF level in blood fluctuate with circadian periodicity (Ramirez and Sawyer, 1965; Meyer et al., 1974). Ovarian secretion of estrone and 20α-hydroxypregn-4-en-3-one was shown to undergo the influence of circadian rhythm (Baranczuk and Greenwald, 1973). This periodicity does not depend essentially upon negative feedback mechanism. Removal of ovaries did not disrupt a 24-hr pattern of LH and FSH release (Lawton and Schwartz, 1968; Lawton and Smith, 1970).

Although the CNS circadian rhythm is intrinsic, and independent of feedback effect, some modification by target hormone was indicated. Lawton and Smith (1970) demonstrated that in the ovariectomized rat duration of peak secretion is shorter than that in the intact rat. Similarly, the peak value of CRF content in the SME showed an advancement or shift toward the earlier period and the amplitude increased in the adrenalectomized rats (Hiroshige and Sakakura, 1971). These facts led us to presume that certain kinds of ovarian steroids, other than estradiol, should exert a certain influence on the manifestation of MPO circadian rhythm, and tested the effect of estrone on the MPO activity.

The results are demonstrated graphically in Fig. 1. The higher concentration of serum LH which should be observed in the afternoon, compared with that in the morning, of the ovariectomized rat was found to be depressed in the afternoon of the day of single s.c. injection of estradiol benzoate (EB, 20 µg in sesame oil) at 12:00, and this was more evident in the afternoon of the following day. On the second and third day, the afternoon LH levels become higher than those in the morning, indicating a circadian LH release by the stimulatory feedback action of estradiol. Implantation of crystalline estrone into the MPO through chronically implanted outer cannula at 12:00 on the third day after EB injection induced an increase in LH release in the afternoon. Implantation of estradiol into the MPO was not effective to increase LH release. The effect of estrone implanted in the MPO was rather inhibitory to LH release in the afternoon of the day of EB injection and of the day following. Similarly, in the EB pretreated ovariectomized rat bearing the

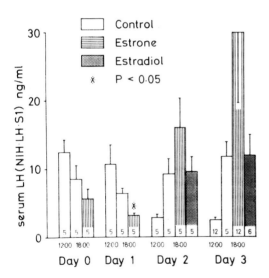

Fig. 1. Effect of implantation of estrone or estradiol into the MPO upon serum LH concentration in the ovariectomized rat. Estradiol benzoate (20 µg) was injected subcutaneously at 12:00 (day 0) and implantation of estrogen was done at 12:00 on the same day (day 0), the first (day 1), the second (day 2) or the third (day 3) day after estrogen injection. On day 0 the decrease in LH concentration at 18:00 after estrogen injection was further decreased by estrone implantation into the MPO, and on day 3 LH concentration at 18:00 was further increased by estrone implantation into the MPO compared with that in non-implanted control, but not by estradiol implantation.

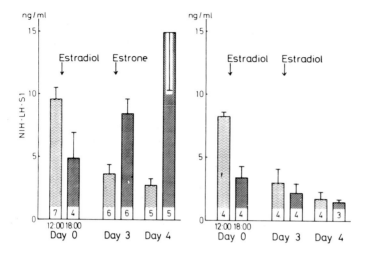

Fig. 2. Effect of subcutaneous injection of estrone or estradiol upon serum LH concentration in the ovariectomized rat bearing the MPO-roof cut on the third day (day 3) after the first estradiol injection (day 0). While the second injection of estradiol on day 3 could not induce an increase in LH concentration at 18:00 on both day 3 and day 4, injection of estrone on day 3 increased LH at 18:00 on the same day and more remarkably on day 4.

MPO roof cut, the second injection of EB (20 μg) at 12:00 on the third day after the first injection could not induce an increase in serum LH concentration, at 18:00 on the day of injection and on the following day which should have been observed in intact rats, but rather decreased further. While, when the second estrogen injected was estrone, serum LH increased at 18:00 on the day of injection and on the following day. Therefore, it seems that estradiol may act on some areas out of the MPO, possibly on the limbic system (Fig. 2), and estrone may act directly on the MPO to facilitate gonadotropin release. Results from the electrophysiological studies supported the above findings. That is, only estrone was effective to elevate multiunit activity in the ARC remarkably within about 3 hr of recording under the same procedure of implantation into the MPO. It is clearly indicated that estrone implanted in the MPO is able to augment the release of LH going with circadian periodicity, whereas the estradiol in the MPO is not.

The afternoon increase of estrone secretion in the 5-day cycling rat was confirmed in our experiment. Serum estrone was determined by the method described by Abraham (1969) (Fig. 3). On the day of either diestrus 3 or proestrus, the levels at 12:00 and 16:00 were significantly higher than that at 8:00, although the morning level on the day of proestrus was higher than the peak on the day before. This daily rhythm is supposedly a reflection of the circadian release of LH and/or FSH. However, there is a possibility that via the MPO as one of the stimulatory feedback sites, estrone prolongs the circadian secretion of LH and/or FSH, which may further stimulate the steroid secretion of estrone. This claim may be supported by the studies of Hilliard et al. (1967), who reported the importance of progestin in prolongation of preovulatory release of LH in the copulated rabbit.

It should be noted, from the unsuccessfulness of inducing ovulation in the cycling rat, to be described further, as well as from the fact described, that the excitation of the MPO only is not enough for inducing ovulatory LH release. The implantation of estrone in the MPO at 12:00 on diestrus 2 advanced vaginal cycle in 6 out of 14 animals, but induced early ovulation in only 2 of the 6 animals. Similarly, the same implantation on diestrus 3 was effective to advance vaginal cycle, inducing cornified smear on the day following, in 4 out of 8 rats, but ovulation occurred only in 2 of the 4 rats. The incidence of the advancement of vaginal cycle and ovulation is higher than that observed in the implantation of estradiol in the MPO, but is much less than that in the implantation of estradiol in the m-AMYG or the BST. Therefore, it is indicated that the implantation of estrone in the MPO in mid- or late-diestrus is effective in inducing the release of a certain amount of gonadotropin but this is not enough to induce early ovulatory gonadotropin release, unlike the estradiol implantation in the m-AMYG or the BST and subsequent excitatory impulses from those areas may be needed.

In this regard, the ARC seems to behave like the MPO. The release of LH by the implantation of estradiol in the ARC, on the third day after estradiol priming in the ovariectomized rat, occurred in the afternoon of the same day, whereas in the case of m-AMYG and the BST the release occurred on the day following. Estrone implantation in the ARC also induced the LH release in the afternoon of the same day. Additionally,

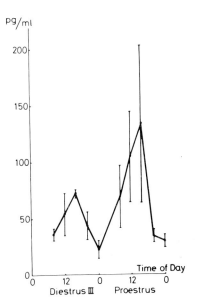

Fig. 3. Diurnal fluctuation of estrone on the days of diestrus 3 and proestrus in the 5-day cycling rat. The levels at 12:00 and 16:00 were significantly higher than that at 8:00 on the day of either diestrus 3 or proestrus, although even the level at 8:00 on the day of proestrus was higher than the peak on the day before.

simultaneous anterior deafferentation with estradiol implantation in the ARC prevented estradiol effect on induction of early ovulation (Kawakami et al., 1975a). Thus, the ARC requires co-operative facilitatory inputs possibly from the MPO, either of circadian or of 4- or 5-day nature, for stimulating LH surge in response to estrogen. However, the MPO−ARC system itself seems to manifest rather a circadian nature than a 4- or 5-day cyclic one. Because, the rat with MPO-roof cut revealed more significant circadian fluctuations of MPO and ARC multiunit activity than those of intact ones. Another important fact that the value of estrone is highest on the day of proestrus may indicate that circadian excitation of the MPO−ARC system is also maximal on that day (Kawakami et al., 1973e). Therefore, it may be said that the preovulatory gonadotropin release undergoes two complex rhythms, one based upon 4- or 5-day periodical excitation of the m-AMYG and the BST and the other on circadian excitation of the MPO−ARC system.

As Halász (1969) previously demonstrated, when the MPO was disconnected rostrally and laterally so as to leave this area connected to the tuber through the anterior hypothalamic area, the female rats continued to have an estrous cycle and to luteinize instead of developing persistent estrus as when cuts are made caudally. It seemed to indicate that the hypothalamic area including the MPO can carry out the control of cyclical gonadotropin release, without influences from the limbic structures. However, the following experimental result (Kawakami and Kimura, 1975) suggests that such maintained cyclical gonadotropin release is due to the functional alteration of the hypothalamic areas, i.e., of the MPO.

In ovariectomized rats we severed the limbic afferents to the hypothalamus by an L-shaped knife with a 2.0 mm long horizontal blade rotating through rostrally 180° at the level of anterior commissure (MPO-roof cut). This type of transection interrupted the fornical and stria terminalis components including the BST and a part of diagonal bundle of Broca and damaged the medial septal nucleus. Estradiol benzoate (20 μg s.c.), given 3 weeks postoperatively into such rats, increased serum LH at day 4 after treatment, second application of EB on the day preceding the increment had neither effect while estrone (200 μg s.c.) application brought about 3-fold higher serum LH. Twice administration of estrone in such preparation never altered serum LH level. The implantation of EB directly into either the AMYG or the BST of the intact rats induced ovulation but whenever the MPO-roof cut was performed ovulation did not result (Kawakami et al., 1975d, 1975e). The above mentioned facts possibly imply the requirement of estradiol primed AMYG as well as estradiol primed MPO for normal functional response of the MPO itself towards estrone on the release of LH.

The non-ovariectomized MPO-roof cut rats resumed 4- or 5-day vaginal cycles within 5−7 weeks after showing variable period of diestrus or irregular cycles. In these rats, ovarian and uterine weights, the number of ova in the oviducts and the timing of LH surge were not significantly different from those in the intact rats. The ovulatory response to EB administered on diestrus 2 in 5-day cyclers was observed in 4 out of 6 rats. This effect of estradiol was mimicked by the implantation of estradiol in the MPO, although the number of ova ovulated being relatively few. This implantation dominantly increased the multiunit activity in the ARC with a latency of about 2 hr. Therefore, one should consider that the restoration of estrous cycle after the limbic deafferentation is, partly, resultant from the functional compensation of the hypothalamus including the MPO, i.e., the acquirement of responsiveness to estradiol by the MPO (Fig. 4).

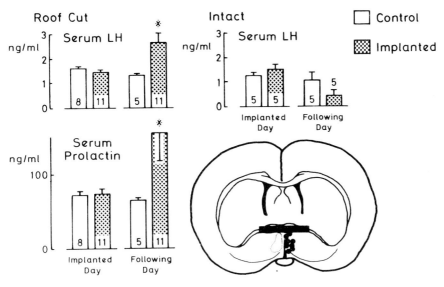

Fig. 4. Effect of implantation of estradiol into the MPO upon LH release in the rat bearing MPO-roof cut 5—7 weeks after the operation. Note that in these animals the implantation on diestrus 2 in 5-day cyclic rat induced LH release at 18:00 on the day following, while this implantation could not induce LH release in intact rat.

III. MODES OF NOREPINEPHRINE AND DOPAMINE IMPLICATION IN THE MECHANISM FOR CYCLIC GONADOTROPIN RELEASE

Early evidence for the participation of adrenergic mechanism in the regulation of ovulatory gonadotropin release was demonstrated by Sawyer and his colleagues, who showed that dibenamine would block ovulation (Sawyer et al., 1947), and reserpine could also block ovulation (Barraclough and Sawyer, 1957) and led to the development of pseudopregnancy (Barraclough and Sawyer, 1959). More recent works showed that ovulation or LH surge could be triggered by injecting epinephrine or norepinephrine into the third ventricle, but could not and rather be inhibited by dopamine (Rubinstein and Sawyer, 1970; Tima and Flerkó, 1974; Kimura et al., 1975). The importance of norepinephrine as putative transmitter for ovulatory mechanism in the hypothalamus has been generally realized. On the other hand, it was demonstrated that dopamine was more effective than norepinephrine in stimulating LH release in vivo and in vitro systems (Schneider and McCann, 1969; Kamberi et al., 1969; Kordon and Glowinski, 1969).

In order to know why such inconsistent results happened, the following experiments were designed. The m-AMYG and BST—MPO—ARC—ME system and the PVA—ARC—ME system, which we propose to be involved in facilitatory neural mechanisms responsible for ovulatory gonadotropin release, should be analyzed from the aspect of chemical substances involved in the activation of each indvidual neuronal component. Theoretically, facilitatory or inhibitory effects of those monoamines upon either part of these systems would modify the gonadotropin release.

III.1. Norepinephrine as a possible activator of the BST and nucleus preopticus pars suprachiasmatica (POSC)

From the facts described before, the BST may not only be involved in the mechanisms triggering LH surge on the day of proestrus, but also in the gonadotropin secretion that stimulates follicular growth and steroid genesis preceding proestrus. The BST is postulated as an important structure in conducting the amygdala impulses for stimulating gonadotropin release to the hypothalamus. However, electrical stimulation of the BST through chronically implanted electrodes induced an increase of LH release either on the day of proestrus or estrus, while the amygdalar stimulation with the same parameter as given to the BST could not induce an increase of LH release on neither of the two days. Multiunit activity in this area, different from that in the m-AMYG, showed a characteristic elevated pattern during the critical period on the day of proestrus. These observations suggest that the BST may possess its own function somewhat independent from the m-AMYG.

Intending to determine the physiological activator of the BST neuron in relation to gonadotropin release, a minute amount of dopamine (dopamine hydrochloride, Wako Purechemical Ind.) or noradrenaline (Sankyo Co.) was implanted through chronically implanted double cannula into the BST of the ovariectomized rat at 12:00 on the third day after a single s.c. injection of EB (20 μg). As shown in Fig. 5, on that day before any implantation, serum concentration of LH at 18:00 was high compared with that at 12:00.

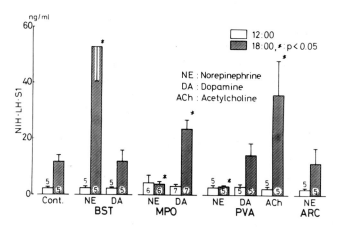

Fig. 5. Effect of implantation of catecholamines or acetylcholine into the brain at 12:00 on the third day after estradiol benzoate (20 μg) injection in the ovariectomized rat. Norepinephrine implantation into the BST, dopamine into the MPO and acetylcholine into the PVA further facilitated afternoon LH release which observed on the third day after single estrogen injection. Norepinephrine implantation into the MPO rather inhibited it.

When norepinephrine was implanted, serum concentration of LH was significantly increased at 18:00 compared with those in non-implanted rats. On the other hand, implantation of dopamine could not induce significant changes in serum LH concentration. Other evidence is that 20 min after the implantation of norepinephrine into the BST of the ovariectomized rat at 12:00 on the third day after EB injection multiunit activity in the ARC was elevated and lasted for several hours. In addition, with the same scheduled experimental work with BST, the implantation of norepinephrine in POSC produced a similar result as in the BST. Therefore, some noradrenergic input to the BST might be involved in the mechanism of LH release.

III.2. Acetylcholine as a possible activator for the PVA

As described in detail in a later section, the PVA is indicated as one of the components controlling gonadotropin release. To determine whether catecholaminergic or cholinergic mechanisms participate in mobilizing the activities of the PVA, the effects of the catecholamine and acetylcholine implantation into the periventricular zone including the PVA on serum LH concentration were examined. These materials were implanted into the PVA at 12:00 on the third day after a single s.c. injection of EB (20 μg) in ovariectomized rats and blood was collected at 18:00. While the implantation of dopamine could not induce any significant change in serum LH concentration, the implantation of acetylcholine elevated several times as much as that in the non-implanted animals. On the other hand, implantation of norepinephrine tended to decrease serum LH concentration. According to our electrophysiological study, the implantation of acetylcholine into the PVA elevated multiunit activity in the ARC with latency of about 15 min and maintained a high level for more than one hour. Considering that increasing activity in the ARC was correlated with LH release, this fact may support the above findings and some cholinergic input to the PVA, therefore, might be involved in the mechanism of LH release.

III.3. Dopamine as a possible activator for the MPO

Same procedure as on the BST and PVA was performed on the MPO neurons. In the animals implanted with dopamine in the MPO, serum concentration of LH at 18:00 was significantly increased compared with that in the control. However, the implantation of norepinephrine did exert an inhibitory effect on the LH release; the concentration at 18:00 was significantly lower than that in the non-implanted animals. It is clear that dopamine facilitates the MPO activity and stimulated LH release, while norepinephrine inhibits the activity and suppresses the LH release.

Results from electrophysiological studies supported the above findings. Implantation of dopamine in the MPO in the ovariectomized rat on the third day after EB injection induced an increase of multiunit activity of the ARC with a latency of several hours and a high level (Fig. 6). By contrast, after implantation of norepinephrine the electrical activity of the ARC showed almost no changes, within about 24 hr of recording.

There are some possibilities concerning the mode of dopamine action on the MPO. One is that dopamine acts as a mediator for the stimulatory action of estrogen. For instance, Rosenfeld and O'Malley (1970) reported that the stimulatory effect of estradiol on uterine adenyl cyclase activity could be prevented by D,L-propranolol. Although similar observation has not been done in the brain tissues, we confirmed that the stimulatory effect of dopamine implantation in the MPO on the ARC multiunit activity occurred more acutely and sharply when it was implanted with estradiol than without. Catecholamines and estradiol cause an increase in the cyclic AMP level of rat hypothalamus in vitro (Gunaga and Menon, 1973) and the estradiol-induced accumulation of hypothalamic cyclic AMP was abolished by prior treatment with α- and β-adrenergic blockers in vitro and in vivo (Gunaga and Menon, 1973; Gunaga et al., 1974; Weissman and Skolnick, 1975). Furthermore, conversion of estradiol and estrone to catechol estrogens by rat hypothalamic tissue but not by the cerebral cortex was demonstrated (Fishman and Norton, 1975). Therefore, the concept that a certain catecholamine may serve as a mediator for steroid hormone action on the brain tissue seems quite interesting and fascinating.

The other possibility is that dopamine acts as like the neurotransmitter on the MPO neurons. According to Palkovits's experiment (1974), dopaminergic cell bodies located in the ventricular portion of the MPO nearby the periventricular zone and their axons terminate in the middle part of the nucleus. Also the reduction of dopamine in the MPO was very slight, approximately 20%, after the lesion of substantia nigra. He also demonstrated that the BST is extremely high in dopamine content, as also is the ME, although the cell bodies could not be ascertained in this area. Those findings seem to indicate that the MPO receives the dopaminergic inputs from the BST and others. If that is the case, taking into account our present findings, the CNS mechanisms responsible for ovulatory gonadotropin release seem to require at least dopamine as a neurotransmitter for the MPO.

On the other hand, norepinephrine acts probably at the MPO to suppress gonadotropin release. The preoptic area is extensively innervated by the terminals from the ascending ventral noradrenergic pathway orginating in the lower brain stem (Ungerstedt, 1971). Palkovits et al. (1974) demonstrated the possibility, as in the case of dopamine, that a certain amount of norepinephrine in the MPO is supplied by the fibers from an adjacent area, such as the BST or a part of the septal nucleus, which has a high content of norepinephrine that could not be absolutely reduced by the lesion of the locus coeruleus. Therefore, some noradrenergic pathways may suppress the MPO activity. Recently, we proposed the possibility that the inhibitory influence of hippocampal electrical stimulation on the ovulatory gonadotropin release is probably conducted to that certain hypothalamic area which contains many cholinergic cell bodies through the hypothesized noradrenergic fornical pathway (Reivich and Glowinski, 1967; Kawakami et al., 1975b). That the MPO is rich in nerve cells containing cholinesterase was reported by Meszaros et al. (1969) and Jacobowitz and Palkovits (1971).

In the persistent estrous rat made anovulatory by exposure to constant illumination, norepinephrine implantation in the MPO induced a biphasic effect on the ARC multiunit activity (Fig. 6_1). The level started to increase sharply 10 min after the implantation and reached the maximum at about 30 min. Activity became lower than the control level after the high level continued for several hours. It was confirmed that this increase in the ARC multiunit activity was correlated with LH release (Fig. 6_2 and unpublished observation). Implantation of dopamine also induced an increase of the ARC activity, although the increase being not so significant as by norepinephrine. It was speculated, thus, that the lack of cyclical gonadotropin release in those animals is, in part, depending on the abnormal response of the MPO neurons to noradrenergic inputs. We hypothesized that cyclical gonadotropin release is normally controlled by push-pull activities of the facilitatory and inhibitory mechanisms in the brain (Kawakami and Terasawa, 1972b; Kawakami

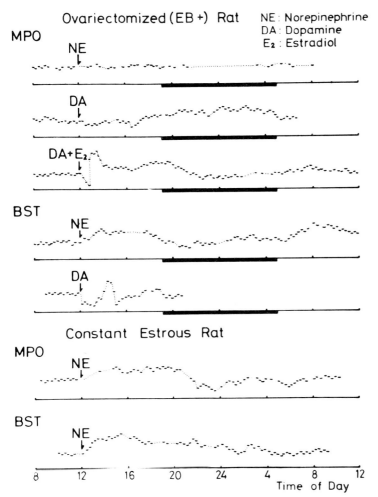

Fig. 6_1. Effect of implantation of norepinephrine and dopamine into the MPO and BST upon multi-unit activity in the ARC in ovariectomized rats and in constant estrous rats induced by continuous illumination. Ovariectomized rats were pretreated with estradiol benzoate (20 μg) at 12:00 and catecholamines were implanted at 12:00 on the third day after estrogen injection. Norepinephrine was effective to elevate multiunit activity in the ARC when implanted into the BST in the ovariectomized rat and into the BST or MPO in the constant estrous rat. Dopamine implantation into the MPO in the ovariectomized rat elevated multiunit activity in the ARC and this effect appeared more acutely and sharply when it was implanted with estradiol than without (Kawakami and Konda, unpublished observation).

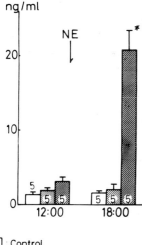

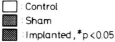

Fig. 6₂. The effects of norepinephrine implanted into the medial preoptic area on the release of LH in the constant estrous rats induced by constant illumination.

et al., 1975d).

Kalra and McCann (1973), however, proposed that the noradrenergic fibers originating in the brain stem are involved in the stimulation of LH-RH neurons whose cell bodies lie in the MPO, observing that the release of LH by the MPO stimulation in the ovulatory period and after administration of estrogen and progesterone was blocked by the administration of diethyldithiocarbamate which depletes norepinephrine levels selectively.

III.4. Norepinephrine as a possible activator of the ARC

Electrophysiologically, the nerve cells in the ARC and its adjacent periventricular region exhibit antidromic responses to electrical shocks in the ME, indicating the projection of their axons to the ME (Kawakami and Sakuma, 1974) (Fig. 7). Those cells are recognized to be neurosecretory based on the specific features of antidromic response such as slow conductive velocity of axons and fractionation of activated A- and B-spike (Dyer, 1973; Kawakami and Sakuma, 1976a) (Fig. 8). Furthermore, microiontophoretic infusion of LH-RH antiserum of blocked spontaneous activity of antidromically identified ARC neurons at a high percentage, the suppression of such activity recovered to the control level by 10—30 sec after the end of infusion. Distribution of axon terminals (Calas et al., 1973; Sétáló et al., 1975) has been confirmed in the ARC and ME, respectively.

In the experimental study into what is the activator for those ARC cells using proestrous rats which were blocked ovulatory surge by pentobarbital anesthesia, it was demonstrated that the anti-LH-RH suppressed cells which showed facilitatory response to the infusion of norepinephrine. Norepinephrine was effective in inducing facilitation in 48

(61.8%) of 78 antidromically identified cells (Kawakami and Sakuma, 1976b). Norepinephrine-facilitated cells tended to be non-responsive to dopamine, and norepinephrine was ineffective on the cells which were activated by dopamine. Furthermore, statistically significant relation was observed between the facilitation induced by norepinephrine and by the MPO stimulation, implying that the MPO induced facilitation might be mediated by noradrenergic transmitter. No relation was seen between the effects of dopamine

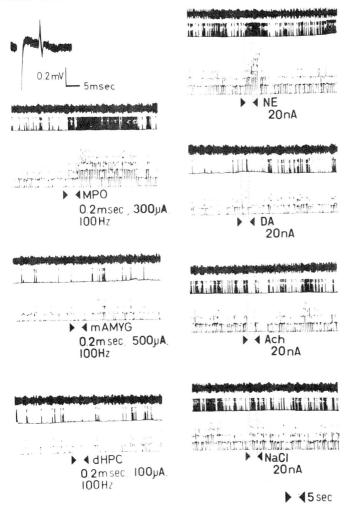

Fig. 7₁. Responses of an antidromically identified arcuate neuron to consecutive electrical stimulation of the brain and microiontophoresis of drugs. In this example, made in a proestrous female rat, in which medial preoptic stimulation facilitated the neuronal activity, microiontophoresis of norepinephrine mimicked the effect of the stimulation (Kawakami and Sakuma, 1976a).

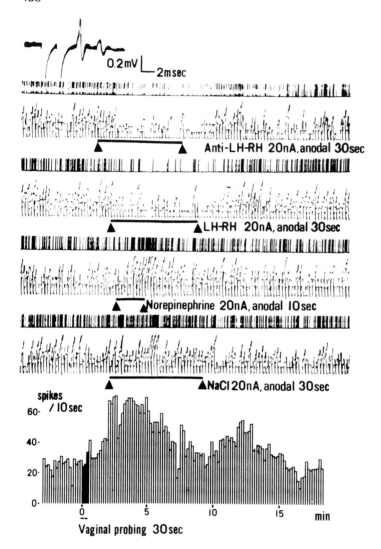

Fig. 7₂. Responses of an antidromically identified arcuate neuron to the microiontophoresis and vaginal probing. Recordings were made in a female rat made anovulatory by constant illumination. Microiontophoretic infusion of antiserum anti-LH-RH suppressed neuronal activity, in contrast to the facilitation induced by norepinephrine. Current application by the saline filled barrel was used for control purpose. Vaginal probing resulted in a marked facilitation in the activity (Kawakami and Sakuma, 1976b).

infusion and the electrical stimulation of the MPO. These findings seem to support the concept that norepinephrine possibly acts as a facilitatory neurotransmitter at the level of the ARC—ME. Because it has been presented that the ARC neuronal activation was correlated with gonadotropin release (Kawakami et al., 1971), norepinephrine may stimulate the cell to cause the release of LH-RH.

Fuxe and Hökfelt (1969) hypothesized that the tuberoinfundibular neurons, whose cell bodies, localized mainly in the ARC, send their axons to the internal layer of the ME, where they inhibit the release of LH-RH and FSH-RH probably by an axo-axonic influence. We noticed that threshold for antidromic invasion of the ARC cells by the ME stimulation showed significant fluctuation according to the endocrine environments of the animal. Elevated threshold was seen in the ovariectomized and progesterone treated

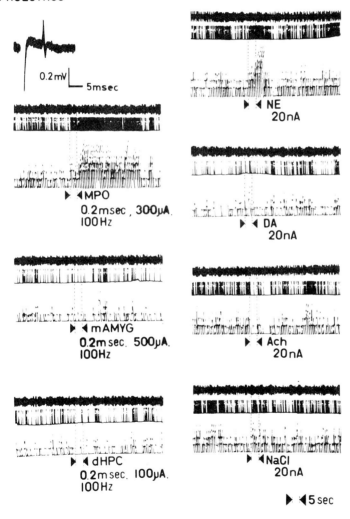

Fig. 8. Responses of an antidromically identified arcuate neuron to consecutive electrical stimulation of the brain and microiontophoresis of drugs. In this example, made in a proestrous female rat, in which medial preoptic stimulation facilitated the neuronal activity, microiontophoresis of norepinephrine mimicked the effect of the stimulation (Kawakami and Sakuma, 1976a).

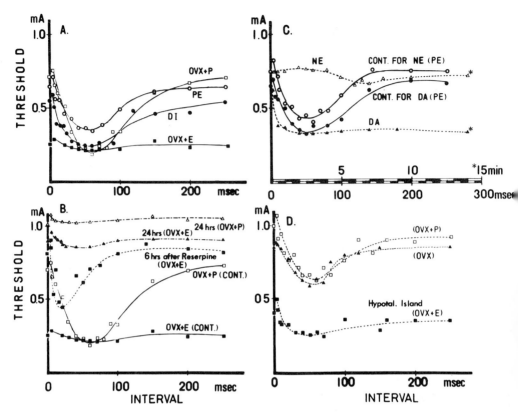

Fig. 9. Effects of the "conditioning" pulse on the threshold for antidromic invasion by the "test" pulse under various sex hormone environments. The vertical axis represents the threshold intensity for antidromic invasion by the "test" pulses, and the horizontal bar indicates intervals between the "conditioning" and "test" pulses. The horizontal axis in C with an asterisk (*) indicates time lapse after application of dopamine (DA) or norepinephrine (NE). A: threshold for antidromic activation showed significant decrease following the conditioning. The decrease was significant in the ovariectomized progesterone primed (OVX + P) and proestrus, whereas the conditioning was effectless in the ovariectomized-estrogen primed (OVX + E). B: threshold-decreasing effect of the conditioning which disappeared after systemic application of reserpine (5 mg/kg). In the OVX + E, the threshold for the conditioning was seen to increase by reserpine treatment. C: the decreasing effect was seen in almostly the same time-course after total deafferentation of the hypothalamus. D: the decreasing effect of the conditioning was mimicked by local application of crystalline DA in the ME, but not by NE.

animals when compared with those treated with EB and was also seen in the ovariectomized-progesterone primed animal when compared with the ovariectomized-EB primed animal. Since preceding conditioning pulse significantly decreased the threshold for activation by the test pulse, it seemed that this decrease might be caused by some neural mechanism (Fig. 9). The time-course of the decrease of the threshold showed a similar pattern with that of spinal presynaptic inhibition. Therefore, it was hypothesized that this fluctuation must be caused by heterosynaptic mechanisms in the ME, which induce partial depolarization of the axon terminals in a manner similar to the presynaptic inhibition, since decreased threshold was observed in preceding pulses when mediated by dopaminergic transmitter, which was indicated from the finding that the application of dopamine on the exposed surface of the ME mimicked the decrease in the effect of electrical shocks, whereas norepinephrine was ineffective. Therefore, dopamine was regarded as an inhibitory transmitter for the release of hypothalamic releasing hormone in the ARC–ME.

IV. A POSSIBLE GONADOTROPIN CONTROL MECHANISM INVOLVING THE ORGANUM VASCULOSUM LAMINAE TERMINALIS AND SUPRA-RETROCHIASMATIC AREA

Most recent studies have demonstrated the sites concentrating LH-RH in the extrahypothalamus other than the medial basal hypothalamus. The sites included the septal and preoptic area (Barry et al., 1973), the suprachiasmatic nucleus (Krulich et al., 1971), and the periventricular area at the rostral part of the third ventricle including the organum vasculosum laminae terminalis (PVA) (Kawakami et al., 1973a, 1974; Zimmerman et al., 1974). Barry and Dubois (1974) could map the perikarya of LH-RH-producing cells in the albino guinea pig and traced their axons as forming a "preoptic-infundibular neurosecretory pathway" to terminate at different levels of the ME. Flerkó (personal communication) hypothesized a supra-retrochiasmatic infundibular pathway in rats which, supposedly, terminates on or in the immediate vicinity of the capillary loops penetrating the ME.

Our preliminary observations have shown a dramatic change of electrical activity in this area, around the critical period on the day of proestrus (Kawakami et al., 1973c), which was similar to that found in the ARC–ME (Kawakami et al., 1970). Furthermore, electrical stimulation of this area could increase serum concentrations of LH and prolactin, and the threshold of electrical stimulation was lower than that of the MPO stimulation, which was known to have a low threshold to induce LH release among the several brain areas. The isolation of the PVA prevented an increase of serum LH and FSH which should follow bilateral ovariectomy (Kawakami et al., 1973a). From the subsequent experiments, an implication of the PVA–ME system in the gonadotropin control mechanism is indicated (Kawakami et al., 1975d).

Electrophysiological study demonstrated the neurosecretory nature of the PVA neurons and the terminations of their axons in the ME without a synapse in between. Extracellular recordings were made from neurons in the PVA and 55 neurons were activated antidromically by the single shock of the ME. The mean latency of the antidromic activated units was 9.1 ± 2.8 (S.D.) msec. The conduction velocity was estimated to be 0.3–0.8 m/sec, implying that the nerve fibers of the neurons were unmyelinated. Some of them (i.e., 30.2%) had clear fluctuation of activated A- and B-spikes. In addition, 26 neurons in the PVA were activated transsynaptically by the ME stimulation. Among those recorded from the ARC–ME region, the neurons activated antidromically by the stimulation of the PVA were also found. Those findings strongly suggested that activities in those

Fig. 10. PVA units (A, B and E–H) activated by ME stimulation and ARC units (C and D) evoked by PVA stimulation. Each tracing is 10–15 superimposed traces. A: antidromically activated PVA unit; B: transsynaptically activated PVA unit: C: antidromically activated ARC unit; D: transsynaptically activated ARC unit; E–H: the same PVA unit activated antidromically with different stimulus frequencies (E, 1 Hz; F, 10 Hz; G, 20 Hz and H, 40 Hz). Note that as the stimulus frequency increases, the peak latency of B potential fluctuates and fractionation of A and B potential enhances. Calibration bars, 5 msec and 1 mV (Kawakami et al., 1975).

two areas, the PVA and the ARC–ME, influence each other and might be coordinated with regard to the release of pituitary hormone by the neural network between them.

The neurons in the BST, SRC and the area medial to the suprachiasmatic nuclei also showed antidromic activation following the ARC–ME stimulation. The mean latency of 104 activated units was 12.6 ± 4.5 (S.D.) msec. Some of them (i.e., 34.6%) had clear fluctuation of activated A- and B-spikes. The conduction velocity was estimated to be 0.4–0.9 m/sec. These observations are almost in agreement with the concept that the axons of the neurosecretory cells in the BST and the SRC run through or aside the ARC to terminate in the ME.

In order to examine the significance of those neurons, the effect of PVA stimulation on LH release was tested in the ovariectomized and EB primed rat with anterior deafferentation. The knife cut extended to the paraventricular nucleus severing afferent and efferent fibers of the medial basal hypothalamus anteriorly. Increase in LH concentration in serum was elicited by the PVA stimulation in the rat in which deafferentation was not performed or incomplete, leaving the basal part intact. However, no increase in serum LH level was observed following PVA stimulation in the rat in which the deafferentation was complete, cutting fibers running through the basal part of the brain. The results seem to support that the neurosecretory neurons in the PVA release their hormone at the axon terminals in the ME, but not into the ventricular cerebrospinal fluid as suggested by Knigge et al.(1973). Furthermore, such axons were suggested to run through the base of the brain along the ventral border of the third ventricle, in that way the fibers of the SRC converge.

In contrast was the ARC, which was responsive to stimulatory action of estradiol. Implantation of EB into the PVA on the day of diestrus 2 in the 5-day cyclic rat could not advance the ovulation, and the implantation in the ovariectomized rat on the third day after EB injection could not increase the afternoon release of LH and rather inhibited it. In addition, implantation of estradiol in the ovariectomized rat depressed the increase of LH and FSH release (Kawakami et al., 1974). Stumpf (1971) reported the organum vasculosum laminae terminalis among the estrogen–neuron system as well as the ME, the

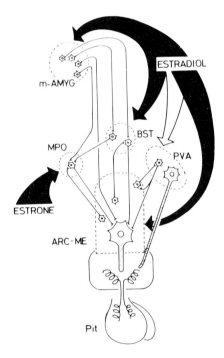

Fig. 11. Schematic illustration of estrogen feedback mechanisms which regulate gonadotropin release. The mechanisms consist of m-AMYG : BST : ARC–ME system which may be mobilized by positive feedback of estradiol and relates in cyclic release of gonadotropin; PVA : ARC–ME system which may be mobilized by negative feedback of estradiol and relates in cyclic release; and MPO : ARC–ME system which is affected by estrone and relates in tonic and 24-hr periodic release. Illustration is made exclusively on presumed shortest pathways, without referring on possible transsynaptic connections. Projections of ARC neurons to the PVA are omitted.

MPO and other limbic structures. The PVA was indicated as one of the negative feedback sites of estradiol, but not of the positive feedback sites, unlike the ARC—ME.

V. DISCUSSION

Although many questions remain unanswered, our data presented in this chapter may be available for giving a profile of the CNS mechanisms responsible for cyclical gonadotropin release. Data strongly suggested the initiative roles of the limbic structures for cyclic gonadotropin release. The m-AMYG and the BST are responsive to stimulatory action of estradiol which increase in blood before the critical period of proestrus. The MPO is not responsive to this estradiol action but is to estrone which increases during the everyday time. The successive excitation of the m-AMYG—BST—MPO neural axis is considered to be a main stimulatory mechanism for LH-RH release with both 4- or 5-day and 24-hr periodicity. Therefore, such periodical ovulatory gonadotropin release is prepared by both the estradiol and estrone sensitive neurons in the brain.

The characteristic of the limbic structures as sites of intimate correlation with steroid hormones has been demonstrated by the high uptake of tritiated estrogens and androgens in the septum and AMYG as well as in the preoptic-hypothalamic areas (McEwen et al., 1970; McEwen and Pfaff, 1973; Stumpf, 1971; Pfaff, 1973), and by the existence of aromatizing enzyme systems which convert androstenedione to estradiol and estrone in the fetus limbic structure as well as in the hypothalamus (Naftolin et al., 1971). Such nature of the limbic system has given a ground attributing sexual differentiation to this system. Dysfunction of the m-AMYG and the hippocampus concerning the control of LH release was reported on the neonatally androgenized anovulatory rat (Terasawa, 1971; Kawakami and Terasawa, 1972d).

It was surprising at first to find that the MPO is not responsive to estradiol concerning gonadotropin release in spite of a high uptake of tritiated estradiol. Kato and Villee (1967) and Luttge and Whalen (1970) have demonstrated, however, that after the injection of estradiol, significant amounts of estrone can be found in selected regions of the brain. Furthermore, only 9—10% of chromatographic radioactive estrone was found in the preoptic anterior hypothalamus, whereas over 20% was found in the ME-posterior hypothalamic areas. This fact seems to indicate a possibility that estradiol taken up in the brain tissue becomes effective after it is converted into estrone. If the MPO is less functional in this process, as shown by Luttge and Whalen (1970), the unappreciable or non-response to estradiol can be reasonable. In the subsequent work by Luttge and Whalen (1972), the preoptic anterior hypothalamic area can retain a considerable amount of estrone, although being less than the cerebral peduncle or the posterior hypothalamus. Thus, it is indicated that the MPO has the estrone receptor, which could not be saturated by estradiol (Luttge and Whalen, 1972).

The finding suggesting that dopamine is implicated as the possible transmitter at the level of the MPO is quite interesting. Experimental results, on one hand, supported that norepinephrine is a neurotransmitter for stimulating LH-RH release, but on the other hand indicate the implication of dopamine in this process. However, those results are not necessarily incompatible with each other, if one assumes the implication of at least two neurotransmitters, the one at the site of the MPO neurons and the other at the site of LH-RH neurons. The injection of drugs which deplete catecholamine, both norepinephrine and dopamine, into the brain inhibited spontaneous and induced ovulation (Sawyer et al., 1947, 1949; Coppola et al., 1966; Rubinstein and Sawyer, 1970). The LH release could be induced, if the LH-RH neurons were directly activated by intraventricular application of norepinephrine (Sawyer et al., 1974; Tima and Flerkó, 1974). The abolishment of LH release by drugs which lower norepinephrine and dopamine levels could be restored by drugs that selectively raise norepinephrine, but not dopamine levels (Kalra and McCann, 1973; Kalra et al., 1972), because the LH-RH neurons, supposedly, cannot be activated by the deficiency of norepinephrine in spite of the activation of the MPO neurons by dopamine. The blockade of induced ovulation in the immature rat by α-methyltyrosine was reversed by DOPA which restores the levels of both dopamine and norepinephrine but not by dihydroxyphenyserine which restores norepinephrine levels only (Kordon and Glowinski, 1969 and 1970), and pimozide, a dopamine-receptor blocker, was inhibitory to the release of LH-RH in the hypophysectomized rats (Corbin and Upton, 1973).

The site where the effect of norepinephrine, applied intraventricularly, was proposed, by Weiner et al. (1971) and Sawyer (1975), to lie in the region of the ME, in the inner layer of which the norepinephrinergic nerve endings have been revealed (Björklund et al., 1970, 1973). However, Tima and Flerkó (1974) have shown that in the rats made anovulatory by a frontal cut the percentage of ovulated rats by intraventricular norepinephrine injection was much lower than in the rats made anovulatory by the anterior hypothalamic lesions. The difference between the cut and lesion was in the numbers of LH-RH nerve endings degenerated in the ME; the cut induced more degeneration than the lesion. It was suggested that the intraventricularly injected norepinephrine stimulated the LH-RH producing neurons in the preoptic and supra- and retrochiasmatic areas as well as the BST. In relation to this problem, we implanted norepinephrine directly into the ARC—ME in the rat in diestrus 2 or pentobarbital-blocked proestrus, expecting it to be effective in inducing ovulation. The result was that the norepinephrine implant could not induce early or expected ovulation, respectively. Craven and McDonald (1973) also reported that norepinephrine placed in the ARC was also not able to overcome the anti-ovulatory effect of pentobarbitone or reserpine. However, they observed that in the rats pretreated with a monoamine oxidase inhibitor, the norepinephrine infusion in that region could advance the critical period. Furthermore, in the experiment of norepinephrine implantation directly into the BST at 12:00 on the third day after EB injection in the ovariectomized rat, we observed a remarkable increase in serum LH at 18:00. Those findings seem to support that the major site of norepinephrine stimulation for phasic release of LH-RH is in the anterior preoptic area including the BST and the site in the basal hypothalamus including the ARC seem to be implicated in the further modulation of LH-RH release induced by stimulation of the former site. The modulation of dopamine neurons in the latter site may also be involved. Intraventricular injection of dopamine inhibits LH release, whose effect is assumed to be induced through this mechanism.

As evidence supporting the stimulatory implication of norepinephrine at the level of basal hypothalamus, we have observed an alteration of norepinephrine levels in this area after effective stimulation of the m-AMYG on LH release. The proestrous rat chronically implanted with concentric bipolar electrode in the m-AMYG was electrically stimulated between 14:00 and 14:30 with 100 μA of 100 Hz and 0.1 msec duration train pulses. Norepinephrine content in the brain tissue from the caudal border of the optic chiasm to the caudal end of the infundibulum, including the ARC and ME widely, measured by the method described by Sandler (1968) with a slight modification, was significantly lower just after stimulation than in the non-stimulated animals. Taking into account the fact that electrical stimulation in the m-AMYG was effective to induce the rise of ARC electrical activity which correlated with LH release, it was indicated that the stimulation might increase the norepinephrine turnover, which in turn may account for the decrease of the content in that region. The hippocampal stimulation, which was confirmed to be inhibitory to the ARC electrical activity as well as to LH release, was effective to increase the content.

From an electrophysiological point of view, electrical activities of facilitatory and inhibitory systems related in the mechanisms controlling gonadotropin might differ under various hormonal environments. For example, electrical stimulation in the l-AMYG in the ovariectomized rat lowered the multiunit activity in the ARC, while the stimulation in the same site elevated it after EB injection. Neuronal connections between one site and another in the brain seem to have a number of complex pathways in which different aminergic structures are possibly involved, and changes in excitability of these pathways probably occur in different hormonal environments. This might be one of the reasons why disagreement still exists regarding the overt effects of the various catecholamines on pituitary secretion, and why the chemical identity of the principal participating catecholamine remains a controversial issue. Anatomically, it is well-known that main aminergic pathways arise in the brain stem (Dahlström and Fuxe, 1964; Fuxe, 1965; Ungerstedt, 1971), though the existences of noradrenergic cell bodies in the posterior hypothalamus and dopaminergic cell bodies in the ARC—ME are demonstrated. However, in respect to the aminergic pathways relevant for gonadotropin regulation, the precise origin of these and the relationship between ascending aminergic fibers and the hypothalamic functions on gonadotropin release are not well defined. Further future experimentation is required to identify the aminergic pathway relating in the gonadotropin regulation and to determine their role in the selective regulation of reproductive functions.

VI. SUMMARY

The subject in this chapter was to cover some aspects of the limbic structures in relation to those matters which have been observed in neuroendocrine tissues at the subcellular levels. Emphasis has been placed on the actions of ovarian hormones and catecholamines with special reference to cyclic gonadotropin controlled by the limbic forebrain and organum vasculosum laminae terminalis.

It was postulated that the preovulatory gonadotropin release is stimulated by those neural stimuli, which orginate in the estradiol sensitive neurons in the medial amygdala and the bed nucleus of stria terminalis with 4- or 5-day periodicity, and converge to the estrone sensitive neurons in the medial preoptic area, to be integrated and transformed into circadian nature.

There could be a dopaminergic synapse for the transmission of those limbic inputs to the medial preoptic area, noradrenergic to the bed nucleus of stria terminalis, as well as to the nucleus preopticus pars suprachiasmatica and cholinergic to the organum vasculosum laminae terminalis, with regard to the facilitatory input for the gonadotropin release. Furthermore, noradrenergic transmitter might exert influences on the LH-RH producing cells located in the supra-retrochiasmatic areas as well as at the level of final common pathway in the hypothalamus to release gonadotropin.

ACKNOWLEDGEMENTS

We are grateful to our colleagues, Drs. E. Terasawa, Y. Sakuma, N. Konda and M. Manaka for their help in the preparation of this manuscript and permission to describe their data. The authors wish to thank Drs. A. Kambegawa and M. Nagase for determinations of estrone in serum and norepinephrine in brain tissues, respectively. The authors also wish to express their thanks to the National Institute of Arthritis, Metabolism and Digestive Diseases, U.S.A. for supplying materials for the radioimmunoassay. Antiserum Anti-LH-RH was a generous gift of Dr. Y. Wakabayashi.

The experimental work that was performed in the authors' laboratory and described in this chapter was supported by a grant from the Ministry of Education, Japan.

REFERENCES

Abraham, G.E. (1969) Solid-phase radioimmunoassay of estradiol-17β. *J. clin. Endocr.*, 29: 866–870.

Baranczuk, R. and Greenwald, G.S. (1973) Peripheral levels of estrogen in the cyclic hamster. *Endocrinology*, 92: 805–812.

Barraclough, C.A. and Sawyer, C.H. (1957) Blockade of the release of pituitary ovulating hormone in the rat by chlorpromazine and reserpine: possible mechanisms of action. *Endocrinology*, 61: 341–351.

Barraclough, C.A. and Sawyer, C.H. (1959) Induction of pseudopregnancy in the rat by reserpine and chlorpromazine. *Endocrinology*, 65: 563–571.

Barry, J. and Dubois, M.P. (1974) Immunofluorescence study of the preoptico-infundibular LH-RH neurosecretory pathway of the guinea pig during the estrous cycle. *Neuroendocrinology*, 15: 200–208.

Barry, J., Dubois, M.P. and Poulain, P. (1973) LRF producing cells of the mammalian hypothalamus. *Z. Zellforsch.*, 146: 351–366.

Björklund, A., Falck, B. and Hromek, F. (1970) Identification and terminal distribution of the tuberohypophyseal monoamine fiber systems in the rat by means of stereotaxic and microspectrofluorimetric techniques. *Brain Res.*, 17: 1–23.

Björklund, A. and Nobin, A. (1973) Fluorescence histochemical and microspectrofluorimetric mapping of dopamine and noradrenaline cell groups in the rat diencephalon. *Brain Res.*, 51: 193–205.

Bunn, J.P. and Everett, J.W. (1957) Ovulation in persistent estrous rats after electrical stimulation of the brain. *Proc. Soc. exp. Biol. (N.Y.)*, 96: 369–371.

Calas, A., Kerdelhué, B., Assenmacher, I. et Jutisz, M. (1973) Les axones à LH-RH de l'éminence médiane. Mise en évidence chez le canard par une technique immunocytochimique. *C.R. Acad. Sci. (Paris)*, 277: 2765–2768.

Coppola, J.A., Leonardi, R.G. and Lippmann, W. (1966) Ovulatory failure in rats after treatment with brain norepinephrine depletors. *Endocrinology*, 78: 225–228.

Corbin, A. and Upton, G.V. (1973) Effect of dopaminergic blocking agents on plasma luteinizing hormone releasing hormone activity in hypophysectomized rats. *Experientia (Basel)*, 29: 1552–1553.

Craven, R.P. and McDonald, P.G. (1973) The effect of intrahypothalamic infusions of dopamine and noradrenaline on ovulation in reserpine-treated female rats. *J. Endocr.*, 58: 319–326.

Dyer, R.G. (1973) An electrophysiological dissection of the hypothalamic regions which regulate the preovulatory secretion of luteinizing hormone in the rat. *J. Physiol. (Lond.)*, 234: 421–442.

Everett, J.W. (1948) Progesterone and estrogen in the experimental control of ovulation time and other features of the estrous cycle in the rat. *Endocrinology*, 43: 389–405.

Everett, J.W. and Sawyer, C.H. (1950) Twenty-four hour periodicity in "LH-release apparatus" of female rats, disclosed by barbiturate sedation. *Endocrinology*, 47: 198–218.

Everett, J.W. and Sawyer, C.H. (1953) Estimated duration of spontaneous activation which causes release of ovulating hormones from rat hypophysis. *Endocrinology*, 52: 83–92.

Everett, J.W., Sawyer, C.H. and Markee, J.E. (1949) Neurogenic timing factor in control of ovulatory discharge of luteinizing hormone in cyclic rats. *Endocrinology*, 44: 234–250.

Ferin, M., Tempone, A., Zimmering, P.E. and Van de Wiele, R.L. (1969) Effect of antibodies to 17β-estradiol and progesterone on the estrous cycle of the rat. *Endocrinology*, 85: 1070–1078.

Fishman, J. and Norton, B. (1975) Catechol estrogen formation in the central nervous system of the rat. *Endocrinology*, 96: 1054–1069.

Fuxe, K. (1965) Evidence for the existence of monoamine neurones in the central nervous system. IV. Distribution of monoamine nerve terminals in the central nervous system. *Acta physiol. scand.*, 64, Suppl. 247: 36–85.

Fuxe, K. and Hökfelt, T. (1969) Catecholamines in the hypothalamus and the pituitary gland. In *Frontiers in Neuroendocrinology*, W.F. Ganong and L. Martini (Eds.), Oxford Univ. Press, New York, pp. 47–96.

Gay, V.L., Midgley, A.R., Jr. and Niswender, G.D. (1970) Patterns of gonadotropin secretion associated with ovulation. *Fed. Proc.*, 29: 1880–1887.

Gunaga, K.P. and Menon, K.M.J. (1973) Effect of catecholamines and ovarian hormones on cyclic AMP accumulation in rat hypothalamus. *Biochem. biophys. Res. Commun.*, 54: 440–448.

Gunaga, K.P., Kawano, A. and Menon, K.M.J. (1974) In vivo effect of estradiol benzoate on the accumulation of adenosine 3′,5′-cyclic monophosphate in the rat hypothalamus. *Neuroendocrinology*, 16: 273–281.

Halász, B. (1969) The endocrine effects of isolation of the hypothalamus from the rest of the brain. In *Frontiers in Neuroendocrinology*, W.F. Ganong and L. Martini (Eds.), Oxford Univ. Press, New York, pp. 307–342.

Harris, G.W. (1961) The pituitary stalk and ovulation. In *Control of Ovulation*, C.A. Villee (Ed.), Pergamon Press, Oxford, pp. 56–78.

Hilliard, J., Penardi, R. and Sawyer, C.H. (1967) A functional role for 20α-hydroxypregn-4-en-3-one in the rabbit. *Endocrinology*, 80: 901–914.

Hiroshige, T. and Sakakura, M. (1971) Circadian rhythm of corticotropin-releasing activity in the hypothalamus of normal and adrenalectomized rats. *Neuroendocrinology*, 7: 25–36.

Hohlweg, W. (1934) Veränderungen des Hypophysenvorderlappens und des Ovariums nach Behandlungen mit grossen Dosen von Follikelhormon. *Klin. Wochschr.*, 13: 92–95.

Horikoshi, H. and Suzuki, Y. (1974) On circulating sex steroids during the estrous cycle and the early pseudopregnancy in the rat with special reference to its luteal activation. *Endocr. jap.*, 21: 69–79.

Jacobowitz, D.M. and Palkovits, M. (1971) Topographic atlas of catecholamine and acetylcholinesterase-containing neurons in the rat brain. I. Forebrain (telencephalon, diencephalon). *J. comp. Neurol.,* 157: 13–28.

Kalra, P.S., Kalra, S.P., Krulich, L., Fawcett, C.P. and McCann, S.M. (1972) Involvement of norepinephrine in transmission of the stimulatory influence of progesterone on gonadotropin-release. *Endocrinology,* 90: 1168–1176.

Kalra, S.P. and McCann, S.M. (1973) Effect of drugs modifying catecholamine synthesis on LH release from preoptic stimulation in the rat. *Endocrinology,* 93: 356–362.

Kamberi, I.A., Mical, R.S. and Porter, J.C. (1969) Luteinizing hormone-releasing hormone activity in hypophysial stalk blood and elevation by dopamine. *Science,* 166: 388–390.

Kato, J. and Villee, C.A. (1967) Preferential uptake of estradiol by the anterior hypothalamus of the rat. *Endocrinology,* 80: 567–575.

Kawakami, M. and Kimura, F. (1974) Study on the bed nucleus of stria terminalis in relation to gonadotropin control. *Endocr. jap.,* 21: 125–130.

Kawakami, M. and Kimura, F. (1975) Possible roles of CNS estrogen-neuron systems in the control of gonadotropin release. In *Anatomical Neuroendocrinology,* W.F. Stumpf (Ed.), Karger, Basel, pp. 13–32.

Kawakami, M. and Sakuma, Y. (1974) Responses of hypothalamic neurons to the microiontophoresis of LH-RH, LH and FSH under various levels of circulating ovarian hormones. *Neuroendocrinology,* 15: 290–307.

Kawakami, M. and Sakuma, Y. (1976a) Electrophysiological evidences for possible participation of periventricular neurons in anterior pituitary regulation. *Brain Res.,* 101: 79–94.

Kawakami, M. and Sakuma, Y. (1976b) Chemical sensitivity of hypothalamic immuno-reactive cells to antiserum against LH-RH. *Brain Res.,* submitted for publication.

Kawakami, M. and Terasawa, E. (1972a) A possible role of the hippocampus and the amygdala in the androgenized rat: effect of electrical or electrochemical stimulation of the brain on gonadotropin secretion. *Endocr. jap.,* 19: 349–358.

Kawakami, M. and Terasawa, E. (1972b) Electrical stimulation of the brain on gonadotropin secretion in the female prepuberal rat. *Endocr. jap.,* 19: 335–347.

Kawakami, M. and Terasawa, E. (1972c) Acute effect of neural deafferentation of timing of gonadotropin secretion before proestrus in the female rat. *Endocr. jap.,* 19: 449–459.

Kawakami, M. and Terasawa, E. (1974) Role of limbic forebrain structures on reproductive cycles. In *Biological Rhythms in Neuroendocrine Activity,* M. Kawakami (Ed.), Igaku-Shoin, Tokyo, pp. 197–219.

Kawakami, M., Terasawa, E. and Ibuki, T. (1970) Changes in multiple unit activity of the brain during the estrous cycle. *Neuroendocrinology,* 6: 30–48.

Kawakami, M., Terasawa, E., Seto, K. and Wakabayashi, K. (1971) Effect of electrical stimulation of the medial preoptic area on hypothalamic multiple unit activity in relation to LH release. *Endocr. jap.,* 18: 13–21.

Kawakami, M., Kimura, F., Negoro, H. and Yanase, M. (1973a) Possible role of the organum vasculosum laminae terminalis in the release of LH and prolactin in rats. *Int. Soc. Psychoneuroendocrinology IVth Int. Congr.,* Berkeley.

Kawakami, M., Manaka, M., Hiroto, S., Higuchi, T. and Konno, T. (1973b) Cyclic changes in electrical activity of the brain. *Saishin-Igaku,* 28: 1096–1105 (in Japanese).

Kawakami, M., Terasawa, E., Kimura, F., Higuchi, T. and Konda, N. (1973c) Changes in multiple electrical activity (MUA) in rat brain during the estrous cycle and after administration of sex steroids. In *Progr. Brain Res., Drug Effects on Neuroendocrine Regulation, Vol. 39,* E. Zimmermann, W.H. Gispen, B.H. Marks and D. De Wied (Eds.), Elsevier, Amsterdam, pp. 125–134.

Kawakami, M., Terasawa, E., Kimura, F. and Kubo, K. (1973d) Correlated changes in gonadotropin release and electrical activity of the hypothalamus induced by electrical stimulation of the hippocampus in immature and mature rats. In *Hormones and Brain Function,* K. Lissák and E. Endröczi (Eds.), Akadémiai Kiadó, Budapest, pp. 347–374.

Kawakami, M., Terasawa, E., Kimura, F., Shinohara, Y. and Hiroto, S. (1973e) A role of the septal complex in gonadotropin release in the rat. In *Excerpta Med. Int. Congr. Ser. 273,* Excerpta Medica, Amsterdam, pp. 67–72.

Kawakami, M., Terasawa, E., Kimura, F. and Wakabayashi, K. (1973f) Modulating effect of limbic structures on gonadotropin release. *Neuroendocrinology,* 12: 1–16.

Kawakami, M., Kimura, F. and Yanase, M. (1974) Involvement of the circumventricular organ in the regulation of gonadotropins and prolactin. In *Biological Rhythms in Neuroendocrine Activity,* M. Kawakami (Ed.), Igaku-Shoin, Tokyo, pp. 167–187.

Kawakami, M., Kimura, F. and Higuchi, T. (1975a) Localization and mechanisms of stimulatory feedback action of estrogen: effects of limbic forebrain implantation of estradiol benzoate on advancement of ovulation. *Endocr. jap.,* 22: 319–330.

Kawakami, M., Kimura, F. and Kawagoe, S. (1975b). Cholinergic and/or serotonergic neural links are required for the inhibition of hippocampus, lateral amygdala and central gray matter on gonadotropin release. *Endocr. jap.,* in press.

Kawakami, M., Kimura, F. and Konda, N. (1975c) Role of forebrain structures in the regulation of gonadotropin and prolactin secretion. In *Neuroendocrine Regulation of Fertility,* T.C. Anand Kumar (Ed.), Karger, Basel, pp. 10–13.

Kawakami, M., Negoro, H., Kimura, F., Higuchi, T. and Asai, T. (1975d) Neural control of LH-release in anterior periventriculo-median eminence-pituitary system. *Neuroendocrinology,* 17: 410–420.

Kazama, N. and Longcope, C. (1972) Metabolism of estrone and estradiol-17β in sheep. *Endocrinology,* 91: 1450–1454.

Kimura, F., Sakuma, Y. and Kawakami, M. (1975) Monoaminergic control of gonadotropin release. *Nihon-Rinsho,* 33: 534–538 (in Japanese).

Knigge, K.M., Joseph, S.A. and Silverman, A. (1973) Further observations on the structure and function of the median eminence, with reference to the organization of RF-producing elements in the endocrine hypothalamus. In *Drug Effects on Neuroendocrine Regulation, Progr. Brain Res., Vol. 39,* E. Zimmermann, W.H. Gispen, B.H. Marks and D. De Wied (Eds.), Elsevier, Amsterdam, pp. 7–20.

Kobayashi, F., Hara, K. and Miyake, T. (1969) Effects of steroids on the release of luteinizing hormone in the rat. *Endocr. jap.,* 16: 251–260.

Kordon, C. and Glowinski, J. (1969) Selective inhibition of superovulation by blockade of dopamine synthesis. *Endocrinology,* 85: 924–931.

Kordon, C. and Glowinski, J. (1970) Role of brain catecholamines in the control of anterior pituitary function. In *Neurochemical Aspects of Hypothalamic Function,* L. Martini and J. Meites (Eds.), Academic Press, New York, pp. 85–100.

Krey, L.C. and Everett, J.W. (1973) Multiple ovarian responses to single estrogen injections early in rat estrous cycles: impaired growth, luteotropic stimulation and advanced ovulation. *Endocrinology,* 93: 377–384.

Krulich, L., Quijada, M., Illner, P. and McCann, S.M. (1971) The distribution of hypothalamic hypophysiotropic factors in the hypothalamus of the rat. *25th Int. Congr. Physiol., Munich, Proc. int. Union Physiol. Soc.,* 9: 326.

Lawton, I.E. and Schwarz, N.B. (1968) A circadian rhythm of luteinizing hormone secretion in ovariectomized rats. *Amer. J. Physiol.,* 214: 213–217.

Lawton, I.E. and Smith, S.W. (1970) LH secretory patterns in intact and gonadectomized male and female rats. *Amer. J. Physiol.,* 219: 1019–1022.

Luttge, W.G. and Whalen, R.E. (1970) Regional localization of estrogenic metabolites in the brains of male and female rats. *Steroids,* 15: 605–612.

Luttge, W.G. and Whalen, R.E. (1972) The accumulation, retention and interaction of oestradiol and oestrone in central neural and peripheral tissues of gonadectomized female rats. *J. Endocr.,* 52: 379–395.

Marshall, F.H.A. and Verney, E.B. (1936) The occurrence of ovulation and pseudopregnancy in the

rabbit as a result of central nervous stimulation. *J. Physiol. (Lond.)*, 86: 327–336.
McCann, S.M. (1962) A hypothalamic luteinizing hormone-releasing factor (LH-RF). *Amer. J. Physiol.*, 202: 395–400.
McEwen, B.S. and Pfaff, D.W. (1973) Chemical and physiological approaches to neuroendocrine mechanisms: attempts at integration. In *Frontiers in Neuroendocrinology*, W.F. Ganong and L. Martini (Eds.), Oxford Univ. Press, New York, pp. 267–335.
McEwen, B.S., Pfaff, D.W. and Zigmond, R.E. (1970) Factors influencing sex hormone uptake by rat brain regions. III. Effects of competing steroids on testosterone uptake. *Brain Res.*, 21: 29–38.
Meszaros, T., Csuri, I.J., Hazas, J. and Palkovits, M. (1969) Esterase activity in the hypothalamus. *Acta morph. Acad. Sci. hung.*, 17: 201–215.
Meyer, M.H., Masken, J.F., Nett, T.M. and Niswender, G.D. (1974) Serum levels of gonadotropin-releasing hormone (Gn-RH) during the estrous cycle and in pentobarbital-treated rats. *Neuroendocrinology*, 15: 32–37.
Naftolin, F., Ryan, K.J. and Petro, Z. (1971) Aromatization of androstenedione by limbic system tissue from human foetuses. *J. Endocr.*, 51: 795–796.
Pfaff, D.W. (1973) In *The Regulation of Mammalian Reproduction*, S.J. Segal, R. Cozier, P.A. Corfman and P.G. Condliffe (Eds.), Thomas, Springfield, Ill., pp. 5–22.
Pfaff, D.W. and Keiner, M. (1973) Atlas of estradiol-concentrating cells in the central nervous system of the female rat. *J. comp. Neurol.*, 151: 121–157.
Ramirez, V.D. and Sawyer, C.H. (1965) Fluctuations in hypothalamic LH-RF (luteinizing hormone-releasing factor) during the rat estrous cycle. *Endocrinology*, 76: 282–289.
Reivich, M. and Glowinski, J. (1967) An autoradiographic study of the distribution of C^{14}-norepinephrine in the brain of the rat. *Brain*, 90: 633–647.
Richter, C.P. (1971) Inborn nature of the rat's 24-hour clock. *J. comp. physiol. Psychol.*, 75: 1–4.
Rosenfeld, M.G. and O'Malley, B.W. (1970) Steroid hormones: effects on adenylcyclase activity and adenosine 3',5'-monophosphate in target tissues. *Science*, 168: 253–255.
Rubinstein, L. and Sawyer, C.H. (1970) Role of catecholamines in stimulating the release of pituitary ovulating hormone(s) in rats. *Endocrinology*, 86: 988–995.
Sandler, R. (1968) Concentration of norepinephrine in the hypothalamus of the rat in relation to the estrous cycle. *Endocrinology*, 83: 1383–1386.
Sawyer, C.H. (1975) First Geoffrey Harris Memorial Lecture. Some recent developments in brain-pituitary-ovarian physiology. *Neuroendocrinology*, 17: 97–124.
Sawyer, C.H., Markee, J.E. and Hollinshead, W.H. (1947) Inhibition of ovulation in the rabbit by the adrenergic-blocking agent dibenamine. *Endocrinology*, 41: 395–402.
Sawyer, C.H., Markee, J.E. and Townsend, B.F. (1949) Cholinergic and adrenergic components in the neurohumoral control of the release of LH in the rabbit. *Endocrinology*, 44: 18–37.
Sawyer, C.H., Hilliard, J., Kanematsu, S., Scaramuzzi, R. and Blake, C.A. (1974) Effects of intraventricular infusions of norepinephrine and dopamine on LH release and ovulation in the rabbit. *Neuroendocrinology*, 15: 328–337.
Schneider, H.P.G. and McCann, S.M. (1969a) Possible role of dopamine as transmitter to promote discharge of LH-releasing factor. *Endocrinology*, 85: 121–132.
Schneider, H.P.G. and McCann, S.M. (1969b) Stimulation of release of luteinizing hormone-releasing factor from hypothalamic tissue by dopamine in vitro. *J. Reprod. Fertil.*, 18: 178.
Sétáló, G., Vigh, S., Schally, A.V., Arimura, A. and Flerkó, B. (1975) LH-RH-containing neural elements in the rat hypothalamus. *Endocrinology*, 96: 135–142.
Stumpf, W.E. (1971) Hypophysiotropic neurons in the periventricular brain: topography of estradiol concentrating neurons. In *Steroid Hormones and Brain Function*, C.H. Sawyer and R.A. Gorski (Eds.), Univ. Calif. Press, Los Angeles, pp. 215–227.
Szentágothai, J., Flerkó, B., Mess, B. and Halász, B. (1962) *Hypothalamic Control of the Anterior Pituitary*, Akadémiai Kiadó, Budapest.
Terasawa, E. (1971) Changes in electrical activity in the androgenized rat and the anterior deafferented rat. *Endocr. jap.*, 46: 1081–1098.

Terasawa, E. and Kawakami, M. (1974) Positive feedback sites of estrogen in the brain on ovulation: possible role of the bed nucleus of stria terminalis and the amygdala. *Endocr. jap.*, 21: 51–60.

Tima, L. and Flerkó, B. (1974) Ovulation induced by norepinephrine in rats made anovulatory by various experimental procedures. *Neuroendocrinology*, 15: 346–354.

Ungerstedt, U. (1971) Stereotaxic mapping of the monoamine pathways in the rat brain. *Acta physiol. scand.*, Suppl. 367: 1–48.

Van Rees, G.P. (1972) Control of ovulation by the anterior pituitary gland. In *Topics in Neuroendocrinology, Progr. Brain Res., Vol. 38,* J. Ariëns-Kappers and J.P. Schadé (Eds.), Elsevier, Amsterdam, pp. 193–210.

Velasco, M.E. and Taleisnik, S. (1969) Release of gonadotropins induced by amygdaloid stimulation in the rat. *Endocrinology*, 84: 132–139.

Velasco, M.E. and Taleisnik, S. (1971) Effects of the interruption of amygdaloid and hippocampal afferents to the medial hypothalamus on gonadotropin release. *J. Endocr.*, 51: 41–55.

Weiner, R.I., Blake, C.A., Rubinstein, L. and Sawyer, C.H. (1971) Electrical activity of the hypothalamus: effects of intraventricular catecholamines. *Science*, 171: 411–412.

Weissman, B.A. and Skolnick, P. (1975) Stimulation of adenosine 3',5'-monophosphate formation in incubated rat hypothalamus by estrogenic compounds: relationship to biologic potency and blockade by anti-estrogens. *Neuroendocrinology*, 18: 27–34.

Ying, S.-Y. and Greep, R.O. (1972) Effect of a single injection of estradiol benzoate (EB) on ovulation and reproductive function in 4-day cyclic rats. *Proc. Soc. exp. Biol.(N.Y.)*, 139: 741–744.

Zimmerman, E.A., Hsu, K.C., Ferin, M. and Kozlowski, G.P. (1974) Presence of LRH in the organum vasculosum laminae terminalis in the mouse was observed with the immuno-peroxidase technique. *Endocrinology*, 95: 1–8.

Subcellular Mechanisms in Reproductive Neuroendocrinology, edited by
F. Naftolin, K.J. Ryan and J. Davies
© 1976 Elsevier Scientific Publishing Company—Amsterdam, The Netherlands

Chapter 24

The adenohypophysis; functional behavior of the gonadotrophs as target cells

S.S.C. YEN

Department of Reproductive Medicine (Obstetrics and Gynecology), School of Medicine, University of California, San Diego, Calif. (U.S.A.)

I. INTRODUCTION

I was assigned the responsibility of synthesizing our discussion on subcellular mechanisms in the neuroendocrine regulation of the hypophysis. From the beginning, I had serious doubts about being able to fulfill this task. After the massive infusion of new knowledge during the past two days, it became crystal clear that I should not be able to come through. On the other hand, I gained the distinct impression that there are more questions than answers relative to the complexities of the regulation of the brain-pituitary system.

I believe that this Symposium represents the first of its kind in that it attempts to cover in a logical fashion our recent knowledge of the cellular mechanisms involved in neuroendocrine communication. As a student in the explosive field of neuroendocrinology, I have witnessed but still not assimilated the recent advances in the morphology and electrophysiology of LRF neurons and in the related role of neurotransmitters. It would appear that we should be able, in the very near future, to break the code and read the message delivered by those central neurons which have endocrine function. In the overall picture which emerges from the presentations I have heard here, some elements appear to be missing. I will do what I can to fill those gaps.

II. CATECHOLAMINES AND THE ADENOHYPOPHYSIS

Noradrenergic and dopaminergic neurons in the median eminence course through the infundibular stem and terminate in the pars intermedia. In addition, dopaminergic neurons terminate in the pars nervosa. While the adenohypophysis is devoid of catecholaminergic fibers, there is evidence that it is capable of actively concentrating norepinephrine and epinephrine several-fold over the plasma levels. This is apparently mediated through a carrier-dependent saturable uptake system. Most of the PAS-positive cells in the anterior pituitary can take up exogenous L-DOPA from the blood and decarboxylate it to form dopamine. When [^{3}H]adrenaline is administered intravenously to the rat, it is rapidly taken up by the pituitary; this is followed by a relatively slow rate of disappearance with a $t_{\frac{1}{2}}$ of 3.5 hr. The pituitary seems to possess the enzymatic machinery for inactivating catecholamines using pathways similar to those found in the brain, i.e., deamination, O-methylation and conjugation. Since the adenohypophysis lays outside the blood-brain barrier for the catecholamines and in view of its avidity for catecholamines, a subtle relationship between circulating catecholamine and adenohypophysis function may be considered.

III. MECHANISM(S) IN THE REGULATION OF PORTAL BLOOD FLOW

Since anterior pituitary function is thought to be determined, in part, by perfusion characteristics of the gland, the disclosure of the mechanism(s) for the vasomotor control of the superior hypophyseal artery and portal vessels would be of great importance to our understanding of the functional relationship between the secretory activity of endocrine neurons and the response of pituitary cells. Happily, investigation of blood flow dynamics in the portal system is now feasible (Kopaniky and Gann, 1975). More information in this regard should provide important clues in the understanding of the delivery system.

IV. THE POSSIBLE INTERACTION BETWEEN THE NEUROHYPOPHYSIS AND THE ADENOHYPOPHYSIS

The neurophysin story presented by Dr. Zimmerman serves to reemphasize the long-standing questions of a relationship between the neuro- and adeno-components of the hypophysis in the regulation of hormone release. Postulates concerning the relationship between oxytocin and prolactin, ACTH and vasopressin, and estrogen-oxytocin-gonadotropin now appear to be testable.

During the remainder of my time, I'd like to present some of the data concerning the neuroendocrine regulation of gonadotropin secretion in humans. I hope that you may be able to help me plug-in the cellular mechanisms to explain the phenomena I will describe.

V. THE DOMINATING ROLE OF THE CNS IN THE REGULATION OF PITUITARY FUNCTION

The presence of pulsatile fluctuations in LH concentration, in the face of a constant infusion of LRF (Fig. 1), probably represents an expression of pituitary response to pulses of endogenous LRF, although local pituitary mechanisms, such as rhythmic vasodilation of the portal system or an inherent rhythm of the pituitary gonadotrophs, have not been excluded. Similar observations have been reported in the rhesus monkey (Ferin et al., 1973) and the rat (Dowd et al., 1975). Several lines of evidence are now available to support a hypothalamic mechanism for pulsatile gonadotropin release: (1) a pulsatile pattern of LRF concentration in the portal blood of the rhesus monkey has been observed (Carmel et al., 1975); (2) pulsatile secretion of LH and FSH can be abolished by the administration of antiserum to LRF (McCormack and Knobil, 1975); and (3) pulsatile

LH release is observed only with pulsed delivery of LRF but not by constant infusion during superinfusion of isolated rat pituitaries (Osland et al., 1975). Thus, the pulsatile release of gonadotropins appears to be the direct consequence of pulsed LRF secretion.

In view of the involvement of catecholaminergic neuronal input to the hypothalamic LRF secreting mechanisms, it is interesting that changes in LH pulses are found to be associated with certain emotional and psychological states. A marked increase in the amplitude of LH and PRL pulses has been observed in a patient with pseudocyesis (Yen et al., 1975b). Our findings in this case indicate unequivocally the presence of an excessive pituitary secretion of LH as well as PRL in association with evidence of luteal activity (Fig. 2). FSH levels were normal. The enhanced LH-release, comparable to the levels seen during the mid-cycle surge, should provide sufficient luteotropic effect for the persistence of luteal activity. Although a luteotropic role for PRL has yet to be established in humans, the findings in this case raise the possibility of a synergistic effect between LH and PRL in the maintenance of luteal function in humans. Delineation of the effect(s) of

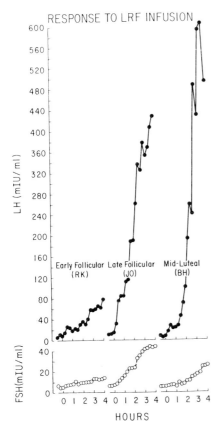

Fig. 1. The presence of spontaneous fluctuation of LH during the rapid gonadotropin release induced by constant LRF infusion (0.2 μg/min x 4 hr) in the early and late follicular and during the mid-luteal phases of the cycle.

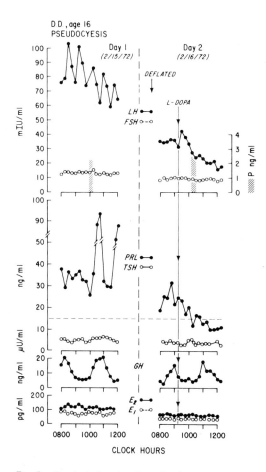

Fig. 2. Basal pituitary and ovarian hormone concentrations and their fluctuating patterns in a patient with pseudocyesis studied before (day 1) and after (day 2) the clinical resolutions (deflation). L-DOPA (500 mg, p.o.) was given on day 2. The hatched bars in the top panels represent progesterone levels. The dotted lines in the second panel denote the mean value of PRL in normal women.

2-brom-α-ergocryptine in this syndrome should shed light on this issue. Whether the modest increase in the levels of progesterone and estrogen are reflections of activity of the corpus luteum or of the luteinized follicles, cannot be resolved in this case. Nonetheless, the dominant role of the CNS in the regulation of pituitary function is evident.

The mechanism(s) for the hypersecretion of PRL and LH, and possibly an aberration in GH release, undoubtedly reside in the CNS. The role of catecholamines in the modulation of mood and behavior, as well as hypothalamic function, is well recognized (Axelrod, 1975). Brown and Barglow have recently suggested that "endogenous" depression is crucially significant in the genesis of pseudocyesis (1971). The fantasy of pregnancy appears to function as a defense against depression. Since preliminary data suggest that mental depression is associated with a decrease in CNS catecholamine activity (Sachar,

1976), one might reason that a reduced activity of the catecholaminergic mechanism could account for the hypersecretion of PRL. The hypersecretion of LH, however, is inconsistent with this thesis, since increased LRF release appears to require an increase in catecholamines. Furthermore, the suppression of both LH and PRL and the absence of a GH rise in response to L-DOPA administration suggests the involvement of additional neuroendocrine mechanisms, other than the catecholamine system. In any case, the role of "anxiety" in causing the hypersecretion of PRL and LH is suggested by the almost immediate fall in the serum concentrations of these hormones after revealing the diagnosis to the patient. The reverse situation is found in patients with anorexia nervosa (Boyar et al., 1974a) and certain hypothalamic dysfunctions (Yen et al., 1975a) where the pulsatile pattern of gonadotropin release is absent.

Evidence for the role of CNS catecholamines in the regulation of gonadotropin release by the pituitary in humans is necessarily limited. Drugs such as reserpine and phenothiazine known to affect CNS catecholamines are frequently associated with the disturbances of cyclic gonadotropin release. Attempts to increase hypothalamic and/or pituitary dopamine by the administration of L-DOPA (Fig. 3) and dopamine infusion (Leblanc et al., 1975) have yielded no demonstrable changes in basal gonadotropin secretion, but were accompanied by marked changes in GH and PRL (Fig. 4). It is interesting to note that the higher the basal PRL secretion, the greater is the inhibitory effect of

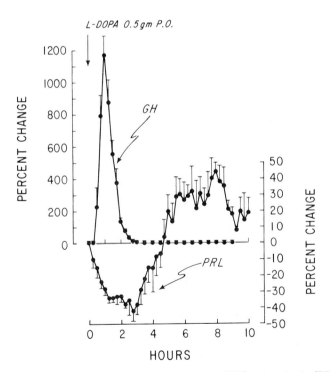

Fig. 3. Per cent change in growth hormone (GH) and prolactin (PRL) levels in response to 500 mg of L-DOPA in three normal women. (From Leblanc and Yen, unpublished.)

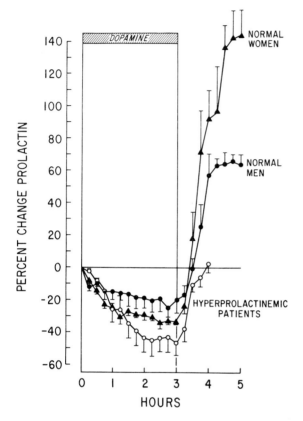

Fig. 4. Changes in serum concentrations of PRL in response to dopamine infusion (4 μg/kg/min) in normal women, normal men and in patients with hyperprolactinemia (expressed as per cent change from basal levels). Gonadotropin and TSH levels were unaffected.

dopamine. Does this mean that hyperprolactinemic state is associated with less dopamine-like substance delivered from the hypothalamus to the pituitary and consequently an increase in dopamine receptors in the lactotrophs? Measurement of catecholamine levels and of enzymatic activities such as tyrosine hydroxylase (for the conversion of tyrosine to DOPA) and dopamine β-hydroxylase (for the conversion of dopamine to norepinephrine) during acute gonadotropin release, either spontaneous or estrogen and/or progesterone induced, has not been reported. On the other hand, plasma monoamine oxidase (MAO) activity (using [^{14}C]tyramine as a substrate in the assay) is found to be significantly greater in the postovulatory than in the preovulatory phase of the cycle. The MAO activity in plasma of amenorrheic and postmenopausal women is significantly greater than in postovulatory subjects. These elevated levels of MAO can be reduced by estrogen treatment, but estrogen plus progestin treatment significantly increased the MAO activity over that found during estrogen treatment alone (Klaiber et al., 1971). These data imply that estrogens tend to inhibit MAO activity in human plasma, a finding which is consistent with the changes in hypothalamic MAO activity in rats under similar hormonal environments (Kobayashi et al., 1964). Since plasma MAO activity is believed to be inversely related to the level of central adrenergic activity, they reasoned that estrogen, which appears to inhibit MAO activity, may also enhance central adrenergic processes. Progesterone may be assumed to effect a reversal of this process. It should be emphasized

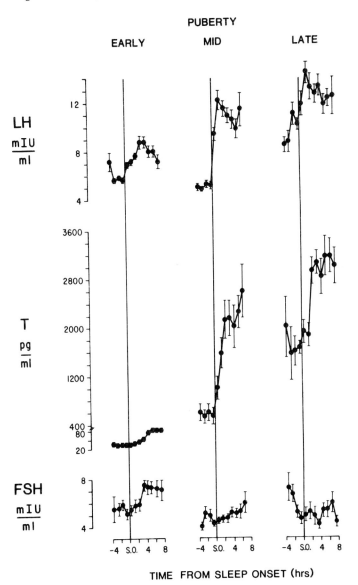

Fig. 5. The sleep-induced gonadotropin release during different stages of puberty development. The concomitant elevation of circulating testosterone (T) is also depicted.

that the real clinical significance of these findings cannot be assessed since measurements of enzymatic activities in circulation may have no relationship to cellular catecholamine metabolism.

The mechanism responsible for the onset of puberty has been, until recently, an unsolved puzzle. The enigma has resided in the apparent development of secondary sex characteristics prior to the detection of an increase in plasma levels of sex steroids. However, the recent disclosure of a sleep-associated increase in pulsatile pituitary LH-release at the time of puberty, which regresses after sexual maturation (Boyar et al., 1972; Parker et al., 1975), has advanced our understanding of the basis for sexual maturation. A progressive increase in the pulsed release of LH during sleep occurs from puberty stage I through stage V, and a temporally associated nocturnal testosterone secretion is also found (Fig. 5) (Boyar et al., 1974b; Judd et al., 1974; Parker et al., 1975). Since this augmented LH secretion during sleep is independent of gonadal activity, as is found in gonadal dysgenesis (Boyar et al., 1973), the concept of a "CNS program" for the pubertal activation of the hypothalamic-pituitary system has been proposed. To what extent the catecholamine and LRF neuronal activities are involved in this pubertal CNS program, as well as in pseudocyesis, anorexia nervosa, and certain hypothalamic dysfunctions remains to be determined. Nonetheless, these are outstanding examples for the dominating role of CNS in the regulation of gonadotropin output.

VI. OVARIAN MODULATION OF THE HYPOTHALAMIC-PITUITARY SYSTEM IN THE RELEASE OF GONADOTROPIN

VI.1. Cyclic gonadotropin release: "CNS clock" vs. "ovarian clock"

Under a variety of experimental conditions, ovarian estradiol has been shown to exert a stimulatory feedback action on gonadotropin release in the rat (Krey and Everett, 1973). The most interesting feature of these observations in the rat is that estrogen-induced surges of LH were observed only in the afternoon of proestrus, suggesting the involvement of a "CNS clock". Since the LH surge can be blocked by the administration of antibodies to 17β-estradiol (Ferin et al., 1969), it was proposed that in the rat, ovarian estradiol determines the day of the surge while the CNS determines its hour of onset (Bogdanove, 1972).

Recent studies of gonadotropin secretion revealed a remarkable similarity between the menstrual cycles in humans and in the rhesus monkey. They appear to be different from the rat in that the evidence supports an "ovarian clock" rather than a "CNS clock" in the regulation of cyclic gonadotropin surge. The background information supporting that ovarian estradiol constitutes the critical feedback signal in the regulation of cyclic gonadotropin release has been extensively reviewed both for the human (Ross et al., 1970; Van de Wiele et al., 1970; Yen et al., 1974a, 1975) and for the rhesus monkey (Knobil et al., 1974).

VI.2. Approaches in the functional analysis of gonadotrophs as target cells

There is good experimental evidence to indicate that both the pituitary and the hypothalamus are involved in the feedback action of estrogen. Further, estrogen receptors have been found in both the hypothalamus and the pituitary (Davies et al., 1975). Although much information has been obtained concerning the pituitary gonadotropin responses to LRF under varied steroid environments (Jaffe and Keys, 1974; Yen et al., 1974b; Yen, 1975), assessments of the quantitative and qualitative relationships between individual controllers (LRF and gonadal steroids) and the status of gonadotrophs have been a recent approach. It should be recognized that these functional dynamics are extremely complex and characterization of gonadotrophs, as target cells, in the human is hampered by the lack of meaningful in vitro information. These studies are based on the premise that the functional capacity of the gonadotrophs is determined by the relative inputs of hypothalamic LRF and ovarian E_2 (Fig. 6). The gonadotropin response of the pituitary to pulse injections of LRF (5 times) at 2-hr intervals with large (150 μg), small (10 μg), incremental (10–300 μg) and decremental (300–10 μg) doses was assessed in eugonadal males with a relatively stable hypothalamic-pituitary-gonadal system (Lasley et al., 1975a). This experimental design was employed (a) in an effort to simulate a pulsatile hypothalamic LRF input assumed to be immediately responsible for the pulsatile pituitary gonadotropin output, and (b) to determine whether the pituitary detects small, incremental and decremental changes of LRF input at 2-hr intervals. At constant doses, the rates at which serum LH increase and decrease appear remarkably similar to the rates characteristic of spontaneous pulses in normal male subjects (Naftolin et al., 1972). It is apparent that the pituitary can detect relatively small increments or decrements of LRF as reflected by corresponding changes in circulating LH concentrations. At no time was there a suggestion of a refractory period, an augmentation or a reduction of sequential responses when tested at 2-hr intervals. These data indicate that graded doses of LRF induce a graded variation in the quantity of LH released and suggest, as indicated earlier, that variations in the amount of LRF delivered represent a significant factor in the control of the pituitary gonadotropin output.

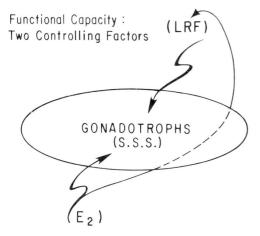

Fig. 6. A diagrammatic illustration of the functional capacity of the gonadotrophs, as determined by the two controllers, LRF and E_2.

The technique of administering "pulses of LRF" and "LRF infusion" were employed in further studying the feedback action of estrogen and progesterone on pituitary responce in women. It is assumed that the magnitude of initial, or first phase response is dependent upon the previous history or "set" of gonadotrophs, which is determined by the relative inputs of LRF and E_2. The rapidity and quantity (Δ) of this initial release will probably reflect the size of the acute releasable pool and may be a reasonable measure of "sensitivity". The second phase of release also appears to depend on the previous "set" of gonadotrophs. It reflects the size of the second pool (pituitary reserve) but in addition, is influenced by the rate of new synthesis which may be augmented by the time-related inputs of both E_2 and LRF at an appropriate ratio. These premises are instrumental in the functional characterization of the gonadotrophs, as target cells, relative to the changes of sensitivity/capacity and the variables of controlling inputs: LRF, estrogen and progesterone.

VI.3. Ovarian modulations of the pituitary capacity

VI.3.1. The concept of two pools of pituitary gonadotropin. The biphasic pattern of serum LH response to continuous LRF stimulation supports the concept of the presence of two components of releasable LH in the anterior pituitary, one immediately releasable and the other requiring continued stimulus input (Fig. 7). Similar data have been reported for man (Bremner and Paulson, 1974) and for pubertal (but not pre-pubertal) children (Reiter et al., 1975). It is likely that the initial rise in serum LH during the first 30 min results from release of the first pool and with continued LRF stimulation, the second or reserve pool is tapped with a secondary rise in serum LH concentration. In the case of pulsed LRF administration, the first pulse would appear to induce LH release from the first pool, and the increased release from subsequent pulses of LRF may be a result of activations of the reserve pool of LH.

VI.3.2. Pituitary sensitivity versus reserve. Although this two-pool concept remains arbitrary, it may nevertheless be useful in the analysis of the functional meaning of the pituitary sensitivity and reserve. When the LRF stimulation is small and brief, such as a 10-μg pulse, the initial increment at peak serum concentration (Δ_i) may be a reasonable measure of pituitary sensitivity and probably reflects the size of the first pool or acutely releasable gonadotropin in the pituitary. On the other hand, estimation of pituitary reserve may require a longer duration of stimulation, and the quantitative release as a function of time is likely an approximation of the size of the second or storage pool which may include a component of, yet unmeasurable amount, newly synthesized gonadotropins. When a large single dose of LRF is used (100–150 μg), the components of sensitivity and reserve cannot be separately appreciated since a large dose provides a longer action than a small dose and thereby, induces a release not only from the acutely releasable pool but also the second pool (Van den Berg et al., 1974; Wang et al., 1975).

VI.3.3. Changes in pituitary sensitivity and reserve as related to E_2 and P. Major changes in the pituitary sensitivity and reserve occur during the menstrual cycle (Fig. 8). In general, these changes are in synchrony with the cyclicity of ovarian steroid levels. During the early follicular phase, both pituitary sensitivity and reserve seem to be at a minimum, but with increasing levels of E_2 a preferential increase in pituitary reserve than

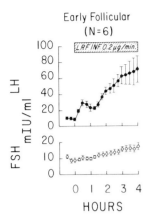

Fig. 7. The two components of LH release elicited by constant LRF infusion in subjects during the early follicular phase of the cycle (mean ± S.E.).

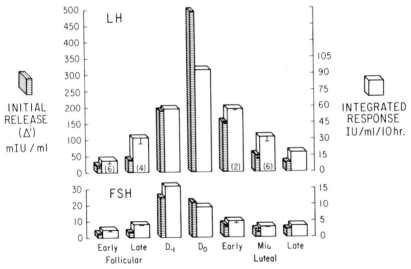

Fig. 8. Analyses of the changes in pituitary sensitivity (Δ_i) and reserve (integrated response) elicited by pulses of LRF during different phases of the menstrual cycle (mean ± S.E.).

in sensitivity ($P < 0.005$) in response to LRF pulses was elicited. The finding that pituitary reserve (but not sensitivity) in hypogonadal women is enhanced by estrogen treatment (Lasley et al., 1975b) and negated by the administration of anti-estrogen (clomiphene) during the late follicular phase (Wang and Yen, 1975) suggests that increasing reserve in the late follicular phase of the cycle is causally related to the rising levels of E_2. This preferentially augmented pituitary reserve requires that the rate of synthesis exceeds the rate of release, thereby, increasing the pituitary gonadotropin store. This build-up in pituitary store, in all probability, constitutes the prequisite for the "development" of the mid-cycle surge.

Pituitary sensitivity and reserve continued to be high during the early luteal phase but reduced progressively thereafter. The presence of progesterone as well as high circulating E_2 levels during the mid-luteal phase was associated with almost identical functional capacity (sensitivity and reserve) of the gonadotrophs, as that found in the presence of high E_2 levels alone. This finding provides good evidence that progesterone at a relatively high level does not impair pituitary sensitivity or reserve. On the contrary, we have recently demonstrated that progesterone at a relatively low circulating level (2—5 ng/ml) enhances both sensitivity and reserve in the estrogen-augmented pituitary (Lasley et al., 1975b). Thus, progesterone is probably responsible, in part, for maintaining pituitary sensitivity and reserve at relatively high levels during the early luteal phase. The pituitary sensitivity and reserve during the late luteal phase are reduced, but remain higher than those observed during the early follicular phase at which time they are at the lowest level found during the entire cycle. It should be noted that although the FSH responses are less obvious, they are remarkably parallel to the pattern of LH responses.

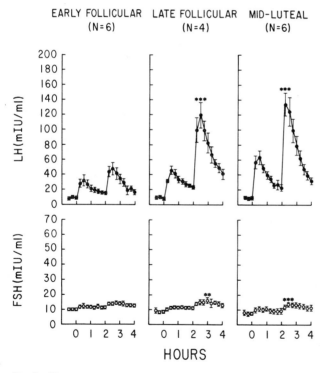

Fig. 9. The quantitative and qualitative comparison between gonadotropin responses to the first and second pulses of LRF during the low (early follicular) and the high (late follicular and mid-luteal) estrogen phases of the cycle (mean ± S.E.). ** = $P < 0.01$; *** = $P < 0.005$.

VI.3.4. Self-priming effect of LRF. In the late follicular and in the early to midluteal phases of the menstrual cycle particularly, the second and subsequent gonadotropin responses to pulsed LRF always exceed the response to the first pulse (Fig. 9). One is tempted to postulate that the first pulse not only induces release from the first pool but also "activates" the reserve pool so that subsequent pulses of LRF may induce release of "stored" or reserve gonadotropin. This "self-priming" effect of LRF occurs only during those phases of the menstrual cycle when the serum E_2 concentration is high and, in fact, can be produced in hypogonadal women through estrogen treatment (Lasley et al., 1975b). Although the physiological significance of the self-priming effect of LRF is not clear, it seems likely that it may serve to activate the reserve pool and render its gonadotropins more readily releasable; this is then revealed as an increase in sensitivity. The finding of a more rapid increase in sensitivity than in reserve from the late follicular to the mid-cycle surge (Fig. 8) is consistent with this postulate and is displayed by a dramatic and stepwise increase in the pituitary sensitivity on the day before and during the mid-cycle surge. This confirmation of changes in pituitary function may play an important role in the development of mid-cycle surge. A remarkably similar finding of self-priming by LRF in the rat has been reported (Aiyer et al., 1974). It is concluded that the functional capacity of the gonadotrophs exhibits a remarkable cyclic change and that the adenohypophysis represents a critical feedback site in the development of preovulatory gonadotropin surge.

VI.4. *The contribution of endogenous LRF*

These dramatic changes in pituitary sensitivity and reserve provide a basis for rationalizing the hypothalamic component in the operating characteristics of the pituitary. During the late follicular and mid-luteal phases of the cycle, when sensitivity and reserve are found to be high and the self-priming effects of LRF found to be at a maximum, the relatively low basal gonadotropin secretion normally found requires that endogenous LRF release be very low. The modest increase in basal LH secretion observed just prior to the onset of the mid-cycle surge (Midgley and Jaffe, 1968) may reflect the beginnings of incremental LRF secretion; increased amounts of LRF have been found at the time of the mid-cycle surge in the portal blood of rhesus monkeys (Carmel et al., 1975) and in the peripheral blood of humans (Malacara et al., 1972; Arimura et al., 1974). The increased pituitary sensitivity and reserve, the development of the self-priming effect of LRF and the increments in LRF release may represent the essential combination required to induce the stepwise release of LH at the time of the mid-cycle surge.

A progressive decrease in sensitivity and reserve characterizes pituitary function from the mid-luteal to late luteal phases and into the early follicular phase of an ensuing cycle. This is probably largely the result of the progressive decline in concentration of ovarian steroids on which sensitivity and reserve appear to be dependent. Thus, a continued low level of endogenous LRF may be speculated. However, the elevation in basal LH and FSH release in the face of reduced pituitary sensitivity and reserve, observed during the early follicular phase, requires that an increased LRF release be postulated for this period of the cycle. Additionally, a decrease in the speculated ovarian inhibitory factor with prefer-

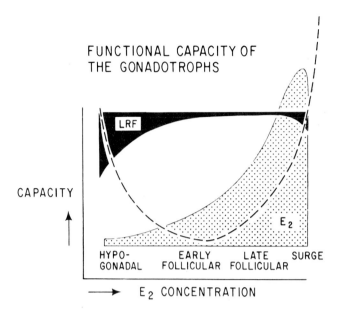

Fig. 10. The relationship between the functional capacity (sensitivity and reserve) of the gonadotrophs and E_2 as a function of concentration and time. The rationalization of the endogenous LRF input is also depicted.

ential action on FSH secretion should be considered in order to account for the higher level of FSH than LH seen at this time.

VII. THE PARADOX OF THE NEGATIVE AND POSITIVE FEEDBACK CONTROL SYSTEMS ON GONADOTROPIN RELEASE

When the role of estradiol as a feedback modulator of LH release is studied through the administration of this hormone to otherwise normal women, a somewhat puzzling observation is made. Whereas estradiol inhibits the pituitary response to LRF in hypogonadal women (Yen et al., 1974b), in normal women the response of the pituitary to LRF is enhanced (as discussed above). Thus, ovarian steroids are said to exert both a "negative" and "positive" feedback on gonadotropin secretion, and for many years numerous attempts were made to sort out and explain these effects. The positive feedback effect of estradiol on gonadotropin release is viewed as being paradoxical since, under most circumstances, estrogen inhibits gonadotropin secretion. From the results of a series of recent investigations in which estrogen was employed and circulating gonadotropin concentrations measured, it is proposed that the negative and positive feedback action of estradiol is not discontinuous but represents an overlapping event (Yen et al., 1974a). When changes in pituitary capacity (sensitivity and reserve) are being analyzed, the feedback event is represented by a continuum in the form of a U-shaped curve (Yen et al., 1975a). Fig. 10 describes this U-shaped functional capacity of the gonadotrophs as influenced by the relative inputs of LRF and E_2. Moderate concentrations of estradiol, such as are found in the early follicular phase of the menstrual cycle, inhibit gonadotropin output but a change in estradiol concentration *in either direction* tends to diminish that inhibition and permit increased activity of the hypothalamic-hypophyseal complex. The normal, unopposed hypothalamic-pituitary system (as in the hypogonadal and ovariectomized woman) produces gonadotropin at a relatively high rate under the influence of increased hypophysiotropic effect of LRF; small amounts of estradiol suppress the gonadotropin output (via inhibition of LRF release) but when the estradiol concentration is further elevated, the ability to suppress the gonadotropin release is no longer operative. Instead, an augmentation of the pituitary capacity occurs principally through the increase in sensitivity and reserve within the population of the gonadotrophs. The concomitant development of the "self-priming" effect of LRF to LRF action may accelerate the movement of the reserve pool to the acute releasable gonadotropin pool and, thereby, be expressed as a rapid increase in pituitary sensitvity when E_2 has reached a critical high level of sufficient duration. These intrapituitary dynamics described provide for the first time a more meaningful understanding of the positive feedback action of E_2. Although these interacting pituitary dynamics by themselves may eventuate an accelerated pituitary gonadotropin release, the initiation of mid-cycle surge should and does involve an increase in endogenous LRF release. The mechanism for triggering this neuroendocrine event in the hypothalamic LRF release is unknown, and will undoubtedly occupy the effort of many in the next few years.

REFERENCES

Aiyer, M.S., Shiappa, S.A. and Fink, G. (1974) A priming effect of LH-releasing factor on the anterior pituitary in the female rat. *J. Endocr.*, 62: 573.

Arimura, A., Kastin, A.J. and Schally, A.V. (1974) Immunoreactive LH-releasing hormone in plasma: mid-cycle elevation in women. *J. clin. Endocr.*, 38: 510.

Axelrod, J. (1975) Relationship between catecholamines and other hormones. *Recent Progr. Hormone Res.*, 31: 1.

Bogdanove, E.M. (1972) Hypothalamic-hypophyseal interrelationships: basic aspects. In *Reproductive Biology*, H. Balin and S. Glasser (Eds.), Excerpta Medica, Amsterdam, p. 5.

Boyar, R.M., Finkelstein, J., Roffwarg, H., Kapen, S., Weitzman, E. and Hellman, L. (1972) Synchronization of augmented LH secretion with sleep during puberty. *New Engl. J. Med.*, 287: 582–586.

Boyar, R.M., Finkelstein, J.W., Roffwarg, H., Kapen, S., Weitzman, E.G. and Hellman, L. (1973) Twenty-four-hour luteinizing hormone and follicle stimulating hormone secretory patterns in gonadal dysgenesis. *J. clin. Endocr.*, 37: 521–525.

Boyar, R.M., Katz, J., Finkelstein, J.W., Kapen, S., Weiner, H., Weitzman, E.G. and Hellman, L. (1974a) Anorexia nervosa; immaturity of the 24-hour LH secretory pattern. *New Engl. J. Med.*, 291: 861.

Boyar, R.M., Finkelstein, J.W., Roffwarg, H., Kapen, S., Weitzman, E.D. and Hellman, L. (1974b) Simultaneous augmented secretion of luteinizing hormone and testosterone during sleep. *J. clin. Invest.*, 54: 609.

Bremner, W.J. and Paulson, A. (1974) Two pools of luteinizing hormone in the human pituitary:

evidence from constant administration of luteinizing hormone-releasing hormone. *J. clin. Endocr.*, 39: 811.

Brown, E. and Barglow, P. (1971) Pseudocyesis a paradigm for psychophysiological interaction. *Arch. gen. Psychiat.*, 24: 221.

Carmel, P.C., Araki, S. and Ferin, M. (1975) Prolonged stalk portal blood collection in rhesus monkeys: pulsatile release of gonadotropin-releasing hormone (Gn-RH). *Endocrinology*, 96: 107A.

Davies, I.J., Naftolin, F., Ryan, K.J. and Siu, J. (1975) A specific high affinity, limited capacity estrogen binding component in the cytosol of human fetal pituitary and brain tissue. *J. clin. Endocr.*, 40: 909.

Dowd, A.J., Barofsky, A.-L., Chanhuri, N., Lloyd, D.W. and Weisz, J. (1975) Patterns of LH and FSH release from perfused rat pituitaries in response to infusions of hypothalamic extracts. *Endocrinology*, 96: 243.

Ferin, M., Tempone, A., Zimmering, P.E. and Van de Wiele, R.L. (1969) Effect of anti-bodies to 17β-estradiol and progesterone on the estrous cycle of the rat. *Endocrinology*, 85: 1070.

Ferin, M., Khalaf, S., Warren, M., Dyrenfurth, I., Jewelwiez, R., White, W. and Van de Wiele, R.L. (1973) Effect of continuous infusions of synthetic LH releasing hormone (LRF) in rhesus monkeys. *Endocrinology*, 93: A-21.

Jaffe, R.B. and Keys, W.R., Jr. (1974) Estradiol augmentation of pituitary responsiveness to gonadotropin-releasing hormone in women. *J. clin. Endocr.*, 39: 850.

Judd, H.L., Parker, D.C., Siler, T.M. and Yen, S.S.C. (1974) The nocturnal rise of plasma testosterone in pubertal boys. *J. clin. Endocr.*, 38: 710.

Klaiber, E.L., Kobayashi, Y., Broverman, D.M. and Hall, F. (1971) Plasma monoamine oxidase activity in regularly menstruating women and in amenorrheic women receiving cyclic treatment with estrogens and progestin. *J. clin. Endocr.*, 33: 630.

Knobil, E. (1974) On the control of the gonadotropin secretion in the rhesus monkey. *Recent Progr. Hormone Res.*, 30: 1.

Kobayashi, T., Kobayashi, T., Kato, J. and Menaguchi, H. (1964) Fluctuations and monoamine oxidase activity in the hypothalamus of rat during the estrous cycle and after castration. *Endocr. jap.*, 11: 283.

Kopaniky, D.R. and Gann, D.S. (1975) Anterior pituitary vasodilation after hemorrhage in dogs. *Endocrinology*, 97: 630.

Krey, L.C. and Everett, J.W. (1973) Multiple ovarian responses to single estrogen injections early in rat estrous cycles: impaired growth, luteotropic stimulation and advanced ovulation. *Endocrinology*, 93: 377.

Lasley, B.L., Wang, C.F. and Yen, S.S.C. (1975a) Assessments of the functional capacity of the gonadotrophs in men; effects of estrogen and an anti-estrogen. *J. clin. Endocr.*, in press.

Lasley, B.L., Wang, C.F. and Yen, S.S.C. (1975b) The effects of estrogen and progesterone on the functional capacity of the gonadotrophs. *J. clin. Endocr.*, 41: 820.

Leblanc, H., Abu-Fadil, S. and Yen, S.S.C. (1975) Effects of dopamine infusion on pituitary hormone secretion in humans. *J. clin. Endocr.*, in press.

Malacara, J.M., Seyler, L.E., Jr. and Reichlin, S. (1972) Luteinizing hormone releasing factor activity in peripheral blood from women during the mid-cycle luteinizing hormone ovulatory surge. *J. clin. Endocr.*, 34: 271.

McCormack, J.T. and Knobil, E. (1975) The suppression by antiserum to LH-RH, of tonic LH and FSH secretion in the rhesus monkey. *Endocrinology*, 96: 108A.

Midgley, A.R., Jr. and Jaffe, R.B. (1968) Regulation of human gonadotropins. IV. Correlation of serum concentrations of follicle stimulating and luteinizing hormones during the menstrual cycle. *J. clin. Endocr.*, 28: 1699.

Naftolin, F., Yen, S.S.C. and Tsai, C.C. (1972) Rapid cycling of plasma gonadotropins in normal men as demonstrated by frequent sampling. *Nature New Biol.*, 236: 92.

Osland, R.B., Gallo, R.V. and Williams, J.A. (1975) *In vitro* release of luteinizing hormone from anterior pituitary fragments superfused with constant or pulsatile amounts of luteinizing hormone-releasing factor. *Endocrinology*, 96: 1210.

Parker, D.C., Judd, H.L., Rossman, L.G. and Yen, S.S.C. (1975) Pubertal sleep-wake patterns of episodic LH, FSH and testosterone release in twin boys. *J. clin. Endocr.*, 40: 1099.

Reiter, E.O., Duckett, G.E. and Root, A.W. (1975) Effect of constant infusion of luteinizing hormone-releasing hormone (LH-RH) upon gonadotropin secretion in children. *Pediat. Res.*, 9: 294 (A-223).

Ross, G.T., Cargille, C.M., Lipsett, M.B., Rayford, P.L., Marshall, J.R., Strott, C.A. and Rodbard, D. (1970) Pituitary and gonadal hormones in women during spontaneous and induced ovulatory cycles. *Recent Progr. Hormone Res.*, 26: 1.

Sachar, E.J. (1976) The use of neuroendocrine techniques in psychopharmacological research. In *Hormones, Behavior and Psychopathology*, E.J. Sachar (Ed.), Raven Press, New York, in press.

Van den Berg, G., DeVane, G. and Yen, S.S.C. (1974) Effects of exogenous estrogen and progestin on pituitary responsiveness to synthetic luteinizing hormone-releasing factor. *J. clin. Invest.*, 53: 1750.

Van de Wiele, R.L., Bogumil, J., Dyrenfurth, I., Ferin, M., Jewelewicz, R., Warren, M., Rixkallah, T. and Mikhail, G. (1970) Mechanisms regulating the menstrual cycle in women. *Recent Progr. Hormone Res.*, 26: 63.

Wang, C.F. and Yen, S.S.C. (1975) Direct evidence of estrogen modulation of pituitary sensitivity to luteinizing hormone releasing factor during the menstrual cycle. *J. clin. Invest*, 55: 201.

Wang, C.F., Lasley, B.L. and Yen, S.S.C. (1975) Gonadotropin secretion in response to low and high doses of LRF in normal and hypogonadal women (functional disparity of the gonadotrophs). *J. clin. Endocr.*, in press.

Yen, S.S.C. (1975) Gonadotropin-releasing hormone. In *Annual Review of Medicine: Selected Topics in the Clinical Sciences, Vol. 26*, W.P. Creger et al. (Eds.), Annual Reviews Inc., Palo Alto, Calif., pp. 403.

Yen, S.S.C., Van den Berg, G., Tsai, C.C. and Siler, T. (1974a) Causal relationship between the hormonal variables in the menstrual cycle. In *Biorhythms and Human Reproduction*, M. Ferin et al. (Eds.), Wiley, New York, pp. 219.

Yen, S.S.C., Van den Berg, G. and Siler, T.M. (1974b) Modulation of pituitary responsiveness to LRF by estrogen. *J. clin. Endocr.*, 39: 170.

Yen, S.S.C., Lasley, B.L., Wang, C.F., Leblanc, H. and Siler, T.M. (1975a) The operating characteristics of the hypothalamic-pituitary system during the menstrual cycle and observations of biological action of somatostatin. *Recent Progr. Hormone Res.*, 31: 312.

Yen, S.S.C., Rebar, R.W. and Quesenberry, W. (1975b) Pseudocyesis: neuroendocrine assessments. *J. clin. Endocr.*, in press.

Chapter 25

Neuroendocrine mechanisms in sexual behavior

CARLOS BEYER

Universidad Autonoma Metropolitana, Unidad Iztapalapa, Mexico, D.F. (Mexico)

I. INTRODUCTION

Psychologists distinguish two components in sexual behavior: the motivational or appetitive and the consummatory (Craig, 1918). In most species, the internal drive which stimulates an animal to search for a sexual partner depends on the presence of gonadal hormones. Moreover, hormones increase the probability for displaying sexual consummatory behavior by "sensitizing" some neural circuits to respond to the appropriate stimulus with copulation. Beach (1967), in a review on the reflexive mechanisms involved in copulatory behavior, proposed that sexual intercourse in the male depends on one mechanism for sexual arousal (AM) and another for the execution of the copulatory pattern (CM), the activity of the latter mechanism being under the control of the former. Recent evidence, including data presented in this symposium, indicates that another modulatory mechanism (MM) is involved in the control of sexual behavior in both sexes. These three mechanisms and its functional interrelationships are schematized in Fig. 1.

II. THE COPULATORY MECHANISM (CM)

CM comprises neural elements, i.e., receptors, interneurons and motoneurons, whose activation results in the movements and postural adjustments involved in copulation. There is evidence that some of the response units involved in copulation are mediated by reflex arcs located in the spinal cord, since males with total spinal cord transection can respond to adequate stimulation with erection of the penis and seminal emission, and spinal females can display some of the motor components of the lordosis response (for review see Beach, 1967). Yet, copulation is a complex, sequential motor pattern which requires delicate motor adjustments to the performance of the sexual pattern, and therefore, the participation of supraspinal motor mechanisms like the pyramidal and extrapyramidal systems. Moreover, the successful performance of a complex motor response always involves the suppression of movements which are inefficient or inadequate, and the CM must include such an inhibitory component.

In contrast to the stereotyped consummatory response (copulation), the appetitive component of sexual behavior is variable, unspecific and often it is only associated with an increase in the exploratory activities of the animal (see Meyerson and Lindström, 1973;

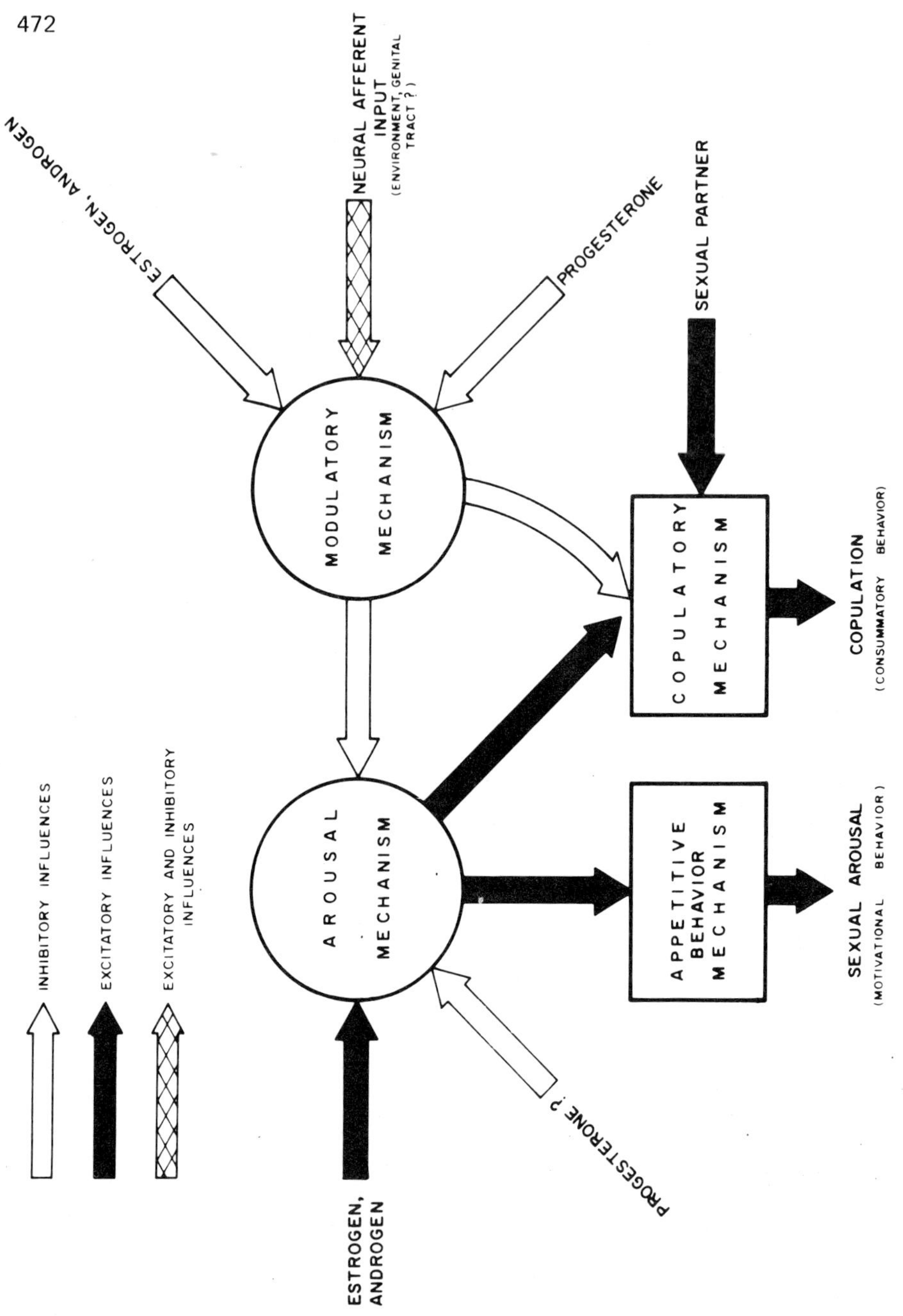

Fig. 1. Model for the possible mechanism of action of hormones and environmental factors on male and female sexual behavior. It is proposed that sexual behavior results from the interaction of 3 mechanisms: (1) a steroid-dependent arousal mechanism; (2) a modulatory mechanism exerting basically inhibitory influences through monoaminergic transmitters; and (3) a neuromotor mechanism, comprising the neuromotor elements involved in either appetitive behavior or in copulation. As shown in this figure, facilitatory effects of estrogen can result from a stimulatory effect on the arousal mechanism by interacting with steroid-dependent neurons, or by decreasing the tone of the modulatory mechanism. Environmental or internal factors resulting in inhibition of sexual behavior probably act by stimulating the activity of the basically inhibitory modulatory mechanism.

Meyerson, 1975). It appears that the neural substrate mediating such motor changes is not specific for sexual behavior, but probably involves the brain stem noradrenergic neurons proposed to mediate the activation of a broad sequence of exploratory movements (Margules and Margules, 1973).

III. THE AROUSAL MECHANISM (AM)

AM includes neurons whose activation enhances both sexual drive and facilitates the display of copulation. This mechanism contains steroid-dependent neurons, and interneurons linking this mechanism with the CM. Knowledge on the location of the neurons constituting part of the AM has been acquired through lesioning experiments, stereotaxic implantation of gonadal steroids in various regions of the brain, and studies of selective uptake and accumulation of radioactive steroids by the brain. All these experimental approaches coincide to indicate that the hypothalamus and preoptic area are regions indispensable for the display of copulatory behavior in male and female mammals (for review see Lisk, 1973), and therefore that these regions are part of the AM. Thus, lesions in the preoptic area suppress male sexual behavior in several species, without interfering with motor behavior, and gonadal steroid implantation in this same region restores copulatory behavior in castrated males. Furthermore, electrical stimulation of the medial preoptic area in male rats facilitates ejaculation (Malsbury, 1971) or induces highly exaggerated stimulation-bound sexual behavior (Merari and Ginton, 1975). Similar studies have also delineated certain areas in the hypothalamus that are essential for the activation of lordosis behavior in female animals (Lisk, 1973). These results correlate well with autoradiographic studies (see Pfaff, 1968; Tuohimaa, 1971; Pfaff and Keiner, 1973; McEwen et al., 1974), demonstrating that preoptic and hypothalamic neurons incorporate and retain gonadal steroids (estrogen and androgen).

IV. MODULATORY MECHANISM (MM)

In some species of seasonal breeders, periods of intense sexual activity are correlated with an actively secreting gonad, and periods of sexual quiescence with an atropic gonad (Asdell, 1946). In this case, sexual inactivity is probably due to the lack of gonadal hormones. However, in most species rapid fluctuations in sexual drive and copulatory activity cannot be explained by concomitant variations in the secretion of gonadal hormones, but to the intervention of active inhibitory neural mechanisms. Moreover, sexual behavior is not only modulated by hormones, but also by afferent impulses caused by environmental changes. In some cases, copulation in certain conditions might not only be inappropriate, but risky, and under these circumstances in spite of the existence of all prerequisites for mating, i.e., presence of an adequate sexual partner and high sexual motivation, copulation must be inhibited. Several lines of research point to the existence of an MM for sexual behavior. This mechanism comprises neurons having neural connections capable of deactivating sexual behavior, and sensory pathways transmitting information to these neurons. Evidence for the existence of brain regions exerting inhibitory effects on sexual behavior comes mainly from lesion studies. Destruction of the septum, habenula, preoptic area, interpeduncular nucleus, amygdala and neocortex has been reported to enhance sexual activity in both female and male animals (see Beach, 1967; Komisaruk, 1974). Conversely, in some of these brain areas hyperactivity by electrical stimulation (Zasorin et al., 1975) or by an irritative process (Vigouroux and Naquet, 1961) might result in a marked depression of sexual behavior.

It appears that besides exerting a tonic inhibitory effect on sexual behavior, some of these structures participate in the phasic inhibitory effect on sexual drive observed after ejaculation in most mammalian males, including man. This phenomenon, i.e., the postejaculatory refractory period, is an active process mediated by nervous connections, since lesions in the mesodiencephalic area shorten considerably the postejaculatory refractory period in male rats (Heimer and Larsson 1964; Barfield et al., 1975).

There is some evidence that some of the monoaminergic neurons and pathways, described by Fuxe and his colleagues in this symposium (Fuxe, 1976), are part of the neural substrate for the modulation of sexual behavior in mammals (Ahlenius et al., 1975). In spite of some contradictory results, many studies indicate that pharmacological manipulations, decreasing serotonin and probably dopamine concentration, result in an enhancement of sexual activity. Conversely, increasing the activity of these monoaminergic pathways usually results in an inhibition of sexual behavior (Dewsbury, 1975; Everitt et al., 1975b; Gessa and Tagliamonte, 1975; Meyerson, 1975).

Monoaminergic neurons might modulate sexual activity by inhibiting the firing of neurons either located in the AM or in the CM (see Fig. 1). Iontophoretic administration of serotonin and dopamine in the neighborhood of spinal or hypothalamic neurons usually inhibits their firing rate (Bloom 1969; Bloom et al., 1973). An alternative mode of action of the MM on the neurons of the AM would be the transsynaptic control of some biosynthetic processes related to the excitability of these neurons. For example, impulses carried on by serotonergic fibers could alter some of the mechanisms involved in the uptake or intracellular transport of the steroid to its active site of action in the nucleus by influencing the rate of synthesis of proteins involved in this process. The primary inhibitory effect of the monoaminergic neurons appears to be exerted through its ascending connections with the diencephalic AM, since monoamines not only inhibit copulatory behavior, but also sexual motivation (Meyerson, 1964). Moreover, as previously mentioned, lesions in the mesodiencephalic area disconnecting some of the monoaminergic projections to the diencephalon reduce the refractory period following ejaculation (Heimer and Larsson, 1964; Barfield et al., 1975).

As previously mentioned, the activity of the MM is partially regulated by stimulation originating in the periphery. Thus, the abrupt inhibition of sexual motivation following ejaculation cannot be explained by hormonal factors, but it must be the result of afferent signals originating during the act of ejaculation. Moreover, diurnal changes in sexual behavior occur in some species in relation to changes in illumination. In this context, the recent observation by Mosko and Jacobs (1974), that the discharge of raphe serotonergic neurons in the rat is slower during the dark phase than during the light phase of the cycle, is a finding of interest since sexual activity in the rat is lower during the light phase.

V. NATURE OF THE HORMONAL STIMULUS

Although there is evidence that sexual behavior can be manifested in some species, and by some individuals in the total absence of gonadal hormones (see Beyer et al., 1969a), there is no doubt that gonadal steroids are the most important factors in regulating sexual behavior. Behavioral and biochemical studies permit us to make some inferences on the possible modes of action of gonadal steroids (androgen and estrogen) on neurons. Sexual behavior is facilitated in most species by androgen or estrogen through a rather slow process. Thus, restoration of lordosis behavior in castrated females usually takes more than 24 hr, even when large amounts of estrogen are given (Michael and Scott, 1964; McDonald et al., 1970b; Beyer et al., 1971), and castrated males copulate in response to androgen administration with even more prolonged latencies (Davidson, 1972). These prolonged latencies suggest that the neurons responding to estrogen or androgen do not act like chemoreceptors, i.e., transducing chemical energy into electrical changes, but that they rather act as steroid target cells by responding to a chemical signal with biochemical events like protein synthesis. This idea is supported by the finding that neurons from the AM bind and retain estrogen and androgen as other target cells (Pfaff and Keiner, 1973; McEwen et al., 1974). Therefore, it appears likely that the delay in hormone action involves intracellular processes such as transport of the steroid to the nucleus, interaction with the genome, synthesis of proteins (enzymes, steroidophilic molecules, postsynaptic receptors, etc.) and transport of these molecules to their sites of action. These "slow" biochemical events eventually result in changes in the excitability of the neurons of the AM, changes that can explain the alterations in both spontaneous and induced neuronal firing in hypothalamic and preoptic areas reported after the administration of estrogen or androgen (for review see Beyer and Sawyer, 1969). Short-term alterations in neuronal firing following estrogen or androgen administration may not be relevant for the regulation of sexual behavior, though it is also theoretically possible that part of the chemical message encoded within the steroid is initially translated by the neuron into electrical events which secondarily would induce biosynthetic changes.

It appears also probable that sex hormones besides acting on the AM can also influence sexual behavior by depressing or increasing the activity of the monoaminergic component of the MM, or other structures influencing the lower, reflexive mechanisms directly involved in copulation. There is evidence that gonadal steroids (estrogen, androgen, progesterone) can affect the uptake (Wirz-Justice et al., 1974), synthesis or turnover rates of some of the monoamines presumably related to the control of sexual behavior and gonadotropin secretion (Everitt et al., 1975a, b; Gradwell et al., 1975; Luine et al., 1975; Brownstein, 1976). The cellular mechanism through which steroids influence the biochemical activity of monoamine neurons is not known with certainty, since serotonergic neurons apparently do not bind gonadal steroids (Pfaff and Keiner, 1973).

A third possibility through which steroids might influence sexual behavior is by acting directly on the neuronal membrane. Thus, it appears possible that progesterone or some of its metabolites exerts its short-latency facilitatory effects on sexual behavior by acting directly on the neuronal discharge of the monoaminergic inhibitory system. Recently Kow et al. (1974) have proposed that progesterone facilitates sexual behavior by depressing the activity of a serotonergic neural loop, interconnecting the septum and the raphe nuclei and having an inhibitory effect on the lower neural substrate for lordosis behavior. An alternative possibility would be that progesterone would release the AM by inhibiting the neuronal activity of the monoaminergic component of the MM. At any rate, it appears that in this case progesterone does not act at a nuclear level, but directly inhibits neuronal discharge by acting at the membrane. In this connection it is interesting to mention that administration of extremely small amounts of some brain metabolites of progesterone, like pregnanolone and pregnanedione, induces with very short latencies, marked depressory effects on the firing of brain stem and diencephalic neurons (Kubli et al., 1976). On the other hand, the inhibitory effect on sexual behavior of progesterone described in some species like the rabbit (Beyer et al., 1969b) and the guinea pig (Zucker, 1966) could be due to a depression of neurons of the AM or the CM. An alternative possibility would be that progesterone, by increasing serotonin turnover, indirectly inhibited the activity of the AM or CM (Everitt and Herbert, 1971; Gradwell et al., 1975).

There is evidence that gonadal hormones can also directly influence sexual responses integrated at spinal levels (treading, tail deviation, etc.) at least in some species (see Hart, 1968, 1970, 1971). Moreover, recent studies (Komisaruk et al., 1972; Kow and Pfaff, 1973; Komisaruk, 1974) have shown that estrogen can increase the genital sensory field of ovariectomized rats treated with estradiol benzoate. The mechanism through which estrogen elicits this most interesting effect has not been well established, but it suggests that gonadal steroids might modulate sexual responses by acting, through a variety of mechanisms.

Hormones appear to influence female or male sexual behavior not only through different hormones or combinations of hormones, but also through the activation of different cellular mechanisms. Therefore, it will be convenient to deal separately with the neuroendocrine control of female and male sexual behavior.

VI. FEMALE SEXUAL BEHAVIOR

The hormonal basis of female sexual behavior appears to be relatively well established, though few systematic comparative studies on the efficiency of diverse estrogens have been made (Beyer et al., 1971; Feder and Silver, 1974). These studies have shown that estradiol is the most potent estrogen to stimulate lordosis behavior, followed by estrone. Estriol had also a stimulatory effect on lordosis behavior in ovariectomized rats (Beyer et al., 1971) and guinea pigs (Feder and Silver, 1974), though no clear correlation between dose and effect can be established, probably due to its peculiar pattern of nuclear binding. Interestingly, sulfoconjugates can also elicit sexual behavior in the rat and guinea pig, though in the ovariectomized rabbit estrone sulfate is much less effective than free estrone (Beyer and McDonald, 1973). Estrophilic molecules in brain tissue have been well characterized (see McEwen, 1976), and there is good evidence to believe that they are involved in the expression of female sexual behavior. Thus, antiestrogens (MER-25, CI-628, cis-clomifene), which compete for estrogen binding in these areas, interfere with estradiol-induced sexual behavior (Meyerson and Lindström, 1968; Komisaruk and Beyer,

1972; Ross et al., 1973; Sodersten and Larsson, 1974). It appears likely that estrone and estriol exert their stimulatory effect on female sexual behavior by binding to the same system of "receptors" than estradiol, since reduction of estrone to estradiol appears to be minimal, and estriol cannot be converted to estradiol. Recently, Vreeburg et al. (1975) have demonstrated that both estrone and estriol show good affinity for a brain protein which strongly binds estradiol.

It is well known that some androgens can induce lordosis behavior in females (see Young, 1961), though the doses required are much larger than those of estrogen. We have studied the effect of androgens with different stereochemical characteristics on the sexual behavior of ovariectomized rabbits (Beyer et al., 1970) and rats (Beyer and Komisaruk, 1971). Aromatizable androgens possessing the Δ4-3-keto- or the Δ5 structure (testosterone, androstenedione, 19-hydroxy-androstenedione, dehydroepiandrosterone) were effective to stimulate lordosis behavior, while non-aromatizable androgens like ring A reduced androgens (dihydrotestosterone, 5α-androstanediol, 5α-androstanedione) or 11β-hydroxy-androgens were not. From these studies it was suggested that aromatization, i.e., conversion of androgen to estrogen, was essential for the stimulation of lordosis behavior in subprimate females by androgens. This hypothesis received support from the observation that antiestrogens (MER-25, CI-628) block testosterone-induced lordosis behavior in ovariectomized rabbits (Beyer and Vidal, 1971) and rats (Whalen et al., 1972).

In summary, data on the neuroendocrine events related to female sexual behavior suggest that the following steps are involved in the production of lordosis in subprimate species: (a) estrogen secreted by the ovaries or resulting from peripheral aromatization binds to a well characterized system of estrophilic molecules (cytosol receptor) present in hypothalamic neurons; (b) estrogen is transported to the nucleus where the steroid binds to another receptor, and estrogen or an estrogen receptor complex activate the genome to initiate events resulting in the production of molecules related either to the capacity of the neuron to receive information or to transmit information. The necessity of protein synthesis for the induction of female sexual behavior in response to hormones is supported by the inhibition of lordosis behavior, obtained with local implantation or infusion of chemicals interfering with protein synthesis (see Whalen et al., 1974; Quadagno and Ho, 1975).

The hormonal stimulus for female sexual behavior in primates is probably different from that of subprimate mammals (see Michael, 1971). Thus, ovariectomy in women and monkeys does not result in a sharp decline in sexual activity as in subprimate species (Money, 1961; Everitt et al., 1975b). This observation, however, cannot be considered as conclusive evidence against the participation of gonadal steroids in the regulation of sexual behavior in primates, since adrenal steroids are an important source of both androgen and indirectly estrogen in women (McDonald et al., 1967). Moreover, adrenalectomy suppresses sexual behavior in female rhesus monkeys (Everitt et al., 1975b). There is evidence that systemic administration of androgen increases sexual behavior in female primates, but estrogen does not (Money, 1961; Everitt et al., 1975b). Therefore, it is possible that primates have developed a different neuroendocrine mechanism for female sexual behavior regulation, using androgen rather than estrogen as the "erotic" stimulus.

It is also possible that androgen is aromatized in brain tissue as has been demonstrated by Flores et al. (1973) in the primate brain, and that estrogen resulting from intracellular aromatization would have access to cellular compartments that circulating estrogen had not. In this connection, it is particularly interesting that the distribution of nuclear estrogen, resulting from testosterone aromatization, varies somewhat from the distribution obtained with administration of estrogen (McEwen, 1976).

VII. MALE SEXUAL BEHAVIOR

The endocrine control of male sexual behavior appears to be more complicated than it was previously believed. It is well known that testosterone initiates or restores sexual behavior in castrated animals of all species so far studied. Moreover, other androgens such as androstenedione and androstenediol can initiate or maintain copulatory behavior in male rats (Whalen and Luttge, 1971; Moralí et al., 1974).

On the other hand, the effect of non-aromatizable 5α reduced androgens on sexual behavior appears to depend on the species studied. Thus, ring A reduced androgens are ineffective to stimulate or maintain sexual behavior in castrated rats (Beyer et al., 1973; Parrot, 1975), hamsters (Clemens, 1974) and in some strains of mice (Luttge and Hall, 1973). Dihydrotestosterone, a potent non-aromatizable androgen, initiates sexual activity in some castrated rabbits (Beyer and Rivaud, 1973; Agmo and Sodersten, 1976), though its potency is significantly less than that of testosterone. A similar situation appears to occur in the castrated male rhesus monkey in which dihydrotestosterone propionate can restore full copulatory behavior, though it takes twice as long as testosterone to achieve this result (Phoenix, 1974). Finally, dihydrotestosterone is as effective as testosterone to initiate sexual behavior in prepuberally castrated pigs (Alsum and Goy, 1974).

The finding that potent non-aromatizable androgens do not restore or maintain normal copulatory behavior in the rat led some investigators (McDonald et al., 1970a; Beyer et al., 1973) to propose that estrogen resulting from androgen aromatization was responsible for stimulating male sexual behavior. This idea received support from the following findings: (a) estrogen per se can restore copulatory behavior in castrated male rats (Sodersten, 1973) and hamsters (Clemens, 1974); (b) direct implantation of estrogen in the preoptic area is more effective than testosterone to restore copulatory behavior in castrated male rats (Christensen and Clemens, 1974); (c) some antiestrogens, CI-628 (Luttge, 1975a) and ICI-46474 (Beyer et al., 1976b), suppress or depress the copulatory behavior that normally follows testosterone administration to castrated male rats; (d) aromatase inhibitors either systemically injected (Beyer et al., 1976b) or directly infused into the preoptic area (Christensen and Clemens, 1975) inhibit the stimulatory action of testosterone on the sexual behavior of castrated male rats; (e) limbic tissue can aromatize androgen to estrogen (Naftolin et al., 1971, 1976); and (f) estradiol is the main nuclear metabolite of testosterone in limbic areas of the rat brain (Lieberburg and McEwen, 1975).

However, some experimental data apparently do not fit in the aromatization hypothesis. Thus, non-aromatizable androgens can restore normal copulatory behavior in castrated males of several species, e.g., rabbit (Beyer and Rivaud, 1973), guinea pig (Alsum and Goy, 1974) and rhesus monkey (Phoenix, 1974) and estrogen administration even in large dosages usually fails to initiate or maintain normal sexual behavior in castrated males of several species (see Young, 1961). It must be stated, however, that in those experiments in which dihydrotestosterone has been found to stimulate sexual behavior, it is possible that estrogens were produced from aromatization of adrenal androgens (McDonald et al., 1971). On the other hand, the relative failure of exogenous estrogen to stimulate sexual behavior in castrated males can be due to the fact that estrogen passing from plasma to the brain does not have the same pattern of distribution in brain tissue than that resulting from androgen aromatization (McEwen, 1976).

It appears, however, that under normal conditions male sexual behavior results from the interaction of both estrogen and androgen. Thus, the effect of systemic injection of estrogen on male sexual behavior is enormously enhanced by the simultaneous administration of dihydrotestosterone in the castrated rat (Baum and Vreeburg, 1973; Larsson et al., 1973, 1976; Feder et al., 1974) and rabbit (Beyer et al., 1975), and estrogen enhances the stimulatory effect of testosterone on the sexual behavior of the pig (Joshi and Raeside, 1973) and man (Foss, 1939).

Moreover, a detailed analysis of the effects of estrogen on male sexual behavior suggests that these hormones are specifically related to behavioral responses associated to sexual arousal. Thus, Parrot (1975) has found that estradiol can maintain the duration of the postejaculatory refractory period within normal values in castrated rats, though it fails to prevent the gradual decrease in the proportion of copulating rats that normally follows castration. Similarly, implants of estradiol in the preoptic area of castrated male rats were significantly more effective than testosterone implants in stimulating mean number of mounts and intromissions, but not in restoring the percentage of tests during which animals presented ejaculatory responses (Christensen and Clemens, 1974). Moreover, Joshi and Raeside (1973) have found that the main behavioral effect induced by estrogen on testosterone-treated castrated boars is to significantly reduce the reaction time to mount a dummy in the collection area. This estrogen effect is clearly related to an increase in sexual arousal.

Several hypotheses can be proposed to explain the synergistic effect of estrogen and androgen on sexual behavior.

(a) Androgen maintains the morphological and physiological integrity of the sexual accessories (penis, seminal vesicles, etc.), while estrogen activates brain mechanisms. This hypothesis is based on the idea that afferent inflow from the androgen stimulated male genital area is important for male sexual behavior. Yet, this idea does not explain the finding that dihydrotestosterone and estradiol synergize to activate mounting behavior in female rats, even when their genital area was anesthetized (Baum et al., 1974).

(b) Estrogen acts on the brain mechanisms related to sexual arousal while androgen facilitates the responsivity of the spinal reflexes involved in copulation.

(c) Estrogen acts on the AM and androgen on the MM, disinhibiting either the AM or the CM. In this connection, it is important to recall that androgen can decrease the turnover of serotonin (Gradwell et al., 1975) and increase the level of monoamine oxidase (Luine et al., 1975), both effects probably resulting in a decrease in the tone of the inhibitory component of the MM. That a decrease in the activity of the serotonergic system of fibers can facilitate the response to low doses of estrogen by the castrated male rat has been recently demonstrated by Luttge (1975b).

(d) Androgen activates in the castrated animal the synthesis of molecules involved in either uptake, metabolism or transport of estrogen to its site of action. In this connection it has been reported by Larsson et al. (1975) that pretreatment of castrated male rats with dihydrotestosterone shortens the latency to the initiation of copulation induced by testosterone. Moreover, estradiol elicits intense sexual behavior with very short latencies in castrated rabbits pretreated with dihydrotestosterone (Beyer et al., 1976a).

(e) Androgen facilitates the intracellular transport of estrogen to its site of action at the nuclear level. In this connection it is of interest that Rochefort et al. (1972) found in an in vitro preparation of rat uteri that androgen can transfer the estrogen cytosol receptor to the nucleus without binding to this molecule, and suggested that androgens modify membrane permeability for the estrogen cytosol receptor. Although, this process has been observed in peripheral effector tissues, it is theoretically possible that it occurs at the neuronal level and may have some relevance for sexual behavior. From these highly speculative considerations it is obvious that much experimental work remains to be done to establish the mechanism through which androgen and estrogen facilitate sexual behavior.

REFERENCES

Agmo, A. and Sodersten, P. (1976) Sexual behavior in castrated rabbits treated with testosterone, oestradiol, dihydrotestosterone or oestradiol in combination with dihydrotestosterone. *J. Endocr.*, in press.

Ahlenius, S., Engel, J., Erikson, H., Modigh, K. and Sodersten, P. (1975) Involvement of monoamines in the mediation of sexual behavior. In *Sexual Behavior. Pharmacology and Biochemistry*, M. Sandler and G.L. Gessa (Eds.), Raven Press, New York, pp. 137–145.

Alsum, P. and Goy, R.W. (1974) Actions of esters of testosterone, dihydrotestosterone or estradiol on sexual behavior in castrated male guinea pigs. *Hormone Behav.*, 5: 207–217.

Asdell, A.S. (1946) *Patterns of Mammalian Reproduction*, Constable and Co., London.

Barfield, R.J., Wilson, C. and McDonald, P.G. (1975) Sexual behavior: extreme reduction of postejaculatory refractory period by midbrain lesions in male rats. *Science*, 189: 147–149.

Baum, M.J. and Vreeburg, J.T.M. (1973) Copulation in castrated male rats following combined treatment with estradiol and dihydrotestosterone. *Science*, 182: 283–285.

Baum, M.J., Sodersten, P. and Vreeburg, J.T.M. (1974) Mounting and receptive behavior in the ovariectomized female rat: influence of estradiol, dihydrotestosterone, and genital anesthetization. *Hormone Behav.*, 5: 175–190.

Beach, F.A. (1967) Cerebral and hormonal control of reflexive mechanisms involved in copulatory behavior. *Physiol. Rev.*, 47: 289–316.

Beyer, C. and Komisaruk, B.R. (1971) Effects of diverse androgens on estrous behavior, lordosis reflex and genital tract morphology in the rat. *Hormone Behav.*, 2: 217–225.

Beyer, C. and McDonald, P.G. (1973) Hormonal control of sexual behaviour in the female rabbit. In *Advances in Reproductive Physiology, Vol. 6,* G. Bishop (Ed.), Logos Press, London, pp. 185–219.

Beyer, C. and Rivaud, N. (1973) Differential effect of testosterone and dihydrotestosterone on the sexual behavior of prepuberally castrated male rabbits. *Hormone Behav.*, 4: 175–180.

Beyer, C. and Sawyer, C.H. (1969) Hypothalamic unit activity related to control of the pituitary gland. In *Frontiers in Neuroendocrinology*, W.F. Ganong and L. Martini (Eds.), Oxford Univ. Press, New York, pp. 255—287.

Beyer, C. and Vidal, N. (1971) Inhibitory action of MER-25 on androgen induced oestrous behavior in the ovariectomized rabbit. *J. Endocr.*, 51: 401—402.

Beyer, C., Cruz, M.L. and Rivaud, N. (1969a) Persistence of sexual behavior in ovariectomized-adrenalectomized rabbits treated with cortisol. *Endocrinology*, 85: 790—793.

Beyer, C., Vidal, N. and McDonald, P.G. (1969b) Interaction of gonadal steroids and their effect on sexual behavior in the rabbit. *J. Endocr.*, 45: 407—413.

Beyer, C., Vidal, N. and Mijares, A (1970) Probable role of aromatization in the induction of estrous behavior by androgens in the ovariectomized rabbit. *Endocrinology*, 87: 1386—1389.

Beyer, C., Moralí, G. and Vargas, R. (1971) Effect of diverse estrogens on estrous behavior and genital tract development in ovariectomized rats. *Hormone Behav.*, 2: 273—277.

Beyer, C., Larsson, K., Pérez-Palacios, G. and Moralí, G. (1973) Androgen structure and male sexual behavior in the castrated rat. *Hormone Behav.*, 4: 99—108.

Beyer, C., de la Torre, L., Larsson, K. and Pérez-Palacios, G. (1975) Synergistic actions of estrogen and androgen on the sexual behavior of the castrated male rabbit. *Hormone Behav.*, 6: 301—306.

Beyer, C., Cruz, M.L., Olivera, J. and Contreras, J.L. (1976a) Estrogen-induced sexual behavior in the castrated-adrenalectomized male rabbit pretreated with dihydrotestosterone. In preparation.

Beyer, C., Moralí, G., Naftolin, F., Larsson, K. and Pérez-Palacios, G. (1976b) Effect of some antiestrogens and aromatase inhibitors on androgen induced sexual behavior in castrated male rats. *Hormone Behav.*, in press.

Bloom, F.E. (1969) Serotonin neurons: localization and possible physiological role. In *Advances in Biochemical Psychopharmacology, Vol. 1,* E. Costa and P. Greengard (Eds.), Raven Press, New York, pp. 28—47.

Bloom, F.E., Hoffer, B.J., Nelson, C., Sheu, Y.S. and Siggins, G.R. (1973) The physiology and pharmacology of serotonin mediated synapses. In *Serotonin and Behavior*, J. Barchas and E. Usdin (Eds.), Academic Press, New York, pp. 249—261.

Brownstein, M.J. (1976) Hormonal regulation of the synthesis and metabolism of neurotransmitters. In *Subcellular Mechanisms in Reproductive Neuroendocrinology*, F. Naftolin, K.J. Ryan and I.J. Davies (Eds.), Elsevier, Amsterdam, Ch. 10.

Christensen, L.W. and Clemens, L.G. (1974) Intrahypothalamic implants of testosterone or estradiol and resumption of masculine sexual behavior in long term castrated male rats. *Endocrinology*, 95: 984—990.

Christensen, L.W. and Clemens, L.G. (1975) Personal communication.

Clemens, L.G. (1974) Neurohormonal control of male sexual behavior. In *Advances in Behavioral Biology, Vol. 11, Reproductive Behavior,* W. Montagna and W. Sadler (Eds.), Plenum Press, New York, pp. 23—53.

Craig, W. (1918) Appetites and aversions as constituents of instincts. *Biol. Bull.*, 34: 91—107.

Davidson, J.M. (1972) Hormones and reproductive behavior. In *Hormones and Behavior*, S. Levine (Ed.), Academic Press, New York, pp. 63—103.

Dewsbury, D.A. (1975) The normal heterosexual pattern of copulatory behavior in male rats: effects of drugs that alter brain monoamine levels. In *Sexual Behavior. Pharmacology and Biochemistry*, M. Sandler and G.L. Gessa (Eds.), Raven Press, New York, pp. 169—179.

Everitt, B.J. and Herbert, J. (1971) The effect of dexamethasone and androgens on sexual receptivity of female rhesus monkeys. *J. Endocr.*, 51: 575—588.

Everitt, B.J., Fuxe, K., Hökfelt, T. and Jonsson, G. (1975a) Studies on the role of monoamines in the hormonal regulation of sexual receptivity in the female rat. In *Sexual Behavior. Pharmacology and Biochemistry*, M. Sandler and G.L. Gessa (Eds.), Raven Press, New York, pp. 147—159.

Everitt, B.J., Gradwell, P.B. and Herbert, J. (1975b) Humoral and aminergic mechanisms regulating sexual receptivity in female rhesus monkeys. In *Sexual Behavior. Pharmacology and Biochemistry*, M. Sandler and G.L. Gessa (Eds.), Raven Press, New York, pp. 181—191.

Feder, H.H. and Silver, R. (1974) Activation of lordosis in ovariectomized guinea pigs by free and esterified forms of estrone, estradiol-17β and estriol. *Physiol. Behav.*, 13: 251–255.

Feder, H.H., Naftolin, F. and Ryan, K.J. (1974) Male and female sexual responses in male rats given estradiol benzoate and 5α-androstan-17β-ol-3-one propionate. *Endocrinology*, 94: 136–141.

Flores, F., Naftolin, F., Ryan, K.J. and White, R.J. (1973) Estrogen formation by the isolated perfused rhesus monkey brain. *Science*, 180: 1074–1075.

Foss, G.L. (1939) Clinical administration of androgens. *Lancet*, 236: 502–505.

Fuxe, K. (1976) This symposium.

Gessa, G.L. and Tagliamonte, A. (1975) Role of brain serotonin and dopamine in male sexual behavior. In *Sexual Behavior. Pharmacology and Biochemistry*, M. Sandler and G.L. Gessa (Eds.), Raven Press, New York, pp. 117–128.

Gradwell, P.B., Everitt, B.J. and Herbert, J. (1975) 5-Hydroxytryptamine in the central nervous system and sexual receptivity of female rhesus monkeys. *Brain Res.*, 88: 281–293.

Hart, B.L. (1968) Sexual reflexes and mating behavior in the male rat. *J. comp. physiol. Psychol.*, 65: 453–460.

Hart, B.L. (1970) Mating behavior in the female dog and the effects of estrogen on sexual reflexes. *Hormone Behav.*, 1: 93–104.

Hart, B.L. (1971) Facilitation by estrogen of sexual reflexes in female cats. *Physiol. Behav.*, 7: 675–678.

Heimer, L. and Larsson, K. (1964) Drastic changes in the mating behaviour of male rats following lesions in the junction of diencephalon and mesencephalon. *Experientia* (Basel), 20: 460–461.

Joshi, H.S. and Raeside, J.I. (1973) Synergistic effects of testosterone and oestrogens on accessory sex glands and sexual behaviour of the boar. *J. Reprod. Fertil.*, 33: 411–423.

Komisaruk, B.R. (1974) Neural and hormonal interactions in the reproductive behavior of female rats. In *Advances in Behavioral Biology, Vol. 11, Reproductive Behavior*, W. Montagna and W.A. Sadler (Eds.), Plenum Press, New York, pp. 97–129.

Komisaruk, B.R. and Beyer, C. (1972) Differential antagonism by MER-25, of behavioral and morphological effects of estradiol benzoate in rats. *Hormone Behav.*, 3: 63–70.

Komisaruk, B.R., Adler, N.T. and Hutchinson, J. (1972) Genital sensory field: enlargement by estrogen treatment in female rats. *Science*, 178: 1295–1298.

Kow, L.M. and Pfaff, D.W. (1973) Estrogen effect on pudendal nerve receptive field size in the female rat. *Anat. Rec.*, 175: 362–363.

Kow, L.M., Malsbury, C.W. and Pfaff, D.W. (1974) Effects of progesterone on female reproductive behavior in rats: possible modes of action and role in behavioral sex differences. In *Advances in Behavioral Biology, Vol. 11, Reproductive Behavior*, W. Montagna and W.A. Sadler (Eds.), Plenum Press, New York, pp. 179–210.

Kubli, C., Cervantes, M. and Beyer, C. (1976) Changes in multiunit activity and EEG induced by the administration of natural progestins to Flaxedil immobilized cats. *Brain Res.*, in press.

Larsson, K., Sodersten, P. and Beyer, C. (1973) Sexual behavior in male rats treated with estrogen in combination with dihydrotestosterone. *Hormone Behav.*, 4: 289–299.

Larsson, K., Pérez-Palacios, G., Moralí, G., and Beyer, C. (1975) Effects of dihydrotestosterone and estradiol benzoate pretreatment upon testosterone-induced sexual behavior in the castrated male rat. *Hormone Behav.*, 6: 1–8.

Larsson, K., Beyer, C., Moralí, G., Pérez-Palacios, G. and Sodersten, P. (1976) Effects of estrone, estradiol and estriol combined with dihydrotestosterone on mounting and lordosis behavior in castrated male rats. *Hormone Behav.*, in press.

Lieberburg, I. and McEwen, B.S. (1975) Estradiol-17β: a metabolite of testosterone recovered in cell nuclei from limbic areas of adult male rat brains. *Brain Res.*, 91: 171–174.

Lisk, R.D. (1973) Hormonal regulation of sexual behavior in polyestrous mammals common to the laboratory. In *Handbook of Physiology, Section 7, Endocrinology. Vol. 11, Female Reproductive System, Part I*, R.O. Greep (Ed.), American Physiological Society, Washington, D.C., pp. 223–260.

Luine, V.N., Khylchevskaya, R.I. and McEwen, B.S. (1975) Effect of gonadal steroids on activities of monoamine oxidase and choline acetylase in rat brain. *Brain Res.*, 86: 293–306.

Luttge, W.G. (1975a) Effects of anti-estrogens on testosterone stimulated male sexual behavior and peripheral target tissues in the castrate male rat. *Physiol. Behav.,* 14: 839–846.

Luttge, W.G. (1975b) Stimulation of estrogen induced copulatory behavior in castrate male rats with the serotonin biosynthesis inhibitor p-chlorophenylalanine. *Behav. Biol.*, 14: 373–378.

Luttge, W.G. and Hall, N.R. (1973) Differential effectiveness of testosterone and its metabolites in the induction of male sexual behavior in two strains of albino mice. *Hormone Behav.*, 4: 31–43.

Malsbury, C.W. (1971) Facilitation of male rat copulatory behavior by electrical stimulation of the medial preoptic area. *Physiol. Behav.*, 7: 797–805.

Margules, D.L. and Margules, A.S. (1973) The development of operant responses by nonadrenergic activation and cholinergic suppression of movements. In *Efferent Organization and the Integration of Behavior*, J.D. Maser (Ed.), Academic Press, New York, pp. 203–228.

McDonald, P.C., Rombaut, R.P. and Siiteri, P.K. (1967) Plasma precursors of estrogen. I. Extent of conversion of plasma Δ4-androstenedione to estrone in normal males and non pregnant normal, castrate and adrenalectomized females. *J. clin. Endocr.*, 27: 1103–1111.

McDonald, P., Beyer, C., Newton, F., Brien, B., Baker, R., Tan, H.S., Sampson, C., Kitching, P., Greenhill, R. and Pritchard, D. (1970a) Failure of 5α-dihydrotestosterone to initiate sexual behavior in the castrated male rat. *Nature (Lond.)*, 227: 964–965.

McDonald, P., Vidal, N. and Beyer, C. (1970b) Sexual behavior in the ovariectomized rabbit after treatment with different amounts of gonadal hormones. *Hormone Behav.*, 1: 161–172.

McDonald, P.C., Grodin, J.M. and Siiteri, P.K. (1971) Dynamics of estrogen secretion. In *Gonadal Steroid Secretion*, D.T. Baird and J.A. Strong (Eds.), Edinburgh Univ. Press, Edinburgh, pp. 158–174.

McEwen, B.S. (1976) Steroid receptors in neuroendocrine tissues: topography, subcellular distribution, and functional implications. In *Subcellular Mechanisms in Reproductive Neuroendocrinology*, F. Naftolin, K.J. Ryan and I.J. Davies (Eds.), Elsevier, Amsterdam, Ch. 14.

McEwen, B.S., Denef, C.J., Gerlach, J.L. and Plapinger, L. (1974) Chemical studies of the brain as a steroid hormone target tissue. In *Neurosciences, 3rd Study Program*, F.O. Schmitt and F.G. Worden (Eds.), MIT Press, Cambridge, Mass., pp. 549–620.

Merari, A. and Ginton, A. (1975) Characteristics of exaggerated sexual behavior induced by electrical stimulation of the medial preoptic area in male rats. *Brain Res.*, 86: 97–108.

Meyerson, B.J. (1964) Central nervous monoamines and oestrous behaviour in the spayed rat. *Acta physiol. scand.*, 63, Suppl. 241: 1–32.

Meyerson, B.J. (1975) Drugs and sexual motivation in the female rat. In *Sexual Behavior, Pharmacology and Biochemistry*, M. Sandler and G.L. Gessa (Eds.), Raven Press, New York, pp. 21–31.

Meyerson, B.J. and Lindström, L. (1968) Effect of an oestrogen antagonist ethamoxy-triphetol (MER-25) on oestrous behaviour in rats. *Acta endocr. (Kbh.)*, 59: 41–48.

Meyerson, B.J. and Lindström, L. (1973) Sexual motivation in the female rat. *Acta physiol. scand.*, Suppl. 389: 1–80.

Michael, R.P. (1971) Neuroendocrine factors regulating primate behavior. In *Frontiers in Neuroendocrinology*, L. Martini and W.F. Ganong (Eds.), Oxford University Press, London, pp. 359–398.

Michael, R.P. and Scott, P. (1964) The activation of sexual behaviour in cats by the subcutaneous administration of oestrogen. *J. Physiol. (Lond.)*, 171: 254–274.

Money, J. (1961) Components of eroticism in man. The hormones in relation to sexual morphology and sexual desire. *J. nerv. ment. Dis.*, 132: 239–248.

Moralí, G., Larsson, K., Pérez-Palacios, G. and Beyer, C. (1974) Testosterone, androstenedione and androstenediol: effects on the initiation of mating behavior of inexperienced castrated male rats. *Hormone Behav.*, 5: 103–110.

Mosko, S.S. and Jacobs, B.L. (1974) Midbrain raphe neurons: spontaneous activity and response to light. *Physiol. Behav.*, 13: 589–593.

Naftolin, F., Ryan, K.J. and Petro, Z. (1971) Aromatization of androstenedione by the diencephalon.

J. clin. Endocr., 33: 368—370.

Naftolin, F., Ryan, K.J. and Davies, I.J. (1976) Androgen aromatization by neuroendocrine tissues. In *Subcellular Mechanisms in Reproductive Neuroendocrinology*, F. Naftolin, K.J. Ryan and I.J. Davies (Eds.), Elsevier, Amsterdam, Ch. 17.

Parrot, R.F. (1975) Aromatizable and 5α-reduced androgens: differentiation between central and peripheral effects on male rat sexual behavior. *Hormone Behav.*, 6: 99—108.

Pfaff, D.W. (1968) Uptake of ^{3}H-estradiol by the female rat brain. An autoradiographic study. *Endocrinology*, 82: 1149—1155.

Pfaff, D. and Keiner, M. (1973) Atlas of estradiol-concentrating cells in the central nervous system of the female rat. *J. comp. Neurol.*, 151: 121—158.

Phoenix, C.H. (1974) The role of androgens in the sexual behavior of adult male rhesus monkeys. In *Advances in Behavioral Biology, Vol. 11, Reproductive Behavior*, Plenum Press, New York, pp. 249—258.

Quadagno, D.M. and Ho, G.K.W. (1975) The reversible inhibition of steroid-induced sexual behavior by intracranial cycloheximide. *Hormone Behav.*, 6: 19—26.

Rochefort, H., Lignon, F. and Capony, F. (1972) Formation of estrogen nuclear receptor in uterus: effect of androgens, estrone and nafoxidine. *Biochem. biophys. Res. Commun.*, 47: 662—670.

Ross, J.W., Paup, D.C., Brant-Zawadski, M., Marshall, J.R. and Gorski, R.A. (1973) Effects of cis- and trans-clomiphene in the induction of sexual behavior. *Endocrinology*, 93: 681—685.

Sodersten, P. (1973) Estrogen activated sexual behavior in male rats. *Hormone Behav.*, 4: 247—256.

Sodersten, P. and Larsson, K. (1974) Lordosis behavior in castrated male rats treated with estradiol benzoate or testosterone propionate in combination with an estrogen antagonist, MER-25, and in intact male rats. *Hormone Behav.*, 5: 13—18.

Tuohimaa, P. (1971) The radioautographic localization of exogenous tritiated dihydrotestosterone, testosterone and oestradiol in the target organs of female and male rats. In *Basic Actions of Sex Steroids on Target Organs*, P.O. Hubinot, F. Leroy and P. Galand (Eds.), Karger, Basel, pp. 208—214.

Vigouroux, R. et Naquet, R. (1961) Essai d'interprétation de la physiologie du rhinencéphale. In *Les Grandes Activités du Rhinencéphale. Vol. II. Physiologie et Pathologie du Rhinencéphale*, Masson, Paris, pp. 245—262.

Vreeburg, J.T.M., Schretlen, P.J.M. and Baum, M.J. (1975) Specific, high-affinity binding of 17β-estradiol in cytosols from several brain regions and pituitary of intact and castrated adult male rats. *Endocrinology*, 97: 969—977.

Whalen, R.E. and Luttge, W.G. (1971) Testosterone, androstenedione, and dihydrotestosterone: effects on mating behavior of male rats. *Hormone Behav.*, 2: 117—125.

Whalen, R.E., Battie, C. and Luttge, W.G. (1972) Anti-estrogen inhibition of androgen induced sexual receptivity in rats. *Behav. Biol.*, 7: 311—320.

Whalen, R.E., Gorzalka, B.B., De Bold, J.F., Quadagno, D.M., Ho, G.K. and Hough, J.C. (1974) Studies on the effects of intracerebral actinomycin D implants on estrogen-induced receptivity in rats. *Hormone Behav.*, 5: 337—343.

Wirz-Justice, A., Hackman, C. and Lichsteiner, M. (1974) The effect of oestradiol dipropionate and progesterone on monoamine uptake in the rat brain. *J. Neurochem.*, 22: 187—189.

Young, W.C. (1961) The hormones and mating behaviour. In *Sex and Internal Secretions*, W.C. Young (Ed.), Williams and Wilkins, Baltimore, Md., pp. 1173—1239.

Zasorin, N.L., Malsbury, C.W. and Pfaff, D.W. (1975) Suppression of lordosis in the hormone-primed female hamster by electrical stimulation of the septal area. *Physiol. Behav.*, 14: 595—599.

Zucker, I. (1966) Facilitatory and inhibitory effects of progesterone on sexual responses of spayed guinea pigs. *J. comp. physiol. Psychol.*, 62: 376—381.

Chapter 26

Control of gonadotropin secretion

K. BROWN-GRANT

Medical Research Council External Scientific Staff, A.R.C. Institute of Animal Physiology, Babraham, Cambridge CB2 4AT (Great Britain)

I. INTRODUCTION

Over the two preceding days, and during many excellent presentations, attention has been focussed on events taking place at the subcellular and molecular level and it is a truism that we shall not fully understand any biological system until we understand it at this level. At the same time, it is the implications and consequences of these findings and their integration into what we know or suspect about the functioning of the hypothalamus-pituitary-gonad system in the whole animal, in both health and disease, that is the basic interest of many of the participants at this meeting and also the justification for the expenditure of scarce research talent and, in these hard times, scarce research funds on work of this type. Research, like light or electron microscopy, seems to follow an invariable rule that the higher the degree of resolution desired, the greater the expense. In a meeting centred around the application of sophisticated and costly techniques to problems in reproductive neuroendocrinology, a discussion of some well documented animal model systems and the possibilities they offer for highly reproducible steroid input—gonadotropin output manipulations seemed to be of possible practical value and to offer a framework to which some of the new information forthcoming at the meeting could be related. Equally, some superficially attractive model systems where input—output relationships and their physiological basis are far less simple and far less clear cut than was once thought to be the case will be discussed.

In many instances some assurance of the physiological state of the animal or evidence of a biologically meaningful response is an essential part of a biochemical experiment. It is a depressing experience, fortunately less common these days, to come across a report of a painstaking, detailed and costly biochemical study the results of which could be very valuable and of great relevance to one's work, only to find that an elementary misconception about or ignorance of basic reproductive physiology has vitiated the entire study. A classic example is the assumption that all rats showing ovarian cycles and having a fully cornified morning vaginal smear are therefore in a comparable physiological state. I have may own list of such papers. Other examples are given by Schwartz (1969) in an excellent review that gives the only recent detailed account of the vagaries of the 5 day estrous cycle of the rat from which this type of error can arise. A word of caution, also about the mouse, where in my experience a fully cornified vaginal smear may persist up to the equivalent of the day of metestrus in the rat cycle and be a potential source of confusion (Brown-Grant, 1966a). The mouse cycle can be sorted out and any biochemist in need of guidance should consult the papers of Bingel and Schwartz (1969a, b, c) and Bingel (1974).

A second relatively common error concerns the effects of steroid hormones given to female rodents in the neonatal period on spontaneous ovulation when adult. This simple manoeuvre, offering a model system for the study of the biochemical basis of the sexual differentiation of the brain, has been used extensively as the basis for the studies of the inhibition and modification of steroid action (Gorski, 1971; Salaman and Birkett, 1974). However, though the phenomenon was first reported by Swanson and Van der Werff ten Bosch (1964), the fact that animals treated in this way may show fairly regular ovulations up to about 100 days of age but cease to ovulate by 150 days of age is frequently ignored. An end point of success or failure in the blockade of ovulation determined at 80 or 90 days of age is inadequate as a guide as to whether a putative antagonist to the process of masculinization has been effective or an observed biochemical change induced by steroid administration can or cannot be reliably linked to the process of masculinization (Brown-Grant, 1974b, c).

One can also rather easily convince oneself that having had the forethought and cunning to incorporate the assessment of a physiologically meaningful end point into the experimental design, one can then relax and assume that responses are being adequately monitored. Ovulation (the culminatory event of the female reproductive process for the neuroendocrinologist, as it was described some years ago by a disgruntled biologist) can be easily detected and quantified in terms of the number of tubal ova in the rat. Such a procedure does provide assurance, by definition, that an amount of luteinising hormone (LH) sufficient to induce ovulation has been released. It is of no value, however, as a guide as to whether the *normal* preovulatory discharge of LH has occurred or not. A plasma concentration of LH reaching only about 25% of the normal maximum was found to be sufficient to induce ovulation in one study (Naftolin et al., 1972) and an estimate of 14% of the normal as the necessary minimum was obtained in another study (Greig and Weisz, 1973). Moreover, egg numbers were not consistently related to the maximal LH concentration observed, at least in Nembutal-blocked rats receiving synthetic decapeptide LH-releasing factor (LRF) in the detailed study of Aiyer et al. (1974b).

II. A GOOD MODEL

If an animal preparation is to be used to provide the tissues or cells for a costly biochemical investigation of the mechanisms involved in a particular response, then the response being studied should be unequivocal, reliably inducible, of a sufficient magnitude for progressive experimental attenuation to be detectable without the response falling to the level where background "noise" makes interpretation difficult and, if possible, there should be natural controls. The long term gonadectomized, estrogen-primed rat comes close to being such a preparation for the study of steroid hormone effects on gonadotropin secretion. It was originally described by Caligaris et al. (1968) and is known colloquially, at least in Oxford, as the Taleisnik rat. Two or more months after ovariectomy, the animals are primed with 20 µg of estradiol benzoate (EB) at 12.00 hr and 72 hr later given a second injection of steroid, commonly 2.5 mg of progesterone, at 12.00 hr. Over the next few hours there is a dramatic rise in plasma LH concentration, sufficiently great, in fact, to have been originally quantified by bioassay (Caligaris et al., 1968) though, of course, later and more detailed studies have utilized radioimmunoassay (Taleisnik et al., 1971). There is an associated, though much less marked, rise in plasma FSH concentration. This preparation has, as the estate agents say of house properties, many desirable features. The maximal increase seen 5 hr after progesterone administration is 15–20 times the pre-injection value in rats of the Wistar strain though the increase may be much less in other strains both with this experimental procedure (Brown-Grant, 1974a) and in other situations involving responses to estrogen priming (Legan et al., 1973a). The response can be inhibited or aborted by Nembutal given at the time of or after progesterone administration in the normal female and is completely lacking in the male rat castrated when adult and given the same steroid treatment. Parallel studies of biochemical changes in the hypothalamus under the conditions of such an experiment in the responding female and the totally non-responding male can be used to try to pinpoint those directly linked to the stimulation of gonadotropin secretion. For the purist, mindful of the X-Y chromosomes, it can be added that these responses are also absent in genetic females that have been rendered anovulatory by the administration of high doses of aromatizable androgen or estrogen in the neonatal period (Brown-Grant, 1973, 1974a).

The priming effect of the estrogen is also susceptible to experimental modification, offering the possibility of contrasting post-estrogen biochemical changes in normal females under circumstances where one group can be assumed to be progressing toward the primed state while the other is not. The trick is to surround the estrogen with progesterone, injections of 2.5 mg being given 2 hr before and 2, 6 and 21 hr after the injection of EB; under these conditions no gonadotropin release follows a second injection of progesterone or estrogen 72 hr after the first injection of EB (Tapper et al., 1974).

The Taleisnik rat is not a completely foolproof preparation, however. At the usual level of priming (20 µg EB) it is extremely sensitive to progesterone, so that a moderate increase in plasma LH may follow a clumsily given control injection of oil rather than progesterone (see Table 3 of Brown-Grant, 1974a), an increase that is probably related to the stimulation of adrenal progesteron secretion (Fajer et al., 1971) There is also evidence that minor "spontaneous" peaks of LH may occur at the time of day when, under the prevailing lighting conditions, the ovulatory surge of LH would be expected to take place at proestrus (Legan et al., 1973b, 1975; Burnet and MacKinnon, 1975). Whether these "spontaneous" peaks are due to a diurnal increase in adrenal progesterone secretion is not clear. Adrenalectomy has been observed to abolish (F.R. Burnet, personal communication) or reported to have no effect on these peaks (Coon et al., 1975). Parenthetically, peak is not a precise quantitative term. There are Pennine/Appalachian peaks and Himalayan/Rocky Mountain peaks and the magnitude of the "spontaneous" peaks of LH are in the former category and the induced peaks in the latter so that confusion in work on the Taleisnik preparation is unlikely. Size is not all important, of course, and the spontaneous peaks are, indeed, of very considerable physiological interest and widespread occurrence (Norman et al., 1973; Norman and Spies, 1974).

Modification of estrogen action by progesterone has not attracted the degree of attention at the biochemical level that it perhaps merits. There are, of course, technical problems raised by the fact that this steroid, in contrast to estradiol, is extensively metabolized by many target tissues (Armstrong, 1970; Karavolas and Herf, 1971) but the alternative of using a non-metabolizable synthetic analogue is available (John and Rogers, 1972).

The artificial blockade of estrogen-priming by progesterone in the rat was mentioned above. Physiological evidence suggests that the absence of an LH surge in early pregnancy in the rat despite a normal rate of follicle maturation (Brown-Grant, 1969a) and a relatively high circulating estradiol concentration (Brown-Grant et al., 1972) may be due to the high endogenous progesterone level. Studies on the rhesus monkey have also shown that endogenous or exogenous progesterone will impair the facilitatory effect of estrogen on LH release (Spies and Niswender, 1972; Dierschke et al., 1973). Similar effects have been demonstrated in the sheep (Pelletier and Signoret, 1969; Scaramuzzi et al., 1971) though for the cow the evidence is equivocal (Short et al., 1973; Lemon et al., 1975).

There are also many intriguing aspects to the effects of progesterone on sexual behaviour that make it surprising to a physiologist that its actions on neuroendocrine systems have not been more intensively studied. There is evidence for the rat (Brown-Grant, unpublished) that sexual receptivity in ovariectomized rats, given a single (2.5 µg) dose of EB and tested 72 hr later either with no further treatment or 4 hr after an injection of 1.25 mg of progesterone, is depressed or abolished if progesterone is given with the initial priming dose of estrogen. This effect has also been demonstrated for the sheep whether the estrogen is given systemically (Pelletier and Signoret, 1969; Scaramuzzi et al., 1971) or by intrahypothalamic implantation (Domański et al., 1972). In this species there is a further interaction to be considered; if fairly prolonged treatment with progesterone precedes but is terminated 48 hr or more before estrogen administration the effect of the estrogen on behaviour is facilitated or enhanced, rather than inhibited (Robinson, 1954, 1955). These behavioural effects must surely be at the neural level and amenable to investigation by methods currently available.

Just as the male, female, neonatally castrated male and neonatally "androgenised" female rats provide a range of preparations in which the facilitation of gonadotropin release by progesterone after estrogen priming is clearly present or totally absent, these same types of rats show a striking sexually dimorphic response to progesterone after estrogen-priming with respect to female sexual receptivity as evaluated by the lordosis response. Contrary to popular belief, the male and the "androgenised" female rat retain the capacity to exhibit the lordosis response when adequately tested (Brown-Grant, 1973, 1975) as is probably true of species other than the ramstergig (Beach, 1971) but, in the rat, the ancillary effect of progesterone is clearly abolished in animals exposed to high levels of androgen or estrogen neonatally. Curiously, the ancillary effect of luteinising hormone releasing factor (LRF; specifically a synthetic decapeptide with LRF-like properties) on receptivity (Moss and McCann, 1973; Moss, 1974) is not, in contrast to the action of progesterone, absent in the masculinized rat (Brown-Grant, 1975), suggesting a biochemically or neurally different mode of action. Both morphological and neurophysiological evidence of hypothalamic sexual dimorphism in the masculinized rat of either genetic sex is now available (Raisman and Field, 1973; Dyer et al., 1976). The correlation of these findings with currently available physiological data and the future results of the applications of the techniques described at this meeting should provide exciting new concepts in the next few years.

Rodent-oriented investigators may have been guilty of over-selling the concept of sexual differentiation of the brain. Currently, they seem to be receiving some support from more agricultural fields of research (Short, 1974; Karsch and Foster, 1975) though in the sheep, as in the rat, it may be premature to describe the female pattern of reproductive behaviour as having been abolished, rather than taking the more considered view, as exemplified by Beach (1971), that the probability of eliciting female sexual behaviour has been altered. Primatologists are currently somewhat ambivalent to the concept. Clearly the male (or masculinized female) hypothalamo-pituitary system is not incapable of exhibiting a positive feedback response to estrogen or estrogen plus progesterone (Goy, 1970; Karsch et al., 1973a; Stearns et al., 1973) but it seems likely to me that in view of the complexity of the response in the rat (Aiyer et al., 1974a, 1976) that a form of test input can and will be found that does demonstrate a sexual dimorphism in the control of gonadotropin secretion in the primate that is related to prenatal exposure to androgens. Ample evidence for such effects on non-sexual, sexually dimorphic behaviour is already available (Money and Ehrhardt, 1968; Goy, 1970; Goy and Resko, 1972).

III. A FAIRLY GOOD MODEL

So far the input side of the animal models for the study of the control of gonadotropin secretion that have been considered has been humoral. The manipulation of steroid hormone levels or the utilization of the known pattern of endogenous steroid hormone secretion when this is fairly closely tied to the light—dark cycle, as in the rat, is the essence of this approach. However, there are situations where a rapid and predictable response to an accurately timed physiological stimulus, neural rather than humoral in nature, may provide a more appropriate experimental situation. The rat, though normally a spontaneous ovulator, can also ovulate in response to the stimuli provided by mating. A number of preparations involving pharmacological blockade of the expected ovulation in normal cycling rats or exogenous hormone treatment have been described (Aron et al., 1966, 1968; Everett, 1967; Harrington et al., 1967; Zarrow and Clarke, 1968) but there has not been uniform success in adapting these procedures to other strains in other laboratories, at least at the 90% plus level of reliability that appears desirable if they are to be utilized in connection with expensive investigations of the underlying mechanisms. The female rat exposed to constant light has long been known to become anovulatory but to be capable of ovulating after mating (Dempsey and Searles, 1943). Hardy (1970) demonstrated that such animals were in persistent behavioural estrus as well as showing persistent vaginal cornification and Brown-Grant et al. (1973) established that ovulation could consistently follow mating and determined the level of coital stimulation necessary to induce ovulation. The increases in plasma LH (and follicle-stimulating hormone (FSH)) concentration following mating are prompt (less than 10 min) and striking. Here, one presumes, is a clear case of a neural input acting to release LRF in fairly large amounts over a limited period of time, particularly as the responsiveness of the pituitary gland itself to exogenous LRF is not high (Davidson et al., 1973; Fink, 1975). The decrease in hypothalamic LRF content, though definite, is not, however, very striking (Smith and Davidson, 1974).

This model system is not without complications, however. Non-specific stimulation may also induce gonadotropin release and ovulation, though the time course of the changes in plasma gonadotropin concentration differs from those seen after mating and strongly suggests that the release of progesterone from the adrenal cortex is the underlying mechanism (Brown-Grant et al., 1973). The pineal, happily, does not seem to be involved (Brown-Grant and Östberg, 1974). The response to mating would, on first principles (or rather on the basis of established prejudice), be expected to be quite susceptible to barbiturate blockade. In fact, attempts to prevent or curtail mating-induced rises in plasma gonadotropin concentration by the administration of Nembutal gave rather variable and disappointing results (Brown-Grant et al., 1973; K. Brown-Grant, G. Fink and S. Henderson, unpublished results, 1975). This aspect of the earlier work merits further investigation. The other general precautionary note is that strain susceptibility to the ovulation suppressing effect of exposure to constant light is quite variable. The majority of the early studies were made on albino rats which are now known to undergo extensive retinal photoreceptor degeneration when exposed to constant light (O'Steen, 1970; O'Steen and Lytle, 1971; Glickstein et al., 1972). Anovulation can occur in pigmented rats exposed to constant light (Davidson et al., 1973) and can do so without any evidence of retinal damage at the light microscopic level (Brown-Grant, 1974d) but, on the whole, albino strains are more responsive. Another important factor is the age and early history in terms of light exposure and the extensive studies of Hoffman in this field should be borne in mind (Hoffman, 1967, 1969, 1970, 1973).

The potentialities of the light-exposed anovulatory female rat are not limited to the exploration of the mechanism of mating-induced LH release. It has recently been demonstrated that, in contrast to the rabbit where mating-induced ovulation (the normal pattern for this species) is invariably followed by the development of functional corpora lutea and pseudopregnancy or pregnancy, in the light-exposed rat ovulation and the development of functional luteal tissue following mating are not necessarily associated and in fact require different levels and patterns of coital stimulation (Brown-Grant, 1976). Ejaculation per se, though more effective than an intromission, is neither necessary, nor, alone, sufficient in the rat in contrast to the findings in the mouse (McGill and Coughlin, 1970). This preparation should enable a further dissection to be made of the biochemical processes involved in LH and prolactin release.

One finding of particular interest concerned the "storage" of copulatory information. With respect to the induction of pseudopregnancy, coital stimuli delivered over hours rather than minutes seem to be as effective in the light-exposed female as in the normal female and results closely comparable to those of Edmonds et al. (1972) for the normal female were obtained. However, in contrast to the normal females in whom adequate LH release to ensure ovulation had already occurred spontaneously before the coital stimuli necessary to induce luteal function were delivered, coital stimulation was also necessary to induce ovulation in the light-exposed females. The limiting factor turned out to be the effectiveness in inducing ovulation; when intromissions were too dispersed in time, the proportion of animals ovulating fell to impracticably low values, though the incidence of pseudopregnancy in those that did ovulate remained high (Brown-Grant, 1976). The females' "memory" with respect to LH release seems to be shorter than her memory with regard to establishing a pattern of continuing prolactin secretion, however, bizarre this may prove to be in constant light (Everett, 1968; Freeman and Neill, 1972; Freeman et al., 1974; Beach et al., 1975). Occasional failures to induce ovulation by mating in females exposed to constant light (J.D. Neill, personal communication, 1975; K. Brown-Grant and G. Fink, unpublished, 1975) may be related to the short memory as a leisurely performance by the male was ineffective. The small print of Brown-Grant et al. (1973) describes how a mating procedure giving very high rates of delivery of coital stimulation was adopted in order to facilitate the timing of postcoital samples; in retrospect we were probably fortunate to have carried out these experiments in the way that we did.

Finally, lest again this section may be found too rodent-oriented, it should be noted that there is circumstantial evidence for mating-induced stimulation of LH release in sheep or at least the advancement of LH release (Goding et al., 1969) and that the question of coitus-induced ovulation in the human is by no means finally settled (Jöchle, 1973).

IV. TWO BAD MODELS (BUT SOME GOOD PROBLEMS)

Even physiologists feel the urge to perform physiological experiments. The Taleisnik rat (see Section II) has many advantages but the levels of hormone treatment are grossly high (20 µg of estradiol benzoate, 2.5 mg of progesterone). Attempts to scale down these doses were unrewarding; 2.5 µg of estradiol benzoate was quite ineffective as a priming dose (Brown-Grant, 1974a). One can argue that the peripheral plasma levels are of less significance than tissue levels and that the fact that the concentration of receptors in the tissues of the estrogen-deprived rat may be low and that regeneration may have to occur before significant uptake can take place provides an explanation for the very large doses required. But this is unsatisfactory (and frustrating) when compared with the elegant studies of Knobil and his group on the monkey where plasma steroid hormone concentrations were monitored and manipulated to establish that their experimental input was clearly comparable to that seen under physiological conditions (Yamaji et al., 1971; Dierschke et al., 1973; Karsch et al., 1973a, b, c, d).

In the hope of being able to reproduce the normal ovulatory surge of gonadotropins in the rat by the administration of physiological amounts of steroids shortly after ovariectomy during the cycle, a fairly lengthy series of experiments along these lines was performed (Tapper et al., 1974). The results were disappointing; even in the acute situation the necessary doses of estrogen were shown to result in plasma concentrations far above the modest maxima of about 30 pg estradiol-17β/ml seen on the morning of proestrus in the normal cycle (Brown-Grant et al., 1970a; Naftolin et al., 1972) and quite incommensurate with the known production rates by the ovary (Tapper and Brown-Grant, 1975). The probable explanation was thought to be interference by progesterone of adrenal origin, released in response to the stress of surgical removal of the ovaries; despite various manipulations, including suppression of adrenal function by exogenous synthetic glucocorticoid, we were not able to devise a useful protocol for the acute study of steroid input versus gonadotropin output in this particular rodent at the periovulatory stage of the cycle. Our experience was not unique (see Mann and Barraclough, 1973). The gonadotropin release mechanism appears to be particularly susceptible to experimental disturbance; our findings and those of Nequin and Schwartz (1971) with respect to sexual receptivity were more satisfactory as were the studies of Aiyer and Fink (1974) on the role of the ovarian steroids in modulating pituitary sensitivity to LRF.

The acutely ovariectomized or otherwise surgically stressed rat is not a preparation that can be recommended (see also Krulich et al., 1974). Nor is it a valid assumption that the evils of acute stress associated with blood sampling can be circumvented by use of previously implanted catheters; see the sad case of the disappearing diurnal prolactin peak of pregnancy (Butcher et al., 1972; Freeman and Neill, 1972; Freeman et al., 1974). The ovarian cycle of the hamster, possibly because of a low adrenal progesterone contribution (Lukaszewska and Greenwald, 1970; Ridley and Greenwald, 1975) is far less susceptible to unwanted experimental perturbation than that of the rat (Brom and Schwartz, 1968). The hamster is probably the species of choice for acute studies in rodents, particularly in view of the by now considerable accumulation of data on plasma steroid and gonadotropin concentrations (Goldman and Porter, 1970; Leavitt and Blaha, 1970; Goldman et al., 1971b; Bosley and Leavitt, 1972; Baranczuk and Greenwald, 1973).

The second bad model is the prepuberal rat. The steady base line, on which any experimental manoeuvre must be superimposed and subsequent changes in gonadotropin secretion established before any investigation of mechanisms can be meaningful, appears to be absent in these animals over a prolonged period of development as far as plasma LH concentrations are concerned. The LH levels were reported as quite variable in the early studies of Goldman and Gorski (1970) and Goldman et al. (1971a) but until recently this point has been tacitly ignored or obscured by the use of large numbers of rats per group or by the pooling of plasma from several animals for each determination. In this way, fairly meaningful trends with time have been established for the developmental changes (see Brown-Grant et al. (1975) for references to the many studies on males and Brown-Grant (1974c) for females). When individual values, particularly for females between 10 and 25 days of age, were determined, however, it was found that 10—20% of values from a presumed homogeneous group of littermates could be many times higher than the group mean, 20 or 30 ng S-13 LH/ml against a group mean of 2—3 ng LH/ml (unpublished observations).

No explanation in terms of handling or killing procedure could be found. A far more detailed study on rats of the same strain will be published (MacKinnon et al., 1976). In this study a very important finding is reported, namely that these sporadic episodes of LH secretion can be eliminated by disturbance of the animals 20—60 min before killing or by the implantation of cardiac catheters. This latter finding is reminiscent of the effect on the diurnal prolactin surge in the adult pregnant rat referred to earlier. In summary, beware the immature rat, male or female. If you are careful, your basal, pre-experimental values for plasma LH will be all over the place; if you decide to study the variation for its own sake after catheterization, it will disappear!

The instability of plasma gonadotropin levels under other conditions and in other animal preparations presents a range of highly intriguing problems for investigation. It is not possible to discuss or even list the many instances of episodic secretion that have come to light over the last few years. It appears to be characteristic of both male and female mammals at some stage of prepuberal development (see Grumbach et al. (1974) and also Forest and Bertrand (1974) for many reports of this in many species). Initially one might be tempted to suggest that this is the characteristic open loop behaviour of the hypothalamo-pituitary-gonad system in the immature animal where gonadal secretion of steroids is not adequate quantitatively or qualitatively to control the pituitary. The frequent occurrence of episodic secretion in gonadectomized animals (Dierschke et al., 1970; Butler et al., 1972; Gay and Sheth, 1972) might be considered to be in favour of this view but against this is a mass of evidence for episodic secretion of LH in intact males of very many species (Katongole et al., 1971; Naftolin et al., 1973; Sanford et al., 1974; Wilson and Sharp, 1975). Moreover, the older idea that the prepuberal gonad is inactive is in the process of rapid revision (for references see Grumbach et al., 1974; Forest and Bertrand, 1974). Generalization about the phenomenon of episodic secretion of pituitary tropic hormones is not possible at present. There are too many unexplained differences; why should the intact human female secrete in an episodic pattern (Kapen et al., 1973) but not the rhesus female (Weick et al., 1973)? Why is the relationship of episodic secretion to sleep stage so variable for different tropic hormones (Rubin, 1975) and what is the basis of age related changes in this relationship (Boyar et al., 1972; Rubin et al., 1972; Kapen et al., 1974)? Clearly, the phenomenon of episodic secretion has altered the order of magnitude of the inevitable complexity of the mechanisms that control gonadotropin secretion.

V. A PLEA FOR THE PITUITARY

Much of the emphasis at this meeting has been on supradiaphragmatic events but any consideration of the integrated control of gonadotropin secretion must take into account the many reports of changes occurring at the pituitary level. Historically, it has been an uphill battle until very recently for the reality of these changes, with respect to gonadotropin secretion, to gain general acceptance. It is a curious commentary on specialization within an already narrowly specialized field that feedback at the pituitary level of thyroid hormones to regulate thyroid-stimulating hormone (TSH) secretion was well documented and generally accepted by the early sixties (Brown-Grant, 1966b) and that evidence of modulation of pituitary responsiveness to a hypothalamic releasing factor, the then unidentified thyrotropin-releasing factor (TRF), was already available (Reichlin, 1964). Even Bogdanove, who preached to the unconverted reproductive physiologists for many years (Bogdanove, 1963, 1964; Kingsley and Bogdanove, 1973), probably did not anticipate the central importance that later work has established for variation in pituitary responsiveness to gonadotropin releasing factor(s). Two fairly recent symposium volumes contain much information (Gual and Rosemberg, 1973; Motta et al., 1975). It is not possible to go into detail here. A highly selective guide to work in this area might reasonably include papers giving evidence for a reduced responsiveness following certain steroid treatments (Yen et al., 1974), an increased sensitivity in response to endogenous and exogenous steroids, specifically estrogen (Aiyer and Fink, 1974), and well documented effects of LRF itself, acting on an estrogen-primed gland, to increase the responsiveness to its own action (Aiyer et al., 1974a). Current work suggests that this phenomenon may be dependent upon newly synthesized, freely releasable pools (Bremner and Paulsen, 1974; Pickering and Fink, 1975). Such is the rate of advance, however, that these effects are almost ancient history by now, latecomers having only the satisfaction of tidying up work (Aiyer et al., 1976).

Two more recent findings seem to me to be of prime importance. One is the observation in both sheep (Chakraborty et al., 1974) and in rats (Gnodde and Schuiling, 1975; Schuiling and Zurcher, 1975) that continuous infusions of LRF over several hours give an initial mono- or a biphasic increase in plasma LH concentration but that LH concentration subsequently falls to very low levels despite the continued infusion of LRF and despite the presence of large amounts of unreleased immunoassayable LH within the pituitary gland. This should provide the physiological basis for some interesting biochemical investigations. The second is the fascinating evidence that in women and adolescent girls who have undergone self-imposed pathological weight loss, the pituitary response to exogenous LRF reverts to the prepuberal pattern (these patients are, of course, amenorrhoeic). On refeeding, however, the normal post-menarchal response returns when a critical level of body weight is regained (Sherman et al., 1975; Warren et al., 1975). The mechanism of these dramatic changes is unknown but the significance of a body weight—pituitary responsiveness relation is obviously very great, not only in the pathogenesis of anorexia nervosa but also in relation to the timing and physiological basis of human female puberty (Grumbach et al., 1974).

VI. ON MODELLING AND A SUMMARY

As succinctly pointed out by Davidson (1969), the use of the concepts (or at least the jargon) of control systems theory has become widespread among reproductive neuroendocrinologists. He also to some extent indicates his own position by prefacing his comments with "For better or for worse ...". The philosophy behind the use of models in research on endocrine, and indeed physiological systems in general, has been expounded many times (Stear and Kadish, 1969; Yates et al., 1969, 1971). A favourable and very relevant viewpoint is that expressed by Schwartz (1969). More recent excursions into modelling in relation to the control of gonadotropin and corticoid secretion include the studies of Bogumil et al. (1972a, b), Jusko et al. (1975) and Shotkin (1974a, b).

An earlier criticism of the modelling approach (Davidson, 1971), that frequent simultaneous measurements of the circulating levels of the relevant hormones were not available, has by now either been met or can be met. The problem, indeed, is probably now one of an excess of data at this level rather than a shortage. My own experience (Brown-Grant, 1969b; Brown-Grant et al., 1970b) with the hypothalamus-pituitary-thyroid system was that neither the input nor the output side of the system presented insuperable problems to adequate or at least personally gratifying simulation studies (but see footnote to page 358 of Davidson (1971) with reference to this). The total breakdown in the realism of the model came at the stage of undertaking the representation of the central control elements in mathematical terms when none of the underlying physico-chemical processes involved were known (Brennan and Brown-Grant, 1970). It is relatively simple to produce a few punched cards incorporating one's fantasies or prejudices about proportional or derivative error detectors that the computer would accept as the truth and act on to generate printouts and graphs of system performance (a West Coast, not a cynical Yankee computer, I should say). But the whole exercise did degenerate into science fiction at this point. It has been lack of definite evidence about the physico-chemical basis of the regulatory process at the hypothalamic and pituitary level that has formed the basis of the really valid criticism of any modelling so far undertaken. This has been a major deterrent to anyone who has once been bitten returning to this type of work. The preceding sentence has been deliberately written in the past tense because my overall impression of this most enjoyable meeting has been that a major breakthrough in this crucial area of the detailed study of the mechanisms of the control process at the physico-chemical level is now occurring.

REFERENCES

Aiyer, M.S. and Fink, G. (1974) The role of sex steroid hormones in modulating the responsiveness of the anterior pituitary gland to luteinizing hormone releasing factor in the female rat. *J. Endocr.*, 62: 553–572.

Aiyer, M.S., Chiappa, S.A. and Fink, G. (1974a) A priming effect of luteinizing hormone releasing factor on the anterior pituitary gland in the female rat. *J. Endocr.*, 62: 573–588.

Aiyer, M.S., Fink, G. and Greig, F. (1974b) Changes in the sensitivity of the pituitary gland to luteinizing hormone releasing factor during the oestrous cycle of the rat. *J. Endocr.*, 60: 47–64.

Aiyer, M.S., Sood, M.C. and Brown-Grant, K. (1976) The pituitary response to exogenous luteinizing hormone releasing factor in steroid-treated gonadectomized rats. *J. Endocr.*, in press.

Armstrong, D.T. (1970) Regulation of progesterone-5α-reductase activity in the rat uterus. In *Abstracts of papers presented at the 3rd int. Congr. on Hormonal Steroids, Hamburg, 1970, Int. Congr. Ser. 210*, L. Martini (Ed.), Excerpta Medica, Amsterdam, p. 133.

Aron, C., Asch, G. and Roos, J. (1966) Triggering of ovulation by coitus in the rat. *Int. Rev. Cytol.*, 20: 139–172.

Aron, C., Roos, J. and Asch, G. (1958) New facts concerning the afferent stimuli that trigger ovulation by coitus in the rat. *Neuroendocrinology*, 3: 47–54.

Baranczuk, R. and Greenwald, G.S. (1973) Peripheral levels of estrogen in the cyclic hamster. *Endocrinology*, 92: 805–812.

Beach, F.A. (1971) Hormonal factors controlling the differentiation, development, and display of copulatory behavior in the ramstergig and related species. In *The Biopsychology of Development*, E. Tobach, L.R. Aronson and E. Shaw (Eds.), Academic Press, New York, pp. 249–296.

Beach, J.E., Tyrey, L. and Everett, J.W. (1975) Serum prolactin and LH in early phases of delayed versus direct pseudopregnancy in the rat. *Endocrinology*, 96: 1241–1246.

Bingel, A.S. (1974) Timing of LH release and ovulation in 4- and 5-day cyclic mice. *J. Reprod. Fertil.*, 40: 315–320.

Bingel, A.S. and Schwartz, N.B. (1969a) Pituitary LH content and reproductive tract changes during the mouse oestrous cycle. *J. Reprod. Fertil.*, 19: 215–222.

Bingel, A.S. and Schwartz, N.B. (1969b) Timing of LH release and ovulation in the cyclic mouse. *J. Reprod. Fertil.*, 19: 223–229.

Bingel, A.S. and Schwartz, N.B. (1969c) Timing of LH release and ovulation in the post partum mouse. *J. Reprod. Fertil.*, 19: 231–237.

Bogdanove, E.M. (1963) Direct gonad-pituitary feedback: an analysis of effects of intracranial estrogenic depots on gonadotrophin secretion. *Endocrinology*, 73: 696–712.

Bogdanove, E.M. (1964) The role of the brain in the regulation of pituitary gonadotropin secretion. *Vitam. u. Horm.*, 22: 205–260.

Bogumil, R.J., Ferin, M., Rootenberg, J., Speroff, L. and Van de Wiele, R.J. (1972a) Mathematical studies of the human menstrual cycle. I. Formulation of a mathematical model. *J. clin. Endocr.*, 35: 126–143.

Bogumil, R.J., Ferin, M. and Van de Wiele, R.L. (1972b) Mathematical studies of the human menstrual cycle. II. Simulation performance of a model of the human menstrual cycle. *J. clin. Endocr.*, 35: 144–156.

Bosley, C.G. and Leavitt, W.W. (1972) Dependence of preovulatory progesterone on critical period in the cyclic hamster. *Amer. J. Physiol.*, 222: 129–133.

Boyar, R., Finkelstein, J., Roffwarg, H., Kapen, S., Weitzman, E.D. and Hellman, L. (1972) Synchronization of augmented luteinizing hormone secretion with sleep during puberty. *New Engl. J. Med.*, 287: 582–586.

Bremner, W.J. and Paulsen, C.A. (1974) Two pools of luteinizing hormone in the human pituitary: evidence from constant administration of luteinizing hormone-releasing hormone. *J. clin. Endocr.*, 39: 811–815.

Brennan, R.D. and Brown-Grant, K. (1970) Simulation of the human thyroid-pituitary system. In *Proceedings of the 1970 Summer Computer Simulation Conference, Denver*, pp. 849–858.

Brom, G.M. and Schwartz, N.B. (1968) Acute changes in the estrous cycle following ovariectomy in the hamster. *Neuroendocrinology*, 3: 366–377.

Brown-Grant, K. (1966a) The relationship between ovulation and the changes in thyroid gland activity that occur during the oestrous cycle in rats, mice and hamsters. *J. Physiol. (Lond.)*, 184: 402–417.

Brown-Grant, K. (1966b) The action of hormones on the hypothalamus. *Brit. med. Bull.*, 22: 273–277.

Brown-Grant, K. (1969a) The induction of ovulation during pregnancy in the rat. *J. Endocr.*, 43: 529–538.

Brown-Grant, K. (1969b) Regulation and control of the thyroid-pituitary system. *J. Basic Engng*, 91: 313–320.

Brown-Grant, K. (1973) Recent studies on the sexual differentiation of the brain. In *Foetal and Neonatal Physiology*, K.S. Comline, K.W. Cross, G.S. Dawes and P.W. Nathanielsz (Eds.), Cambridge University Press, Cambridge, pp. 527–545.

Brown-Grant, K. (1974a) Steroid hormone administration and gonadotrophin secretion in the gonadectomized rat. *J. Endocr.*, 62: 319–332.

Brown-Grant, K. (1974b) Failure of ovulation after administration of steroid hormones and hormone antagonists to female rats during the neonatal period. *J. Endocr.*, 62: 683–684.

Brown-Grant, K. (1974c) On "critical periods" during the post-natal development of the rat. In *Endocrinologie Sexuelle de la Période Périnatale*, M.G. Forest and J. Bertrand (Eds.), INSERM, Paris, pp. 357–373.

Brown-Grant, K. (1974d) The role of the retina in the failure of ovulation in female rats exposed to constant light. *Neuroendocrinology*, 16: 243–254.

Brown-Grant, K. (1975) A re-examination of the lordosis response in female rats given high doses of testosterone propionate or estradiol benzoate in the neonatal period. *Hormone Behav.*, in press.

Brown-Grant, K. (1976) The induction of pseudopregnancy and pregnancy by mating in albino rats exposed to constant light. *Hormone Behav.*, in press.

Brown-Grant, K. and Östberg, A.J.C. (1974) Lack of effect of pineal denervation on the responses of the albino female rat to exposure to constant light. *J. Endocr.*, 62: 45–50.

Brown-Grant, K., Exley, D. and Naftolin, F. (1970a) Peripheral plasma oestradiol and luteinizing hormone concentrations during the oestrous cycle of the rat. *J. Endocr.*, 48: 295–296.

Brown-Grant, K., Brennan, R.D. and Yates, F.E. (1970b) Simulation of the thyroid hormone-binding protein interactions in human plasma. *J. clin. Endocr.*, 30: 733–751.

Brown-Grant, K., Corker, C.S. and Naftolin, F. (1972) Plasma and pituitary luteinizing hormone concentrations and peripheral plasma oestradiol concentration during early pregnancy and after the administration of progestational steroids in the rat. *J. Endocr.*, 53: 31–35.

Brown-Grant, K., Davidson, J.M. and Greig, F. (1973) Induced ovulation in albino rats exposed to constant light. *J. Endocr.*, 57: 7–22.

Brown-Grant, K., Fink, G., Greig, F. and Murray, M.A.F. (1975) Altered sexual development in male rats after oestrogen administration during the neonatal period. *J. Reprod. Fertil.*, 44: 25–42.

Burnet, F.R. and MacKinnon, P.C.B. (1975) Restoration by oestradiol benzoate of a neural and hormonal rhythm in the ovariectomized rat. *J. Endocr.*, 64: 27–35.

Butcher, R.L., Fugo, N.W. and Collins, W.E. (1972) Semicircadian rhythm in plasma levels of prolactin during early gestation in the rat. *Endocrinology*, 90: 1125–1127.

Butler, W.R., Malven, P.V., Willett, L.B. and Bolt, D.J. (1972) Patterns of pituitary release and cranial output of LH and prolactin in ovariectomized ewes. *Endocrinology*, 91: 793–801.

Caligaris, L., Astrada, J.J. and Taleisnik, S. (1968) Stimulating and inhibiting effects of progesterone on the release of luteinizing hormone. *Acta endocr. (Kbh.)*, 59: 177–185.

Chakraborty, P.B., Adams, T.E., Tarnavsky, G.K. and Reeves, J.J. (1974) Serum and pituitary LH concentrations in ewes infused with LH-RH/FSH-RH. *J. Anim. Sci.*, 39: 1150–1157.

Coon, G.A., Legan, S.J. and Karsch, F.J. (1975) Persistence of a daily surge of luteinizing hormone (LH) following adrenalectomy in estrogen treated ovariectomized rats. *Fed. Proc.*, 34: 287 (Abstract).

Davidson, J.M. (1969) Feedback control of gonadotropin secretion. In *Frontiers in Neuroendocrinology, 1969*, W.F. Ganong and L. Martini (Eds.), Academic Press, New York, pp. 343–388.

Davidson, J.M. (1971) Overview and summary of conference. In *Steroid Hormones and Brain Function*, C.H. Sawyer and R.A. Gorski (Eds.), University of California Press, Los Angeles, pp. 355–371.

Davidson, J.M., Smith, E.R. and Bowers, C.Y. (1973) Effects of mating on gonadotropin release in the female rat. *Endocrinology*, 93: 1185–1192.

Dempsey, E.W. and Searles, H.F. (1943) Environmental modification of certain endocrine phenomena. *Endocrinology*, 32: 119–128.

Dierschke, D.J., Bhattacharya, A.N., Atkinson, L.E. and Knobil, E. (1970) Circhoral oscillations of plasma LH levels in the ovariectomized Rhesus monkey. *Endocrinology*, 87: 850–853.

Dierschke, D.J., Yamaji, T., Karsch, F.J., Weick, R.F., Weiss, G. and Knobil, E. (1973) Blockade by progesterone of estrogen-induced LH and FSH release in the rhesus monkey. *Endocrinology*, 92: 1496–1501.

Domański, E., Przekop, F. and Skubiszewski, B. (1972) Interaction of progesterone and estrogens on the hypothalamic center controlling estrous behavior in sheep. *Acta Neurobiol. exp.*, 32: 763–766.

Dyer, R.G., MacLeod, N.K. and Ellendorff, F. (1976) Electrophysiological evidence for sexual dimorphism and synaptic convergence in the preoptic and anterior hypothalamic areas of the rat. *Proc. roy. Soc. B*, in press.

Edmonds, S., Zoloth, S.R. and Adler, N.T. (1972) Storage of copulatory stimulation in the female rat. *Physiol. Behav.*, 8: 161–164.

Everett, J.W. (1967) Provoked ovulation on long-delayed pseudopregnancy from coital stimuli in barbiturate-blocked rats. *Endocrinology*, 80: 145–154.

Everett, J.W. (1968) Delayed pseudopregnancy in the rat, a tool for the study of central neural mechanisms in reproduction. In *Perspectives in Reproductive and Sexual Behavior*, M. Diamond (Ed.), Indiana University Press, Bloomington, pp. 25–31.

Fajer, A.B., Holzbauer, M. and Newport, H.M. (1971) The contribution of the adrenal gland to the total amount of progesterone produced in the female rat. *J. Physiol. (Lond.)*, 214: 115–126.

Fink, G. (1975) The responsiveness of the anterior pituitary gland to luteinizing hormone releasing factor in rats exposed to constant light. *J. Endocr.*, 65: 439–445.

Forest, M.G. et Bertrand, J. (Eds.) (1974) *Endocrinologie Sexuelle de la Période Périnatale*, INSERM, Paris.

Freeman, M.E. and Neill, J.D. (1972) The pattern of prolactin secretion during pseudopregnancy in the rat: a daily nocturnal surge. *Endocrinology*, 90: 1292–1294.

Freeman, M.E., Smith, M.S., Nazian, S.J. and Neill, J.D. (1974) Ovarian and hypothalamic control of the daily surges of prolactin secretion during pseudopregnancy in the rat. *Endocrinology*, 94: 875–882.

Gay, V.L. and Sheth, N.A. (1972) Evidence for a periodic release of LH in castrated male and female rats. *Endocrinology*, 90: 158–162.

Glickstein, M., Brown-Grant, K. and Raisman, G. (1972) Light-induced retinal degeneration in the rat and its implications for endocrinological investigations. *J. Anat. (Lond.)*, 111: 515 (Abstract).

Gnodde, H.P. and Schuiling, G.A. (1975) The pituitary response to continuous infusions of synthetic LH-RH in the ovariectomized rats. Effects of oestradiol and progesterone. *Acta endocr. (Kbh.)*, Suppl. 199: 197 (Abstract).

Goding, J.R., Catt, K.J., Brown, J.M., Kaltenbach, C.C., Cumming, I.A. and Mole, B.J. (1969) Radioimmunoassay for ovine luteinizing hormone. Secretion of luteinizing hormone during estrus and following estrogen administration in the sheep. *Endocrinology*, 85: 133–147.

Goldman, B.D. and Gorski, R.A. (1970) Effects of gonadal steroids on the secretion of LH and FSH in neonatal rats. *Endocrinology*, 89: 112–115.

Goldman, B.D. and Porter, J.C. (1970) Serum LH levels in intact and castrated golden hamsters. *Endocrinology*, 87: 676–679.

Goldman, B.D., Grazia, Y.R., Kamberi, I.A. and Porter, J.C. (1971a) Serum gonadotropin concentrations in intact and castrated neonatal rats. *Endocrinology*, 88: 771–776.

Goldman, B.D., Mahesh, V.B. and Porter, J.C. (1971b) The role of the ovary in control of cyclic LH release in the hamster, *Mesocricetus auratus. Biol. Reprod.*, 4: 57–65.

Gorski, R.A. (1971) Gonadal hormones and the perinatal development of neuroendocrine function. In *Frontiers in Neuroendocrinology, 1971*, L. Martini and W.F. Ganong (Eds.), Oxford University Press, New York, pp. 237–290.

Goy, R.W. (1970) Experimental control of psychosexuality. *Phil. Trans. B*, 259: 149–162.

Goy, R.W. and Resko, J.A. (1972) Gonadal hormones and behavior of normal and pseudohermaphroditic non-human female primates. *Recent Progr. Hormone Res.*, 28: 707–733.

Greig, F. and Weisz, J. (1973) Preovulatory levels of luteinizing hormone, the critical period and ovulation in rats. *J. Endocr.*, 57: 235–245.

Grumbach, M.M., Grave, G.D. and Mayer, F.E. (Eds.) (1974) *The Control of the Onset of Puberty*, Wiley, New York.

Gual, C. and Rosemberg, E. (Eds.) (1973) *Hypothalamic Hypophysiotropic Hormones*, Excerpta Medica, Amsterdam.

Hardy, D.F. (1970) The effect of constant light on the estrous cycle and behavior of the female rat. *Physiol. Behav.*, 5: 421–425.

Harrington, F.E., Eggert, R.G. and Wilbur, R.D. (1967) Induction of ovulation in chlorpromazine-blocked rats. *Endocrinology*, 81: 877–881.

Hoffman, J.C. (1967) Effects of light deprivation on the rat estrous cycle. *Neuroendocrinology*, 2: 1–10.

Hoffman, J.C. (1969) Light and reproduction in the rat: effect of lighting schedule on ovulation. *Biol. Reprod.*, 1: 185–188.

Hoffman, J.C. (1970) Light and reproduction in the rat: effects of photoperiod length on albino rats from two different breeders. *Biol. Reprod.*, 2: 255–261.

Hoffman, J.C. (1973) Light and reproduction in the rat. Effects of early lighting on responses measured in adult females. *Biol. Reprod.*, 8: 473—480.

Jöchle, W. (1973) Coitus-induced ovulation. *Contraception*, 7: 523—564.

John, P.N. and Rogers, A.W. (1972) The distribution of [^{3}H]progesterone and [^{3}H]megestrol acetate in the uterus of the ovariectomized rat. *J. Endocr.*, 53: 375—387.

Jusko, W.J., Slaunwhite, W.R., Jr. and Aceto, T., Jr. (1975) Partial pharmacodynamic model for circadian-episodic secretion of cortisol in man. *J. clin. Endocr.*, 40: 278—289.

Kapen, S., Boyar, R., Hellman, L. and Weitzman, E.D. (1973) Episodic release of luteinizing hormone at mid-menstrual cycle in normal adult women. *J. clin. Endocr.*, 36: 724—729.

Kapen, S., Boyar, R.M., Finkelstein, J.W., Hellman, L. and Weitzman, E.D. (1974) Effect of sleep-wake cycle reversal on luteinizing hormone secretory pattern in puberty. *J. clin. Endocr.*, 39: 293—299.

Karavolas, H.J. and Herf, S.M. (1971) Conversion of progesterone by rat medial basal hypothalamic tissue to 5α-pregna-3,20-dione. *Endocrinology*, 89: 940—942.

Karsch, F.J. and Foster, D.L. (1975) Sexual differentiation of the mechanism controlling the preovulatory discharge of luteinizing hormone in sheep. *Endocrinology*, 97: 373—379.

Karsch, F.J., Dierschke, D.J. and Knobil, E. (1973a) Sexual differentiation of pituitary function: apparent difference between primates and rodents. *Science*, 179: 484—486.

Karsch, F.J., Dierschke, D.J., Weick, R.F., Yamaji, T., Hotchkiss, J. and Knobil, E. (1973b) Positive and negative feedback control by estrogen of luteinizing hormone secretion in the Rhesus monkey. *Endocrinology*, 92: 799—804.

Karsch, F.J., Weick, R.F., Butler, W.R., Dierschke, D.J., Krey, L.C., Weiss, G., Hotchkiss, J., Yamaji, T. and Knobil, E. (1973c) Induced LH surges in the rhesus monkey: strength-duration characteristics of the estrogen stimulus. *Endocrinology*, 92: 1740—1747.

Karsch, F.J., Weick, R.F., Hotchkiss, J., Dierschke, D.J. and Knobil, E. (1973d) An analysis of the negative feedback control of gonadotropin secretion utilizing chronic implantation of ovarian steroids in ovariectomized rhesus monkeys. *Endocrinology*, 93: 478—486.

Katongole, C.B., Naftolin, F. and Short, R.V. (1971) Relationship between blood levels of luteinizing hormone and testosterone in bulls, and the effects of sexual stimulation. *J. Endocr.*, 50: 457—466.

Kingsley, T.R. and Bogdanove, E.M. (1973) Direct feedback of androgens: localised effects of intrapituitary implants of androgens on gonadotrophic cells and hormone stores. *Endocrinology*, 93: 1398—1409.

Krulich, L., Hefco, E., Illner, P. and Read, C.B. (1974) The effects of acute stress on the secretion of LH, FSH, prolactin and GH in the normal male rat, with comments on their statistical evaluation. *Neuroendocrinology*, 16: 293—311.

Leavitt, W.W. and Blaha, G.C. (1970) Circulating progesterone levels in the golden hamster during the estrous cycle, pregnancy and lactation. *Biol. Reprod.*, 3: 353—361.

Legan, S.J., Coon, G.A. and Midgley, A.R. (1973a) Evidence for a 24-hour periodicity of LH and FSH release in estrogen-treated ovariectomized rats. *Fed. Proc.*, 32: 240 (Abstract).

Legan, S.J., Gay, V.L. and Midgley, A.R. (1973b) LH release following steroid administration in ovariectomized rats. *Endocrinology*, 93: 781—785.

Legan, S.J., Coon, C.A. and Karsch, F.J. (1975) Role of estrogen as initiator of daily LH surges in the ovariectomized rat. *Endocrinology*, 96: 50—56.

Lemon, M., Pelletier, J., Saumande, J. and Signoret, J.P. (1975) Peripheral plasma concentrations of progesterone, oestradiol-17β and luteinizing hormone around oestrus in the cow. *J. Reprod. Fertil.*, 42: 137—140.

Lukaszewska, J.H. and Greenwald, G.S. (1970) Progesterone levels in the cyclic and pregnant hamster. *Endocrinology*, 86: 1—9.

MacKinnon, P.C.B., Mattock, J.M. and Ter Haar, M.B. (1976) Serum gonadotrophin levels during development in male, female and androgenized female rats and the effect of acute general disturbance on episodic LH release. *J. Endocr.*, in press.

Mann, D.R. and Barraclough, C.A. (1973) Role of estrogen and progesterone in facilitating LH release in 4-day cyclic rats. *Endocrinology*, 93: 694—699.

McGill, T.E. and Coughlin, R.C. (1970) Ejaculatory reflex and luteal activity induction in *Mus musculus. J. Reprod. Fertil.*, 21: 215–220.

Money, J. and Ehrhardt, A.A. (1968) Prenatal hormonal exposure: possible effects on behavior in man. In *Endocrinology and Human Behaviour*, R.P. Michael (Ed.), Oxford University Press, London, pp. 32–48.

Moss, R.L. (1974) Relationship between central regulation of gonadotropins and mating behavior in female rats. In *Reproductive Behavior*, W. Montagna and W.A. Sadler (Eds.), Plenum, New York, pp. 55–76.

Moss, R.L. and McCann, S.M. (1973) Induction of mating behavior in rats by luteinizing hormone-releasing factor. *Science*, 181: 177–179.

Motta, M., Crosignani, P.G. and Martini, L. (Eds.) (1975) *Hypothalamic Hormones: Chemistry, Physiology, Pharmacology and Clinical Uses*, Academic Press, New York.

Naftolin, F., Brown-Grant, K. and Corker, C.S. (1972) Plasma and pituitary luteinizing hormone and peripheral plasma oestradiol concentrations in the normal oestrous cycle of the rat and after experimental manipulation of the cycle. *J. Endocr.*, 53: 17–30.

Naftolin, F., Judd, H.L. and Yen, S.S.C. (1973) Pulsatile pattern of gonadotropins and testosterone in man: the effects of clomiphene with and without testosterone. *J. clin. Endocr.*, 36: 285–288.

Nequin, L.G. and Schwartz, N.B. (1971) Adrenal participation in the timing of mating and LH release in the cyclic rat. *Endocrinology*, 88: 325–331.

Norman, R.L. and Spies, H.G. (1974) Neural control of the estrogen-dependent twenty-four-hour periodicity of LH release in the golden hamster. *Endocrinology*, 95: 1367–1372.

Norman, R.L., Blake, C.A. and Sawyer, C.H. (1973) Estrogen-dependent twenty-four-hour periodicity in pituitary LH release in the female hamster. *Endocrinology*, 93: 965–970.

O'Steen, W.K. (1970) Retinal and optic nerve serotonin and retinal degeneration as influenced by photoperiod. *Exp. Neurol.*, 27: 194–205.

O'Steen, W.K. and Lytle, R.B. (1971) Early cellular disruption and phagocytosis in photically-induced retinal degeneration. *Amer. J. Anat.*, 130: 227–233.

Pelletier, J. et Signoret, J.P. (1969) Contrôle de la décharge de LH dans le sang par la progestérone et le benzoate d'oestradiol chez la brébis castrée. *C.R. Acad. Sci. (Paris)*, 269: 2595–2598.

Pickering, A.J.M.C. and Fink, G. (1975) The priming effect of LH-RF: in vitro and in vivo evidence consistent with its dependence upon protein and RNA synthesis. *J. Endocr.*, Submitted.

Raisman, G. and Field, P.M. (1973) Sexual dimorphism in the neuropil of the preoptic area of the rat and its dependence on neonatal androgen. *Brain Res.*, 54: 1–29.

Reichlin, S. (1964) Function of the hypothalamus in regulation of pituitary-thyroid activity. Brain-thyroid relationships. *Ciba Found. Study Group*, 18: 17–32.

Ridley, K. and Greenwald, G.S. (1975) Progesterone levels measured every two hours in the cyclic hamster. *Proc. Soc. exp. Biol. (N.Y.)*, 149: 10–12.

Robinson, T.J. (1954) The necessity for progesterone with estrogen for the induction of recurrent estrus in the ovariectomized ewe. *Endocrinology*, 55: 403–408.

Robinson, T.J. (1955) Quantitative studies on the hormonal induction of oestrus in spayed ewes. *J. Endocr.*, 12: 163–173.

Rubin, R.T. (1975) Sleep-endocrinology studies in man. In *Hormones, Homeostasis and the Brain, Progr. in Brain Res., Vol. 42*, W.H. Gispen, Tj. B. van Wimersma Greidanus, B. Bohus and D. de Wied (Eds.), Elsevier, Amsterdam, pp. 73–80.

Rubin, R.T., Kales, A., Adler, R., Fagan, T. and Odell, W. (1972) Gonadotropin secretion during sleep in normal adult men. *Science*, 175: 196–198.

Salaman, D.F. and Birkett, S. (1974) Androgen-induced sexual differentiation of the brain is blocked by inhibitors of DNA and RNA synthesis. *Nature (Lond.)*, 247: 109–112.

Sanford, L.M., Winter, J.S.D., Plamer, W.M. and Howland, B.E. (1974) The profile of LH and testosterone in the ram. *Endocrinology*, 95: 627–631.

Scaramuzzi, R.J., Tillson, S.A., Thorneycroft, I.H. and Caldwell, B.V. (1971) Action of exogenous progesterone and estrogen on behavioral estrus and luteinizing hormone levels in the ovariectomized ewe. *Endocrinology*, 88: 1184–1189.

Schuiling, G.A. and Zurcher, A.F. (1975) An investigation into the contribution of the central nervous system and the pituitary gland to the characteristics of pre-ovulatory LH-surges in the rat; the role of progesterone. *Acta endocr. (Kbh.)*, Suppl. 199: 198 (Abstract).

Schwartz, N.B. (1969) A model for the regulation of ovulation in the rat. *Recent Progr. Hormone Res.*, 25: 1–53.

Sherman, B.M., Halmi, K.A. and Zamudio, R. (1975) LH and FSH response to gonadotropin-releasing hormone in anorexia nervosa: effect of nutritional rehabilitation. *J. clin. Endocr.*, 41: 135–142.

Short, R.V. (1974) Sexual differentiation of the brain of the sheep. In *Endocrinologie Sexuelle de la Période Périnatale*, M.G. Forest and J. Bertrand (Eds.), INSERM, Paris, pp. 121–142.

Short, R.E., Howland, B.E., Randel, R.D., Christensen, D.S. and Bellows, R.A. (1973) Induced LH release in spayed cows. *J. Anim. Sci.*, 37: 551–557.

Shotkin, L. (1974a) A model for LH levels in the recently castrated adult rat and its comparison with experiment. *J. theor. Biol.*, 43: 1–14.

Shotkin, L.M. (1974b) A model for the effect of daily injections of gonadal hormones on LH levels in recently-castrated adult rats and its comparison with experiment. *J. theor. Biol.*, 43: 15–28.

Smith, E.R. and Davidson, J.M. (1974) Luteinizing hormone releasing factor in rats exposed to constant light: effects of mating. *Neuroendocrinology*, 14: 129–138.

Spies, H.G. and Niswender, G.D. (1972) Effect of progesterone and estradiol on LH release and ovulation in Rhesus monkeys. *Endocrinology*, 90: 257–261.

Stear, E.B. and Kadish, A.H. (Eds.) (1969) Hormonal control systems. *Math. Biosci.*, Suppl. 1.

Stearns, E.L., Winter, J.S.D. and Faiman, C. (1973) Positive feedback effect of progestin upon serum gonadotropins in estrogen-primed castrate men. *J. clin. Endocr.*, 37: 635–638.

Swanson, H.H. and Van der Werff ten Bosch, J.J. (1964) The "early-androgen" syndrome; its development and the response to hemi-spaying. *Acta endocr. (Kbh.)*, 45: 1–12.

Taleisnik, S., Caligaris, L. and Astrada, J.J. (1971) Feedback effects of gonadal steroids on the release of gonadotropins. In *Proc. of 3rd int. Congr. on Hormonal Steroids, Int. Congr. Ser. 219*, V.H.T. James and L. Martini (Eds.), Excerpta Medica, Amsterdam, pp. 699–707.

Tapper, C.M. and Brown-Grant, K. (1975) The secretion and metabolic clearance rates of oestradiol in the rat. *J. Endocr.*, 64: 215–227.

Tapper, C.M., Greig, F. and Brown-Grant, K. (1974) Effects of steroid hormones on gonadotropin secretion in female rats after ovariectomy during the oestrous cycle. *J. Endocr.*, 62: 511–525.

Warren, M.P., Jewelewicz, R., Dyrenfurth, I., Ans, R., Khalaf, S. and Van de Wiele, R.L. (1975) The significance of weight loss in the evaluation of pituitary response to LH-RH in women with secondary amenorrhea. *J. clin. Endocr.*, 40: 601–611.

Weick, R.F., Dierschke, D.J., Karsch, F.J., Butler, W.R., Hotchkiss, J. and Knobil, E. (1973) Periovulatory time courses of circulating gonadotropic and ovarian hormones in the rhesus monkey. *Endocrinology*, 93: 1140–1147.

Wilson, S.C. and Sharp, P.J. (1975) Episodic release of luteinising hormone in the domestic fowl. *J. Endocr.*, 64: 77–86.

Yamaji, T., Dierschke, D.J., Hotchkiss, J., Bhattacharya, A.N., Surve, A.H. and Knobil, E. (1971) Estrogen induction of LH release in the rhesus monkey. *Endocrinology*, 89: 1034–1041.

Yates, F.E., Brennan, R.D. and Urquhart, J. (1969) Adrenal glucocorticoid control system. *Fed. Proc.*, 28: 71–83.

Yates, F.E., Russel, S.M. and Maran, J.W. (1971) Brain-adenohypophysial communication in mammals. *Ann. Rev. Physiol.*, 33: 393–444.

Yen, S.S.C., Vandenberg, G. and Siler, T.M. (1974) Modulation of pituitary responsiveness to LRF by estrogen. *J. clin. Endocr.*, 39: 170–177.

Zarrow, M.X. and Clarke, J.H. (1968) Ovulation following vaginal stimulation in a spontaneous ovulator and its implications. *J. Endocr.*, 40: 343–352.

Subcellular Mechanisms in Reproductive Neuroendocrinology, edited by
F. Naftolin, K.J. Ryan and J. Davies
© 1976 Elsevier Scientific Publishing Company—Amsterdam, The Netherlands

Chapter 27

Special aspects — rhythms

ROGER SHORT

MRC Reproductive Biology Unit, Department of Obstetrics and Gynaecology, Edinburgh EH3 9ER (Great Britain)

Dr. Naftolin, you remember that when you wrote to invite me to this meeting, you asked me if I would review, and I quote, "Yanks, jerks, rhythms, and puberty." So I would like to start on Yanks.

I think you Yanks are fantastic. Your hospitality this week has been incredible, and one can't come to this country without feeling overcome by your friendship and your genuine kindness and sympathy. So much for Yanks. I could say so much more, and it would all be flattering.

Jerks — those are the people that left the meeting early!

And so, to the meat of my talk — rhythms. I would like to summarize all that we have heard about neuroendocrine mechanisms in the last three days by quoting Hamlet: "There are more things in heaven and on earth, Horatio, than are dreamt of in your philosophy." We haven't even begun to consider the complexity of rhythms, and the manifold ways in which they can control reproductive and neuroendocrine functions.

Let's start with a definition: a rhythm is an event with a predictable frequency. So I am going to exclude from my discussion, at least initially, these episodic bursts of gonadotrophin secretion, and talk about basic rhythms.

Now, Fred Naftolin was a student of Geoffrey Harris, and I think it would be nice to remember Geoffrey at this meeting. I will remind you of what he wrote in 1964 in his article in *Endocrinology*. He posed the following question: "Does the normal female nervous system regulate a rhythmic discharge of gonadotrophin in the absence of an ovarian (estrogen) feedback?" (Harris, 1964). I think that question is as relevant today as when he first posed it, and none of us have really attempted to answer it during the course of this meeting. Are these rhythms of gonadotrophin discharge that we talk about entirely dependent on an input from the gonad before they can be expressed, or are they produced by an inherent hypothalamic clock, just ticking away?

My personal belief is that the so-called "reproductive rhythms in the hypothalamus" are driven by the mainspring of gonadal steroid feedback; I find it difficult to believe that there are hypothalamic rhythms operating in the absence of gonadal feedback (Short, 1974). But in his opening remarks to us, Fred Naftolin said that the male has tonic rhythms of gonadotrophin release and the female has cyclic rhythms; in his closing remarks, he said that "rhythms live in the limbic system." So I suspect that Fred believes that there are differences in the limbic system between male and female, and that the female has inherent limbic rhythms and the male doesn't.

I will discuss this problem from the standpoint of three different types of rhythms, and I will begin by talking about the rhythm of the estrous or menstrual cycle, which is a fairly basic reproductive rhythm — or isn't it?

I would like to remind all of you that if you take a male rat and a female rat and put them together in a cage and observe them for their lifetime, I doubt very much whether you would ever see a single 4- or 5-day estrous cycle. You would possibly see one or two cycles of pseudopregnancy, and I don't think that I have heard the pseudopregnant rat mentioned during this meeting; you would also see a great many pregnancies. So although the 4- or 5-day cyclic rat is a very convenient animal to study, let's remember that it is an artifact, and that the "rhythm of the cycle", or even as John Challis would have it, "the normally cycling sheep", represents an abnormal situation, because repeated cycles do not normally occur in Nature. It is fair enough to study them, but let's get them in perspective and remember them for what they are — artifacts of man's making.

I said that it seemed to me that it was likely to be the gonad that was the mainspring of hypothalamic rhythms, and I would like to give you 4 examples in 4 different species as evidence of the fact that the brain is dancing to a tune played upon it by the gonads, so that the rhythm we see is inherently gonadal.

This morning Dr. Klein eulogized about the chicken as an experimental animal, and I share that enthusiasm. Macroscopically, the ovary of the hen contains a very large number of small primordial follicles, a smaller population of growing secondary follicles, and usually about half a dozen large tertiary follicles at different stages of development. The hen offers a beautiful experimental situation because it is possible to mark the yolk of growing follicles by daily intramuscular injections of different colored fat-soluble dyes (Gilbert, 1972), and thus study the sequence of follicular development leading up to ovulation. You will appreciate that it is normal for a bird to lay an egg about every 24 hr for a sequence of several days before having a 24-hr pause, and then starting again. It seems that the secondary follicles are recruited from a large pool of primary follicles, and several months may elapse before they enter the phase of rapid yolk deposition as they enlarge to become tertiary follicles. But once a tertiary follicle has started to develop and accumulate yolk, it appears to be committed to a rather rigid program, leading to eventual ovulation 5—14 days later. If a tertiary follicle is missing from the sequence, the bird will miss a day in its ovulatory cycle, hence the laying "pause" we have just mentioned.

I think this is a good example of how the follicle has to go through a rather immutable programmed course of development before it can be ovulated. Thus the rhythm of egg laying in the hen is to an extent predetermined by the hierarchy of tertiary follicles in the ovary. So I would commend the bird to you for studies of ovulatory rhythms. It is cheap, expendable and tasty.

The next example I would like to take is very different; it concerns a genetic lesion that one can insert in the germ cell line of the ovary in order to study the rhythm of the estrous cycle.

If you mate a jack donkey to a horse you produce a mule. These animals are particularly interesting because the chromosomal makeup of the two parental species is completely different. The karyotypes are widely dissimilar, both with respect to chromosome shape and total chromosome number. So in the mule, although mitosis and somatic growth is perfectly normal, meiosis is almost impossible, because of the extreme difficulty in pairing homologous chromosomes prior to the first reduction division (Chandley et al., 1974). So the mule is really God's gift of man's making as an experimental animal; it allows us to introduce a genetic lesion specifically into the germ cell line, which impedes gamete formation.

The consequences of this are that the ovary of the mule just after birth contains a severely depleted stock of oocytes (Taylor and Short, 1973). However, a very small number of oocytes do manage to survive, and may even develop to the point of ovulation, so that it is possible to recover eggs from the fallopian tube (Short, 1975).

How can all this be relevant to the study of reproductive rhythms? Nishikawa in Japan (1959) and Bielanski in Poland (1968, 1972) have made a careful study of the length of the estrous cycle in the mule, and they found occasional normal 22-day cycles, interspersed with cycles of up to more than 900 days in length. This seems to me strong evidence to suggest that there may be a hierarchy of primordial follicles in the ovary that spontaneously progress to a gonadotrophin-sensitive stage. Severe oocyte depletion leaves major gaps in the hierarchy, with corresponding gaps in the length of the estrous cycle. The hypothalamus and pituitary may be ready, willing and able to induce follicular development, estrus and ovulation, but they have to bide their time until the ovary becomes responsive. So I feel that the mule is a spectacular, if somewhat unusual, example of the remarkable degree of control that the ovary can exert on cycle length. If the ovary was entirely at the mercy of the pituitary, then mules would have a brief series of 22-day cycles until the entire stock of oocytes was depleted, followed by anestrus. This clearly does not happen.

The next example that I would like to take is another rather exotic one, the chimpanzee. As you know, the chimpanzee has a perineum which shows a marked swelling at the time of maximal follicular development, and which is known to be estrogen dependent. So, by studying the backsides of our nearest ancestors, we can actually learn a great deal about cyclical ovarian function without the tedium or trauma of measuring hormone levels.

I would like to show you some very remarkable information that comes from the Gombe Stream Reserve in Tanzania, where Caroline Tutin has been studying the swelling cycles in free-living wild chimpanzees as they go through puberty (Fig. 1). Those of you who have read Jane Goodall's books will recognize the name Fifi; you can see that in 1967, Fifi started to show cycles of vaginal swelling *about 9 months in advance of menarche*. Following menarche, menstruation occurred fairly regularly for 2.5 years. Even though she was being mated repeatedly, she did not conceive until 1970, continued to have one or two swelling cycles *after* conception, and eventually gave birth in mid-1971.

This is one of the most convincing pieces of evidence to show that there are cycles ("rhythms") of follicular development and estrogen secretion in the ovary for a considerable period of time *before* menarche, and probably for an even longer time before the first ovulation. It would be most interesting to know what is timing these cycles. Is it some hypothalamic or pituitary rhythm? If so, it is a rhythm that is operating by an oscillating negative feedback, because the positive feedback mechanism whereby estrogen provokes an ovulatory discharge of LH will not mature until some time after menarche, if

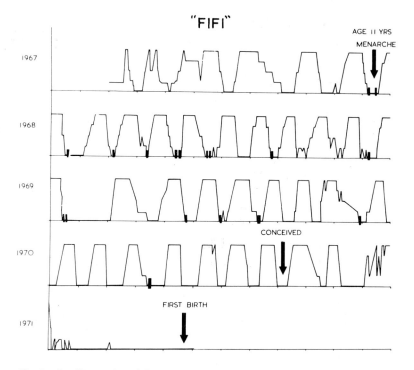

Fig. 1. Swelling cycles of the chimpanzee Fifi in the Gombe Stream Reserve, Tanzania. Data kindly supplied by Dr. Caroline Tutin.

the chimpanzee is anything like the human and the rhesus monkey. This should give us much food for thought about pubertal mechanisms.

The last of my 4 cycle rhythms concerns the human menstrual cycle. If we look at Treloar's data on the length of the menstrual cycle from menarche to the menopause (Fig. 2), we see that a spectacular increase in cycle length *precedes* the menopause (Treloar et al., 1967).

Even post-menopausal human ovaries contain a few oocytes, so the story that the menopause is caused by a *total* depletion of oocytes is incorrect. Instead, I would like to suggest that a woman entering the menopause is probably in a very analogous situation to the female mule. She has a very depleted stock of oocytes, and even though the gonadotrophin levels may have started to rise, she will have to wait longer and longer for the spontaneous maturation of a gonadotrophin-sensitive follicle before ovulation can take place. So once again, this is yet another example of how it may be the gonad that is controlling cycle length.

I would now like to say something about circadian and circannual rhythms. Thinking of Ken Ryan's remarks last night about the way in which the human appears to have escaped from many of the environmental controls that beset other animals, I am sure that he is right. Almost all mammals are seasonal breeders, whereas for some peculiar reason, I suppose simplicity, we have selected for detailed study those few mammals, like the rat, which are *not* seasonal breeders.

In temperate regions of the world, it is essential for most mammals to give birth to their young in due season. So they must be able to determine with accuracy where they are in the circannual cycle (Short, 1973). It seems probable that mammals, like birds, sense the circannual cycle by accurate perception of the light/dark rhythms of the circadian cycle. There is now good evidence from the study of men confined in underground caves for long periods of time, that in the absence of any photoperiodic cues, we have a free-running circadian rhythm of 25 or 26 hr; normally, we entrain our circadian rhythm to 24 hr, by means of a photoperiodic stimulus.

The same is true of bird, as Menaker (1971) has shown so beautifully, and it seems that they perceive the circadian cycle in a very subtle way. The brain is only able to respond to light for a very brief period at a set time in the circadian cycle. In the winter, when the days are short, this photosensitive period falls in darkness, and the bird effectively perceives no light. As the days get longer, this photosensitive period comes to lie in the daytime, and so the bird now responds to the light. It is this changing day length that in turn controls gonadal activity in so many birds and mammals.

So here, if you think about it, we have a "circadian multiplier" which can be used to time the circannual cycle very accurately. If we applied to mammals the level of sophistication which has been applied to circadian rhythms in birds, we might be a deal wiser. And I would like to think that Dr. Klein's studies of the pineal gland will reveal much more about the nature of the mammalian timing mechanism. This morning, I was impressed by his demonstration of the effect of the pineal in controlling seasonal testicular regression in the hamster. It is no accident that he happened to show us a hamster; the

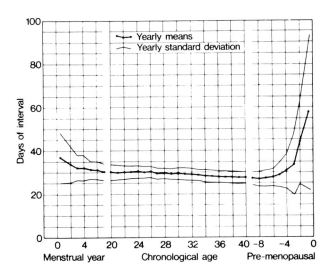

Fig. 2. Mean length ± S.D. of the human menstrual cycle from menarche to menopause. (Data from Treloar et al. (1967); drawing from Austin and Short (1972), *Reproduction in Mammals*, Book 5, *Artificial Control of Reproduction*, Cambridge University Press.)

effect is not present in the rat. This serves to introduce a general plea that I would make to everybody at this conference; we ought to pay much more attention to the hamster, and less to the rat. The hamster is a superb seasonal breeder, and absolutely ideal if we are looking for models in the laboratory where we can study both circadian and circannual rhythms.

And so I come to my final topic, circannual rhythms, and I would like to talk specifically about male seasonal breeders. I would like to bring to your attention some work that Dr. Gerald Lincoln has recently done in my laboratory on the ram (Lincoln, 1976). In fact, he has used a rather unusual sheep, the Soay, which is a very primitive, diminutive, unimproved breed that comes from the remote Atlantic island of St. Kilda, far off the west coast of Scotland. These animals were probably brought to Britain by early settlers, and they appear to be closely related to the wild European mouflon. Their particular attraction from our point of view was that even the males are marked seasonal breeders (Jewell et al., 1974). Lincoln has established that these animals show a marked increase in testicular size in the autumn, in response to decreasing day length. When he looked at the plasma LH levels in these animals, bled daily, there was a clear increase in LH which preceded the autumnal rise in testis size, and increased blood testosterone levels. The sequence of events appeared to be: daylight change → increased LH discharge (FSH was not measured) → increased testis size → increased testosterone secretion → behavioral change. Lincoln went on to analyze these changes in more detail, by bleeding at 15-min intervals during certain critical periods. During the summer, when the testes were regressed, there was only about one episodic discharge of LH, and a very poor testosterone response, during the 8-hr bleeding period. When day length began to decrease, the magnitude of the LH discharges appeared to increase, as did the testosterone response. By the time that the testis was at its maximum size, the LH pattern was again altered, with elevated basal levels, an increased frequency of discharge, but diminished amplitude. This was accompanied by major increases in the testosterone level, much as we had found in our earlier study of domestic rams (Katongole et al., 1974). These findings are summarized in Fig. 3.

Lincoln went on to test the pituitary reserve of LH in these rams by giving a standard injection of LHRH at different times in the circannual cycle, and Fig. 4 summarizes his results. It seems that the differing patterns of LH discharge that occur spontaneously are associated with marked variations in the pool size of releasable LH in the pituitary.

These wildly fluctuating levels of LH and testosterone that occur from hour to hour and month to month we could call the "Language of Love", and I submit that we just don't understand it. It is as if Dr. Kawakami had delivered his lecture in Japanese. We would have known that he was speaking, we could hear him, but we wouldn't understand what he was saying. I think we should realize that although we can measure pituitary hormone levels, and define changing episodic discharge patterns, we have no clue as to the significance of these "yanks and jerks" for the gonad. Sam Yen touched on this point this morning, and I agree with him that if the gonadotrophin remains bound to a gonadal receptor for a reasonable length of time, you only need to top up the blood hormone level occasionally, when the receptor becomes depleted. That seems sensible, but there is also no doubt that a rapid increase in the amplitude and/or frequency of the gonadotro-

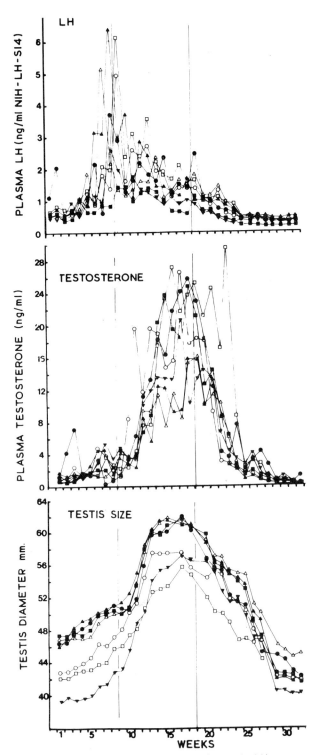

Fig. 3. Diagrammatic summary of the changes in LH, testosterone and testis size as Soay rams enter the autumn breeding season. (Data from Lincoln (1976), *J. Endocr.*, in press.)

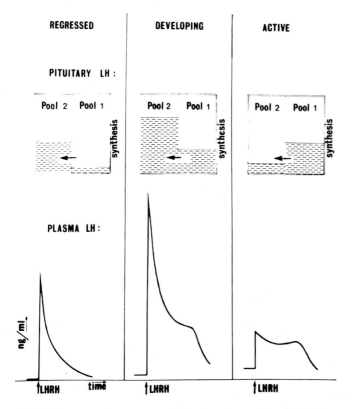

Fig. 4. Diagrammatic summary of the changing pool sizes of LH in the pituitary, and the response to a standard injection of LHRH, in Soay rams as they enter the autumn breeding season. (Data from Lincoln (1976), *J. Endocr.*, in press.)

phin firing rate can certainly produce biological effects. And we must also ask ourselves how seasonal influences can change pool sizes in the pituitary. I think that we endocrinologists need to be able to explain a great deal more about the true biological significance of the hormone levels we measure, because at the moment we really do not understand the language.

Finally, I would like to touch briefly on the behavioral consequences of circannual rhythms, and in so doing link up with what Dr. Beyer was telling us today.

As some of you may know, I have spent a great deal of time in the last 8 years working on wild red deer; red deer, because they are superb seasonal breeders; wild red deer, because, as Carlos Beyer said, to study reproductive behavior we really need to obtain information on more than just intromission frequencies. Because red deer confine their sexual activity to a circumscribed time of the year, we can use the rest of the year to investigate the *social* effects of sex hormones, and this is a most valuable attribute.

One of the things that has intrigued us is the way in which the circannual cycle appears to control the ability of the deer to respond behaviorally to the sex hormones (Lincoln et al., 1972; see Table I). In deer, we can quantitate social behavior by assessing an animal's position in the social hierarchy; the expression of male or female sexual behavior is self-evident. Testosterone in the male will always induce social aggression, at any time of the year, and the response appears to be dose dependent. But testosterone has a very different effect on male sexual behavior. Here it only has a permissive action; you cannot heighten libido by giving more testosterone, but *some* testosterone is essential for the full expression of male libido, and castration rapidly abolishes it. Even more important is the fact that male libido can *only* be shown at the appropriate time of the year, namely the autumn. Thus if a castrate is implanted with testosterone in the spring, there will be an immediate aggressive response, but no sexual response until the following autumn.

In the hind, testosterone will always induce social aggression, provided that she is not pregnant. Thus the effects of testosterone on social behavior are identical in male and female. Testosterone is also very effective at producing estrous behavior, but again, only at the appropriate time of year.

Estradiol-17β has no effect whatsoever on male social behavior apart from making the animals roar continuously, and yet minute doses (10 μg/day) are very effective at restoring male libido in castrates (Fletcher and Short, 1974). Estrogen also has no effect on female social behavior, although it is effective in inducing estrus at the appropriate time of year. These results seem to be in accord with the hypothalamic aromatization concept of Fred Naftolin and Ken Ryan, at least as far as sex behavior is concerned. But note that estradiol and testosterone have very different effects from one another on social behavior.

I hope that this evidence has persuaded you that not only do circannual rhythms induce great changes in pituitary and gonadal activity, but they also regulate some behavioral responses to the hormones. So here is a whole new level of control mechanisms that we have not even begun to think about; we are even beginning to suspect that circannual rhythms of gonadotrophin secretion may even persist in the castrate. Thus, circannual rhythms can clearly have a very dominant effect on hypothalamic activity, even in the absence of gonadal feedback.

TABLE I

Summary of effects of testosterone and estradiol-17β on the social and sexual behavior of castrate and intact male and female red deer

Data from Lincoln et al. (1972), Fletcher and Short (1974), and Fletcher (unpublished).

	Stags		Hinds	
	Social behavior	Sexual behavior	Social behavior	Sexual behavior
Testosterone	Inductive aggression any time	Permissive male libido in autumn only	Inductive aggression any time	Permissive estrus in autumn** only
Estradiol-17β	No effect*	Permissive male libido in autumn only	No effect	Permissive estrus in autumn only

* Estradiol-17β implants do make stags roar all the year round.
** Testosterone implants also eventually induce components of male libido.

I must end on a note of speculation. Why on earth should Upjohn finance a meeting such as this? What can we contribute to them? I would like to make a few suggestions.

One of the world's major meat-producing animals, the sheep, is a seasonal breeder. For half the year, ewes will not come into heat. In spite of all our sophisticated studies of the maturation of positive and negative feedback mechanisms at puberty, we still have no understanding of the mechanisms that bring about this photoperiodically controlled summer anestrus. If only we could understand how a seasonal breeder can switch on and off its seasonal puberty (for seasonal breeding can be regarded as a form of recurrent puberty), we could double agricultural productivity overnight. This would more than justify the expense of the puny research effort made so far in this area.

I hope my talk has illustrated some of the advantages of whole animal biology; but we reproductive biologists also need to think much more about what you neurophysiologists are doing. I hope that if we are impressed by your science, you will occasionally stop and listen to our biology, since your brain is after all only a servant of our gonads.

REFERENCES

Bielanski, W. (1972) Clinical observations on sexual behaviour and function of reproductive organ in she-mules. In *Riproduzione Animale e Fecondazione Artificiale,* Edagricole, Bologna, pp. 45–55.

Bielanski, W. and Zapletal, Z. (1968) Ovulation in she-mules: a report of two cases. *Proc. VI Int. Congr. Anim. Reprod. Art. Insem., Paris,* II: 1555–1558.

Chandley, A.C., Jones, R.C., Dott, H.M., Allen, W.R. and Short, R.V. (1974) Meiosis in interspecific equine hybrids. 1. The male mule *(Equus asinus x Equus caballus)* and hinny *(E. caballus x E. asinus). Cytogenet. Cell Genet.,* 13: 330–341.

Fletcher, T.J. and Short, R.V. (1974) Restoration of libido in castrated red deer stag *(Cervus elaphus)* with oestradiol-17β. *Nature (Lond.),* 248: 616–618.

Gilbert, A.B. (1972) The activity of the ovary in relation to egg production. In *Egg Formation and Production,* B.M. Freeman and P.E. Lake (Eds.), British Poultry Science Ltd., Edinburgh, pp. 3–21.

Harris, G.W. (1964) Sex hormone, brain development and brain function. *Endocrinology,* 75: 627–648.

Jewell, P.A., Milner, C. and Boyd, J.M. (1974) *Island Survivors: the Ecology of the Soay Sheep of St. Kilda,* Athlone Press, London.

Katongole, C.B., Naftolin, F. and Short, R.V. (1974) Seasonal variations in blood luteinising hormone and testosterone levels in rams, *J. Endocr.,* 60: 101–106.

Lincoln, G.A. (1976) Seasonal variation in the episodic secretion of LH and testosterone in the ram. *J. Endocr.,* in press.

Lincoln, G.A., Guinness, F. and Short, R.V. (1972) The way in which testosterone controls the social and sexual behavior of the red deer stag *(Cervus elaphus). Hormone Behav.,* 3: 375–396.

Menaker, M. (1971) Rhythms, reproduction and photoreception. *Biol. Reprod.,* 4: 295–308.

Nishikawa, Y. (1959) *Studies on Reproduction in Horses,* Japan Racing Association, Tokyo, p. 320.

Short, R.V. (1973) A general discussion of the effects of the environment on reproduction. In *The Environment and Reproduction in Mammals and Birds,* I.W. Rowlands and J.S. Perry (Eds.), *J. Reprod. Fertil.,* Suppl. 12: ix–xii.

Short, R.V. (1974) Rhythms of ovulation. In *Chronobiological Aspects of Endocrinology,* J. Aschoff, F. Ceresa and F. Halberg (Eds.), Schattauer Verlag, New York, pp. 221–228.

Short, R.V. (1975) The contribution of the mule to scientific thought. *J. Reprod. Fertil.*, Suppl. 23: 359–364.

Taylor, M.J. and Short, R.V. (1973) Development of the germ cells in the ovary of the mule and hinny. *J. Reprod. Fertil.*, 32: 441–445.

Treloar, A.E., Boynton, R.E., Behn, B.G. and Brown, B.W. (1967) Variation of the human menstrual cycle throughout reproductive life. *Int. J. Fertil.*, 12: 77–126.

Subject Index

Acetylcholine, 37, 48, 162, 163, 433
    biosynthesis of, 162
    -containing systems, increase of ACTH secretion, 239
    metabolism, 295
    neuron systems, stimulatory effect on LRF secretion, 239
Acetylcholinesterase stainings, 205
ACTH, 189
    biosynthesis, 290
    release, effects of prostaglandins on, 408
    secretion, 290
        increase by acetylcholine-containing systems, 239
        inhibition of, 239
        inhibitory effect of hypothalamic NA terminals on, 236
        involvement of 5-HT pathways in, control of, 238
Actinomycin D, 110, 292
Adenohypophyseal hormone secretion, control by prostaglandins, 407
Adenohypophysis, 453
    possible interaction with neurohypophysis, 454
Adenosine 3′, 5′-cyclic phosphate (cyclic AMP), 391
Adenylate cyclase, 392
Adenylate cyclase system, 393
Adenylyl cyclase, 383, 385
Adrenalectomy, effect on NA turnover, 236
Adrenaline, 193
    neuron systems, 199
Adrenal medulla, 45–50, 53
Adrenal progesterone, 256
Adrenergic cells, 36
Adrenergic mechanism, 431
Afferent pathways, 167
Affinity column technique, 113
Age, 73
Albumin, tryptophan binding to, 159
Amine interactions, 170
Amine metabolism, 168
Amine–hormone interaction, 173, 174
Aminergic innervation, pathway of, 171
Amino acids, 64
    compartmentation of, 66
    incorporation rate, 67, 293
    metabolism, 152
        daily rhythms, 152

    postprandial overflow of, 152
    precursors, 152
    release of, 132
Aminopeptidases, 140, 141
Aminophylline, 49
Amphetamine, 49
Amygdala, 19, 37, 278, 280, 281, 283, 368, 374, 377
    corticomedial, 282
    medial, 424
Androgens, 38, 267, 268, 347, 350, 475, 478
    administration, 329
    adrenal, 479
    aromatizable, 280, 477
    binding, 269
    binding components, 269
    mechanism of action, 350
    metabolic pathways of, 348
    metabolites, 267
    non-aromatizable, 477
    receptors, 267, 268, 280
Androgenic molecules, aromatization to estrogen, 338
Androstan-3α,17β-diol, conversion into, 328
Androstenediol, 267
    effects on copulatory behavior, 478
Androstenedione, 252, 253, 281
    concentration of, 257
    effects on copulatory behavior, 478
Angiotensin, 119, 140, 143
    actions of, 141
    cleavage by brain enzymes, 138
Angiotensinases, 141
Annulate lamella, 10
Anorexia nervosa, 457
Anterior hypothalamus, 364
Anterior periventricular hypothalamic nucleus, 195
Anterior pituitary, 358
    gland, 392
    hormones, release of, 397
    in vivo effect of prostaglandins on, 400
Antidromic activation, 442
Antidromic identification, 34
Antiestrogenic compounds, 292
Antiestrogens, 476
    non-steroidal, 278
Antipsychotic drugs, 385
Arcuate median eminence, 426
Arcuate neuron, 9, 12

of castrate, 11
Arcuate nucleus (ARC), 2, 4–6, 9, 13, 195, 199, 209, 424
    development of, 13
    modified by neonatal testosterone, 13
    multiunit activity, 430, 434
    neurons, antidromically identified, 436
    norepinephrine as activator of, 436
    ventromedial region, 2
Aromatizable androgens, 280
Aromatization, 347
    hypothesis, 478
    in NET, 350
    in vitro by rat brain tissues, 352
    in vitro studies, 348
    of in vivo NET, 351
    system, 282
        in NET, 350
Arousal mechanism (AM), 473
Aspirin, 417
Astrocyte, 12
Astrocytic process, 12
ATP, 50, 51, 121
ATPase, 120
Autoradiography, 270, 277
Axo-axonic interaction, 209
Axo-axonic synapses, 179
Axons
    degeneration, 13
    flow of proteins, 71
    intraventricular, 23
    terminals, depolarization of, 441
    transport, 110
Axoplasmic transport, 185

Bed nucleus of stria terminalis (BST), 424
    norepinephrine as activator of, 432
Behavior, 277, 456
    estrogen effects on, 291
    patterns of, 384
    in rats, 159
    reproductive, 510
    sexual, 21
    social, 510
Bioenergetic processes, 364
Biosynthesis, 150
Biosynthetic enzymes, 186
Biotransformations, 357
Birds, 504
Body weight–pituitary responsiveness, 494
Bradykinin, 142
Brain, 249, 256, 263, 270, 349
    acetylcholine synthesis, 162

catecholamine synthesis, 160
choline levels, 162
cytosol, estrogen binding by, 352
cytosol receptor, 284
deamidase, 136
differentiation, 349
enzyme activities, 293
enzymes, 129
function, biochemistry of, 386
glucocorticoid receptors, 291
neurons, serotonin synthesis in, 150
neurotransmitter, 151
programming, 349
protein synthesis, rates of, 67
regions, 291
serotonin, 153, 154, 160
    concentration, diet-induced changes in rat, 158
    diet-induced changes in, 160
    levels, 159
    synthesis, 151, 159
structures, 331
    oxidative activity of, 363
tissues, 264, 277, 424, 434
    estrophilic molecules in, 476
tryptophan, 156
tryptophan levels, diet-induced changes in, 159
tyrosine, 160
Bundle of Schutz, 171
2-Brom-α-ergocryptine, 456
Burster neurons, 36

Caffeine, 48, 49
Calcium, 49, 50, 54, 187
    extracellular, 46
    influx, 47
    intracellular, 47, 48
    ionophores, 47
    –microtubule interaction, 50
    requirement for LH-RH action, 395
    role in hormone storage, 45
    role in transport, 45
Carbohydrate consumption, 154
Carboxyamide peptidase, 139, 143
Carboxypeptidase N (kininase I), 142
Carboxypeptidase (kininase II), 142
Castrated animals, effect of testosterone on sexual behavior in, 478
Castration, 38
    cell, 9
    effect on oxidative metabolism, 374
Catalytic mechanisms, 139

Catecholamines, 168, 185, 294, 383, 423, 434, 453, 456
   biosynthesis of, 186
   mechanisms, 433
   neuronal input, 455
   neurons, 149
   release, 46
Catechol estrogens, 266, 357, 359, 360
Catechol-O-methyltransferase, 189, 360
Cathepsin A, 143
Cathepsin M, 143
Celite microcolumns, 252
Cells
   adrenergic, 36
   continuously active, 36
   fractionation, 277
   labeling of, 278
   mechanisms, 174
   nuclear binding, 284
   nuclear estrogen–receptor complex, 278
   nuclear labeling, 277, 284
   nuclear labeling pattern, 286
   nuclear retention, 282
   silent, 36
Central aromatization, substrates and products in, 351
Central nervous system (CNS), 357, 454
   circadian rhythm, 426
   clock, 460
   mechanisms, 423
      critical period, 423
   role of, 456
Central tissues, transformations in, 359
Cerebral cortex, 280, 358, 364, 367, 376
   during development of, 334
Cerebral proteins, half-life of, 67
Cerebral tissue, 363
Cerebrospinal fluid (CSF), 256
   content of, 23
   transependymal pathway, 24
Chemical signal, 475
Chimpanzee, 505
Chlorimipramine, 160
p-Chlorophenylalanine, 161
Chlorpromazine, 49
Choline acetylase, 205
Choline content, dietary, 162
Cholinergic mechanisms, 433
Cholinergic neuron systems, 205
Cholinergic regulation, 170
Chromaffin granules, 50
Chromatography, 111
Circadian gonadotropic release, 170

Circadian multiplier, 507
Circadian rhythmicity, 424, 426, 506
Circannual rhythms, 506, 510
Cleavage
   of analogs, 132
   rate-limiting, 132, 137, 140
Clomiphene, 292
Clonidine, 234
Coital stimulation, patterns of, 490
Colchicine, 110
Competitive interaction, 265
Conjugated free estrogens, 358
Constant light, rat exposed to, 489
Contractile activity, regulation by calcium, 54
Contractile process, 55
Copulatory behavior, 474
   effects of androgens on, 478
Copulatory information, storage of, 490
Copulatory mechanism, 471
Cortex, 250, 253, 257
Corticosteroids, 188
Corticosterone, 255, 290
[^{3}H]Corticosterone, 283
Corticotropin-releasing factor (CRF), 4, 15, 290, 408
   release, inhibition of, 240
   secretion, inhibitory effect of NA on, 236
Corticotropin-releasing hormone (CRH), 176
Cortisol, 255, 283
Cranial tissue, 248
C-terminal fragments, release of, 138
Cycle length, 505
Cyclical reproductive activity, 423
Cyclic AMP, 49, 54, 187, 189, 381, 382, 391, 392
   accumulation in anterior pituitary gland, 394
   accumulation in anterior pituitary tissue, 396
   change in, 383
   concentration, 394
   functions, 383
   inhibitory effect of somatostatin on, 394
   intracellular, 411
   neuronal role of, 384
   role, 383
   stimulatory effect of LH-RH on, 392
   stimulatory effect of purified GH-RH on, 393
   stimulatory effect of TRH on, 393
Cyclic GMP, 383
   role, 383
Cyclic gonadotropin release, 423, 431, 460
Cyclic nucleotide phosphodiesterase, 392
Cyclic nucleotides, 53, 381

Cycloheximide, 292
Cyproterone, 282
   inhibitor of the aromatizing enzyme, 282
Cysteine-^{35}S, 109
Cytoimmunochemistry, 174
Cytology
   changes, 11
   neurosecretory, 7
Cytopathology, 12
Cytophysiological organization of the MPOA, 22
Cytoplasmic filamentous bodies (CFB), 7
Cytoplasmic fraction, 264, 271
   radioactivity in, 264
Cytosol, 279
   binding, 284
   binding pattern, 286
   in hippocampus, 288
   receptor, 278

Daily rhythm, 428
Deafferentation, 442
Deamidation, 135, 140
Deamido LRH, 119
Deamido TRH, 119
Decarboxylase enzyme, 150
Decarboxylase inhibitor, 161
Degradation system, 119
Dendritic spines, 21
   non-strial synapses on, 21
Desulfation, 359
Dexamethasone, 284, 290
Dietary choline content, 162
Dietary protein, 156
Diethylstilbestrol, 280
Diet-induced changes, 159, 160
5α-Dihydroprogesterone (5α-DHP), 307, 320
   in vivo uptake of, 315
   metabolism of, 315
   progesterone-like effects of, 320
20α-Dihydroprogesterone (20α-DHP), 307, 312
   in vitro metabolism of, 306
5α-Dihydrotestosterone (DHT), 252, 265, 281, 327, 339, 352
   conversion of testosterone into, 258, 330
   transformation of testosterone into, 330
Dihydroxyphenylalanine (DOPA), 150
Dimorphism, sexual, 21, 22
3α-Diol, 339
   transformation of testosterone into, 330
Disulfide bridge, 139
Diurnal rhythmicity, 426
DNP-TRH, 117
DOPA, 186, 234

   synthesis, 161
L-DOPA, administration of, 457
Dopadecarboxylase immunofluorescence, in DA cell bodies, 195
Dopamine (DA), 173, 186, 193, 410, 431, 434, 457
   cell bodies
      dopadecarboxylase immunofluorescence in, 195
      mesencephalic periaqueductal, 198
      tyrosine immunofluorescence, 195
   -containing nerve terminals, 209
   control, 213
   effect on GH secretion, 236
   fluorescence, induced by H44/68, 221
   implantation of, 433
   mechanism, 221
   microelectrophoretically applied, 37
   mode of action, 433
   nerve cells, 195
   neurons, 4
      in control of GH and TRF secretion, 230
      in control of LRF and prolactin secretion, 230
      in control of sexual behavior, 230
   neuron systems, 193, 215
   pathways
      lateral controlling the LRF secretion, 215
      medial controlling prolactin secretion, 215
      two types of, 215
   receptor agonists
      apomorphine, 230
      ergocornine, 225
      ET-495, 225, 230
      CB 154, 225
      pimozide, 225
   terminals, mediate LRF secretion, 219
   transmitter, 441
   turnover, 215, 216, 234
      increase by ovine prolactin, 226
      study with H44/68, 217
Dopamine-β-hydroxylase, 186
Dopamine-β-hydroxylase inhibitors, 234
Dopaminergic tuberoinfundibular neurons, 173
Double antibody method, 120
Drug treatments, 160

Effector neurons, 167
Electrical stimulation, 46, 424, 474
Electrolytic lesions, effects on oxidative metabolism, 376
Electrophoresis, 39, 111

Electrophysiological correlates, 33
Electrophysiological localization, 37
Electrophysiological study, 433, 441
Endocrine influences, 75
Endocrine neurons, 33, 40
    electrophysiological analysis of, 40
Endogenous 5-HT, 173
Endogenous steroid levels, in made rat, 250
Endogenous steroids, in neuroendocrine tissues, 247
Endopeptidases, 129, 130, 134, 140
Endoplasmic reticulum, 7
Energy mobilization, 382
Energy, sources of, 382
Enzymatic conversions, 307
Enzyme
    activation of, 189
    cleavage, 110
    degradation, 188
    distribution, 133
    inhibition of, 189
Ependymal tanycytes, 23
Epinephrine, 151, 186
    release of, 45
Ergotropic mediator, 383
Ergotropic system, 382
Estradiol, 249, 253, 265, 267, 295, 358, 424, 426, 428, 466, 511
    binding capacity, 278
    binding by neural and pituitary tissues, 257
    -concentrating neurons, 425
    concentration, 250
    implantation of, 430, 443
    negative feedback sites of, 444
    physiological amounts of, 249
    potentiating, 425
    stimulatory action of, 423, 425
    treatment, 13
    uptake of, 264
Estradiol-^{3}H, 249
17β-Estradiol, 280
    secretion of, 424
Estradiol 17β-oxidoreductase, 358
Estriol, 424
Estrogens, 38, 263, 264, 266, 280, 347, 350, 423, 475, 476
    actions
        central, 349
        genomic involvement in, 292
    antagonists, 266
    binding, 357
    binding protein, 268, 280
    binding sites, 268
    biotransformations, 361
    dynamics, 268
    effects
        on behavior, 291
        on male sexual behavior, 479
    feedback action of, 461
    feedback action on LH secretion, 216
    formation in NET, 353
    metabolism, 357
    and NA neuron activity, 233
    neurochemical effect of, 292
    in neuroendocrine control, 353
    -neuron system, 443
    production rate of, 248
    receptors, 264, 277, 278, 360, 461
        intracellular, 293
        levels, 267
        neonatal, 280
        physicochemical characterization of, 265
        sites, 279, 291
        translocation, 267
    replenishment, 266
    -stimulated neurophysin, 83
        mid-cycle, 92
    treatment, 458
Estrone, 253, 357, 358, 424, 426
    diurnal fluctuation of, 429
    implantation of, 427
    implanted in MPO, 427, 428
    stimulatory action of, 423
Estrone—estradiol interconversion, 359
Estrone-^{3}H, 249
Estrophilic molecules, in brain tissues, 476
Estrophilic neurons, 278
Estrous cycle, 364, 367, 424, 426, 430
Estrus
    behavioral, 292
    persistent, 430
ET-495, DA receptor agonist, 221
Exocytosis, 55, 187
Exogenous progesterone, 256
Exogenous steroid administration, in NET, 249
Exopeptidases, 129, 130
Extracellular recordings, 34
Extrahypothalamic areas, metabolic activity of, 367
Extrahypothalamic control, 36
Extrahypothalamic limbic structures, 20

False transmitter, 189
Feedback effect, 426
Feeding, response to, 382
Female rats, 359

Female sheep, 357
Fetal brain, 350
Fetoneonatal, 280
α-Fetoprotein, 268
Fluorescence intensity, 197
Follicle-stimulating hormone (FSH), 39, 391, 424
    circadian release of, 428
    pulsatile secretion of, 454
    release, 170, 320, 414, 418, 465
        effects of systemic injections of testosterone and of its 5α-reduced metabolites, 336
    secretion of, 221, 339, 373
    serum levels of, 309
Follicular development, 504
Food consumption, 149, 152
Free amino acid pool, 65, 66
Fusion process, 52

GABA, 66
    -containing neurons
        inhibition of ACTH secretion, 239
        inhibition of CRF release, 240
        mechanisms, 205
        stimulation of LH secretion, 239
Genomic activity, 289
Geoffrey Harris, 503
Glia, 6
Glial cells, 288
Glucocorticoid, 283, 285, 290
    multiple binding sites, 287
    receptors, 287
        in adult neuroendocrine tissues, 283
    synthetic, 284
    —target cell interactions, 284
Glucose, biotransformation of, 363
Glucose-6-phosphate dehydrogenase, 293
Glutamic acid, 66
Glutathione, 115
Glycine, 66
Golgi apparatus, 15
Golgi cisternae, 7
Gomori stains, 81
Gonadal steroids, 38, 461, 475
    effects, time dependence of, 294
    feedback, 11
    organizational actions of, 295
Gonadectomy, 13
Gonadotrophs
    functional behavior of, 453
    functional capacity of, 461, 464, 467
    as target cells, 453, 460

Gonadotropic pathway, final common, 1
Gonadotropic secretion, 173
Gonadotropins, 38, 411
    control, 350
        electrophysiological analysis of, 36
        mechanisms, 349, 360, 441
    cyclic release, 430, 431
    feedback control of, 338
    normal ovulatory surge of, 491
    output, 492
    pulsatile release, 454
    regulation, 305
        role of preoptic area, 21
    release, 294, 367, 425, 432
        negative and positive feedback control systems on, 466
        pulsatile pattern of, 457
    -releasing factor (GRF) systems, 4
    -releasing hormone (Gn-RH), 81, 97, 99
        localization of, 95
    -releasing neurohormones, 167
    rhythmic release of, 424
    secretion, 215, 221, 432
        animal models for study of the control of, 489
        control of, 20, 377, 485, 493
        steroid hormone effects on, 486
Growth hormone (GH), 124, 391, 457
    release, 394, 409, 418
    secretion
        increase by NA, 236
        inhibition by DA, 236
    -release inhibiting hormone (GH-RIH), 391
    -releasing hormone (GH-RH), 391
        effect on cyclic AMP, 391
        stimulatory effect of, 393

H44/68, 216, 221
Hamster, 507
Head, 248
Hen, 504
Hippocampal cytosol, transcortin-like binding in, 288
Hippocampal tissue, 286
Hippocampus, 281, 283, 368
    of female rat, 367
Histamine containing neurons, 199
Hormonal control, 188
Hormonal regulation of the synthesis and metabolism of neurotransmitters, 185
Hormonal stimulus
    for female sexual behavior in primates, 477
    nature of, 475

for sexual behavior, 475
Hormone—receptor interactions, 383
Hormones
    calcium for storage of, 49
    concentrations in NET in sheep, 249
    radiolabeled, 263
    release of, 45, 46
    storage, 45
    transport of, 50
Human ovaries, post-menopausal, 506
Humoral inputs, 167
2-Hydroxyestradiol, 189, 360
2-Hydroxyestrone, 357, 360
5-Hydroxyindole, 156, 160
5-Hydroxyindole acetic acid (5-HIAA), 154–156, 163
Hydroxylase enzymes, 151
2-Hydroxylation, 359
    of estrogens by human fetal neuroendocrine tissues, 360
20α-Hydroxysteroid dehydrogenase enzyme system, 249
5-Hydroxytryptamine (5-HT), 168, 193
    inhibition of TRF secretion, 238
    neuron systems, 199, 238
    neurotransmission, 238
    pathways, 238
        involvement in control of ACTH secretion, 238
    turnover, effects of estrogen on, 238
5-Hydroxytryptophan (5-HTP), 150, 163
Hyperprolactinemic state, 458
Hypophysectomized animals, oxygen uptake of the hypothalamus, 371
Hypophysectomy, effect on oxidative metabolism, 373
Hypophysiotropic area, 2, 4, 33, 281
Hypophysiotropic factors, hypothalamic, 109
Hypophysiotropic hormones, biosynthesis of, 110
Hypophysiotropic substances, 40, 41
Hypophysiotropic system, 40
Hypophysiotropins, 94
Hypophysis, neuroendocrine regulation of, 453
Hypothalamus, 167, 171, 193, 198, 199, 250, 253, 257, 278, 283, 350, 358, 359, 363, 374, 376, 408, 431, 432, 461
    anterior, 18, 172, 265
    area, 430, 475
        neuronal firing in, 475
    aromatization concept, 511
    basal, 38
    culture, 111

cyclic AMP, in vivo effects of prostaglandins, 400
    dysfunctions, 457
    extracts, 121, 124
        TRH biosynthesis in, 115
    function, 456
    hormones
        cellular localization of, 193
        interactions in hypothalamic and extra-hypothalamic control of pituitary function and sexual behavior, 193
    hypophysiotropic factors, 109
        biosynthesis of, 109
        degradation of, 109
    –limbic system, 352
    LRF-containing neurons in, 205
    medial, 364
    medial basal (MBH), 1, 36, 40, 174, 176
        nuclei in, 2
    NA terminals, inhibitory effect on stress-induced increases in ACTH secretion, 236
    neonatal, 280
    neurons, electrical activity of, 39
    nucleus, 5, 210, 215
    oxidative activity, 368
    peptides, cleavage of, 131
    periventricular, 1
    -pituitary axis of the rat, sexual differentiation of, 332
    and pituitary oxidative metabolism of male rats, 370
    -pituitary system, 467
    posterior, 364
    production of releasing hormone, 365
    5α-reductase activity, 307
    regulatory hormones, 392
    releasing factors, 36
        biosynthesis, 179
        inactivation of, 133
    sexual dimorphism, 488
    single-unit activity, 33
    tissue, 115, 360, 416
Hypothalamo-pituitary-gonad system, in immature animal, 493

Immunohistochemical studies, 5, 199
Immunoprecipitation, 120
Immunoreactive cells, 5
Implantation experiment, 425
Inactivation, 137, 140, 188
Incorporation studies, 112
Incubation studies, 120
Indomethacin, 417

Infundibular stalk, 210
Infundibular tract, MPOA, 23
Insulin, 154
Integrator structures, 167
Interpeptide, 111
Intracellular processes, 475
Intracellular recordings, 34
Intrahypothalamic and intrapituitary implants of testosterone and of its 5α-reduced metabolites, 339
Intramammary pressure, changes in, 35
Intrapituitary dynamics, 467
Intraventricular axons, 23
3-Iodotyrosine, 189

Kallikrein, 142
Kininase I, 142
Kininase II, 142
Kininogenin, 142
Kinins, 141, 143
    cleavage by brain enzymes, 138
    release of, 142

Lactate dehydrogenase (LDH), 175
Lamina terminalis, 5, 99
    organum vasculosum of, 25
Lateral palisade zone, 219
Libido, male, 511
Limbic deafferentation, 430
Limbic forebrain, 423
Limbic neuronal impulses, 426
Limbic structures, 19, 430, 444
    extrahypothalamic, 20
Limbic system, 367, 376, 381, 383, 384
    nuclei of, 198
Limbic tissues, 350
Lipofuscin, 13
Long feedback, 370
Lordosis behavior, 475, 476
Luliberin (LH-RF), 131, 132
    breakdown of, 133
    inactivation of, 134
Luteinizing hormone (LH), 170, 391
    circadian release of, 428
    ovulatory surge of, 424
    pulsatile secretion of, 454
    pulses, 455
    releasable LH, two components of, 462
    release, 170, 320, 411, 418, 424, 434, 442, 465, 490
        during sleep, 460
        effects of systemic injections of testosterone and of its 5α-reduced metabolites, 336
        mating-induced, 490
    secretion, 24, 221, 234, 339, 373
        feedback action of estrogen on, 216
        pulsatile, 24
        stimulation of, 239
    serum concentration of, 424
    serum levels of, 309
    spontaneous pulses in, 461
    surge, 292
Luteinizing hormone-releasing factor (LRF), 38, 39, 461
    cells, localization of, 5
    cellular organization, 1
    cerobrospinal fluid (CSF), release, 23
    concentrations of, 5
    -containing cell bodies, 210
    -containing nerve terminals, 209
    -containing neurons, in hypothalamus, 205
    in control of sexual behavior, 210
    delivery structures, 1
    delivery systems, 1
        location of, 1
    endogenous release, 465
    facilitates sexual behavior, 210
    immunofluorescence, 202
    infusions of, 494
    infusion into the 3rd ventricle, 23
    intraventricular, 24
    neurosecretory cells, 23
    as a neurotransmitter, 210
    neurovascular release of, 25
    pulsatile pattern of, 454
    pulsatile secretion, 455
    release, 489
        from intraventricular terminals, 23
    secretion, 215, 219, 224, 225, 238
        inhibitory estrogen feedback on, 219
        stimulation by acetylcholine neuron systems, 239
    self-priming effect of, 465
    tuberoinfundibular delivery system, 6
    tuberoinfundibular neurons, 14, 19
        source of, 14
    tuberoinfundibular system, 9, 18
Luteinizing hormone-releasing hormone (LRH, LH-RH), 293, 391, 392, 415
    activity, 124
    biosynthesis, 112
    carrier, 114
    distribution of, 174
    effect on cyclic AMP, 391
    labeled, 114

producing neurons, 170
radioactive, 179
release, effect of prostaglandins on, 399
-secreting neurons, 416
separation, 120
stimulatory effect of, 392
Lysosomes, 15

Magnesium, 47
Magnocellular system, 33, 40
Male rat, 256
Malnutrition, effects of, 74
Mammalian brains, 149
Mechanisms of inactivation, 135, 138
Medial forebrain bundle, 171
Medial palisade zone (MPZ), 219
Medial preoptic area (MPOA), 20, 36, 219, 426
    activity, 434
    circadian fluctuations of, 430
    cytophysiological organization of, 22
    dopamine as activator for, 433
    infundibular tract, 23
    LRF delivery system, 22
    neurons, 434
Median eminence (ME), 2, 4, 5, 81, 193, 205, 210, 215, 216, 224, 234, 270, 271, 400, 440
    DA mechanism in, 221
    DA nerve terminals, 224
    neurovascular junction of, 36
Medulla oblongata, 199
Melanostatin (MIF), 119, 131, 136
Membrane hyperpolarization, 385
Menstrual cycle, follicular phase of, 467
    human, 506
MER-25, 292
Mesencephalon, 199
Metabolic activity, 365, 367
Metabolic clearance rate (MCR), 248, 249
Metabolic transformation, 258
Metabolites, 281
15-Methyl PGE$_2$, 414
α-Methyl-tyrosine methylester (H44/68), 217
Microelectrophoresis, 35
Microelectrophoretically applied drugs, 38
Microfilaments, 51, 187
Microfluorometry, 216
Microiontophoretic infusion, 436
Microtubules, 50, 51, 187
Midbrain, 37, 281
Mid-cycle surge, 465
Mineralocorticoid receptors, 289
Mineralocorticoids, 285

Mitochondria, 15
Mitochondrial fraction, 124
Modulatory mechanism (MM), 473
    mode of action of, 474
Monoamines, 150, 173, 432
    biosynthesis of, 185
    fluorescence, 200
    neurons, biochemical activity of, 475
    neurons and pathways, 474
    projections, 168
    regulation, 168
    release of, 187
    storage of, 187
    system, neuronal discharge of, 475
Monoamine oxidase (MAO), 187
    activity, 458
    inhibitors, 162
Morphine-like factor, 386
Morphine research, 385
Mule, 504
Multi-unit background activity technique, 38
Multi-unit technique, 38
Muscarinic excitation, 35
Myelin figures, 13

Neonatal androgenization
    effect on amygdala QO$_2$, 368
    in female rats, 488
    effect on hippocampus QO$_2$, 368
Neonatal critical period, 280
Neonatally castrated male rats, 488
Neonatal tissues, 350
Neural activity, 290
Neural inputs, 167, 489
Neural stimuli, 425
Neuroendocrine control, 169, 391
    role of cyclic AMP in, 391
    of sexual behavior, 476
Neuroendocrine function, 277, 291, 294
Neuroendocrine mechanisms, 457
    in sexual behavior, 471
Neuroendocrine regulation, subcellular mechanisms in, 453
Neuroendocrine secretion, 45
Neuroendocrine system, 361
Neuroendocrine tissues, 81, 167, 253, 305, 347, 357, 363
    adult, 280
    androgen aromatization by, 347
    androgen reduction by, 327
    aromatization, in vivo, 351
    endogenous steroids in, 247
    formation of estrogens in, 353

hormone concentrations in, 249
protein metabolism in, 63
specific binding of steroids by, 263
steroid receptor function in, 289
steroid receptors in, 277
Neuroendocrinology, 364
Neurohemal organ, 25
Neurohormones
   in vitro release of, 176
   subcellular distribution of, 174
   synaptosomal distribution of, 178
Neurohypophysial system, 40
   electrophysiological analysis of, 33
Neurohypophysis, 55
   possible interaction with adenohypophysis, 454
Neuronal activity, 170
Neuronal membrane, 475
Neuronal perikarya, 5
Neuronal uptake, 190
Neurons, 288
   arcuate, 12
   burster, 36
   catecholaminergic, 149
   endocrine, 33
   hypothalamic, 38
   neurohormone-containing, 173
   preoptic, 38
   serotoninergic, 149
   whorled body containing, 12
Neurophysins, 81, 82
   estrogen-stimulated, 83
   localization of, 83
   nicotine-stimulated, 83
   in the suprachiasmatic nucleus (SCN), 85
   and the ventricular system, 86
   in the zona externa of the median eminence, 87
Neurophysin I, 82
Neurophysin II, 82
Neurosecretory cells, 16, 40
Neurosecretory cytology, 7
Neurosecretory material, 81
Neurosecretory neurons, 443
Neurosecretory peptides
   localization of, 81
   turnover of, 129
Neurosecretory theory, 82
Neurosecretory tissue, 34
Neurotransmission, serotoninergic, 160
Neurotransmitters, 35, 149, 185, 188, 434
   cellular localization of, 193
   facilitatory, 438

interactions in hypothalamic and extrahypothalamic control of pituitary function and sexual behavior, 193
interactions with neuroendocrine tissue, 167
metabolism and action, 294
synthesis, 149
   control by precursor availability, 149
   food consumption, 149
systems
   hypothalamic and extrahypothalamic, 215
   mapping of, 193
Newborn rat, 268
Newt brain, 117
Nicotine, 224
   reduction of FSH, LH and prolactin secretion, 221
   -stimulated neurophysin, 83
Nicotinic excitation, 35
Nissl bodies, 7
Nissl body—myelin figure complex, 16
Non-neuronal cells, 188
Non-ribosomal mechanism, 110
Non-steroidal antiestrogens, 278
Noradrenaline (NA), 193
   effect on GH secretion, 236
   neuron systems, 198, 232
   turnover, 215, 216, 234, 294
      after adrenalectomy, 236
Noradrenergic input, 433
Noradrenergic pathway, 434
Norepinephrine, 35, 48, 431, 432, 434, 436, 438
   implantation of, 433
   microelectrophoretically applied, 37
   release of, 45
19-Nor progestin, 265
Nuclear binding capacity, 278
Nuclear envelope, 16
Nuclear fraction, 264
Nuclear labeling, 352
Nuclear translocation process, 264
Nucleolus, 15, 16
Nucleus, 16
   arcuatus, 2, 4–6, 9, 13, 193
   caudatus, 219
   paraventricular (PVN), 34, 40, 81
   preopticus pars suprachiasmatica (POSC), 432
   raphe dorsalis, 199
   raphe medianus, 199
   suprachiasmaticus, 171, 199
   supraoptic, 34, 40, 81

ventromedial, 37

6-OH-DA-induced lesion, 234, 236
Ontogeny, 268
Opiate receptors, 386
Optic crest, 210
Orchidectomy, 374
Organum vasculosum, 5, 99
Organum vasculosum laminae terminalis, 25, 423, 441, 443
Ovarian estradiol, 460
Ovarian steroids, 427
    cyclicity of, 462
Ovariectomized rats, 409
Ovine prolactin, increase of DA turnover, 226
Ovulation, 170, 424, 425
    pharmacological studies on, 224
Ovulatory LH surge, 426
Ovulatory mechanism, transmitters for, 431
Ovulatory rhythms, 504
Oxidative metabolism, 363
    effects of castration on, 374
    effects of electrolytic lesions on, 376
    effects of hypophysectomy on, 373
17β-Oxidoreductase system, 249
Oxygen consumption (QO_2), 363, 364
    of anterior hypothalamus of androgenized females, 366
Oxygen uptake, 363, 367, 374
    of hypothalami and pituitaries of castrated animals, 371
    of the hypothalamus of hypophysectomized animals, 371
Oxytocin, 34, 40, 119, 138
    cleavage by brain enzymes, 138
    localization of, 90
Oxytocinase, 138

Parolfactory area, 1
Parvicellular system, 33, 40, 94
Pathological changes, 76
Peptidase, 131
Peptide hydrolases, 140
Peptides
    biodegradation of, 129
    hormonally active, 129
    labeled, 112
Pericapillary neuron, 16
Periventricular hypothalamic nucleus, 193
Periventricular zone, 433
Persistent estrous rat, 434
Persistent vaginal cornification, 489
Phenylethanolamine-N-methyltransferase, 188

Phosphorylation, cyclic AMP-activated, 51
Photoperiodic stimulus, 507
Physiological stimulus
    neural, 489
    humoral, 489
Pimozide, a DA receptor blocking agent, 221
Pineal gland, 100
Pituitary, 176, 250, 253, 256, 257, 263–265, 271, 277, 278, 281, 282, 292, 349, 409, 461, 493
    capacity, 462
    cell nuclei, 287
    cells, 408
    function, 193, 465
        regulation of, 454
        role of CNS in regulation of, 456
    glands, 290
    posterior, 45–50, 53
    gonadotropin, two pools of, 462
    hormone release, 407, 418
    oxygen consumption, 371
    5α-reductase activity, 307
    reserve, 462, 463
    sensitivity and reserve, 465
    stalk, electrical stimulation of, 40
    tropic hormones, episodic secretion of, 493
Plasma, 257
    LH concentration, mono- or biphasic increase in, 494
    prolactin, 410
PMS, in prepuberal rats, 221
Polycarboxypeptidase, 141
Pons, 199
Porcine, 121
Portal blood flow, mechanism(s) in the regulation of, 454
Positive feedback, 116
Postejaculatory refractory period, 474
Postprandial overflow, 152
Postsynaptic receptors, 187
Precursor availability, 149
Precursor molecule, 110
Precursor pool, 65
Preoptic area, 1, 5, 19, 167, 198, 210, 278, 280, 434
    neuronal firing in, 475
    role of, 21
    tuberoinfundibular pathway, 21
Preoptic-hypothalamic system, 425
Preoptic-infundibular neurosecretory pathway, 441
Preoptic neurons, electrical activity of, 39
Preoptic-suprachiasmatic region, 5

Preoptic system, 423
Prepuberal rat, 492
Preservation of the species, 381
Presynaptic receptors, 187
Proangiotensin, 141
Progesterone, 38, 256, 265–267, 282, 284, 289, 464, 476
  effects on sexual behavior, 488
  exogenous, 256
  in vitro metabolism of, 306, 310
  in vivo uptake of, 315
  -like effects, 320
  metabolism, 305, 315
  metabolites of, 307, 316
  organ clearance, 248
  organ extraction, 248
  physiological amounts of, 249
  production rate of, 248
  radioactive, 248, 252
  receptor-like components, 270
  receptor-translocation mechanism for, 270
  5α-reductase, tissue localization of, 309
  -sensitive neural functions, 305
  serum levels of, 309
  tissue concentration of, 250
Progestins, 266, 289
Prohormone, 110, 179
Prolactin (PRL), 457
  inhibiting activity, 176, 391
  inhibiting factor (PIF) activity, 175, 410
  pulses, 455
  release, 171, 173, 409, 490
  releasing activity, 124
  secretion, 215, 221, 224, 226, 234, 238
Prostaglandin (PG), 383, 391, 407, 411, 418
  control of adenohypophyseal hormone secretion, 407
  derivative, 413
  effect of, 396
  effects on ACTH release, 408
  effect on cyclic AMP, 391
  effect on LH-RH release, 399
  effect on the release of gonadotropins, 411
  effects on the release of prolactin, 409
  effects on the release of TSH, 408
  in vivo effects on anterior pituitary, 400
  in vivo effects on hypothalamic cyclic AMP, 400
  role in control of LH release, 418
  stimulatory effect on the release of anterior pituitary hormones, 397
  synthesis, inhibitors of, 417
Prostaglandin E_1 (PGE_1), 408, 410

PGE_2, 408, 415
  intraventricular, 413
$PGF_{1\alpha}$, 408
$PGF_{2\alpha}$, 408
Prostate cytosol, 269
Proteinase activity, 140
Proteins
  axonal flow of, 71
  breakdown of, 129
  estrogen-binding, 280
  metabolism, 63, 74
    alterations of, 74
    measuring rates of, 64
  secretion of, 71
  synthesis, 65
    inhibitors of, 124
  turnover, 64, 65
    developmental changes in, 73
    development of, 72
    nutritional influences on, 74
Puromycin, 110
Putative neurotransmitters, 66
Pseudocyesis, 455
Puberty, mechanism for onset of, 460
PVA, acetylcholine as activator for, 433
Pyroglutamyl moiety, 136
Pyroglutamyl peptidase, 136

Radioimmunoassay techniques, 253
Ram, 508
Raphe serotonergic neurons, discharge of, 474
Rat brain, 277, 283
Rat brain tissues, in vitro aromatization by, 352
Rate-limiting enzymes, 152
Receptors, 357
  androgen, 280
  activity, control of, 266
  -blocking agents, 416
  capacity, 288
  characterization of, 265
  function, steroid, 289
  occupancy, 288
  sites, intracellular, 277
Recurrent pathways, 37
Red deer, 510
5α-Reduced metabolites
  conversion of testosterone into, 329
    in the CNS in vivo, 329
    in the anterior pituitary in vivo, 329
  effects of intrahypothalamic and intrapituitary implants, 339
  effects of systemic injections of, 336
  effects of treatment on serum LH and FSH

levels of adult castrated male rats, 337
5α-Reductase activity
   of anterior pituitary, 332, 334
   of cerebral cortex, 334
   effects of androgen administration on, 329
   effects of castration on, 329
   in hypothalamus, 307, 334
   in pituitary, 307
   sex-linked differences in, 335
   substrate specificity of, 313
Regional heterogeneity, 71
Releasing activity
   prolactin, 124
   growth hormone, 124
Releasing factor (RF), delivery systems, 4
Releasing hormones
   biosynthesis of, 124
   cell-free synthesis of, 124
   hypothalamic production of, 365
Renin, 141
Reproductive tract, 270
Reticular formation, 199
Retrograde invasion, 34
Reuptake, 188
Rhesus monkey, 283
Rhesus monkey brains, 359
Rhythms
   of the cycle, 504
   light/dark, 507
   reproductive, 504
   types of, 503
RNA
   concentrations, changes in, 292
   -dependent mechanism, 117
   production of, 289
RNAse, 117
Rough endoplasmic reticulum (RER), 12
   cisternae of, 16

Saturation analysis, 265
Seasonal breeders, 508
Seasonal influences, 510
Secretion of protein, 71
Secretory granules, 49
Self-preservation, 381
Septal region, 210
Septum, 1, 283
Sequential breakdown process, 141
Serotonin, 154–156
   effects of, 171
   receptors, 160
   synthesis, 154
   turnover, 294

Serotoninergic neurons, 149
Serotoninergic neurotransmission, 159, 160, 170
Serum FSH levels, 221, 337
Serum LH concentration, 433
Serum LH levels, 221, 337
Serum tryptophan, 156
Sex differentiation, 172, 268, 279
Sex steroids, 370
Sexual activity, enhancement of, 474
Sexual behavior, 21, 193, 238, 294, 305, 349
   in animals, 350
   diurnal changes in, 474
   effects of estrogen on male, 479
   effects of progesterone on, 488
   female, 476
   hormonal stimulus for female primates, 477
   male, 478
   male and female, 472
   neuroendocrine control of, 476
   neuroendocrine mechanisms in, 471
   synergistic effect of estrogen and androgen on, 479
Sexual differentiation, mechanisms of, 21
Sexual dimorphism, 21, 22, 334
Sexual maturation, 349, 460
Sexual motivation, 474
Sheep, 248
Short feedback, 371
Silent cells, 36
Single neurons
   antidromically identified, 38
   estrogen-sensitive, 38
Single-unit activity, 33, 35
Single units, 40
   antidromic identification of, 34
Social aggression, induced by testosterone, 511
Somatostatin (SRIF), 81, 101, 119, 131, 137, 391, 394
   analogue, 138
   -containing neurons, 210
   effect on cyclic AMP, 391
   release, 213
Specific receptor system, 257
Stalk median eminence (SME), 426
Starvation, 382
Steroids
   activity of, 352
   binding, 258
   concentrations of, 249, 253, 258
   in CSF, 256
   in different tissues, concentrations of, 255
   elution of, 252
   feedback, 9

hormones, 347
    action, 353
    endogenous levels in brain and pituitary, 255
    metabolism, 248
    receptors, 277
    uptake, 248
    input, 492
    metabolic activities, 252
    metabolism, 258
    physiological amounts of, 491
    receptors, 277
        function, 289
        in neuroendocrine tissues, 277
        subcellular distribution, 277
        topography, 277
        —receptor interactions, 263, 277
    reciprocal effects, 39
    specific binding of, 263
    stereotaxic implantation of, 339
    target cells, 475
Δ^4-Steroid 5α-reductase(s), properties of, 311
    rat anterior pituitary, 311
    rat hypothalamic pituitary, 311
    subcellular distribution of, 311
Stria terminalis, 37
    bed nucleus of, 424
Subependymal layer (SEL), 219
Subprimate mammals, 477
Substance P, 131, 139
Substance P-containing neurons, 205
Suckling reflex, 171
Suckling stimulus, 171
Sulpiride, 231
Suprachiasmatic neurons, 19
Suprachiasmatic nucleus (SCN), 6, 18, 171, 442
    neurophysin in, 85
Supra-retrochiasmatic area, 441
Supraspinal motor mechanisms, 471
Surgically stressed rat, 492
Swelling cycles, 505
Synaptic connections, pattern of, 172
Synaptic pattern, 21
Synaptic vesicles, 187
Synaptology, 21
Synthetase, analogous, 116
Synthetically active neuron, 16

Tanycyte, 6, 24
    ependymal, 23
    Gn-RH in, 97
Target tissue, 247
Taurine, 66

Temperature, 48
Testicular regression in hamster, 507
Testosterone (T), 252, 253, 265, 269, 281, 311, 327, 339, 352, 511
    action of, 371
    -binding macromolecule, 269
    binding by neural and pituitary tissues, 257
    concentration of, 253
    conversion of, 328
        into dihydrotestosterone, 328
        into 5α-androstan-3α, 17β-diol, 327
    conversion to DHT, 258
    effect of, 374
    effects of intrahypothalamic and intrapituitary implants, 339
    effects on sexual behavior in castrated animals, 478
    effects of systemic injections of, 336
    effects of treatment on serum LH and FSH levels of adult castrated male rats, 337
    metabolites of, 269
    as a pre-hormone, 258
    transformation of, 327
        in the anterior pituitary, 327
        in the CNS, 327
[^{3}H] Testosterone, distribution of, 280
Tetracaine, local anesthetic, 48
Theophylline, 392
Thymol, 49
Thyroliberin, 131, 135
    inactivation of, 135
Thyrotropin (TSH), 391
    release, 408
Thyrotropin-releasing factor (TRF), 4
    -containing neurons, 210
    nerve terminals, distribution of, 210
    secretion, inhibitory effect of 5-HT on, 238
    transmitter function, 210
Thyrotropin-releasing hormone (TRH), 81, 101, 121, 176, 391, 393, 394, 408
    biosynthesis, 111, 179
        in hypothalamic extracts, 115
    degradation, 117
    effect on cyclic AMP, 391
    secretion, 116
    stimulatory effect on cyclic AMP, 393
    synthesis, 116
Thyroxine, 116, 189
Tissue concentration of progesterone, 250
Tissue fractionation, 264
Tissue localization, 309
Transcortin, 284
Transcortin-like binding, 288

Transfer acyl-tRNA, 117
Triamcinolone acetonide, 284
Trophotropic system, 382, 383
Tryptophan, 149, 156
    binding to albumin, 159
    serum, 156
Tuberoinfundibular dopamine (DA)
    in control of gonadotropin secretion, 215
    neurons, 173, 216, 234
    DA pathway, 224
    system, 193
    turnover of, 294
Tuberoinfundibular LRF delivery system, 6
Tuberoinfundibular neurons, 4, 19
    afferent connections, 19
Tuberoinfundibular pathway, 18
Tuberoinfundibular tract, 39
Turnover rates, 68
Tyrosine, 149, 164, 185
Tyrosine hydroxylase, 162, 186, 188
Tyrosine hydroxylase immunofluorescence, in DA cell bodies, 195
Tyrosine hydroxylase inhibition, 216

Uptake studies, 264
Uterine receptors, 265
Uterus, 249, 250

Vaginal swelling, 505
Vasopressin, 34, 40, 46, 82, 119, 138
    localization of, 90
    biosynthesis of, 109
Ventricular system, 23, 86
Ventrolateral ventromedial nucleus, 14
    neurons in, 14
Ventromedial nucleus (VM), 4, 8, 14, 37
    posterior dorsomedial, 16
Vinblastine, 49

Warburg manometric technique, 363
Whorled body containing neurons, 12